Macromolecular Symposia

Symposium Editor: J. P. Pascault

198

WILEY-VCH

pp. 1–482

August 2003

Macromolecular Symposia publishes lectures given at international symposia and is issued irregularly, with normally 14 volumes published per year. For each symposium volume, an Editor is appointed. The articles are peer-reviewed. The journal is produced by photo-offset lithography directly from the authors' typescripts.
Further information for authors can be obtained from:
Editorial office "Macromolecular Symposia"
Wiley-VCH Verlag GmbH & Co. KGaA,
Boschstrasse 12, 69469 Weinheim,
Germany
Tel. +49 (0) 62 01/6 06-2 38 or -5 81; Fax +49 (0) 62 01/6 06-3 09 or 5 10;
E-mail: macro-symp@wiley-vch.de
http://www.ms-journal.de
Suggestions or proposals for conferences or symposia to be covered in this series should also be sent to the Editorial office at the address above.

Macromolecular Symposia:
Annual subscription rates 2003 (print only or online only)*
Germany, Austria € 1318; Switzerland SFr 2168; other Europe € 1318; outside Europe US $ 1568.
Macromolecular Package, including Macromolecular Chemistry & Physics (18 issues), Macromolecular Bioscience (12 issues), Macromolecular Rapid Communications (18 issues), Macromolecular Theory & Simulations (9 issues) is also available. Details on request.
* For a 5 % premium in addition to **Print Only** or **Online Only**, Institutions can also choose both print and online access.
Packages including Macromolecular Symposia and Macromolecular Materials & Engineering are also available. Details on request.
Single issues and back copies are available. Please inquire for prices.

Orders may be placed through your bookseller or directly at the publishers:
WILEY-VCH Verlag GmbH & Co. KGaA, P.O. Box 10 11 61, 69451 Weinheim, Germany, Tel. +49 (0) 62 01/6 06-400, Fax +49 (0) 62 01/60 61 84. E-mail: service@wiley-vch.de

Macromolecular Symposia (ISSN 1022-1360) is published with 14 volumes per year by WILEY-VCH Verlag GmbH & Co. KGaA, P.O. Box 10 11 61, 69451 Weinheim, Germany. Air freight and mailing in the USA by Publications Expediting Inc., 200 Meacham Ave., Elmont, NY 11003, USA. Application to mail at Periodicals Postage rate is paid at Jamaica, NY 11431, USA. US POSTMASTER please send address changes to: Macromolecular Symposia, c/o Wiley-VCH, III River Street, Hoboken, NJ 07030, USA.

Printing: Strauss Offsetdruck, Mörlenbach. Binding: J. Schäffer, Grünstadt

7th European Symposium on Polymer Blends

held in
Lyon-Villeurbanne, France
May 27–29, 2002

Organized by
«Groupement Français d'Etudes et d'Application des Polymères», GFP
and «Fédération des Polyméristes Lyonnais», FR-CNRS 2151

Symposium Editor (Chairman)
Professor Jean-Pierre Pascault
IMP/LMM, INSA Lyon,
Bat. Jules Verne,
20 Avenue Albert Einstein,
69621 Villeurbanne Cedex
France

Sponsors
CNRS, INSA-Lyon, University Claude Bernard-Lyon I,
University J. Monnet-St-Etienne, University Savoie-Chambery,
Région Rhône-Alpes, Conseil Général du Rhône, City of Villeurbanne, Rhodia,
Atofina, IFP, Hutchinson, Exxon Mobil, Veeco, ThermoRheo, Waters

---

ISBN 3-527-30702-8

The **tables of contents** of the published issues are displayed on the WWW.

This service as well as further information on our journals can be found at the following WWW address:

http://www.ms-journal.de

Contents of Macromolecular Symposia 198

7th European Symposium on Polymer Blends
Lyon-Villeurbanne (France), 2002

* Asterisks indicate the name(s) of the author(s) to whom inquiries should be addressed.

Author Index

Preface

On May 27–29th, 2002, the 7th European Symposium on Polymer Blends took place in Lyon, France. Previous symposia were organized in Bruges (1984), Strasbourg (1987), Cambridge (1990), Capri (1993), Maastricht (1996), and Mainz (1999). The next one, the 8th Symposium is planned to come back to Bruges in 2005. The success of this series of symposia demonstrated that the area of polymer blends is still lively and important from both research and applied viewpoints.

More than 250 participants from 35 countries, and most of them from Europe, were present during the three days meeting in Lyon-Villeurbanne. Main topics and submitted papers selections have been made with the help of the International Advisory Board. The program was composed of 42 lectures (including eight industrial speakers) and 180 posters. A strong emphasis was put on the six hours poster session during one afternoon till the evening, concluded by "wine and spécialités lyonnaises" and four poster prizes.

The key of the success was also rapid positive answers from sponsors, an Organizing committee, and the help during the three days meeting of PhD students from the different laboratories of the "Fédération des Polyméristes Lyonnais" (FR-CNRS 2151).

Most of the lectures and selected posters are collected in this symposium volume divided in four sections:
1. Thermodynamics, Morphologies and Properties of Non-Reactive Blends;
2. In Situ Compatibilisation of Immiscible Thermoplastic Blends;
3. Blends from Reaction-Induced Phase Separation;
4. Blending for Specific Performances.

This volume reflects the three days meeting with a large part devoted to the use of chemistry to control blends morphologies and properties. I hope that readers will enjoy this volume, which is a modest step in the polymer blends story.

J. P. Pascault

Liquid-Liquid Equilibria of Binary Polymer Blends: Molecular Thermodynamic Approach

*Bong Ho Chang, Young Chan Bae**

Division of Chemical Engineering and Molecular Thermodynamics Lab., Hanyang University, Seoul 133-791, Korea

Summary: We extended and simplified the modified double-lattice model to binary polymer blend systems. The model has two model parameters, C_β and C_γ. Those are not adjustable parameters but universal functions. In comparison with Ryu *et al.*'s simulation data for symmetric polymer blend with various chain lengths ($r_1 = r_2 = 8$, 20, 50, 100), C_γ is determined. Our results show that C_β is negligible for symmetric polymer blend systems. The proposed model describes very well phase behaviors of weakly interacting polymer blend systems.

Keywords: binary polymer blend, chain length, liquid-liquid equilibria, molecular thermodynamics, simulation

Introduction

During the last decades, there are various theoretical and experimental approaches to investigate the structures and thermodynamic properties of polymers and their blends. However, most theoretical considerations on the miscibility of polymer-polymer blend systems have developed through the mean-field theory of Flory and Huggins[1~5] due to the simplicity for practical calculations. To overcome some problems of the mean-field theory and pursue a formal "exact" solution to the lattice model, Freed and coworkers[6~8] derived a lattice field theory (or lattice cluster theory) for polymer/solvent systems using advanced statistical-mechanical methods. This theory formally provides an exact mathematical solution for the Flory-Huggins model. However, for practical reasons, the infinite series with respect to coordination number, temperature, and composition are truncated at a certain order. Therefore, this theory still remains deficiency for the real system.

Hu *et al.*[9~10] reported a new and simple model based on Freed's lattice-cluster theory. However, the model has some arbitrary and empirical parameters.

 DOI: 10.1002/masy.200350801

Recently, Oh *et al.*[11] modified Hu *et al.*'s model by introducing new universal parameters and simplified the expression of the Helmholtz energy of mixing to reduce the arbitrariness. The parameters were obtained by comparing simulation data for the polymer solutions with the model (r_1=1 and r_2=100). Therefore, the parameters are not suitable to apply to the polymer blend systems.

The purpose of this study is to extend Oh *et al.*'s model to the polymer blend systems by introducing a new universal parameter, C_γ, as a function of the chain length of a polymer. The coexistence curves are generated and then compared with the experimental data for polymer blend systems. Our model requires only one adjustable binary parameters to describe an upper critical solution temperature (UCST) of weakly interacting polymer blend systems.

Model Development

Helmholtz energy of mixing

Oh *et al.*[11] proposed a new Helmholtz energy of mixing as the form of the Flory and Huggins theory[1~5]. The expression is given by

$$\Delta A / N_r kT = (\phi_1/r_1)\ln\phi_1 + (\phi_2/r_2)\ln\phi_2 + \chi_{OB}\phi_1\phi_2 \tag{1}$$

where N_r is the total number of lattice sites and k is the Boltzmann's constant. ϕ_i and r_i are the volume fraction and the number of segments for component i, respectively. χ_{OB}, a new interaction parameter, is defined and rearranged by

$$\begin{aligned}\chi_{OB} &= C_\beta\left(\frac{1}{r_2}-\frac{1}{r_1}\right)^2 + \left(2+\frac{1}{r_2}\right)\tilde{\varepsilon} - \left(\frac{1}{r_2}-\frac{1}{r_1}+C_\gamma\tilde{\varepsilon}\right)\tilde{\varepsilon}\phi_2 + C_\gamma\tilde{\varepsilon}^2\phi_2^{\,2} \\ &= C_\beta\left(\frac{1}{r_2}-\frac{1}{r_1}\right)^2 + \left(2+\frac{\phi_1}{r_2}+\frac{\phi_2}{r_1}\right)\tilde{\varepsilon} - C_\gamma\tilde{\varepsilon}^2\phi_1\phi_2\end{aligned} \tag{2}$$

where the first term on the right-handed side represents degree of asymmetry between two molecules. However, we assume that in the polymer blend systems, the differences between r_1 and r_2 are not very large ($r_1 \cong r_2$) compared with those of polymer solution systems ($r_1 << r_2$), that is, phase diagrams of polymer blends nearly show the symmetric shapes and critical compositions are near 0.5. Therefore, this term can be neglected. χ_{OB} has only one model parameter, C_γ. $\tilde{\varepsilon}$ is a reduced interaction parameter given by

$$\tilde{\varepsilon} = \varepsilon/kT = \left(\varepsilon_{11} + \varepsilon_{22} - 2\varepsilon_{12}\right)/kT \qquad (3)$$

where $\varepsilon_{11}, \varepsilon_{22}$ and ε_{12} are the corresponding nearest-neighbor segment-segment interactions. χ_{OB} then is a function of temperature, composition and chain length of polymer and has the second order approximation form of $\tilde{\varepsilon}$. Figure 1 shows the simulation results by Ryu *et al.*[19] for various chain lengths of polymers ($r_1 = r_2 = 8, 20, 50, 100$). As shown in Figure 1, the calculation results (dotted line) with the constant $C_\gamma = A_0$ (Case I) are narrower than those of the simulation data. We assume that C_γ is a function of temperature (Case II);

$$C_\gamma = A_0 + A_1(T_r - 1)$$
$$T_r = T/T_c \qquad (4)$$

where T_c is a critical temperature. When $T = T_c$, Case II returns to Case I. Results show that calculated values using Case II give much better agreements than those of Case I. The solid lines are the best fit given by the proposed model. From these calculations, we obtain the values of A_0 and A_1 for the various chain lengths of polymers.

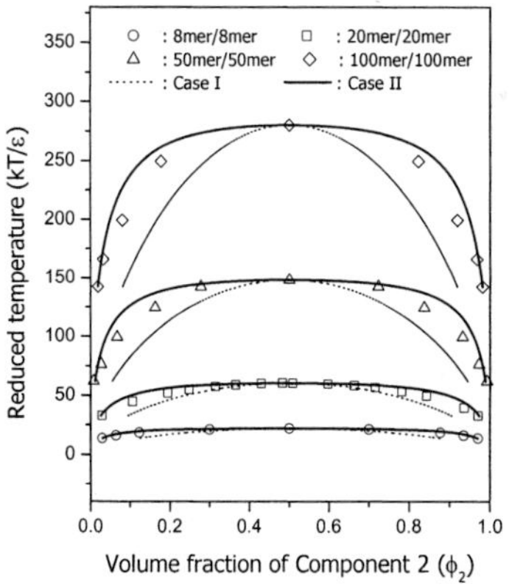

Figure 1. Phase diagram of symmetric polymer blend systems with various chain lengths of polymers

Figure 2 represents the chain length dependence of C_γ. The parameters are appeared to be asymptotic values with respect to r_2. The following mathematical equations are suggested the r-mer dependence of the values of A_0 and A_1 in C_γ, respectively:

$$A_0 = 50.7921 - \frac{30.3247(r_2 - 1)}{1 + 0.0046(r_2 - 1)} \tag{5}$$

$$A_1 = -1124.0506 + \frac{386.4166(r_2 - 1)}{1 + 0.0038(r_2 - 1)} \tag{6}$$

Eqs. (5) and (6) can be applied for large r_2. For r_2>10000, C_γ is very weakly dependent on r_2 according to Figure 2. In our proposed model, parameter C_γ, which is not an adjustable parameter, has fixed value at given temperature and r-mer chain length of each system.

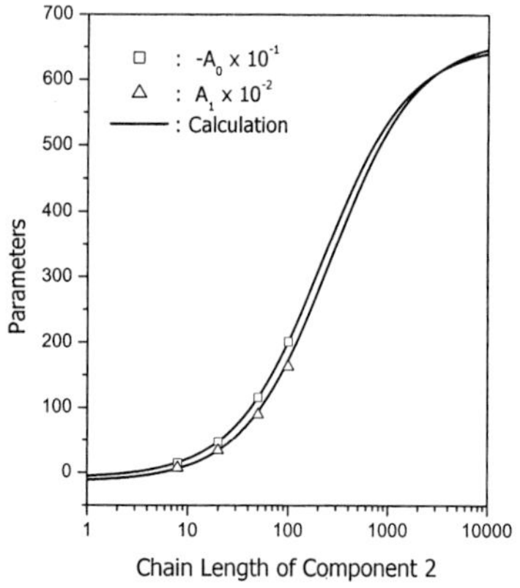

Figure 2. Dependence of parameter C_γ on r-mer chain length. The solid lines are generalized by equation (5) and (6). The open triangles and squres are the results obtained from the previous calculation step

Chemical potentials of component 1 and 2 for calculating the binary coexistence curve are given by

$$\frac{\Delta\mu_1}{kT}=\ln(1-\phi_2)-\left(\frac{r_1}{r_2}-1\right)\phi_2+\left[\left\{\left(\frac{r_1}{r_2}-1\right)+r_1C_\gamma\tilde{\varepsilon}\right\}\tilde{\varepsilon}+r_1\left(2+\frac{1}{r_2}\right)\tilde{\varepsilon}\right]\phi_2^2$$
$$-2\left[\left\{\left(\frac{r_1}{r_2}-1\right)+r_1C_\gamma\tilde{\varepsilon}\right\}\tilde{\varepsilon}+r_1C_\gamma\tilde{\varepsilon}^2\right]\phi_2^3+3r_1C_\gamma\tilde{\varepsilon}^2\phi_2^4 \tag{7}$$

$$\frac{\Delta\mu_2}{kT}=\ln\phi_2+\left(1-\frac{r_2}{r_1}\right)+(2r_2+1)\tilde{\varepsilon}$$
$$-\left[\left(1-\frac{r_2}{r_1}\right)+2\left\{\left(1-\frac{r_2}{r_1}\right)+r_2C_\gamma\tilde{\varepsilon}\right\}\tilde{\varepsilon}+2(2r_2+1)\tilde{\varepsilon}\right]\phi_2$$
$$+\left[\left\{\left(1-\frac{r_2}{r_1}\right)+r_2C_\gamma\tilde{\varepsilon}\right\}\tilde{\varepsilon}+(2r_2+1)\tilde{\varepsilon}+3r_2C_\gamma\tilde{\varepsilon}^2\right]\phi_2^2 \tag{8}$$
$$-\left[6r_2C_\gamma\tilde{\varepsilon}^2+2\left\{\left(1-\frac{r_2}{r_1}\right)+r_2C_\gamma\tilde{\varepsilon}\right\}\tilde{\varepsilon}\right]\phi_2^3+3r_2C_\gamma\tilde{\varepsilon}^2\phi_2^4$$

The coexistence curve is found from the following conditions:

$$\Delta\mu_1'=\Delta\mu_1'' \tag{9}$$

$$\Delta\mu_2'=\Delta\mu_2'' \tag{10}$$

where $\Delta\mu_i$ is the change in chemical potential upon isothermally transferring component i from the pure state to the mixture. Superscripts ′ and ″ denote two phases at equilibrium. For phase equilibrium calculation, we require the experimental coordinates of the critical point. And the critical conditions are given by,

$$\frac{\partial^2(\Delta A/N_r kT)}{\partial\phi_2^2}$$
$$=\frac{-1}{1-\phi_2}+\left(1-\frac{r_1}{r_2}\right)+2\left[\left\{\left(\frac{r_1}{r_2}-1\right)+r_1C_\gamma\tilde{\varepsilon}\right\}\tilde{\varepsilon}+r_1\left(2+\frac{1}{r_2}\right)\tilde{\varepsilon}\right]\phi_2 \tag{11}$$
$$-6\left[\left\{\left(\frac{r_1}{r_2}-1\right)+r_1C_\gamma\tilde{\varepsilon}\right\}\tilde{\varepsilon}+r_1C_\gamma\tilde{\varepsilon}^2\right]\phi_2^2+12r_1C_\gamma\tilde{\varepsilon}^2\phi_2^3=0$$

and

$$\frac{\partial^3(\Delta A/N_r kT)}{\partial \phi_2^{\,3}}$$
$$= \frac{-1}{(1-\phi_2)^2} + 2\left[\left(\left(\frac{r_1}{r_2}-1\right) + r_1 C_\gamma \tilde{\varepsilon}\right)\tilde{\varepsilon} + r_1\left(2+\frac{1}{r_2}\right)\tilde{\varepsilon}\right] \tag{12}$$
$$-12\left[\left\{\left(\frac{r_1}{r_2}-1\right) + r_1 C_\gamma \tilde{\varepsilon}\right\}\tilde{\varepsilon} + r_1 C_\gamma \tilde{\varepsilon}^2\right]\phi_2 + 36 r_1 C_\gamma \tilde{\varepsilon}^2 \phi_2^{\,2} = 0$$

Results and Discussion

We introduce three basic assumptions as follow:

1. Flexibilities and structure effects of polymer chains are neglected. All polymer chains are freely flexible.
2. Volume change of mixing and free volume is not considered, that is, incompressible.
3. Polydispersities of polymers are neglected. All polymers are regarded as monodisperse.

In the proposed model, only one parameter, ε/k, is needed to calculate theoretical coexistence curves. r_i is calculated by

$$r_i = \frac{V^u_{mi(vdW)} \cdot M_{wi}}{10.23(cm^3/mol) \cdot M^u_{wi}} \tag{13}$$

where the constant $10.23(cm^3/mol)$, $V_{mi(vdW)}$ and M_{wi} are the molar van der Waals (vdW) volume of a CH_2 group, which occupies one lattice site, from Bondi[13], the molar vdW volume and molecular weight of component *i*, respectively. Superscript *u* represents a repeating unit of polymers. If r_i's are fixed by Equation (13), the critical composition of each system is numerically determined from critical conditions obtained by Equation (11) and (12).

Figure 3 shows phase diagrams of Polystyrene (PS) / Poly (α - methyl styrene) (PαMeS) systems[14]. This is the ordinary system showing a UCST phase behavior without the oriented interaction.

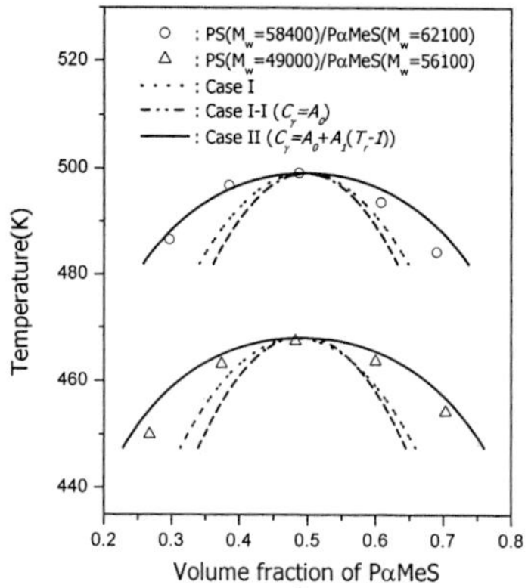

Figure 3. Cloud-point data for PS/Pα MeS systems

For these systems, the values of r_1 and r_2 calculated by Equation (13), respectively. From these values, the numerically determined critical volume fraction of Pα MeS (ϕ_2^c) and ε/k are listed in Table 1, respectively.

Table 1. The values of model parameters for PS/Pα MeS.

r_1	r_2	$\varepsilon/k(K)$	ϕ_c	$T_c(K)$
3641.18	3802.91	0.1019	0.4968	499.2
3055.10	3435.48	0.1071	0.4852	468.1

The very small values of ε/k are due to the similar interaction energies between segments of PS and Pα MeS ($\varepsilon_{11} \cong \varepsilon_{22} \cong \varepsilon_{12}$). In other word, it is due to the structural similarity of segments of PS and Pα MeS. In addition, ε/k is likely to be slightly dependent of molecular weights of polymers. The Case I-I is calculated with C_γ as a fuction of the chain length of polymer at $T_r = 1$. Case II shows a better agreement than thoese of the others.

Figure 4 shows phase diagrams of Polybutadiene (PBD) / PS system[15].

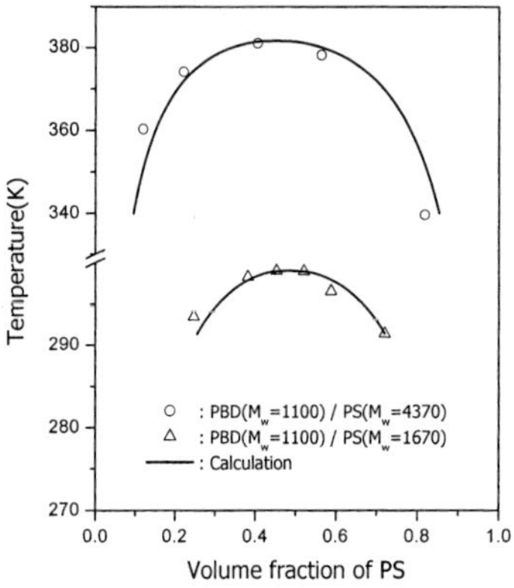

Figure 4. Cloud-point data for PBD/PS systems

r_1, r_2, a critical volume fraction of PS (ϕ_2^c) and ε/k for different molecular weight of PS's are listed in Table 2.

Table 2. The values of model parameters for PBD/PS systems

r_1	r_2	$\varepsilon/k(K)$	ϕ_c	$T_c(K)$
74.26	272.47	0.9827	0.4519	381.6
74.26	104.12	1.1518	0.4860	299.05

The systems having low molecular weight and slightly large asymmetry of polymers show the similar results as those of Figure 3.

In LLE calculation of polymer blends, the critical composition plays a very important role to determine shapes of phase diagrams.

Conclusion

We derived the simple molecular thermodynamic framework for polymer blends. In our model, the critical composition obtained numerically by the asymmetry between two polymers is the most important parameter to determine the shape of calculated coexistence curves. The proposed model describes fairly well phase behavior of polymer blend systems in spite of its simplicity and using only one model parameter.

Acknowledgment

This paper was supported by International Cooperation in Korea Science and Engineering Foundation, 2001 (No 20016-301-03-2).

[1] P. J. Flory, *J. Chem. Phys.* **1942**, *10*, 51.
[2] P. J. Flory, *Principles of Polymer Chemistry.* Cornell Univ. Press, Ithaca. 1953.
[3] P. J. Flory, *J. Am. Chem. Soc.* **1965**, *87*, 1833.
[4] P. J. Flory, *Discuss. Faraday Soc.* **1970**, *49*, 7.
[5] L. Huggins, *J. Phys. Chem.* **1942**, *46*, 151.
[6] K. F. Freed, *J. Phys. A: Math. Gen.* **1985**, *18*, 871.
[7] M. G. Bawendi, K. F. Freed, U. Mohanty, *J. Chem. Phys.* **1988**, *87*, 5534.
[8] M. G. Bawendi, K. F. Freed, *J. Chem. Phys.* **1988**, *88*, 2741.
[9] Y. Hu, S. M. Lambert, D. S. Soane, J. M. Prausnitz, *Macromolecules*, **1991**, *24*, 4356.
[10] Y. Hu, H. Liu, D. S. Soane, J. M. Prausnitz, *Fluid Phase Equilibria*, **1991**, *67*, 65.
[11] J. S. Oh, Y. C. Bae, *Polymer,* **1998**, *39*, 1149.
[12] J. H. Ryu, P. D. Gujrati, *J. Cheml. Phys.* **1997**, *107(10)*, 3945.
[13] D. W. van Krevelen, "*Properties of Polymers*", 3rd edn. Elsevier, Amsterdam, 1990.
[14] J. I. –L. Lin, R. –J. Roe, *Polymer* **1988**, *29*, 1227.
[15] S. Rostami, D. J. Walsh, *Macromolecules*, **1985**, *18(6)*, 1228.

Solubility Islands for Polymer Blends – A New Option to Homogenize Incompatible Polymers?

Attila R. Imre
KFKI Atomic Energy Research Institute, H-1525 Budapest, P.O. Box 49, Hungary
E-mail: imre@sunserv.kfki.hu

Summary: Experimental results of liquid-liquid phase equilibrium in a polydisperse blend of two polyalkylsiloxane are presented here. The UCST has an unusual pressure dependence: pressure induced miscibility at moderate pressures and pressure induced immiscibility at higher pressures, above a double critical point. The cloud point curve has two maxima in (concentration, temperature) as well as in (concentration, pressure) space. Approaching the double critical point, the high and low pressure branches of the cloud point curve merge and in a certain stage of this merging, they form a miscibility island located inside the two-phase region. Islands of this kind can give us a new tool to mix virtually immiscible blends.

Keywords: blends, compatibility, miscibility islands, polysiloxanes, UCST

Introduction

A couple of years ago an unusual kind of phase diagram of polystyrene/ acetone system was published by Rebelo and Van Hook[1] – a small island of miscibility located inside of the hour-glass shaped two-phase region. The existence of these kinds of islands in the (concentration, temperature) space have been predicted a few years earlier[2] and can be expected in other systems. [3-5] In this short paper we would like to report the existence of a similar miscibility island in the (concentration, pressure) space, without giving complete theoretical explanation.

Solubility Island

On Figure 1/a we can see the schematic representation of the liquid-liquid loci in weakly interacting binary homopolymer systems (solutions and blends) at constant pressure. The system is homogeneous in medium temperature range; liquid-liquid phase transition can be induced by cooling or by heating. In some cases the high or the low temperature branch is missing; it can be masked by freezing or degradation, or it might be simply out of our experimental range. The cloud-point curves have two critical points: UCST (Upper Critical Solution Temperature, being the uppermost part of the low -temperature branch) and LCST (Lower Critical Solution Temperature).

 DOI: 10.1002/masy.200350802

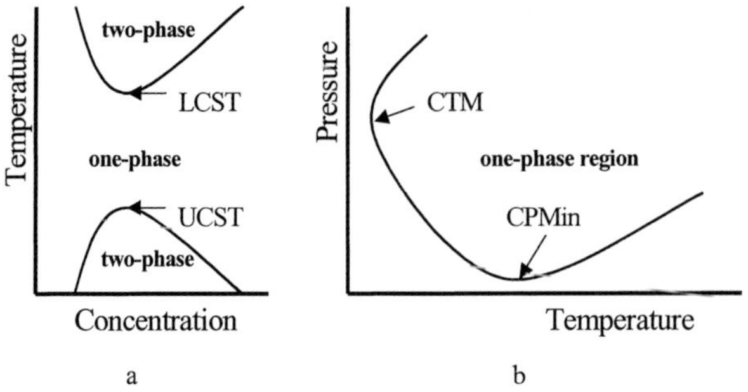

Figure 1.: a.: Schematic representation of the cloud-point curves in weakly interacting binary homopolymer solutions or blends. Critical points, UCST (Upper Critical Solution Temperature) and LCST (Lower Critical Solution Temperature) are marked by arrows. b.: Schematic pressure dependence of the critical liquid-liquid locus in weakly interacting homopolymer solutions and homopolymer blends. The two double-critical points, CTM (Critical Temperature Minimum) and CPMin (Critical Pressure Minimum) are marked by arrows.

In Figure 1/b we can see the schematic representation of the pressure dependence of the critical points[6,7]. The whole curve shown on Figure 1/b can be seen only in a few systems[6,7] ; in most cases only part of this curve can be observed. There are two extrema on this curve: one of them (CTM, Critical Temperature Minimum) can be seen for example in some solutions (like polystyrene/ methylcyclohexane solution[8]) as well as in a few binary blends[9] (like poly(ethylene glycol)/poly(propylene glycol) mixture[10]). The other one (CPMin) can be seen for example in polystyrene/acetone solution[1], but in most cases (like the polystyrene/methylcyclohexane solution[11]) this point is below p=0[12]. Both extrema can be seen for example in polystyrene/1,4-trans-dimethylcyclohexane solution[6] or in polystyrene/ methylcyclohexane/acetone ternary mixture[7].

Decreasing the pressure toward CPMin, the solubility gap between the UCST and LCST branches (in (concentration, temperature) space) is going to shrink and finally the branches touch each other at the UCST/LCST (forming a double critical point, CPMin)[9] and merge, forming a so-called „hourglass"-shaped phase diagram. In most systems, this CPMin can be seen only by decreasing the pressure below zero[6,7,12] ; i.e. by stretching the liquid. Stretched liquids are metastable [13,14] and the generation of negative pressure (high enough to reach the merging point of the UCST and LCST) is difficult, therefore this kind of double critical point can be seen rarely in experiments.

In some systems the cloud-point curve (in (concentration, temperature) space) might be double humped. By decreasing the pressure to approach a double critical point (CPMin) the

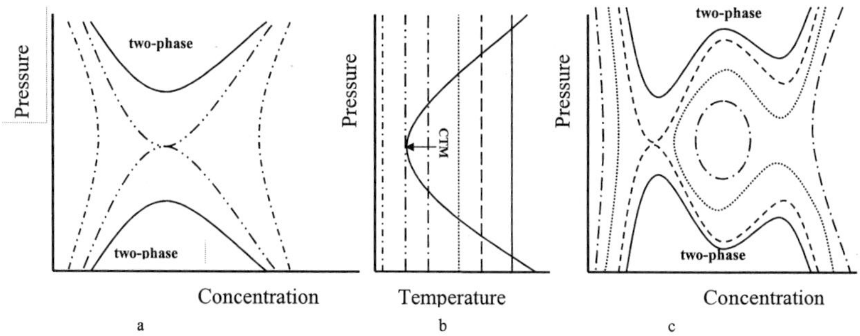

Figure 2.: Schematic representation of the liquid-liquid equilibrium in binary systems with a double critical point (CTM) on the UCST (b) and the merging of the low- and high-pressure branches with sinle (a) and double (c) maxima cloudpoint curves.

two branches will be joined first by the bigger humps, then by the smaller ones, creating first a peninsula, later an island of solubility. The double-humping of the cloud-point curve – which is crucial to get solubility island in weakly interacting binary polymer system – can be explained in two different ways: by the polydispersity of one or both components or by the strong concentration dependence of the Flory-Huggins type of interaction parameter.[1,3,15-20] This kind of solubility island has been found in polystyrene/acetone system[1].

Studying the system at constant temperature (instead of constant pressure) in the vicinity of CTM (Fig.2/b), we can see two cloud-point curves, one at higher pressures and the other one at lower pressures (Fig. 2/a, solid lines), similarly to the constant pressure case (Fig. 1/a). Approaching the CTM by decreasing the temperature, the two branches can merge and form an hourglass shaped phase diagram[8] (Fig. 2/a, dash-dotted lines). Although most theoretical works concerning the shape of the cloud-point curves in polydisperse systems have been done for p=constant case, we can expect that polydispersity causes similar effects (distortion or even double humping) on the cloud-point curve in the (concentration, pressure) space (fig. 2/c, solid lines). In this case, approaching the CTM by decreasing the temperature, we might get a solubility island (Fig.2/c, dash-dotted lines) - similar to the one found by Rebelo and Van Hook in polystyrene/acetone system in the vicinity of the CPMin[1] - but in this case it will be located around the CTM.

In a recent experiment a binary system of two polysiloxanes (already proved to have double humped cloud-point curve in (concentration, temperature) space in a blend of these two polymers with slightly different molecular weight[21]) has been studied by us. We have to emphasize here, that although the cloud-point curve is double humped, the coexistence curve has only a single maximum[9,21], i.e. we do not have three-phase equilibrium or three critical points on the UCST expected in case of double-humped coexistence curve.[18] Polydimethylsiloxane (PDMS, M_w= 1.74 kg/mol, M_w/M_n= 1.3) has been obtained from Wacker, polyhexylmethylsiloxane (PHMS, M_w=90.5 kg/mol, M_w/M_n=6.7) has been obtained from Hüls-Petrarch. Samples have been prepared by weight. Cloud-point temperatures have been measured in a stainless steel cell closed by two sapphyre windows[9,22] by measuring the intensity of the transmitted light with a photodiode, keeping the temperature at constant value and varying the pressure. The critical concentration (at atmospheric pressure) has been determined by the phase-volume ratio method and the obtained concentration was equal (within the experimental error) with the concentration corresponding to the local minimum on the cloud-point curve in the (concentration, temperature) space, as it has been expected in polydisperse systems.[15,23,24] Further details of the experiment is published elsewhere.[9]

Our experimental result obtained from the vicinity of the UCST (i.e. in the region of the local minimum) indicated that the cloud-point curve is double humped in the (concentration, pressure) space[9,21] as well as in the (concentration, temperature) one.

Having a double critical point on the UCST at accessible pressure (here it is around 50-60 MPa), 'separate' low pressure and high pressure cloud-point curves can be seen in (concentration, pressure) space at constant temperature (Fig.2). Without double humping these two curves would merge by forming an hourglass-shape (Fig.2/a). With double humping, first the two bigger humps would merge, then the two smaller ones, forming first a solubility peninsula then a solubility island (Fig.2/c, dotted and dash-dotted lines). Our experimental data focusing on the vicinity of the critical concentration (the local minimum of the lower and the local maximum of the upper curve on Figure 2/c) can be seen on Figure 3. The merging of the two branches and the forming of a solubility peninsula first (above 46 ^{0}C) and a solubility island later (around 44 ^{0}C) is clearly demonstrated, although the exact shape of the island needs further study. The island totally disappeared somewhere between 42 and 41 Celsius.

There are two crucial points which have to be fulfilled to get a solubility island in a weakly interacting binary polymeric system. One of them is the double humping of the cloud-point curves in the (concentration, pressure) space. Probably just like the similar phenomena (in (concentration, temperature) space), this phenomena is also caused by the polydispersity.

The other crucial point (to have a double critical point on the UCST branch) can be explained by a pressure-dependent excess volume of mixing.[22] According to some recent results, the existence of this point – i.e. the turning of dP/dT from negative to positive - is quite general[6,7,22,25-27], but for a lot of polymer blends the experimentally obtained pressure

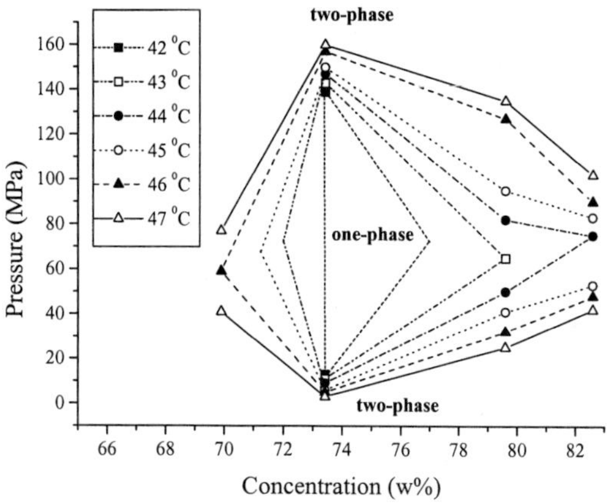

Figure 3.: The appearance of a solubility island in PDMS/PHMS blend. The connecting lines are only guides to the eye. Above 47 Celsius the high- and low-pressure branches are separated. Around 46 Celsius the two higher peaks merge (see Figure 2.), forming a solubility peninsula. At 44 Celsius the lower peaks merge too, forming an island of solubility which disappears between 42 and 41 Celsius.

dependence of the UCST seems to be monotonous[28-30], i.e. the double critical point is usually located in the experimentally not so easily accessible P<0 region.[22]

Solubility islands and compatibilization

The importance of this solubility island is twofold. First, it is theoretically important to show that the system behaves similarly around different kinds of double critical points[6,7] (the one around CPMin) – see the referred polystyrene/acetone system[1] -, and the other one around CTM). The second importance is a practical one. In most cases 'immiscible' refers that the binary system exhibits hourglass shaped cloud-point curves in (concentration, temperature) and/or (concentration, pressure) spaces. In these cases homogeneous system can be prepared only with small amount of component A + big amount of component B (for example 0.1% A+99.9% B) and vice versa, but the two components cannot be homogenized at "medium" concentrations. The existence of a solubility island located inside of the hourglass region shows us, that seemingly incompatible systems might have a small island of compatibility somewhere in the (concentration, pressure) space (Fig. 4). While the other kind of island (in (concentration, temperature) space) is usually located at very high temperatures (over the degradation temperature) and mainly on negative pressures – at the merging of the UCST and

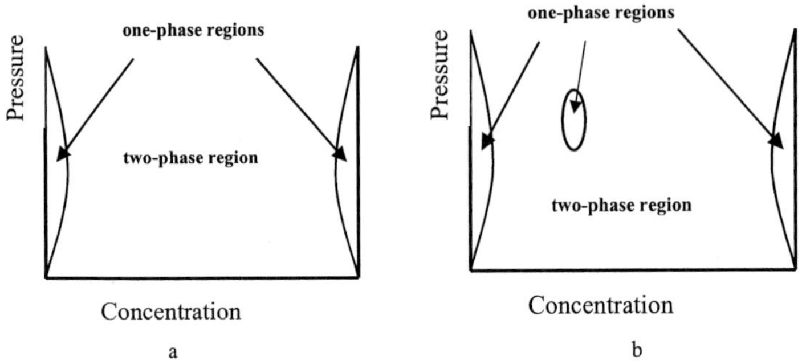

Figure 4.: Schematic representation of the cloud-point curves of two virtually immiscible polymer blends a.: without miscibility island, b.: with miscibility island.

LCST [12,31] -, this one – being in lower temperature - might give us a usable tool to homogenize some virtually incompatible systems.

Conclusions

In polymer blends (similarly to polymer solutions) an extremum (Critical Temperature Minimum) exists on the UCST, where pressure induced miscibility turns to pressure induced immiscibility . For most blends this extremum is located at negative pressure, but in the studied polyalkylsiloxane mixtures it is located on positive pressure; i.e. for these mixtures the UCST - as a function of pressure - is curved and exhibits an extremum. Due to this shape two separate branches can be seen on the concentration-pressure diagram (at constant temperature): a high pressure one and a low pressure one. Approaching the extremum on the UCST –by decreasing the temperature - these two pressure branches can merge, first by the bigger humps (forming a half hourglass-shaped cloud point curve with a miscibility peninsula) then by the smaller humps (forming the "usual" hourglass-shaped cloud point curve, but leaving a small miscibility island inside of the two-phase region). Further approaching the extremum this island shrinks and finally disappears, leaving hourglass-shaped cloud point curve, with two small one-phase parts (at very high and very low concentrations) and a big two-phase regime, located at medium concentrations.

Based on this result, two "virtually" incompatible polymer blends might have a small island of miscibility deep inside of the two-phase region. The existence of these islands might give us a novel way to homogenize virtually incompatible polymers.

Acknowledgement

The author thanks Prof. B.A. Wolf for his helpful suggestions and for providing the experimental setup for this measurement. This work was partially supported by the Hungarian Science Foundation (OTKA) under contract number F 034333. The author also thanks for the Humboldt Foundation for sponsoring his stay in Germany and for the members of the research group of Prof. Wolf (Johannes Gutenberg Univ., Mainz, Germany) – particularly T. Haase and P. Glöckner for the characterisation of the PDMS sample - for their help and hospitality. A.R.I. was also supported by the Bolyai Research Fellowship.

[1] Rebelo, L.P., Van Hook, W.A., *J. Polym. Sci. B.*, **1993**, *31*, 895
[2] Qian, C., Mumby, S.J. and Eichinger, B.E., *J. Polym. Sci. B*, **1991**, *29*, 635
[3] Rebelo, L.P.N., de Sousa, H.C. and Van Hook, W.A., *J. Polym. Sci. B.*, **1997**, *35*, 631
[4] de Sousa, H.C. and Rebelo, L.P.N., *J. Polym. Sci. B.*, **2000**, *38*, 632
[5] de Sousa, H.C. and Rebelo, L.P.N., *J. Chem. Thermodyn.*, **2000**, *32*, 355
[6] Imre, A.R., Melnichenko, G. and Van Hook, W.A., *Phys. Chem. Chem. Phys.*, **1999**, *1*, 4287
[7] Imre, A.R., Melnichenko, G. and Van Hook, W.A., *J. Polym. Sci. B*, **1999**, *37*, 2427
[8] Vanhee, S., Kiepen, F., Brinkmann, D., Borchard, W., Koningsveld, K. and Berghmans, H., *Macromol. Chem. Phys.*, **1994**, *195*, 759
[9] Imre A.R., Kraska, T. and Yelash, L., *Phys. Chem. Chem. Phys.*, **2002**, *4*, 992
[10] Steinhoff, B., Rüllmann, M., Kühne, L. and Alig, I., *J. Chem. Phys.*, **1997**, *107*, 5217
[11] Imre, A. and Van Hook, W.A., *J. Polym. Sci. B.*, **1996**, *34*, 751
[12] Imre, A. and Van Hook, W.A., *Chem.Soc.Rev.*, **1998**, *27*, 117
[13] Debenedetti, P.G., *Metastable Liquids: Concepts and Principles*, Princeton University Press, Princeton, NJ, **1996**
[14] Imre, A., Martinás, K., Rebelo, L.P.N., *J. Non-Equilib. Thermodyn.*, **1998**, *23*, 351
[15] Koningsveld, R., Onclin, M.H. and Kleintjens, L.A., in *Polymer compatibility and incompatibility, Principles and Practices*, MMI Press Symposium Series, vol.2, (Ed.: K. Solc), Harwood Academic Publisher, Chur, **1982**
[16] Kleintjens, L.A., Schoffeleers and H.M., Domingo, L., *B. Polym. J.*, **1976**, *8*, 29
[17] Koningsveld, R., and MacKnight, W.J., *Polym. Int.*, **1997**, *44*, 356
[18] Vanhee, S., Koningsveld, R., Berghmans, H. and Šolc, K. et al, *J. Polym. Sci. B.*, **1994**, *32*, 2307
[19] Šolc, K., Stockmayer, W.H., Lipson, G.E.J. and Koningsveld, in *Multiphase Macromolecular Systems*, B.M. Culberston, Ed., Plenum Press, New York, **1989**, vol. *6*, p.5
[20] Šolc, K., Kleintjens, L.A. and Koningsveld, R., *Macromolecules*, **1984**, *17*, 573
[21] Stammer, A., Wolf, B.A., *Polymer*, **1998**, *39*, 2065
[22] Imre, A.R., Melnichenko, G., Van Hook, W.A. and Wolf, B.A., *Phys. Chem. Chem. Phys.*, **2001**, *3*, 1063
[23] Tsuyumoto, M., Einaga, Y. and Fujita, H., *Polymer J.*, **1984**, *16*, 229
[24] Borchard, W., Frahn, S. and Fischer, V., *Makromol. Chem. Phys.*, **1994**, *195*, 3311
[25] Saeki, S., Kuwahara, N., Nakata, M. and Kaneko, M., *Polymer*, **1975**, *16*, 445
[26] Urbanowicz, P., Rzoska, S.J., Paluch, M., Sawicki, B., Szulc, A. and Ziolo, J., *Chem. Phys.*, **1995**, *201*, 575
[27] Economou, I.G., *Macromolecules*, **2000**, *33*, 4954
[28] Wolf, B.A., *Pure Appl. Chem.*, **1997**, *69*, 929
[29] Rabeony, M., Lohse, D.J., Garner, R.T., Han, S.J., Graessley, W.W. and Migler, K.B., *Macromolecules*, **1998**, *31*, 6513
[30] Kumar, S., *Macromolecules*, **2000**, *33*, 5285
[31] Imre, A., and Van Hook., W.A., *J. Phys. Chem. Ref. Data*, **1996**, *25*, 637 (Erratum: **1996**, *25*, 1277)

Modelling Segmental Dynamics in Miscible Polymer Blends

Eric Leroy,[1,2] *Angel Alegría,*[1] *Juan Colmenero**[1,2]

[1] Departamento de Física de Materiales, Universidad del País Vasco y Unidad de Física de Materiales, Centro Mixto CSIC-UPV/EHU, Facultad de Química, Apdo.1072, 20018 San Sebastián, Spain
[2] Fundación Donostia International Physics Center, Paseo Manuel de Lardizabal, 4 20018, San Sebastián, Spain

Summary: The complex thermorheological behaviour observed in miscible polymer blends is modelled by combining two existing theoretical approaches: The Thermal Concentration Fluctuation model (A. Zetsche, E.W. Fischer, *Acta Polymer.* **1994**, *45*, 168) and the Effective Concentration model (T.P. Lodge, T.C.B. McLeish, *Macromolecules* **2000**, *33*, 5278), giving rise to a simple model with only one adjustable parameter. This model is then tested in the case of two model blends allowing to show its abilities and limitations to describe how the respective segmental dynamics of lowest or highest T_g component are affected by blending.

Keywords: amorphous, blends, glass transition, modelling, thermoplastic polymers

Introduction

Miscible polymer blends tend to present a much more complex thermorheological behaviour than pure polymers, especially when the difference between the glass transition temperatures (ΔT_g) of the blend components is large. The broadening of the relaxation time distribution of each component is one of the key phenomena observed. Fischer and Zetsche[1] showed that it could be formally related to *Thermal Concentration Fluctuations* (TCF). In their model, a blend sample of average composition ϕ_0 is supposed to be divided into small subvolumes V_i in which the local concentration ϕ is different from average composition due to the TCF. In the temperature range where this spatial distribution of compositions is quasi-stationary at the time scale of the segmental α-relaxation, it gives rise to a distribution of local mobilities. Base on these ideas, Fischer and Zetsche propose to describe the broadening of the relaxation time distribution of each blend component by considering that each subvolume relaxes with the same non-exponential behaviour as the pure polymer, but with a different characteristic times τ. The relaxation broadening due to blending is then described by a distribution $H(\log\tau)$ of these characteristic relaxation times, which allows them to calculate the dielectric relaxation of one blend component:

$$\varepsilon''(\log\omega) = \int_{-\infty}^{\infty} H(\log\tau)\cdot \mathrm{Im}\left[\varepsilon^*(\log\omega + \log\tau)\right] d\log\tau \qquad (1)$$

 DOI: 10.1002/masy.200350803

Where ε^* is given by the Havriliak-Negami (HN) function:

$$\varepsilon^*(\omega) = \varepsilon_\infty + \Delta\varepsilon \cdot \frac{1}{\left(1+(i\omega\tau)^\alpha\right)^\gamma} \tag{2}$$

where the relaxation strength $\Delta\varepsilon$ and the exponents α and γ are those experimentally obtained from pure polymer's dielectric relaxation curves analysis.

The key idea of Fischer and Zetsche is to obtain the distribution $H(\log\tau)$ starting from a *gaussian distribution* of concentration $P(\phi)$:

$$P(\phi) \propto \exp\left[\frac{-(\phi-\phi_0)^2}{2\langle(\Delta\phi)^2\rangle}\right] \tag{3}$$

Where the variance $<(\Delta\phi)^2>$ depends of the size of the subvolumes V_i.

They then assume that the composition dependence of blend's macroscopic glass transition temperature (obtained by Dynamic Scanning Calorimetry): $T_g(\phi_0) = F(\phi_0)$, is not modified at the microscopic level, so that one can calculate the glass transition of each subvolume as $T_g(\phi)=F(\phi)$, the function F being for example the Brekner equation:

$$T_g(\phi) = T_{gA} + (T_{gB} - T_{gA})\left[(1+K_1)\cdot\phi - (K_1+K_2)\cdot\phi^2 + K_2\cdot\phi^3\right] \tag{4}$$

Where T_{gA} and T_{gB} are the pure components glass transition temperatures and K_1 and K_2 are fitting parameters determined from Differential Scanning Calorimetry (DSC) measurements.

That leads to a new distribution $N(T_g)$, which in turn can be transformed into $H(\log\tau)$ using the William-Landel-Ferry (WLF) equation:

$$\log\left(\frac{\tau}{\tau_g}\right) = \frac{C_1\cdot[T - T_g(\phi)]}{C_2 + [T - T_g(\phi)]} \tag{5}$$

Where $\log\tau_g$, C_1 and C_2 depend on the polymer component considered, these two latter being in a first approximation linear interpolations of the pure component's parameters. It is noteworthy that in the first formulation of the Fischer's model[1], it is not clear how the Jacobian term $(d\log\tau / d\phi)$ - involved in the transformation from $P(\phi)$ to $H(\log\tau)$ – is taken into account. In the later formulation given by Kumar et al.[3] this Jacobian was explicitly considered and some refinements were also introduced trying to account for another key issue of miscible blends dynamics: the observation of distinctly different segmental mobilities for the two components observed with a wide range of experimental techniques [4-10]. Based on the assumption that the fluctuation subvolumes are equal to the cooperative volumes of Donth, which introduces a strong composition and temperature dependence of $<(\Delta\phi)^2>$, Kumar's model allows a near quantitative description of the behaviour of various miscible polymer blends near T_g. Nevertheless, it seems to fail in explaining the persistence of the different component segmental mobilities well above the glass transition temperature, where the presence of concentration fluctuations does not play a significant role on the dynamic features[5, 9]

This latter point can be addressed by introducing the *Effective Concentration concept*:[2] In a miscible blend of polymers A and B, chain connectivity imposes that the local environment of a segment of polymer A is (on average) necessarily richer in polymer A compared to the *bulk composition* ϕ. The *Effective Concentration* ϕ_{eff} sensed by a polymer segment is thus given by:

$$\phi_{eff} = \phi_{self} + (1 - \phi_{self}) \cdot \phi \qquad (6)$$

where ϕ_{self} is the *Self Concentration* of the considered polymer segment, which was recently expressed by Lodge and McLeish[2] as:

$$\phi_{self} = \frac{C_\infty M_0}{k \rho N_{av} l_k^3} \qquad (7)$$

where: C_∞ is the characteristic ratio, M_0 is the molar mass of repeat unit, k is the number of backbone bonds per repeat unit, ρ is the density, N_{av} is the Avogadro number and l_k is the length of a Kuhn statistical segment. This concept rationalizes the dynamic heterogeneities observed well above T_g where TCF contribution vanishes: As we have shown in a recent experimental study[11], it allows to quantitatively predict the two different effective glass transition temperatures $T_{g\ eff\ A}$ and $T_{g\ eff\ B}$ corresponding to each component in the miscible blend and which are linked to their respective average segmental mobilities

In the present paper, we will describe how the combination of the TCF model with the Effective Concentration concept leads to a simple model where TCF variance $<(\Delta\phi)^2>$ is the only adjustable parameter. As an application, this model will be used to account for the dielectric relaxation behaviour of two model miscible polymer blends in which only one component is dielectrically active, this component being either the one having the lower glass transition temperature (PVME in PVME/PS blends) or the higher one (PoClS in PS/PoClS blends). This particularity will allow us to study selectively the efficiency and limitations of the model to describe the effect of blending on the dynamics of the lower or higher T_g component of the blend.

After discussing the results, we will propose some future improvements of the model.

Modelling

The introduction of the Effective Concentration concept in Fischer's formalism can be carried out as follows. We consider the same local concentration distribution induced by TCF, but inside each relaxing subvolume V_i of composition ϕ. We then introduce two different effective local concentrations $\phi_{eff\,A}$ and $\phi_{eff\,B}$ corresponding to the local environments of polymer segments A and B. Under this assumption we implicitly consider that the size of each subvolume V_i is much larger than l_k^3 (the volume where the self-concentration is relevant). On these grounds, $P(\phi)$ yield two distinct local concentration distributions, depending on the component considered:

$$P(\phi_{eff\,A}) = \frac{1}{1-\phi_{self\,A}} P(\phi) \quad (8a) \quad \text{and} \quad P(\phi_{eff\,B}) = \frac{1}{1-\phi_{self\,B}} P(\phi) \quad (8b)$$

This is illustrated in Figure 1, for a representative example in which the average composition of the blend ϕ_0 would be equal to 0.5 ; TCF would be characterized by $<(\Delta\phi)^2> = 0.1$ and the self concentrations $\phi_{self\,A}$ and $\phi_{self\,B}$ would be both equal to 0.25.

Following the same procedure as Fischer, but now replacing T_g in the WLF equation by T_{geff}, these distributions can be easily transformed into two distinct characteristic relaxation times distributions $H(\log\tau_A)$ and $H(\log\tau_B)$ of each component in the blend, which in turn allow to calculate the dielectric relaxation curves $\varepsilon''(\omega, T)$. The parameters C_1 and C_2 of each subvolume might be different for each polymer component and following Fischer will be taken, in a first approximation, as linear interpolations of pure polymers parameters, at the respective effective concentrations $\phi_{eff\,A}$ and $\phi_{eff\,B}$. In addition, the value of $\tau_g=\tau(T_{geff})$ will be taken fixed to the τ_g value determined for the pure component.

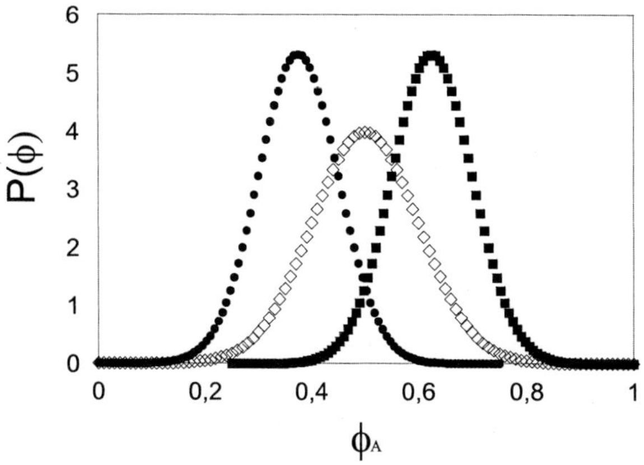

Figure 1. Example of local concentration distributions: Gaussian distribution due to TCF(◊, equation 3) and corresponding local environments of polymer A segments (■, equation 8) and polymer B segments (●, equation 8b) is due to chain connectivity

Experimental

One blend was studied for each polymer pair: a poly(vinyl methyl ether)/Polystyrene blend of weight composition (35/65) and a Poly-o-chlorostyrene/Polystyrene blend of weight composition (50/50). Poly-o-chlorostyrene ($\overline{M}_w$=225000 g/mol; $\overline{M}_w/\overline{M}_n$ =1.7) and poly(vinyl methyl ether) ($\overline{M}_w$=21900 g/mol; $\overline{M}_w/\overline{M}_n$ =3) were obtained from Lark Enterprises and Aldrich Chemical respectively. For PVME/PS blend, a Polystyrene ($\overline{M}_w$=66000 g/mol; $\overline{M}_w/\overline{M}_n$ =1.03), obtained from Polymer Laboratories was used. In contrast, a Polystyrene oligomer ($\overline{M}_w$=740 g/mol; $\overline{M}_w/\overline{M}_n$ =1.06) purchased Polymer Source was used for PS/PoClS blends in order to increase the difference between the T_g of the components. Blends were prepared by solution casting using toluene as solvent. Obtained films were carefully dried under vacuum, above their glass transition temperatures during 72 hours, in order to remove the solvent completely. Reference samples of PVME, PoClS and PS were prepared in a similar way.

Dielectric relaxation measurements were performed on two different set-ups supplied by Novocontrol GmbH. For the frequency range from 10^{-3} to 10^{7} Hz the set-up consisted of a Solartron-Schlumberger frequency response analyser SI 1260 supplemented by a high-impedance preamplifier of variable gain. For the frequency range from 10^{6} to 10^{9} Hz a Hewlett-Packard impedance analyser HP4191A was used. For both ranges the sample was kept between two condenser plates (gold-plated electrodes, 20 and 5 mm diameter, respectively) that were maintained at a fixed distance. Frequency sweeps were performed at constant temperature with temperature stability better than 0.05 K.

Results

In order to have all the parameters required by the model to describe the blends' dielectric relaxation, we first focus on the analysis and modelling of pure components' behaviour. Actually, the dielectric relaxation of PVME, PS and their blends were already studied in detail by our team in a previous paper[6]. The α-relaxation of pure PVME was successfully described by an HN function (Equation 2) with the condition:

$$\gamma(\alpha) = 1 - 0.812 \cdot (1-\alpha)^{0.387} \tag{9}$$

This imposed relationship between the parameters α and γ allows to calculate the corresponding parameters of the Kohlrausch-Williams-Watts (KWW) time function as:

$$\beta = [\alpha \cdot \gamma(\alpha)]^{1/1.23} \tag{10}$$

$$\log \tau_{KWW} = \log \tau_{HN} - 2.6 \cdot (1-\beta)^{0.5} \exp(-3\beta) \tag{11}$$

Therefore the dielectric relaxation of PVME at a given distance from T_g is described by only two parameters, β and $log\ \tau_{KWW}$, plus the relaxation strength which temperature dependence was found to be well described by:

$$\Delta\varepsilon(T) = -0.0125 + 0.23438 \cdot 1000/T \tag{12}$$

Parameter β was found to increase slightly with temperature above T_g, reaching a stable value of 0.5 at higher temperatures. This behaviour has been observed for different polymers[12]. For our calculation of PS/PVME blend's behaviour, we will consider a fixed value β = 0.5. The relaxation time of pure polymers followed the WLF equation with $log\ \tau_{g\ PVME}$ = 0.159; $C_{1\ PVME}$ = 13.20; $C_{2\ PVME}$ = 49.5 K ; $log\ \tau_{g\ PS}$ = -0.769 ; $C_{1\ PS}$ = 11.75 and $C_{2\ PS}$ = 47.0 K.

Another important aspect of this previous work[6] on dielectric behaviour of PS, PVME and their blends was the successful modelling of the *β-relaxation* of PVME which was shown not to be affected by blending. The model obtained allowed us to calculated the contribution of this secondary relaxation in the dielectric curves and to subtract it from experimental curves in order to observe only the *α-relaxation* contribution which we intend to model.

The relaxation behaviour of pure PoClS was studied following the same procedure as for pure PVME. The modified HN function described above successfully fitted the dielectric relaxation curves. The parameter β was found to vary much more near T_g than for PVME. Nevertheless, a few degrees above T_g, β reaches a rather constant value (β = 0.46), that we will use for calculations. Also the relaxation strength $\Delta\varepsilon$ was found to vary strongly with temperature near T_g and reaches at nearly constant value at high temperature. We will consider this high temperature limit value $\Delta\varepsilon$ = 1.2 for the calculations. The relaxation time of pure PoClS and PS oligomer follows the WLF equation with $log\ \tau_{g\ PoClS}$ = -0.132; $C_{1\ PoClS}$ = 12.36; $C_{2\ PoClS}$ = 67.1 K; $log\ \tau_{g\ PSo}$ = -0.632; $C_{1\ PSo}$ = 14.37 and $C_{2\ PSo}$ = 53.7 K.

The self concentration values ($\phi_{selfPVME}$=0.25 and $\phi_{selfPOCLS}$ = 0.22), as well as the pure components T_g and parameters K_1 and K_2 for the two blends were taken from our previous publication[11] . These latter parameters are listed in Table 1.

Let us now focus on the relaxation behaviour of the blends: Figures 2 and 3 show the relaxations of PVME in the (35/65) PS/PVME blend and PoClS in the (50/50) PS/PoClS blend, respectively. One can clearly observe the relaxation broadening due to blending, which is much more important in the case of the PS/PVME blend.

Discussion

The model was fitted to the experimental curves with $<(\Delta\phi)^2>$ as the only adjustable parameter. Figure 2 shows that a near quantitative prediction of dielectric relaxation curves is obtained for PVME/PS blend using a fixed value of $<(\Delta\phi)^2>=0.04$ on the whole temperature range. This result is different from those obtained by Fischer and Zetsche for the same system, whose model, in order to fit experimental curves, needed much lower and clearly temperature dependent values of $<(\Delta\phi)^2>$, together with an adjustment of $\Delta\varepsilon$, C_1 and C_2.

Regarding the accuracy of model predictions, three points have to be enlighten: First, the position of peak maximum and height are reasonably well described despite of the simple assumption made on the values of $\Delta\varepsilon$, C_1 and C_2. The use of $\Delta\varepsilon$ as a free parameter for each temperature and the adjustment C_1 and C_2 via the iterative procedure proposed by Fischer and Zetsche in their paper would of course lead to even better predictions. Second, the description of relaxation broadening toward low frequencies (i.e., slow part of the response) is very good. Nevertheless, the third point we shall stress is that the broadening toward high frequencies (i.e., fast part of the response) is not so well described near T_g. This failure was actually predictable as a previous study of PS/PVME blends[6] showed that close to T_g, the dynamics of the PVME segments in the blend extends to high frequencies more than in pure polymer at the same temperature. This phenomenon, which can be interpreted as the result of a lack of packing of the blend leading to regions with greater "free volume" than in pure PVME, is obviously not taken into account by the simple model we are considering here.

For the PoClS/PS blend (Figure 3), the model seems to describe reasonably well the relaxation broadening using a value of $<(\Delta\phi)^2>=0.0036$. Nevertheless, the prediction of the dielectric curve maximum position is not correct. This evidence that the time scale of the dynamics of PoClS in the blend calculated in the model is far from that in the experiment. A possible reason for this failure is the fact that we fixed τ_g of the component in blend to be the same than that of the pure polymer. This condition is approximately verified for the PVME/PS blend studied (the temperature where $1/\omega_{max}$ equals τ_g (261.5 K) is almost the same that $T_{geff\ PVME}$ = 258.7 K), but clearly it is not for the PoClS/PS blend (in that case the temperature where $1/\omega_{max}$ equals τ_g (345.0 K) is clearly different from $T_{geff\ PoClS}$ = 332.5 K). In order to illustrate the effect of this difference of 12.5 K we have plotted as a dashed line in Figure 3 the curve predicted by the model at T=350 – 12.5 = 337.5 K. One can see that the obtained curve describes quite well the experimental data of the blend at 350 K.

Table 1. Parameters of Brekner's equation used for calculations

Blend A/B	T_{gA}	T_{gB}	K_1	K_2
PVME/PS	249.5	373	-0.707	0.462
PSo/PoClS	280	402	-0.296	0.792

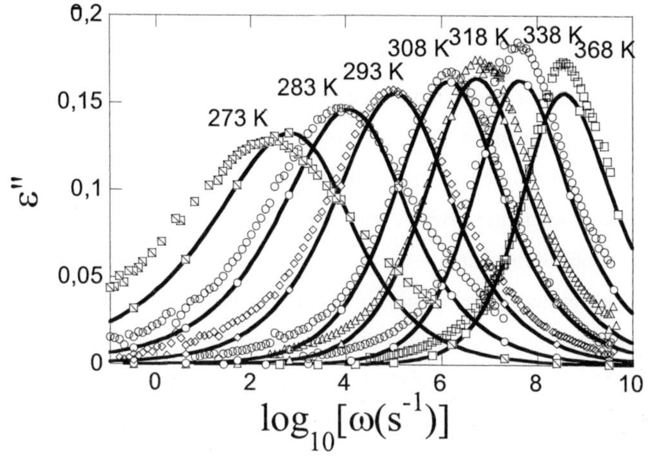

Figure 2. Dielectric relaxation of a PS/PVME (35/65) blend: lines show model predictions

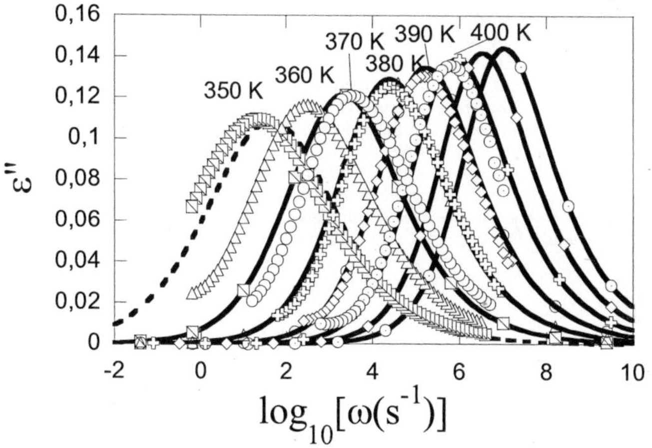

Figure 3. Dielectric relaxation of the PoClS/PSo (50/50) blend: solid lines with points show model predictions at each experimental temperature (solid line with circles on the right side is the prediction for the 400 K experiment in circles), dashed line show model prediction at 337.7 K

Conclusion

The introduction of the *Effective Concentration concept* into the Concentration Fluctuation Model of Fischer lead to a simple model in which *Thermal Concentration Fluctuation* variance $<(\Delta\phi)^2>$ is the only adjustable parameter. As a test, such a model was used to calculate the dielectric α-relaxation curves of two model polymer blends. The results obtained for the

PVME/PS blend, showed that the model seems to give a quite satisfactory description of how the dynamics of the lowest T_g component of the blend (PVME in that case) is affected by blending, except for high frequencies and temperatures close to blend's glass transition, where some additional consideration on blend packing should be introduced. As an opposition with the previous work of Fischer on the same polymer binary blend, we shall point out that, apart from requiring only the adjustment of $<(\Delta\phi)^2>$, our model did not show any apparent need for a strong temperature dependence of this latter.

The results obtained for the second blend studied (PoClS/PS) shown that the model seems to be able to describe the relaxation broadening effect of blending on the highest T_g component of the blend (here PoClS), but not the evolution of relaxation peak maximum with temperature. Apparently, this problem lies in how the isothermal relaxation time is calculated from the glass transition. The refinement of this part of the model is now in progress.

[1] A. Zetsche, E.W. Fischer, *Acta Polymer.* **1994**, *45*, 168.
[2] T.P. Lodge, T.C.B. McLeish, *Macromolecules* **2000**, *33*, 5278.
[3] S.K. Kumar, R.H. Colby, S.H. Anastasiadis, G. Fytas, *J. Chem. Phys.* **1996**, *105*, 3777.
[4] G.C. Chung, J.A. Kornfield, S.D. Smith, *Macromolecules* **1994**, *27*, 5729.
[5] B. Min, X. Qiu, M.D. Ediger, M. Pitsikalis, N. Hadjichristidis, *Macromolecules* **2001**, *34*, 4466.
[6] I. Cendoya, A. Alegria, J. Colmenero, H. Grimm, D. Richter, B. Frick, *Macromolecules*, **1999**, *32*, 4065.
[7] B.H. Arendt, R.M. Kannan, M. Zewail, J.A. Kornfield, S.D. Smith, *Rheologica Acta* **1994**, *33*, 322.
[8] S. Adams, D.B. Adolf, *Macromolecules* **1999**, *32*, 3136.
[9] S. Hoffmann, L. Willner, D. Richter, A. Arbe, J. Colmenero, B. Farago, *Phys. Rev. Let.* **2000**, *85*, 772.
[10] X. Yang, A. Halasa, W.L. Hsu, S.Q. Wang, *Macromolecules* **2001**, *34*, 8532.
[11] E. Leroy, A. Alegria, J. Colmenero, *Macromolecules* **2002**, *35*, 5588
[12] A. Alegria, J. Colmenero, P.O. Mari, I.A. Campbell Phys. Rev. E **1999**, *59*, 6888.

Crystallization and Structure Formation in Polymer Blends with Strong Intermolecular Interactions: Blends of Poly(ethylene oxide) and Styrene-Hydroxystyrene Copolymers

Jaedong Cho,[1] *Ruijian Xu,*[1] *Fengji Yeh,*[2] *Benjamin S. Hsiao,*[2] *James Runt**[1]

[1]Department of Materials Science and Engineering and Materials Research Institute, The Pennsylvania State University, University Park, PA 16802 USA
[2]Department of Chemistry, State University of New York, Stony Brook, NY 11794 USA

Summary: Time-resolved synchrotron wide- and small-angle X-ray scattering experiments were used to investigate crystallization behavior and microstructure development of a nearly monodisperse poly(ethylene oxide) [PEO] (M_w = 53,500), and its melt-miscible blends with two fractionated styrene – hydroxystyrene random copolymers [SHS]. PEO crystallization rates decrease significantly in the presence of the melt-miscible SHS copolymers. All low and high molecular weight SHS blends exhibit a crystallization process at relatively short times characterized by large Avrami exponents (n), followed by a dominant process with n near that of neat PEO. A model for the crystallization of these blends is proposed.

Keywords: crystallization, poly(ethylene oxide), polymer blends, SAXS, WAXD

Introduction

In order to gain greater insight into the influence of blending on crystallization and lamellar organization, we have undertaken a series of experimental studies on polymer blends containing poly(ethylene oxide) [PEO].[1,2,3] For example, time-resolved synchrotron wide- (WAXD) and small-angle X-ray scattering (SAXS) experiments were conducted to investigate the crystallization behavior and microstructure development of PEO and several melt-miscible PEO blends.[3] Blends were prepared with both weakly-interacting poly(methyl methacrylate), and two copolymers that exhibit relatively strong intermolecular interactions with PEO. However, it was difficult to assign specific origins to the different stages of the crystallinity – time behavior of the blends because of uncertainty due to the possible effects of PEO molecular weight fractionation, and contributions from fractionation and/or segregation of polymer diluent molecules. In an attempt to simplify the behavior and facilitate interpretation, the current paper

 DOI: 10.1002/masy.200350804

summarizes our investigation of crystallization and microstructure development in PEO blends, but this time using a narrow PEO molecular weight fraction and two fractionated samples of one of the strongly-interacting random copolymers used in our previous experiments.

Experimental

The PEO molecular weight fraction used in this study was purchased from Polymer Source Inc. and has a weight average molecular weight (M_w) of 53,500 and $M_w/M_n = 1.07$. Two representative 'fractions' of a 50/50 random copolymer of styrene and hydroxystyrene (SHS) were selected for these experiments. The higher molecular weight sample (HMW-SHS) was determined to have a somewhat lower SHS content (44 wt %) and an apparent weight average molecular weight of 72,700 ($M_w/M_n = 2.1$). This molecular weight was determined using GPC with dimethylformamide - 0.05 M LiBr as the mobile phase. Molecular weights are relative to PEO standards. The lower molecular weight SHS fraction (LMW-SHS) contained 50 wt % styrene and has an apparent $M_w = 31,700$ ($M_w/M_n = 1.5$). T_gs for the LWM- and HMW-SHS were determined by differential scanning calorimetry (DSC) to be ~ 150 °C.

PEO/SHS blends were prepared from 2 wt % solutions of 50/50 (w/w) THF/CH_2Cl_2. Films were cast on glass slides and dried in air at room temperature for 48 h and then under vacuum at 90 °C for 6 h. The dried films were then pressed at 69 MPa and 100 °C for 3 min and rapidly cooled in air to room temperature. PEO films were prepared by pressing powdery PEO under the same conditions as the blends. Samples were deposited into a small copper cell for time-resolved experiments.

Time-resolved simultaneous WAXD/SAXS experiments were conducted at the U.S. National Synchrotron Light Source, Brookhaven National Laboratory on the Advanced Polymers Beamline, X27C. The X-ray wavelength (λ) was 0.1309 nm and pinhole collimation was utilized. For SAXS, the sample-to-detector distance was 1.509 m, covering an angular range up to q ~ 1.0 nm^{-1} ($q = \frac{4\pi}{\lambda}\sin(\theta/2)$ and θ is the scattering angle). The WAXD 2θ range was ~8° to 40°, relative to $\lambda = 0.154$ nm. The scattering patterns from a duck tendon and a Lupolen polyethylene standard were used to calibrate the SAXS and WAXD detectors, respectively.

The real-time WAXD/SAXS experiments were conducted using a specially designed sample holder to facilitate a rapid jump between the melt temperature and selected T_cs. All samples were maintained for 5 min in the melt chamber before transferring to the crystallization chamber. SAXS data were first smoothed using a moving window averaging procedure, then background scattering due to thermal density fluctuations was subtracted using the procedure described in ref 4. Crystalline and amorphous layer thicknesses within lamellar stacks were determined using the calculated one-dimensional correlation function, $\gamma(r)$.[5] The relative invariant (Q) was obtained as the ordinate of the linear fit to the self-correlation portion of the correlation function. Defining B as the value of the abscissa where the ordinate first equals zero in $\gamma(r)$, and obtaining the long period (L) from the first maximum of $\gamma(r)$, $x_c(1 - x_c) = B/L$, where x_c is the linear crystallinity. Although one cannot apriori distinguish x_c from x_a ($= 1 - x_c$), it was shown in our earlier work that x_c for a polydisperse PEO was greater than 0.7.[1] Since the PEO under consideration here has lower average molecular weight, lamellar thicknesses for the strongly-interacting blends are greater than those of PEO at the same T_c, and SHS was found previously to reside primarily outside of lamellar stacks,[1] the larger of the two values of x_i for these blends can confidently be taken as x_c. The average lamellar thickness is then calculated from: $l_c = x_c \cdot L$. Likewise, the average amorphous layer thickness (l_a) within lamellar stacks is obtained from: $l_a = (1 - x_c) \cdot L$. Degrees of crystallinity were determined from the WAXD data using a curve fitting program in which a given diffraction profile was separated into three crystalline PEO reflections and an amorphous halo, assuming Gaussian profiles. The apparent degree of crystallinity was defined as the ratio of the area under the resolved crystalline peaks to the total area.

Results and Discussion

1 Poly(ethylene oxide)

Development of the crystallinity and lamellar microstructure of the 53.5K PEO fraction was followed at three crystallization temperatures (T_c): 45, 48 and 52 °C. No systematic changes in the positions of the WAXD crystalline reflections or widths at half peak height were detected during crystallization of either neat PEO or PEO in blends with SHS.

We observe a second order SAXS reflection for PEO crystallized at all T_cs, indicative of the development of relatively well-ordered lamellar stacks. For all PEO samples, the development of the SAXS invariant during crystallization generally mirrors that of the WAXD crystallinity. However, Q is found to consistently develop prior to the appearance of WAXD diffraction peaks, an observation common to many crystallizing polymer systems. The origin of this remains unclear: it has been variously associated with spinodal decomposition,[6] relative detector sensitivities,[7] and primary nucleation.[8]

PEO crystallization data are presented in in Figure 1. The Avrami equation can be written as:[9]

$$1 - X_c = \exp(-zt^n) \qquad (1)$$

where X_c is the fraction of the final crystallinity developed up to time *t*, *z* is a function of the nucleation and growth rates, and *n* is in principle related to the growth geometry and type of nucleation. Although there is some noise in the data, the Avrami plots for PEO crystallized at all three temperatures up to $X_c \sim 92\%$, are best fit by single values of *z* and *n* (see Table 1). This is in contrast to our earlier research on polydisperse PEO in which we observed a distinct transition to a second process (having $n \sim 1$) in the vicinity of $X_c \sim 0.6 - 0.7$.[3] The current results confirm that this longer time process originates from molecular weight fraction during crystallization.[10] Fitted Avrami exponents for the three samples range from 2.6 to 3.8, more or less consistent with spherulitic crystallization (the latter observed in optical microscopy experiments).

Long period and lamellar thickness are frequently observed to decrease early in the crystallization process in time-resolved SAXS experiments of polymer crystallization.[e.g.3,11,12] Such behavior is often associated with crystallization ('insertion') of new lamella in amorphous regions between or within lamellae stacks.[e.g.13,14] Crystallization is rapid at 45 and 48 °C for the 53.5K PEO and the SAXS signal-to-noise ratio is low at short crystallization times (t_c), leading to uncertainty in the analysis of $\gamma(r)$. However, reliable correlation functions can be calculated over the course of crystallization at $T_c = 52$ °C. Both l_c and l_a decrease by ~2 nm with increasing time, becoming constant at longer t_c, consistent with the idea of crystallization of subsidiary lamellae (with restricted lamellar thickness) in non-crystalline regions between dominant lamellae. As expected for homopolymers, l_c and the long period at the conclusion of crystallization increase with increasing T_c (from 17.4 to 20.5 nm, and 23 to28.2 nm, respectively) and the values compare favorably to those determined in our previously experiments on unfractionated PEO.[1,3]

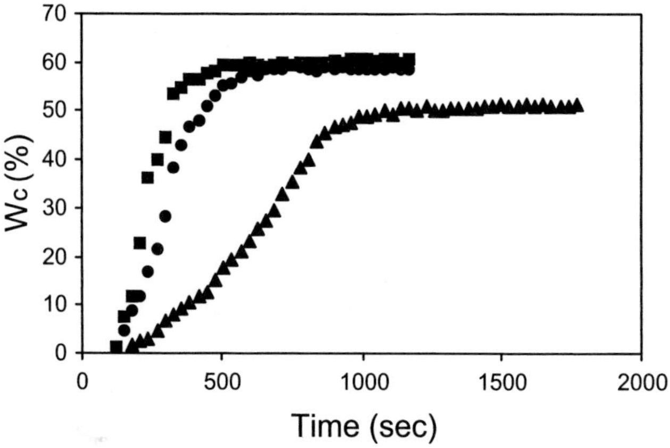

Fig. 1: Development of WAXD crystallinity for PEO as a function of crystallization time at 45 (■), 48 (●) and 52 °C (▲).

Table 1. Avrami parameters and final bulk crystallinities for PEO fractions.

T_c (°C)	$t_{1/2}$ (sec)[a]	W_c (%)	n	z
45	255	68	3.8	3.1×10^{-12}
48	310	62	2.9	7.4×10^{-3}
52	630	54	2.6	5.0×10^{-8}

[a] crystallization half-time

2 PEO/SHS Blends: LMW-SHS

The crystallization behavior of PEO/LMW-SHS blends at T_c = 45 and 48 °C was found to be very similar, and for the sake of brevity only the WAXD results for crystallization at 48 °C will be presented. Bulk crystallization rates for all SHS blends are considerably slower than PEO.

Representative Avrami plots are displayed in Figure 2 and fit parameters in Table 2. In addition to the role of SHS as a miscible polymeric diluent, this arises from a significant reduction in the equilibrium melting point (T_m^0) of PEO in the presence of SHS, due to their relatively strong intermolecular hydrogen bonding.[1] At a given T_c, crystallization proceeds at a significantly smaller degree of supercooling, ΔT, for the SHS blends, compared to neat PEO.

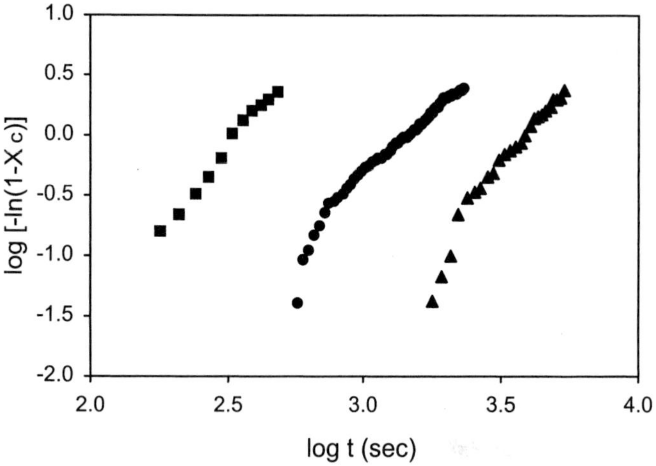

Fig. 2: Avrami plots for PEO and PEO/LMW-SHS blends at T_c = 48 °C. PEO (■), 90/10 LMW-SHS blend (●) and 80/20 LMW-SHS (▲).

The crystallization behavior of the SHS/PEO blends is more complex than that of the neat PEO fraction, even when containing only 10% SHS. As seen in Figure 2, 90/10 and 80/20 LMW-SHS blends appear to exhibit a relatively short time process (through ~20 - 25 % relative crystallinity) with a large apparent *n* (~6 - 7), similar to our previous findings on blends prepared with polydisperse PEO and SHS.[3] Although these large exponents are determined by only a modest number of data points, we consistently observe the same behavior for SHS blends with unfractionated or fractionated PEOs. In fact, similar behavior is observed in DSC studies of crystallization of these same blends.[15] Large Avrami exponents have been predicted for 3D crystallization of branched sheaf-like structures.[16,17] For all SHS blends, the first process is followed by a second dominant one having a 'more typical' Avrami exponent in the range of 2.0 – 2.6. The early time process has no counterpart in the behavior of fractionated (or unfractionated) PEO, crystallized at this or any other T_c investigated, and seems to be uniquely associated with the influence of the polymeric diluent on the crystallization process.

Table 2. Avrami parameters for PEO/LMW-SHS blends crystallized at 48 °C.

Composition	$t_{1/2}$ (s)	W_c (%)	$X_{c,s}$ [a](%)	n	z
PEO	310	62	-	2.9	7.4×10^{-3}
90/10 SHS	1190	54	24	6.3	2.2×10^{-20}
				2.0	6.2×10^{-7}
80/20 SHS	3270	48	20	7.2	1.8×10^{-25}
				2.6	3.9×10^{-10}

[a] relative crystallinity at the end of the initial process

The crystallization time dependence of the SAXS microstructural parameters for SHS blends with <20% LMW-SHS (crystallized at 45 and 48 °C) is similar to that of PEO. However, blends containing 20% LMW-SHS, crystallized at 45 and 48 °C, exhibit a relatively large drop in L and l_c early in the crystallization process: L and l_c both decrease by 5 – 10 nm. An example of this behavior is provided in Figure 3. Some SHS chains become trapped in interlamellar regions, and we have shown previously that the remainder resides between stacks/fibrils.[1] This large drop in L and l_c early in crystallinity development is similar to the behavior we observed previously for polydisperse PEO blends with PMMA and SHS blends.[3] There is a close relationship between the crossover time from high n to more typical n values in the Avrami plots, and the time at which L and l_c become time independent.

Based on the results of the time-resolved WAXD/SAXS studies of blends with fractionated and polydisperse polymers, we propose the following model for crystallization of PEO/SHS blends. Early in crystallization, the process is envisioned as occurring via growth of lamellar stacks organized into a branched sheaf-like structure (that is, one with large n). Booth and Hay estimated that n can be as high as 14 for a highly branched structure.[17] As crystallization proceeds, growth of subsidiary stacks in uncrystallized regions between primary stacks seems likely.[18] To be consistent with the experimental data, crystallization of secondary lamellar stacks must occur with restricted lamellar thickness (and long period) leading to a reduction in mean L and l_c, as observed experimentally. Smaller l_c in regions expected to be enriched in SHS seems at first to be counterintuitive, but could result from restricted lamellar thickening during formation of stable lamellae.[19] At longer times, L and l_c reach a plateau, and we speculate that crystallization of secondary stacks approaches the superstructure boundary. This would lead to a return to a 'normal' Avrami exponent for the remainder of the primary crystallization process.[16]

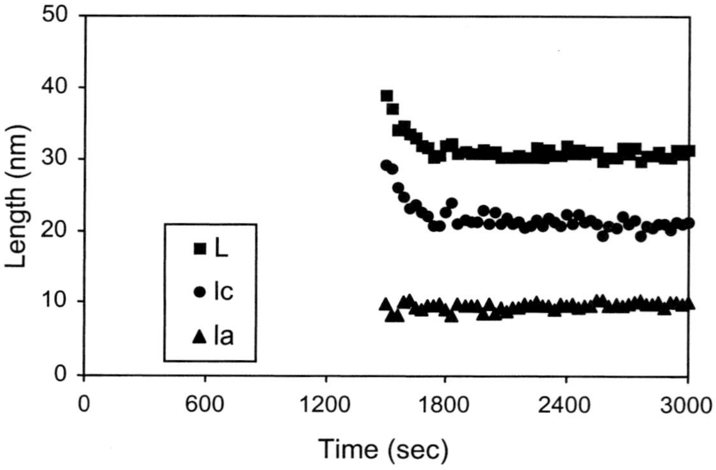

Fig. 3: Dependence of microstructural parameters on crystallization time for the 20 % LMW-SHS/PEO blend crystallized at 45 °C.

Table 3 provides a summary of L, l_c and l_a at the end of the time-resolved experiments for PEO and PEO/LMW-SHS blends crystallized at 45 and 48 °C. In general, the increase in long period results from both an increase in l_a, due to SHS chains trapped in interlamellar regions, and an increase in l_c due to a reduction in ΔT resulting from the presence of the strongly-interacting SHS. The former conclusion is in line with the significantly larger invariant observed for the blends in static SAXS experiments compared to that of neat PEO (SHS has an electron density of 0.573 vs.

0.612 mol e/cc for amorphous PEO).[20] Based on the experimental values for l_c we estimate that the equilibrium melting point is depressed by ~5 °C for the 20 % SHS blend, the same as estimated for 20% blends of polydisperse PEO and SHS.[1] The increase in l_a with SHS content is significantly higher than observed previously for comparable polydisperse blends. This is likely a consequence of more rapid crystallization of the blends in this study compared to the polydisperse PEO/SHS blends, trapping more SHS in interlamellar regions in the present case.

Table 3. Microstructural parameters for PEO and PEO/LMW-SHS blends at the conclusion of time-resolved experiments.

Composition	T_c (°C)	L (nm)	l_c (nm)	l_a (nm)
PEO	45	23.0	17.4	5.6
	48	26.2	20.7	5.5
90/10 SHS	45	26.2	19.3	6.9
	48	28.8	20.8	8.0
85/15	45	28.0	19.3	8.7
	48	30.5	20.8	9.7
80/20	45	31.5	21.3	10.2
	48	34.1	22.4	11.7

3 PEO/SHS Blends: HMW-SHS

Like the LMW-SHS/PEO blends, the behavior at T_c = 48 °C is nearly the same for the HMW-SHS/PEO blends crystallized at 45 °C, and only the former is discussed here. The crystallization behavior is very similar to that of the LMW-SHS blends: a first process with large apparent *n*, followed by a second dominant process with Avrami exponent similar to that of the neat PEO.

Overall crystallization rates are significantly faster for 90/10 HMW-SHS than the comparable LMW-SHS blends. This is likely a consequence of higher mobility and segregation length scale of LMW-SHS, and relative accumulation of LMW-SHS in the uncrystallized phase, inhibiting crystallization. This is also in keeping with the higher degrees of crystallinity observed for 90/10 HMW-SHS blends at the conclusion of the time-resolved experiments. In contrast, however, overall crystallization rates and degrees of crystallinity are comparable for the much more slowly crystallizing 80/20 LMW- and HMW-SHS blends. This implies that crystal growth has been slowed sufficiently (relative to SHS diffusion) so that it dominates the determination of the impurity segregation length scale.

The microstructure parameters at the conclusion of the SAXS experiments are quite similar for the low and high molecular weight SHS blends (see Table 4 for a summary of the PEO/HMW-SHS blend data), implying the same reduction in T_m^0 for the two SHS copolymers. In addition, the crystallization time dependence of the microstructural parameters for the HMW- and LMW-SHS blends is similar.

Table 4. Microstructure parameters for PEO and PEO/HMW-SHS blends at the conclusion of time-resolved experiments.

Composition	T_c (°C)	L (nm)	l_c (nm)	l_a (nm)
PEO	45	23.0	17.4	5.6
	48	26.2	20.7	5.5
90/10 SHS	45	25.6	18.8	6.8
	48	27.6	19.6	8.0
80/20	45	30.9	21.8	9.1
	48	36.7	25.5	11.2

Conclusion

The WAXD crystallinity – time data for the 53.5K PEO molecular weight fraction, crystallized between 45 and 52 °C, are best fit by a single Avrami expression. This is in contrast to our earlier research on a polydisperse PEO in which a distinct transition to n ~ 1 behavior in the vicinity of $X_c \sim 0.6 - 0.7$ was observed.[3] The results in the present paper demonstrate that this longer term process originates from molecular weight fractionation of the polydisperse PEO during crystallization.

PEO crystallization rates decrease significantly in the presence of the melt-miscible SHS copolymers. All LMW- and HMW-SHS/PEO blends exhibit a relatively short-time crystallization process characterized by large n values, followed by a dominant process with n near that of PEO. The magnitude of the initial decrease in mean L and l_c is significantly larger for blends containing either LMW- or HMW-SHS. In addition, the time at which L and l_c become time independent is similar to that at which the Avrami plots crossover from high n to more typical values. Our data leads us to propose that crystallization occurs initially by growth of lamellar stacks organized into a branched sheaf-like structure. As crystallization continues, growth of secondary lamella stacks in uncrystallized regions between primary stacks seems likely. To be consistent with the experimental data, crystallization of secondary stacks must occur with restricted lamellar thickness (and long period). At longer times, L and l_c become constant and crystallization of secondary stacks presumably approaches the superstructure boundary, leading to a typical Avrami exponent for the remainder of the primary crystallization process.

Acknowledgments

The Penn State authors would like to express their appreciation to the National Science Foundation, Polymers Program (DMR – 9900638 and – 0211056) for its support of this research. BH and FY thank a Department of Energy grant (DEFG0299ER45760) for support of the AP-PRT beamline.

[1] S. Talibuddin, L. Wu, J. Runt, J.S. Lin. *Macromolecules* **1996** *29*, 7527.
[2] L. Wu, M.S. Lisowski, S. Talibuddin, J. Runt. *Macromolecules* **1999**, *32*, 1576.
[3] M.S. Lisowski, Q. Liu, J.D. Cho, J. Runt, F. Yeh, B.S. Hsiao. *Macromolecules* **2000**, *33*, 4842.
[4] B.S. Hsiao, R.K. Verma. *J. Synchrontron Rad.* **1998**, *5*, 23.
[5] G.R. Strobl, M. Schneider. *J. Polym. Sci., Polym. Phys.* **1980**, *18*, 1343.
[6] P.D. Olmsted, W.C.K. Poon, T.C.B. McLeish, N.J. Terrill, A.J. Ryan. *Phys. Rev. Lett.* **1998**, *81*, 373.
[7] Z.-G.Wang, B.S. Hsiao, C. Kopp, E.B. Sirota, S. Srinivas. *Polymer* **2000**, *41*, 8825.
[8] E.-Q. Chen, X. Weng, A. Zhang, I. Mann, F.W. Harris, S.Z.D. Cheng, R. Stein, B.S. Hsiao, F. Yeh. *Macromol.Rapid Commun.* **2001**, *22*, 611.
[9] M.J. Avrami. *Chem. Phys.* **1939**, *7*, 1103.
[10] S.Z.D. Cheng, B. Wunderlich. *J. Polym. Sci. Polym. Phys.* **1986**, *24*, 595.
[11] R. Verma, H. Marand, B.S. Hsiao. *Macromolecules* **1996**, *29*, 7767.

[12] D.A. Ivanov, R. Legras, A.M. Jonas. *Macromolecules* **1999**, *32*, 1582.
[13] G. Dreezen, N. Mischenko, M.H.J. Koch, H. Reynaers, G. Groeninckx. *Macromolecules* **1999**, *32*, 4015.
[14] R. Kolb, C. Wutz, N. Stribeck, G. von Krosigk, C. Riekel. *Polymer* **2001**, *42*, 5257.
[15] M. Kanchanasopa, J. Runt. Unpublished results.
[16] B. Wunderlich. *Macromolecular Physics*. New York: Academic Press, **1976** (Vol. 2).
[17] A. Booth, J.N. Hay. *Brit. Polym. J.* **1972**, *4*, 9.
[18] F. Yeh, B.S. Hsiao, B. Chu, B.B. Sauer, E.A. Flexman. *J. Polym. Sci. Polym. Phys.* **1999**, *37*, 3115.
[19] J. Runt. *Macromolecules* **1981**, *14*, 420.
[20] J.D. Cho, J.T. Garrett, J.S. Lin, J. Runt. Manuscript in preparation.

Shear Influence on the Phase Behavior of Systems Containing a Homopolymer A and a Block Copolymer AB

Samy A. Madbouly

Cairo University, Faculty of Science, Department of Chemistry, Orman-Giza 12613, Egypt

Summary: Cloud point temperatures (T_{cp}) and crystallization temperatures ($T_{l/s}$) of the ternary system tetrahydronaphthalene/poly(ethylene oxide)/poly(dimethyl siloxane-*b*-ethylene oxide) have been measured at different constant shear rates using a rheo-optical device and an advanced rheometer. The cloud points temperatures (UCST-type phase diagram) are reduced by several degrees as the system flows; i.e. the shear can suppress the phase separation and enlarge the homogenous region. The crystallization kinetics of PEO in the ternary mixtures has been investigated isothermally and non-isothermally at quiescent state and under shear. The shear could strongly enhance the crystallization i.e. the $T_{l/s}$ shifts to higher temperatures and the induction time, t_0 (the time needs for the onset of crystallization) substantially decreases with increasing shear rate during the non-isothermal and isothermal crystallization processes, respectively. The isothermal crystallization kinetics at quiescent state and at different shear rates was analyzed on the bases of Avrami approach. The Avrami exponent which provides qualitative information about the nature of the nucleation and growth process, was found to be shear rate and temperature dependent. The Avrami exponent increased from ~3 at the quiescent state to as large as 9 at $\dot{\gamma} = 100\ s^{-1}$.

Keywords: Avrami exponent, cloud point, crystallization kinetics, shear rate, ternary mixture

Introduction

The influence of flow on the phase behavior of polymer blends and polymer solutions has been recognized for several decades.[1-3] Knowledge of such behavior would aid in the process design and control the final products. Shear-induced mixing and demixing have been observed, as opposed to the case of small molecular mixtures where only shear-induced mixing reported. In the early study, Silberberg and Kuhn[1] have investigated a ternary polymer mixture of polystyrene/ethyl cellulose in benzene under steady shear flow. They found that the polymer droplets were elongated at moderate shear rates. At higher shear rates the droplets break-up into small sizes leading to homogenization. An increase in the cloud point temperature as large as 28K for solutions of polystyrene in dioctyl phthalate at shear rate

 DOI: 10.1002/masy.200350805

up to 1360 s^{-1} has been observed by Rangel-Nafaile et al.[2] The effects produced by shear have been different perpendicular and parallel to the flow direction for the polystyrene/ polybutadiene/dioctylphthalate (PS/PB/DOP) mixture.[3] Previously we have also initiated systematic experimental and theoretical investigations of shear-induced phase changes in polymer blends and solutions.[4-9] One of the central observation of that work concerns the dominant role the rheological properties of polymers play for shear effects.

A great deal of effort has been devoted to elucidate the structure formation of crystalline polymers under shear flow.[10-15] The earliest work was reported by Trelour and Gent who observed a decrease in the onset of crystallization for cross-linked natural rubber with increasing extension ratio.[10,11] Fritzsche et al. investigated the crystallization kinetics of poly(ethylene oxide), PEO, under varying condition of shear rate and temperature.[12,13] They found that the kinetics of crystallization process was altered by shear, which was seen as a decrease in the induction time and elevation of the Avrami exponent. Eder et al.summarized some of the efforts made in the past to cast more light on the mechanisms of flow-induced crystallization in polymer melts.[15] Similar investigations for other systems, which led to the same conclusion, have recently been reported.[16-18]

In this work we shall investigate the simultaneous effect of shear on the phase separation and crystallization transitions of a ternary mixture of two immiscible polymers in a common solvent which is thermodynamically good for both components. The isothermal and non-isothermal crystallization kinetics will be investigated at quiescent state and at different constant shear rates. The isothermal crystallization kinetics will be also analyzed on the bases of Avrami approach.

Experimental Part

Materials and methods

Poly(ethylene oxide), PEO, and poly(dimethyl siloxane-*b*-ethylene oxide), P(DMS-*b*-EO), COP were obtained from Fluka Chemical Co., Germany. The weight average molecular weight, M_w, and the polydispersity, M_w/M_n, of PEO were 27 kg/mol and 1.3, respectively. The COP contains 75 wt% EO and is relatively low molecular weight (M_w = 0.6 kg/mol), its M_w/M_n = 1.3. PEO was dried for 7 days at 50 °C under vacuum before use. COP was employed without treatment. The solvent 1,2,3,4-tetrahydronaphtalene, THN, was also obtained from Fluka Chemical Co., Germany with 99.8 wt% purity and used as supplied. The boiling point was 205 °C and the density 0.969 g/cm.^{3}

Details of rheo-optical apparatus have been described elsewhere[9]. The apparatus was designed to measure the transmitted light intensity, I/I_0, where I is the twice of light intensity having passed through the solution and I_0 is the intensity of primary laser beam. The measurements were carried out at different constant shear rates ranged from 0 to 500 s^{-1} at 0.25 °C/min cooling rate.

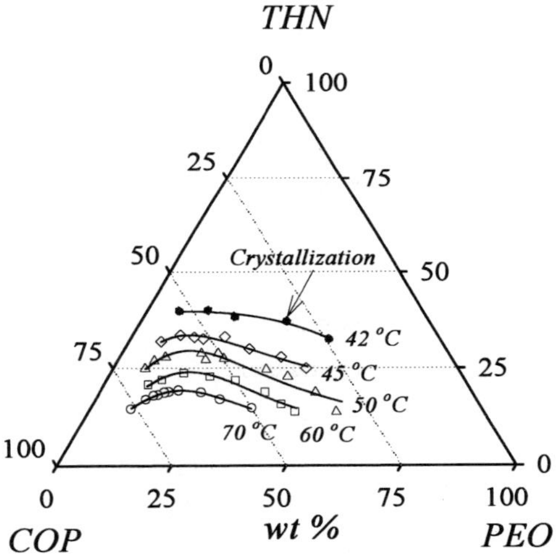

Figure 1. Equilibrium phase diagram of the system THN/COP/PEO at the indicated temperatures as obtained from turbidmetric titration.

The equilibrium cloud point temperatures, T_{cp}, and crystallization temperatures $T_{l/s}$ were estimated in the course of turbidimetric titration. Different solutions of PEO in THN (ca. 5 ml) of typical PEO content were titrated with the block copolymer at constant temperature. Figure 1 demonstrates the equilibrium phase diagram, obviously a typical UCST-type phase diagram was determined at different temperature, however, at 42 °C only $T_{l/s}$ was observed. The existing of only $T_{l/s}$ was conformed by polarized light microscope, which shows only the spherulitic structure of PEO in the mixtures without any indication for the phase separation morphology. The block copolymer may incorporate small amounts of PEO and form mixed crystals, this effect is not considered in the present paper.

The rheological measurements during the crystallization process were carried out using an advanced rheometer, AR 1000 with two 40 mm cone-and-plate geometry with cone angle = 1°. The measurements were carried at different constant shear rates up to 100 s.$^{-1}$

Results and Discussion

Crystallization kinetics at quiescent state

In order to understand the effect of shear on the crystallization kinetics, the crystallization process at quiescent state must be investigated. A typical isothermal crystallization process for THN/COP/PEO = 40/43/17 mixture after melting at T_m = 80 °C for 10 min and rapid quenching to different crystallization temperatures is demonstrated in Figure 2. The figure shows the crystallization time dependence of transmitted light at different crystallization

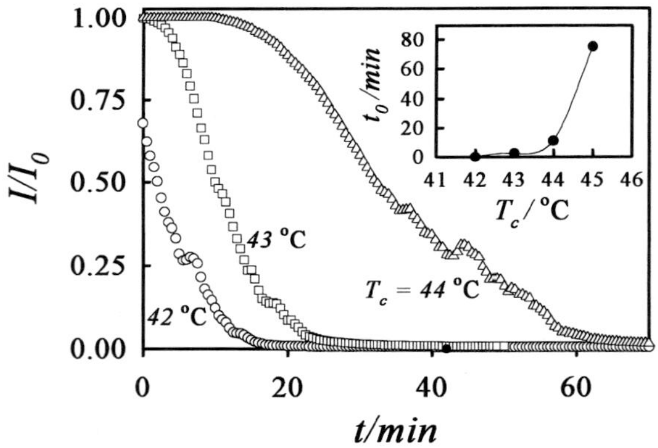

Figure 2. Crystallization time dependence of the transmitted light for THN/COP/PEO = 40/43/17 mixture after melting at 80 °C for 10 min and quenched to different T_c. The inset shows the crystallization temperature, T_c, dependence of the induction time, t_0.

temperatures. The crystallization occurs isothermally at the onset of decreasing the transmitted light of a completely transparent sample (melt sample). The induction time of crystallization, t_0, is defined as the time at which the transmitted light starts to decrease. With the present system the changes in the transmittance of light upon the crystallization time are very sharp and repeated measurements reproduce t_0 within an interval of 30 sec. The crystallization kinetics is substantially affected by the crystallization temperature i.e. t_0 greatly increases with increasing the crystallization temperature; t_0 at 44 °C is approximately four times greater than the corresponding value at 43 °C under the same experimental conditions. Obviously, the crystallization at a temperature lower than 43 °C is very fast and could produce a considerable amount of crystals already during cooling from the melt to the crystallization temperature. However the crystallization process at temperatures higher than 44 °C is very slow and the time needs for complete crystallization is very long.

A general treatment of the crystallization kinetics was developed by Avrami[19] and first applied to polymeric systems by Mandelkern et al.[20] According to this theory the overall crystallization process of PEO can be described by

$$X_t = 1 - \exp(-kt^n) \qquad (1)$$

where X_t is the weight fraction crystallized at a time t, k is the rate constant which is a function of crystallization temperature, and n is the Avrami exponent. The rate constant, k includes the combined effects of nucleation and growth; however, the Avrami exponent provides qualitative information about the nature of the nucleation and growth processes. Linearity between the light transmittance and the degree of crystallinity is assumed (arbitrary hypothesis) for this analysis. The change of the transmitted light during the crystallization process can be converted to the normalized light transmittance, $I_n(t)$:

$$I_n(t) = \frac{I/I_0(t) - I/I_0(0)}{I/I_0(\infty) - I/I_0(0)} \qquad (2)$$

where $I/I_0(t)$ is the light trnsmittance at time t, $I/I_0(0)$ is the melt value, and $I/I_0(\infty)$ is the value at the end of crystallization process. Using this value of the normalized light transmittance, the Avrami-type equation can be rewritten as follows

$$I_n(t) = 1 - \exp(-kt^n) \qquad (3)$$

The kinetics parameters of this equation (the exponent n and the rate constant k) can be obtained by taking the double logarithm of equation 3 and expressing it as

$$\log(-\ln(1-I_n(t))) = \log k + n\log t \qquad (4)$$

The plot of $\log(-\ln(1-I_n(t)))$ *vs.* $\log t$, yields straight line with a slope equal to n and an intercept equal to $\log k$. A typical Avrami-type plots for the THN/COP/PEO = 40/43/17 mixture at different crystallization temperatures are shown in Figure 3. The values of Avrami exponent determined for this mixture was in the range of 1.7 - 3.5 at the temperature range 42 - 44 °C. The rate constant, log *k*, was also found to be temperature dependent and changed from –0.87 at 42 °C to be –5.8 at 44 °C. The sharp decrease in the *k* value with increasing temperature is attributed to the fact that the crystallization rate at 42 °C is several times faster than that at 44 °C (see Figure 2).

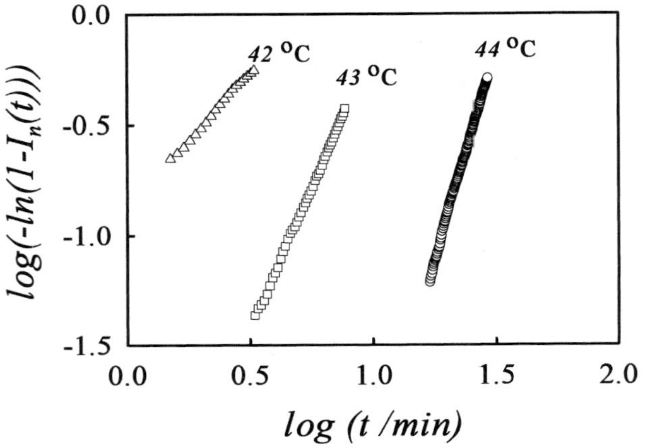

Figure 3. Avrami-type plot for the normalized light transmittance at different crystallization temperatures for THN/COP/PEO = 40/43/17 mixture.

Effect of Shear on the Crystallization Kinetics

The non-isothermal crystallization process of THN/COP/PEO = 40/43/17 mixture under different constant shear rates is shown in Figure 4. The sample is melted at 80 °C for 10 min and then cooled (0.25 °C/min) under different constant shear rates. Obviously as T falls below the crystallization temperature at constant shear rate, the shear stress σ increases dramatically. This substantial increase in the σ during the crystallization process is attributed to the continuos increase of the volume fraction of the crystalline part which may act like crosslinks. The onset of crystallization shifts to higher temperature with increasing shear rate. This behavior is attributed to the fact that the shear increases the orientation in the molecular chains and consequently lowers the conformational entropy. This reduction in the total entropy of fusion allows the crystallization to occur at higher temperatures than in the quiescent state.[21,22]

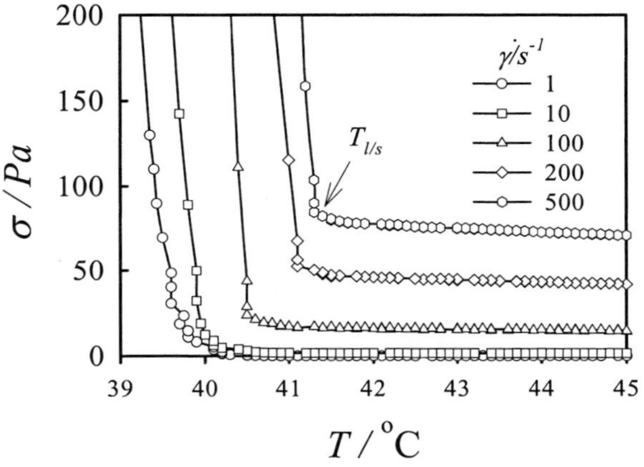

Figure 4. Temperature dependence of the shear stress σ for THN/COP/PEO = 40/43/17 mixture at different shear rates. The crystallization temperature $T_{l/s}$ is marked by a rapid increase in the σ.

Similar behavior (shear-induced crystallization) can be also obtained from the isothermal rheological measurement of the same mixture as shown in Figure 5 which represents the crystallization time dependence of the steady shear viscosity η at different constant shear rates. Once the crystallization process takes place a rapid increase in the sample viscosity was observed. A decrease in the viscosity with crystallization time at high shear rates was also detected as clearly seen in the Figure. This experimental fact is due to the disruption of the crystalline aggregate during shear. The disruption of the crystalline aggregate increases with shear rate. Several authors including us have successfully investigated the kinetics of crystallization for semi-crystalline polymers using rheological measurements.[23-29] The rheological technique for studying the crystallization kinetics was considered to be more sensitive than conventional techniques like DSC.[23,24] Linearity between viscosity and the degree of crystallization is also assumed. Based on this hypothesis one can convert the

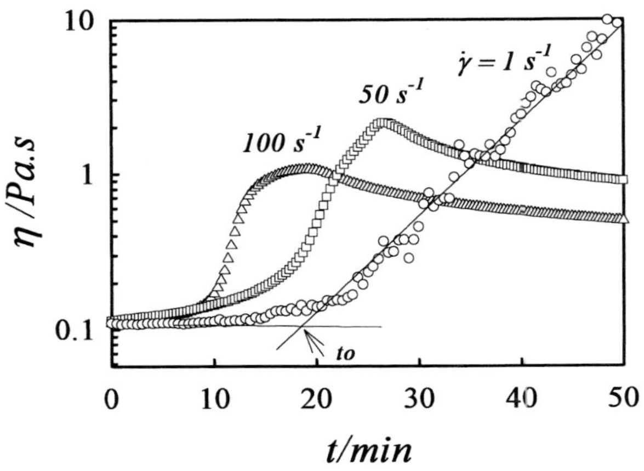

Figure 5. Crystallization time dependence of the steady shear viscosity for THN/COP/PEO = 40/43/17 mixture at 43 °C and different shear rates, the tangent drawn to determine t_0.

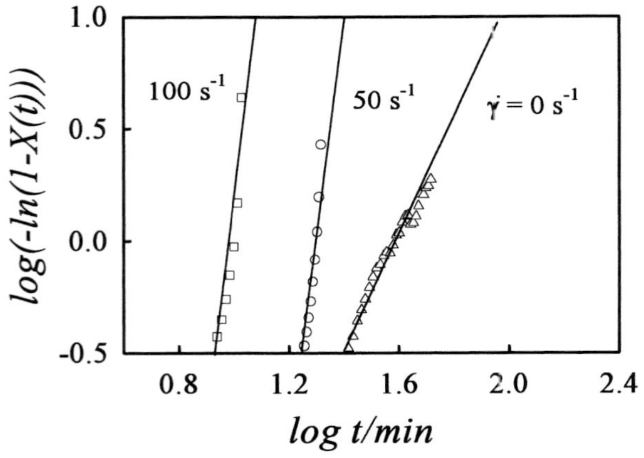

Figure 6. Avrami-type plot for the normalized shear viscosity at different shear rates and for the normalized light transmittance at quiescent state.

steady shear viscosity during the crystallization (Figure 5) to its normalized form as applied for the light transmittance (see equation 2). The Avrami-type plots at 43 °C at different shear rates (the figure includes also the sample at quiescent state obtained from figure 2) are shown in Figure 6. It is apparent that the value of the Avrami exponent increases dramatically with shear rate i.e. the exponent n increases from about 3 at quiescent state to 9 at $\dot{\gamma} = 100\,s.^{-1}$ This behavior was attributed to the fact that, when a crystalline aggregate is growing in a melt under shear, it is subjected to stress. This stress not only causes orientation in the melt but also could cause disruption of the crystalline aggregate. This disruption results in the introduction of more nuclei into the melt as the crystallization proceeds and consequently accelerates the nucleation process. Similar behavior has been observed recently for the crystallization kinetics of some semi-crystalline polymers under shear flow using rheological technique.[28,29]

Simultaneous Shear effect on the T_{cp} and $T_{l/s}$

A typical example for the temperature dependence of the light transmittance for the THN/COP/PEO = 25/58/17 mixture is demonstrated in Figure 7. This composition contains both the T_{cp} (liquid-liquid) and $T_{l/s}$ (liquid-solid or crystallization), so it would be interested to

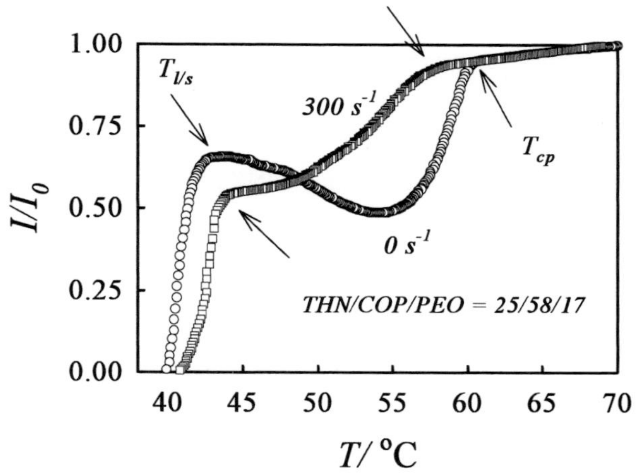

Figure 7. Light transmittance, I/I_0, of THN/COP/PEOP = 25/58/17 mixture as a function of temperature at quiescent state and at $\dot{\gamma} = 300\,s.^{-1}$

check the effect of shear on these two transitions. The sample is phase separated upon cooling (0.25 °C/min) from one-phase regime and then crystallized at a temperature close to the crystallization temperature of PEO. Obviously the shear can significantly influence on both T_{cp} and $T_{l/s}$ transitions. The T_{cp} shifts several degrees to lower temperature, however on the other hand, the $T_{l/s}$ shifts to higher temperature with shear. Based on this experimental fact one can say that the shear can simultaneously induced mixing and crystallization within a single mixture. The data in Figure 7 can be quantitatively represented as shear dependence of the T_{cp} and $T_{l/s}$ transitions (see Figure 8).

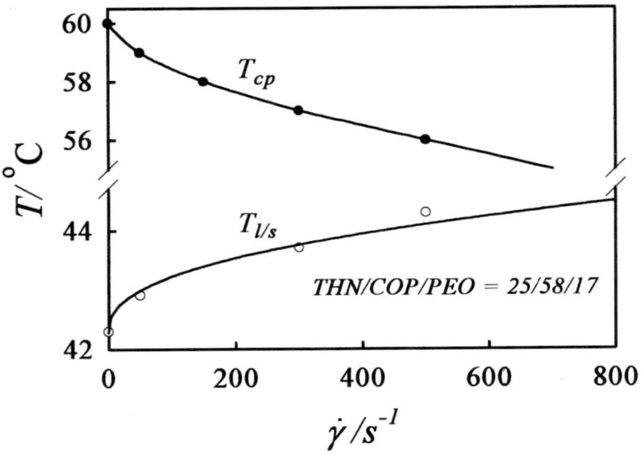

Figure 8. Shear rate dependence of T_{cp} and of $T_{l/s}$ for THN/COP/PEO = 25/58/17 mixture.

Conclusion

Shear effect on the phase behavior of a homopolymer A and a block copolymer A-B in a common solvent has been investigated. The liquid-liquid and liquid-solid transitions were found to be strongly influenced by shear flow. Shear-induced mixing and shear-induced crystallization have been observed in a single mixture. The crystallization kinetics was also changed by shear flow. The induction time, t_0, of the crystallization process decreases dramatically with increasing shear rate. An elevation in the Avrami exponent was also detected under shear, indicating that the shear accelerated the nucleation of crystallization process. These results are interpreted in the terms of the orientation of the molecular chains induced by the shear and consequently an enhancement of the crystallization process.

Acknowledgment

The author is grateful to the Alexander von Humboldt-Stiftung for support of his stay in Germany. He also devotes his sincere appreciation to Professor Bernhard A. Wolf for valuable discussion.

[1] A. Silberberg, W. Kuhn, *Nature* **1950**, *170*, 450.
[2] C. Rangel-Nafaile, A. Metzner, K. Wissbrun, *Macromolecules* **1984**, *17*, 1187.
[3] T. Hashimoto, K. Fujioka, *J. Phys. Soc. Jpn.***1991**, 60, 356.
[4] S.A. Madbouly, M. Ohmomo, T. Ougizawa, T. Inoue, *Polymer* **1999**, *40*, 1465.
[5] S.A. Madbouly, Ougizawa T., T. Inoue, *Macromolecule* **1999**, *32*, 5631 (1999)
[6] S.A. Madbouly, T. Chiba, Ougizawa T., T. Inoue, *Polymer* **2001**, *42*, 1743.
[7] S.A. Madbouly, T. Chiba, Ougizawa T., T. Inoue, *J. Macromol. Sci. Part B-Phys.* **B38** , 79 (1999)
[8] S.A. Madbouly, Ougizawa T. *J. Macromol. Sci. Part B-Phys.* **2002**, *B41*, 629.
[9] C. Krause, R. Horst, B.A. Wolf, *Macromolecules* **1997**, *30*, 890.
[10] L.R. Trelour, *Trans Faraday Soc.* **1941**, *37*, 84.
[11] A.N. Gent, *Trans Faraday Soc.* **1954**, *50*, 521.
[12] A.K. Fritzsche, F.P. Price, *Polym. Eng. Sci.* **1974**, *14*, 401.
[13] A.K. Fritzsche, F.P. Price, R.D. Ullrich, *Polym. Eng. Sci.* **1976**, *16*, 182.
[14] M.H. Theil, *J. Polym. Sci. Polym Phys.* **1975**, *13*, 1097.
[15] G. Eder, H. Janeschitz-Kriegl, S. Liedauer, *Prog. Polym. Sci.* **1990**, *15*, 629.
[16] N.V. Progodina, V.P. Lavrenko, S. Srinivas, H.H. Winter, *Polymer* **2001**, *42*, 9031.
[17] A. Nogales, Hsiao B. S., R.H. Somani, S. Srinivas, A.H. Tsou, F.J. Balta-Calleja, T.A. Ezquerra, *Polymer* **2001**, *42*, 5247.
[18] W.J. Yoon, Myung H. S., B.C. Kim, S.S. Im, *Polymer* **2000**, *41*, 4933.
[19] M. Avrami, *J. Chem. Phys.* **1941**, *9*, 177.
[20] L. Mandelkern, F. A. Quinn, P. J. Flory, *J. Appl. Phys.* **1954**, *25*, 830.
[21] T. W. Hass, B. Maxwell, *Polym. Eng. Sci.* **1969**, *9*, 226.
[22] A. J. Pennings, J. M. Van der Mark, H. C. Booj, *kolloid Z v Z Polym.* **1970**, *236*, 99.
[23] J. W. The, H. P. Blom, A. A. Rudin, *Polymer* **1994**, *35*, 1680.
[24] Y. P. Khanna, *Macromolecules*, **1993**, *26*, 3639.
[25] C. Carrot, J. Guillet, K. Boutahar, *Rheol. Acta* **1993**, *32*, 566-574.
[26] S. A. Vleeshouwers, *Rheo. Acta* **1996**, *35*, 391.
[27] W. Nagatake, T. Takahashi, Y. Masubuchi, J.-I Takimoto, K. Koyama, *Polymer* **2000**, *41*, 523.
[28] Y. Masubuchi, K. Watanabe, W. Nagatake, J.-I Takimoto, K. Koyama, *Polymer* **2001**, *42*, 5023.
[29] S.A. Madbouly, Ougizawa T. *J. Macromol. Sci. Part B-Phys.* **2003**, *42(2)*, in press.

Rheo-Optical Determination of the Interfacial Tension in a Dispersed Blend

Manlio Tassieri,[1] *Nino Grizzuti,**[1] *Francesco Greco*[2]

[1]Dipartimento di Ingegneria Chimica, Università di Napoli "Federico II", Piazzale Tecchio 80, I–80125 Napoli, Italy
[2]Istituto per la Tecnologia dei Materiali Compositi, ITMC, CNR, Piazzale Tecchio 80, I–80125 Napoli, Italy

Summary: In this work, we present two novel methods to determine the interfacial tension of a disperse polymer blend through rheo-optical measurements of flow-induced single drop distortions. A counter-rotating shearing device with transparent plates is used to measure drop distortions. The cell geometry allows for a top view of the deforming drop, i.e., along the velocity gradient direction. Such a view is the only possible option for all currently available commercial rheo-optical instruments. Two different quantities are monitored, namely, the drop axis along the vorticity direction, and the rotation period of the drop surface. We use drops of a polyacrilamide aqueous solution (a shear thinning liquid) immersed in a polyisobutene matrix. Experimental results are interpreted in terms of theories for Newtonian liquids, where the relevant parameter is the Capillary number. If an appropriate viscosity ratio is chosen, that accounts for the shear thinning behaviour of the drop phase, good agreement is found between measurements and theoretical predictions. As a result, a robust estimate of the blend interfacial tension, that makes use only of the information acquired from top view experiments, is obtained.

Keywords: blends; drop deformation; interfacial tension; microstructure; rheo-optics

Introduction

The flow induced deformation of a liquid droplet immersed into a liquid matrix is a widely studied problem[1]. In the Newtonian case, dimensional analysis shows that drop deformation can be described in terms of only two independent dimensionless parameters. One is the ratio between viscous and interfacial stresses, the Capillary number Ca, the other is the drop to matrix viscosity ratio, p. Thus, if a theory for drop deformation is available, comparison with experiments would allow for an estimate of the blend interfacial tension (appearing in Ca), once the viscosity of the

 DOI: 10.1002/masy.200350806

pure components is known[2].

In spite of the reduced number of parameters, shear flow deforms a droplet in a relatively complex way. Experiments as well as theoretical analyses show that the deformed droplet has a non-symmetric, fully three-dimensional shape that, at low Ca, approximates to an ellipsoid[3-7]. As a consequence, such a shape can only be reconstructed by performing optical observations along more than one direction.

Figure 1 depicts a deformed droplet as seen from the two relevant shear flow directions, namely, the vorticity direction (the so-called "side view", Figure1a), and the velocity gradient direction (the "top view", Figure1b). The side view gives direct access to two principal semi-axes, r_{MIN} and r_{MAX}, and to the drop inclination angle with respect to flow, φ. Measurements of side view drop deformation were first performed by Taylor in his pioneering work[3]. In more recent years, several investigators also adopted the same view, by designing their own home-made apparatuses[5,8-11].

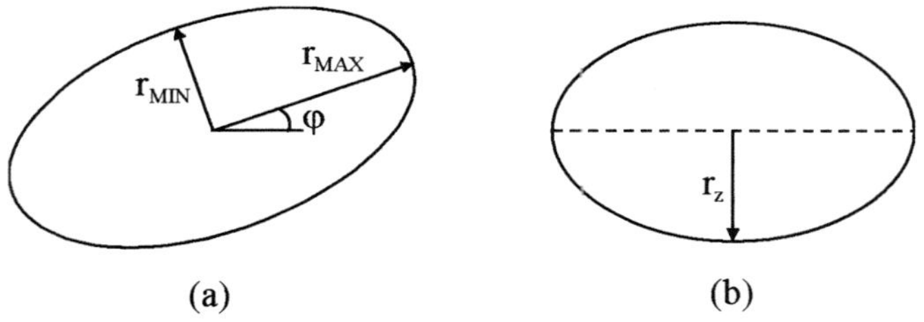

Figure 1. The two classical views of a deformed droplet. (a) side view; (b) top view. The flow direction is horizontal.

In all commercial rheo-optical apparatuses[12-14], however, optical observations can only be made along the velocity gradient direction. In such top view experiments, only the third principal semi-axis can be directly observed, the one along the vorticity direction, r_z (Figure1b). For an affordable determination of the interfacial tension, then, it would be highly desirable to develop a reliable methodology based on top view observations only, *plus* some convenient model describing drop deformation, including predictions of r_z as a function of Ca and p.

Attempts in this direction have already been made. Almusallam *et al*[15], for example, determined the interfacial tension of a Polybutadiene/Polydimethyilsiloxane blend by top view measurements of a droplet subjected to a step shear deformation and subsequent relaxation. In their analysis, however, the drop inclination angle φ was required. This was determined by arbitrarily assuming affine deformation of the drop.

A further difficulty when dealing with drop deformation in polymer systems derives from their non-Newtonian character. While almost all currently available theories for drop deformation are for Newtonian systems, two main non-Newtonian features can affect the drop behaviour under flow. On the one hand, the elasticity of the components is known to produce significant deviations from the Newtonian case[16,17]. On the other hand, the shear rate dependency of the component viscosity, which in turn determines a varying blend viscosity ratio, can also have important consequences on drop evolution under flow. The latter aspect has never been studied in detail.

In this paper we perform top view observations of droplets of aqueous polyacrylamide solutions immersed in a polyisobutene matrix and subjected to a shear flow. The main objective of the work is to determine the blend interfacial tension by only using parameters that can be directly measured by top view experiments. Since the drop phase is characterized by a marked non-Newtonian behaviour, the possibility to adapt models for Newtonian emulsions to the case of a varying viscosity ratio is also considered. The paper is organized as follows. First, the theoretical background for shear-induced drop deformation is set, and the relevant model equations, that are necessary in order to interpret the experimental results, are introduced. Then, after describing the materials used to prepare the blends and the rheo-optical techniques, the experimental results and the methods to determine the interfacial tension are presented and discussed. Conclusions are drawn in the final Section.

Theoretical background

As mentioned in the Introduction, we will examine in this paper two different rheo-optical experiments for obtaining a measure of the interfacial tension under dynamical conditions. In one of these, we look at the drop axis r_z in the vorticity direction of a shear flow, and relate the length variations of this axis to the variation in the imposed shear rate $\dot{\gamma}$. In the other experiment, always under shear flow conditions, the rotation period T of a point at the drop surface is determined, again to relate the variations of this rotation period to $\dot{\gamma}$. In both the just mentioned cases, theoretical predictions do exist, for the r_z-axis and the period T, respectively, which are expressed in terms of the Capillary number and the viscosity ratio:

$$Ca = \frac{\eta_m \dot{\gamma} r_0}{\sigma} \tag{1}$$

$$p = \frac{\eta_d}{\eta_m} \tag{2}$$

In Eqs (1) and (2) η_m and η_d are the matrix and droplet viscosity, respectively, r_0 the radius of the unperturbed droplet, and σ the interfacial tension. Hence, a separate knowledge of r_0, η_d, η_m, and $\dot{\gamma}$ allows one to infer the interfacial tension σ from measurements of r_z and T, and comparison with the theory. In this Section we report the needed formulae from theoretical analysis, and shortly discuss them.

Let us start with the r_z-axis determination. As it is well known, calculating the 3D-shape of a fluid drop immersed in another fluid flowing in an assigned way at infinity is a difficult task. The only available analytic results (for the Newtonian case) come from a perturbative procedure, with the Capillary number as the expansion parameter. For the shear flow case, which is the one experimentally investigated in the present paper, one obtains the following simple expression for the r_z-axis, up to second order in Ca[18]:

$$\frac{r_z}{r_0} = 1 - k(p) Ca^2 \tag{3}$$

with the coefficient $k(p)$ given by:

$$k(p) = \frac{42496 + 191824p + 261545p^2 + 111245p^3}{60480(1+p)^3} \tag{4}$$

(Notice that, from Eq.(3), the r_z-axis only deviates from the equilibrium drop radius r_0 at order Ca^2.) Thus, measuring the maximum drop dimension r_z in the z-direction leaves σ (contained in

Ca^2) as the only unknown quantity of Eq.(3). The interfacial tension is so obtained by performing a "top view" optical experiment, i.e., by looking at the deformed drop from the velocity gradient axis.

The same top view experiment can also lead to another determination of the interfacial tension, if a tracer particle is somehow placed on the drop interface. Indeed, in such a case, it is possible to measure the rotation period of the tracer particle around the drop. For the specific (and easiest) case of a particle moving on the shear plane, we can compare the experimental results to some well-known theoretical predictions available in the literature.

For $Ca \rightarrow 0$, which corresponds to an essentially undeformed droplet, the circulation period is found to be independent of the Capillary number and given by[3]:

$$m_0 = \frac{T\dot{\gamma}}{4\pi} = \frac{p+1}{\sqrt{p(p+2)}} \tag{5}$$

In Eq.(5), m_0 is the so-called "circulation number", and T is the revolution period. For $p \rightarrow \infty$, that is, for a hard spherical particle, Eq.(5) reduces to the well known result m_0=1, i.e., the sphere rotates with a frequency corresponding to one half of the shear rate. It should be stressed that Eq.(5) only holds for very low values of Ca. This point will be exploited later to obtain in any case an estimate for σ from Eq.(5).

An expression for m_0 that explicitly contains the Capillary number is that due to Cox[19]:

$$m_0 = \frac{p+1}{\sqrt{p(p+2)}} + \frac{19p(19p+16)}{4\left[(19p)^2 + (20/Ca)^2\right]}\left[25 - \frac{(p+1)(25p^2+50p-31)}{\sqrt{p^3(p+2)^3}}\right] \tag{6}$$

In this case, therefore, measurements of the rotation period T leave again σ (contained in Ca^2) as the only unknown quantity in Eq.(6). Hence, an "on-line" determination of the interfacial tension is feasible.

For the sake of completeness, let us add some comments on the above reported equations. First, concerning Eq.(6), we want to stress that the two terms in the RHS of the equation have not the same mathematical status. Indeed, the first term (independent of Ca) is the exact result from the first-order theory by Taylor, while the second term, from Cox theory, is not rigorously deduced from the Newtonian fluid-dynamics of the single drop problem. In other words, Eq.(6) does not come from a general analytic solution of the drop problem, nor from a perturbative expansion in Ca (as it is the case, conversely, for Eq.(3) above). On the other hand, Eq.(6) is the only known expression for the rotation period T which accounts for the drop deformation. Actually, Taylor rigorous result (Eq.(5)) simply describes a decrease of T with increasing the shear rate $\dot{\gamma}$, whereas the complete Eq.(6) "corrects" (albeit in an approximate way) for the progressive increase of the interfacial area with $\dot{\gamma}$ itself.

Coming back to Eqs.(3) and (4), we should mention that the quadratic dependence of r_z on Ca has been experimentally shown elsewhere[20] to be in fact preserved up to moderate values of Ca (ca. 0.3, say, with $p\cong 1$), which is a very favourable circumstance, in view of the inherent difficulties of measuring second-order quantities with sufficient accuracy. We should also notice here that the coefficient $k(p)$, as given in Eq.(4), is a rather weakly increasing function of the viscosity ratio, with an asymptote at $p\rightarrow\infty$ less than a factor 3 the value of k at $p=0$. This property of $k(p)$ will prove useful in the experiments reported in the present paper.

Finally, we have to remark that Eqs.(3)-(6) do in fact refer to a Newtonian system, but will be used in this work to deal with a non-Newtonian drop. In this respect, we remind the reader that the only non-Newtonian fluid-dynamic single drop problem that has been (perturbatively) solved is that for the constitutive case of the so-called "second-order fluids". Greco[17] demonstrated how the (order-of-magnitude) condition to have observable non-Newtonian effects under shearing flow condition is given by:

$$(\eta_d \dot{\gamma})^2 \approx \frac{\sigma N_1}{r_0} \qquad (7)$$

with N_1 the first normal stress difference of the droplet fluid. In view of the constitutive choice adopted in [17] (i.e., that of using the second-order fluid constitutive equation), however, Eq.(7) is only valid in a range of shear rates such that η_d stays constant, and N_1 is quadratic in $\dot{\gamma}$. Thus, when using shear-thinning fluids (as we do here, see below), we are in fact in a situation where the only existing non-Newtonian theory cannot be of help, not even in predicting (through Eq.(7))

whether non-Newtonian effects are observable or not. In the lack of anything better, then, we are compelled to use the Newtonian (approximate) formulae, Eqs.(3)-(6), in the rest of this paper.

Materials and methods

Aqueous solutions of polyacrilamide and pure polyisobutene (PIB) were used to prepare the model blend. A 5% wt solution of Separan AP30 (SAP30) was prepared by dissolving the necessary amount of polymer in bidistilled water, followed by gentle mixing with a magnetic stirrer at 50 rpm for several hours. The solution was then kept at rest for at least 24 hours, in order to remove air bubbles, and was sealed in glass containers. SAP30 concentration was periodically checked by dry residue measurements after vacuum oven treatment at 80°C. PIB (Parapol 1300, Exxon Co.) was used as received. For a simpler handling, PIB was preloaded into 10 cc hypodermic syringes, and it was centrifuged at 5000 rpm for 30 minutes in order to remove air bubbles.

The rheological properties of the two components were determined by means of a Rheometric DSR 200 rotational rheometer, equipped with cone and plate fixtures (plate diameter 40mm, cone angle 0.04 rad). A Peltier thermal unit guaranteed constant temperature measurements at 25°C with ±0.1°C accuracy.

Rheo-optical measurements of drop deformation were performed by means of a home-made counter-rotating device. The apparatus consists of two, independently motor-driven glass plates, that can rotate at constant angular speed. The liquid is loaded in the gap between the plates, and can be observed along the velocity gradient direction by means of a video camera connected to an *ad hoc* optical microscopy system. A more detailed description of the apparatus can be found elsewhere[21].

By suitably adjusting the oppositely directed rotation speeds of the two plates, any material point in the liquid can be deformed without being translated with respected to the microscope point of view. In this way, droplets of SAP30 solution in the PIB matrix could be observed for an indefinite time when submitted to a shear flow. In order to obtain well isolated, small SAP30 droplets, two loading procedures were used. In the first procedure, a PIB layer was applied on the bottom glass plate by means of the syringe. A thin thread of liquid SAP30 solution was gently extended on the PIB layer approximately along a plate radius. The thread was then covered by a second PIB layer and the gap between the two plates was slowly set to its working position. A few minutes after loading, the SAP30 filament broke down into droplets by Rayleigh instability. Application of a moderate shear flow was used to displace the droplets from each other, so as to obtain a set of substantially isolated, non interacting SAP30 drops. In the second procedure, the gap between the two plates was completely filled with PIB. Then, a small droplet of SAP30 was injected by means of a microsyringe. If the injected drop had too large a diameter (exceeding about 100 μm), a strong shear flow was applied to the system in order to deform the drop into a filament, that would eventually breakup into smaller drops after cessation of flow.

After the loading procedure was completed, the experiment was started by applying the desired shear rate. Top view images of the deformed drops were video recorded and subsequently computer processed to extract the desired deformation data. All rheo-optical experiments were performed at the controlled ambient temperature of 25±1°C.

Experimental results and discussion

Figure 2 reports the viscosity of PIB and SAP30 as a function of shear rate at 25°C. PIB is a Newtonian liquid with a viscosity of about 95 Pa s. SAP30, conversely, shows a strong shear thinning behaviour, characterized by a power-law exponent n=0.29. The two viscosity curves intersect at a shear rate of about 0.7 s^{-1}. In view of the power law behaviour of SAP30, the viscosity ratio of the system can be cast in simple analytical form as:

$$p = 0.78\,\dot{\gamma}^{-0.71} \qquad (8)$$

Drop deformation experiments were carried out in the shear rate window 0.1÷1 s^{-1}, thus resulting in a viscosity ratio ranging between 4 and 0.8. It should be mentioned that viscoelastic measurements (not reported here) show that PIB does not posses a significant elastic component, which is instead present in the case of SAP30. As pointed out in a previous Section, the effect of component viscoelasticity is not addressed in this work.

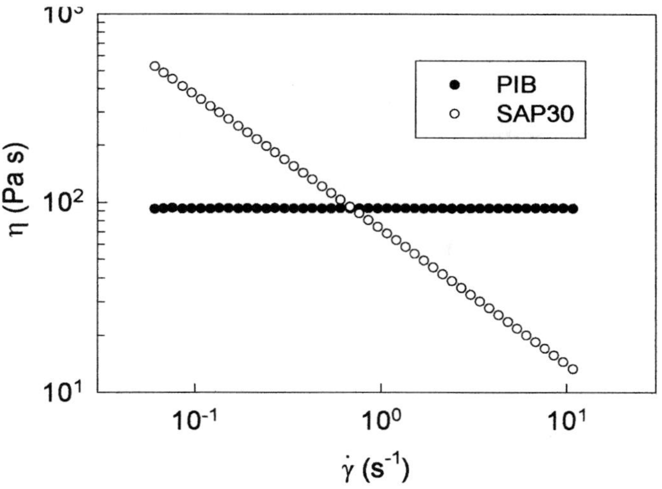

Figure 2. Viscosity of PIB and SAP30 as a function of shear rate at 25°C.

A first set of rheo-optical experiments concerns the measurement of steady-state drop deformation in shear flow. As the counter-rotating apparatus allows for a top view observation, the significant geometrical parameter to be measured is the drop semi-axis r_z. Figure 3 shows the drop deformation along the vorticity axis as a function of the shear rate for two SAP30 droplets of different radius. In Figure 3 the quantity $(r_z - r_0)/r_0^3$ is plotted as a function of the *square* of the shear rate. In fact, according to the theory presented previously, at low values of the capillary number r_z is expected to follow a quadratic dependence upon the shear rate. In particular, Eq.(3) indicates that, in the case of two Newtonian liquids, the following relationship holds:

$$\frac{r_z - r_0}{r_0^3} = \frac{\eta_m^2 k(p)}{\sigma^2} \dot{\gamma}^2 \qquad (9)$$

where k is the analytic function of the viscosity ratio given by Eq.(4).

Equation (9) predicts that, for a given pair of Newtonian liquids, the deformation measure in the LHS of the equation must be a universal function of the shear rate, independent of the droplet size.

Figure 3 confirms the validity of Eq.(9), as the experimental data for both drop sizes fall with good approximation on a single straight line.

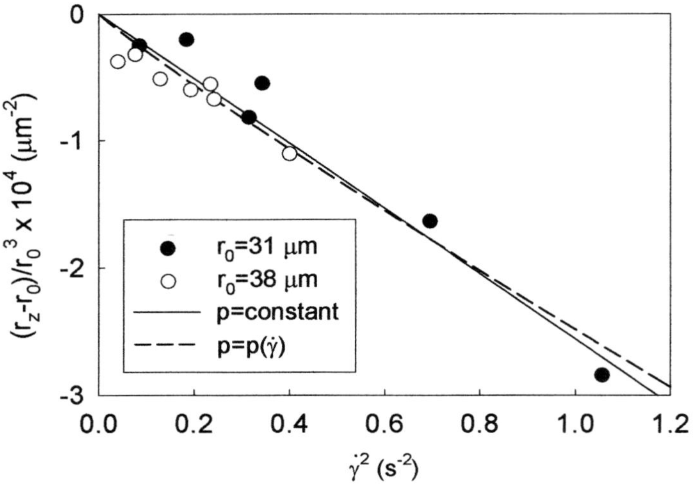

Figure 3. The deformation of two SAP30 droplets in PIB as a function of shear rate. See text for further details.

Figure 3 demonstrates the appropriate procedure to interpret drop deformation results when the optical observations are made along the velocity gradient direction. In particular, if all other quantities in the RHS of Eq.(9) are known, the experimental results can be used to determine the interfacial tension of the system. In the case of the PIB/SAP30 blend, however, this procedure could not be rigorously applied, as SAP30 is characterized by a shear thinning viscosity, in contrast with the assumption of Newtonian fluids dictated by the model. In particular, the non-Newtonian character of SAP30 shows up in the expression for the coefficient k, which is a function of the viscosity ratio.

In order to apply the model to the present situation, we make here the assumption that the model is still valid under a varying viscosity ratio. It should be noticed that, although arbitrary, such a hypothesis does not introduce significant numerical differences in the calculations. In fact, inspection of Eq.(4) shows that k varies only between 1.2 and 1.5 over the shear rate window explored in the experiments.

Two different procedures have been applied in order to estimate the interfacial tension. In one case, a constant viscosity ratio, $\bar{p}$, is used. $\bar{p}$ is chosen as the simple arithmetic mean of the values assumed at the extremes of the experimental window:

$$\bar{p} = \frac{p(\dot{\gamma}_{\min}) + p(\dot{\gamma}_{\max})}{2} \tag{10}$$

where $\dot{\gamma}_{\min}$ and $\dot{\gamma}_{\max}$ are the minimum and maximum experimental attained shear rates, respectively. In this case, i.e., with $p = \bar{p}$, Eq.(9) remains linear in the square of shear rate. The solid line in Figure 3 is indeed the linear fit of Eq.(9). Use of the constant viscosity ratio defined in Eq.(10) yields an interfacial tension of σ=6.8·10^{-3} N/m.

In the second fitting procedure, the viscosity ratio is allowed to change with shear rate. As a consequence, Eq.(9) is no longer simply quadratic in the shear rate, as ***k*** now depends on a varying *p*. The broken line in Figure 3 represents the best fit of Eq.(9) when *k*(*p*) is calculated according to Eq.(4) with a varying *p*. In this case, an interfacial tension of σ=6.4·10^{-3} N/m results. Comparison of the two curves in Figure 3, and of the corresponding interfacial tension values, clearly shows that the agreement between experiments and model is not particularly sensitive to the choice of either a constant or a varying viscosity ratio.

A second set of rheo-optical experiments concerns the measurement of the rotation period of the droplet interface. Figure 4 shows a 38μm radius SAP30 drop in PIB undergoing a shear flow start-up. In Figure 4a the drop is at rest, and an air micro-bubble of about 2μm radius can be clearly seen on the right side of the interface. Such a micro-bubble can be obtained by the drop injection procedure outlined in the Materials and Methods Section. When the flow is started-up (Figure 4b) the air bubble starts traveling along the drop interface. The difference in refractive index between SAP30, PIB and air allows for a clear observation even when the bubble moves along the "hidden" side of the interface. At steady-state (Figure 4c) the drop reaches a constant deformation, and the bubble executes constant period revolutions around the drop interface. It should be also noticed that, at least in the case reported, the air bubble follows essentially the longest trajectory, that is, the one parallel to the flow direction.

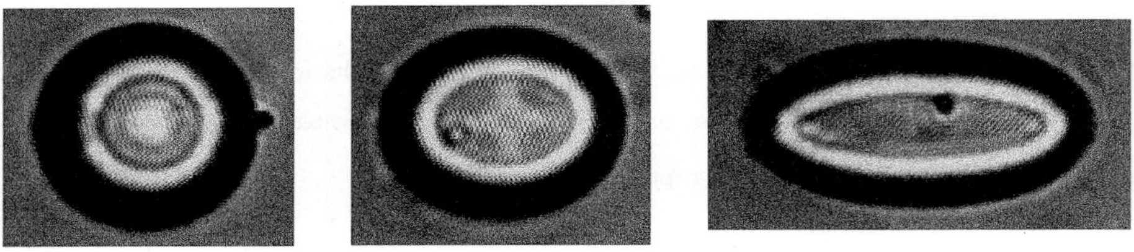

(a) (b) (c)

Figure 4. Top view of a SAP30 drop undergoing a shear flow start-up. The shear rate is 0.6 s^{-1} and the drop size is r_0=38μm. The direction of flow is horizontal. (a) quiescent drop, the air micro-bubble is on the right side; (b) the deforming drop during the transient flow, the air bubble is traveling along the bottom part of the interface; (c) the deformed drop at steady-state, the air bubble is now moving along the top part of the interface.

Assuming that the presence of the air bubble does not alter the flow conditions, and that the bubble itself remains attached to the interface, tracking the bubble motion allows for the measurement of the rotation period of the SAP30 droplet interface.

Figure 5 shows the interface rotation period T as a function of shear rate, as measured by air bubble tracking. T is a decreasing function of shear rate, because the droplet rotates faster as the shear rate is increased. Figure 5 also reports the comparison between the experimental results and the predictions of the available theories for droplet revolution time.

The dashed line in Figure 5 corresponds to Taylor's prediction, Eq.(5). In order to account for the non-Newtonian character of the viscosity ratio, an average value of p has been considered by using the same expression (Eq.(10)) already adopted to describe the drop deformation results. As expected, the prediction is only valid for low shear rates, corresponding to small values of the capillary number. The agreement between theory and experiments for $Ca \to 0$ is anyhow remarkable, as Eq.(5) does not contain adjustable parameters.

Although Taylor's theory does not contain any direct information on the interfacial tension, the observed deviation of the theory from the experimental results at increasing shear rates suggests a procedure to obtain a first, rough estimate of σ. Indeed, as said before, Taylor's theory is only applicable for Ca lower than some threshold value, Ca_t. At the same time, Figure 5 indicates a good quantitative agreement between theory and our data only for shear rates lower than a threshold value, $\dot{\gamma}_t \cong 0.35\, s^{-1}$. Recent work on Newtonian drop deformation for viscosity ratios around p=1[7] suggests $Ca_t \cong 0.1$. Collecting the above information leads to the following approximate relationship between Ca_t and $\dot{\gamma}_t$:

$$Ca_t = \frac{\eta_m \dot{\gamma}_t r_0}{\sigma} \cong 0.1 \tag{11}$$

Equation (11) again contains σ as the only unknown parameter. Calculations yield $\sigma = 1.2 \cdot 10^{-2}$ N/m, a value which is in semi-quantitative agreement with those obtained by the r_z measurements.

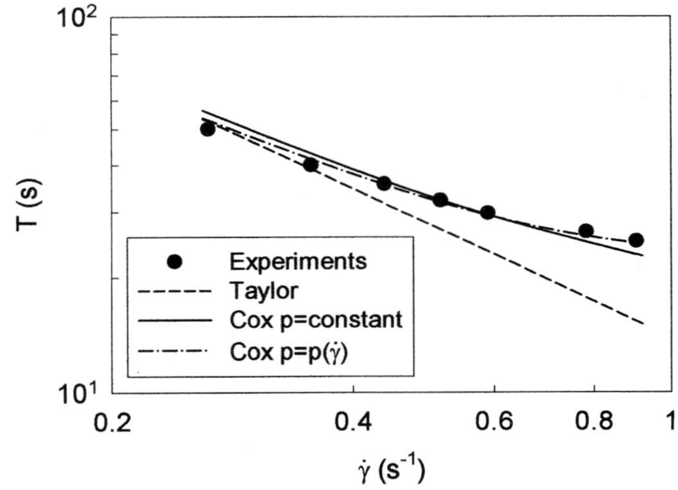

Figure 5. The period of rotation as a function of shear rate for SAP30 in PIB. The lines are the predictions of the Taylor and Cox models (see text).

A direct determination of the interfacial tension from the drop revolution results is obtained by fitting the Cox model, Eq.(6), to the experimental data, with the interfacial tension as the only unknown parameter. The solid line in Figure 5 represents the best fit of Eq.(6) when the average viscosity ratio $\bar{p}$ is used. Good agreement with the experimental results is found. Furthermore, the calculated value of interfacial tension, $\sigma = 6.5 \cdot 10^{-3}$ N/m, is very close to that extracted from the r_z analysis. An even better agreement between theory and experiments is found (see dot-dashed line in Figure 5) when the actual, shear rate dependent viscosity ratio is used in the Cox formula. In this case, a slightly higher interfacial tension results from the model regression results (σ=7.1·10^{-3} N/m), which is however still very similar to those obtained from drop deformation. As shown by Figure 5, the introduction of a varying viscosity ratio produces a slight improvement in the agreement between theory and experiments, and does not change significantly the calculated interfacial tension.

All values of interfacial tension calculated by the different experimental methods and with the different assumptions on the viscosity ratio are summarized in Table 1.

Table 1. Calculated values of interfacial tension from different experimental quantities

	r_z, $\bar{p}$	r_z, $p(\dot{\gamma})$	T (Taylor), $\bar{p}$	T (Cox), $\bar{p}$	T (Cox), $p(\dot{\gamma})$
σ (mN/m)	6.8	6.4	12	6.5	7.1

It can be concluded that, at least for the system investigated in this work, the analysis of top view optical observations leads to a robust estimate of the interfacial tension.

Conclusions

We have shown that top view observations of drop deformation in shear flow, which are typical of all commercial rheo-optical apparatuses, can be successfully used for the quantitative determination of the interfacial tension, once the appropriate deformation parameters are measured and adequately interpreted. In particular, for moderate values of the capillary number, measurements of the drop semi-axis along the vorticity direction, r_z, can be arranged in terms of a universal function of the applied shear rate. This allows for the determination of the blend interfacial tension, once the rheology of the two components is known. As an alternative, top view measurements of the droplet rotation period, which are available if a suitable tracer is deposited on the drop interface, can also be used for a quantitative estimate of the interfacial tension.

We have also shown that, at least for the polymer blend studied in this work, the shear thinning behaviour can be incorporated into models initially elaborated for Newtonian liquids, by allowing the viscosity ratio appearing in such models to depend on the shear rate, or alternatively by defining a suitable mean viscosity ratio as an average over the experimentally accessed shear rate window. It seems feasible, therefore, at least from a practical point of view, to adapt the available Newtonian models to the case of varying viscosity blends.

The results of this work do not have the presumption to be at all exhaustive. As already pointed out before, the authors are aware that elastic phenomena (which are not considered here) could play an important role in the drop deformation of non-Newtonian systems. Efforts must be made to incorporate such effects in the presently available viscous models.

[1] C.L. Tucker III, P. Moldenaers, *Annu. Rev. Fluid Mech.*, **2002**, *34*, 177-210.
[2] I. Sigillo, L. di Santo, S. Guido, N. Grizzuti, *Polym. Eng. Sci.*, **1997**, *37*, 1540-1549.
[3] G.I. Taylor, *Proc. R. Soc. London*, **1932**, *A138*, 41-48.
[4] J. M. Rallison, *Annu. Rev. Fluid Mech.*, **1984**, *16*, 45.
[5] S. Guido, M. Villone, *J. Rheol.*, **1998**, *42*, 395-415.
[6] P.L. Maffettone, M. Minale, *Journal of Non-Newtonian Fluid Mechanics*, **1998**, *78*, 227-241.
[7] S. Guido, F. Greco, M. Villone, *J. Coll. Int. Sci.*, **1999**, *219*, 298-309.
[8] S. Torza, R.G. Cox, S.G. Mason, *J. Coll. Int. Sci.*, **1972**, *38*, 395-411.
[9] H.P. Grace, *Chem. Eng. Commun.*, **1982**, *14*, 225-227
[10] B.J. Bentley, L.G. Leal, *J. Fluid Mech.*, **1986**, *167*, 241-283.
[11] H. Yamane, M. Takahashi, R. Hayashi, K. Okamoto, *J. Rheol.*, **1998**, *42*, 567-580.
[12] Linkam CSS 450 shearing system technical notes, http://www.linkam.co.uk.
[13] Rheometrics Scientific OAMII technical notes, http://www.rheosci.com.
[14] Thermo Haake Rheoscope1 technical notes, http://thermo.com.
[15] A.S. Almusallam, R.G. Larson, M.J. Solomon, *J. Rheol.*, **2000**, *44*, 1055-1083.
[16] L. Levitt, C.W. Macosko, D.S. Pearson, *Polym. Eng. Sci.*, **1996**, *36*, 1647.
[17] F. Greco, *J. Non-Newt. Fluid. Mech.*, **2002**, submitted.
[18] F. Greco, *Phys. Fluids*, **2002**, *14*, 946-954.
[19] R.G. Cox, *J. Fluid Mech.*, **1969**, *37*, 601.
[20] S. Guido, F. Greco, *Rheol. Acta*, **2001**, *40*, 176-184.
[21] N. Grizzuti, O. Bifulco, *Rheol. Acta*, **1997**, *36*, 406-415.

Detection of Co-Continuous Structures in SAN/PA6 Blends by Different Methods

Petra Pötschke,[*1] *Donald R. Paul*[2]

[1] Institut für Polymerforschung Dresden e. V., Hohe Straße 6, D - 01069 Dresden, Germany
E-mail: poe@ipfdd.de
[2] Department of Chemical Engineering and Texas Materials Institute
The University of Texas, Austin, Texas 78712, USA

Summary: This paper describes the 70/30 wt% composition of SAN/PA6 blends having different types of morphology, namely PA6 dispersed in SAN, a co-continuous structures of PA6 and SAN, and a "mixed structure" which exhibits PA6 particles in SAN which themselves form the matrix for smaller SAN particles. These morphologies were achieved by using different processing conditions during extrusion blending in a twin screw extruder, especially variation in the screw speed and by injection molding. Morphological analysis using SEM and TEM, solubility experiments, DMA, and oscillatory rheometry are presented. These methods were shown to be able to distinguish between the different types of morphology. In addition, DSC was used to detect the PA6 crystallization behavior.

Keywords: blends, differential scanning calorimetry (DSC), co-continuous structure, morphology, rheology

Introduction

Co-continuous blends involve the coexistence of at least two continuous structures within the same volume. Blends with co-continuous structures may combine the properties of both components in a favorable way, for example mechanical moduli.

Co-continuous structures are complicated three-dimensional interpenetrating and intertwining structures that are not always isotropic. Compared to dispersed, lamellar or fibrillar structures, it is much more difficult to identify a co-continuous structure unequivocally in a melt mixed blend. Microscopic methods, like light microscopy (LM), scanning electron microscopy (SEM) or transmission electron microscopy (TEM) are often used to visualize blend structures. Microscopic methods based on thin sections which only show a cross-section of the three dimensional structure (TEM, LM) can be problematic for identifying a three dimensional co-continuous structure. Viewing of selectively etched surfaces by SEM can be more informative because it is possible to investigate, at least partially, in the third dimension. Because of the limitations of microscopic methods, indirect methods seem to be more suitable for the unequivocal identification of co-continuous structures. Extraction experiments are shown to be an easy and viable way to check for co-continuity when the components are soluble in specific

 DOI: 10.1002/masy.200350807

solvents. Measurements like dynamic mechanical analysis (DMA), stress-strain-behavior, impact properties, and melt rheology on samples with varying compositions may show a signature indicative of a co-continuous structure composition range; however, such properties generally do not provide an unequivocal indicator of structure.

The aim of the paper is to analyze different methods for their ability to distinguish between co-continuous and dispersed morphology type which were achieved at the same blend composition. Such a comparison is of interest because only then the benefits of co-continuous structures become clear. For this purpose, styrene-acrylonitrile copolymer (SAN) / polyamide 6 (PA6) blends having a composition of SAN/PA6=70/30 wt% were extruded in a twin screw extruder under different conditions, especially variation in the screw speed and by injection molding. Three different types of morphology, namely PA6 dispersed in SAN, co-continuous structures of PA6 and SAN, and a "mixed structure" which exhibits PA6 particles in SAN which themselves form the matrix for smaller SAN particles were obtained. For the comparison, morphological analysis using SEM and TEM, extraction experiments, oscillatory rheometry, and DMA were applied. In addition, DSC was used to detect the PA6 crystallization behavior.

Materials and Methods

Blends with the composition SAN/PA6 = 70/30 wt% were made from SAN 25 (Tyril 100, Dow Chemical) and PA6 (Capron 8207F, Allied Signal). According to Kitayama et al.[1] this SAN exhibiting 25 wt% acrylonitrile has a M_w value of 152,000 g/mol whereas, the PA6 has a molecular weight M_n of 22,000 g/mol. According to the densities of the materials (SAN= 1.07 g/cm^3, polyamide = 1.13 g/cm^3)[1] the volume composition of the blends is SAN/PA6 = 71.1/28.9 vol%. The materials were dried under vacuum at 80°C for at least one day prior to all processing and analyzing operations. A Killion single-screw extruder (L/D=30, D= 25.4 mm) outfitted with an intensive mixing head was used to extrude the blends at 240°C with screw speeds of 40 and 80 rpm. The extrudate was drawn through an ice-water bath before being pelletized or sampled as strand. Injection molding was performed with the material extruded at 40 rpm using an Arburg Allrounder injection molding machine at 240°C to produce standard tensile bars with 3.18 mm thickness (ASTM D638 type 1).

TEM was performed on specimens cryogenically (about -20°C) microtomed perpendicular to the flow direction from the center of the strand or injection molded sample. The thin sections (thickness about 30 nm) were stained with phospotungstic acid by floating the sections on the surface of a 2 % solution for 30 min and washed with distilled water for 20 min. By this procedure, PA6 is stained dark. TEM was carried out on a JOEL 200CX microscope operating at an acceleration voltage of 120 keV.

SEM was performed on cryofractured surfaces (along and perpendicular to the extrusion direction) and cryocut surfaces with and without etching in selective solvents. Formic acid (FA) was used to etch the PA6 phase and tetrahydrofuran (THF) to etch SAN. Investigations were done on sputtered samples using a LEO VP1530 scanning electron microscope with acceleration voltages of 6-10 kV.

Blend solubility was checked by putting one piece of about 0.1 g in 5 ml solvent at room temperature for 1 week. After that the vials were shaken and photos were taken. To dissolve PA6, formic acid (98 % Fluka) was used; to dissolve SAN, THF (99.5 % purity, Merck) was applied. Quantitative analysis was performed with 0.5 g dried blend granules in 50 ml solvent for 96 hours at room temperature. The insoluble part in THF was extracted a second time with fresh THF for 24 h. The sample extracted with formic acid was washed several days in distilled water.

Melt rheological measurements were done using a ARES oscillatory rheometer (Rheometrics Inc.) with parallel plate geometry at 240°C under nitrogen atmosphere. Frequency sweeps were performed at a strain of 5% using a highly sensitive transducer (0.02-200 g cm) on the same sample with increasing frequency from 0.1 to 100 rad/s, with decreasing frequency from 100 to 0.05 rad/s, and after that from 100 to 0.1 rad/s. The second sweep was used for the interpretation. The third sweep was performed as a control to see if changes had occurred in the material during the long annealing time in the melt, which was about 8 min for the first and the third sweep, but about 70 min for the second sweep. The sweeps showed good agreement and all torque values were found to be in the working range of the transducer (> 0.08 g cm). Strain sweeps were performed to prove that the applied strain was within the linear-viscoelastic deformation range.

Dynamical mechanical analysis was performed with the same equipment under torsion rectangular conditions on strands (PA6, SAN, extruded blends) and on the injection molded sample. A heating rate of 5 K/min and a strain of 0.05 % were applied in the temperature range of -100°C to 170°C.

Differential scanning calorimetry was performed using a DSC 7 (Perkin-Elmer) with a Pyris-Version 3.81 under nitrogen atmosphere. The sample weight was about 8 mg taken from the middle of the samples. The temperature range was -10 to 250°C, the cycle consisted of first heating, cooling and second heating at 10 K/min.

Wide angle X-ray diffraction (WAXD) investigations were performed using a X-ray diffractometer P4C (Bruker axs) with monochromatized Cu-Kα radiation in a scattering range $2\Theta = 1.5 - 40°$ in transmission. Reflections at $2\Theta = 20.1°$ and $24.0°$ were assigned to the α-modification of PA6 crystals whereas, reflections at $2\Theta = 21.4°$ correspond to the γ-modification.[2]

Results

Blend Morphology

The selected blend composition is near the phase inversion composition of this blend system as shown by Kitayama for blends extruded under the same conditions[1]. Therefore, it was expected that variations in the extrusion conditions and the subsequent injection molding could change the morphology type.

Figure 1 shows micrographs of the blend extruded at 40 rpm obtained by different techniques.

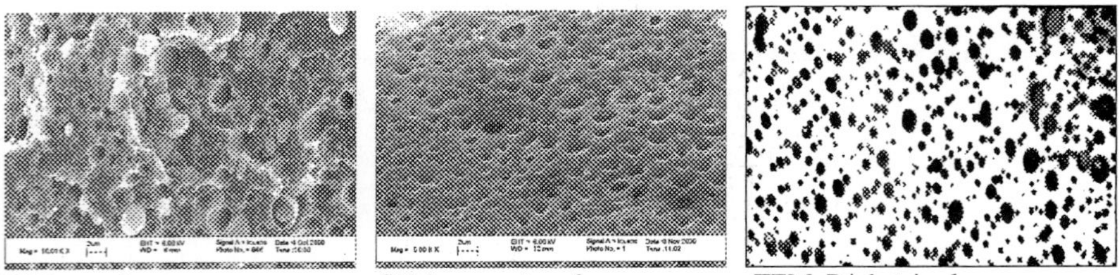

SEM, cryofracture | SEM, cryocut surface, etched with FA | TEM, PA6 stained (image size 34 x 24.5 μm)

Fig. 1. Morphology of PA6/SAN = 70/30 blend extruded at 40 rpm.

SEM cryofracture clearly indicates a matrix-disperse particle structure with nearly spherical particles. During cryofracture some of the dispersed particles dislodge from the matrix indicating poor phase adhesion between PA6 and SAN. SEM on etched surfaces and TEM show that PA6 is the dispersed phase. The particle size ranges between 0.2 and 2 μm.

After injection molding the morphology is changed significantly as shown in Figure 2.

The cryofractured surface also shows dispersed particles. TEM shows that the morphology is characterized by a particle in particle structure which is named in this paper as a "mixed structure". In the continuous phase of SAN small PA6 particles and islands of PA6 filled with small SAN droplets can be found. Etched surfaces also indicate that there are regions with SAN as the matrix component as well as regions with PA as the continuous phase. Such structures typically are found for blend systems which experienced phase inversion during their mixing history.[3] The morphology of the blend extruded at 80 rpm is shown in Figure 3.

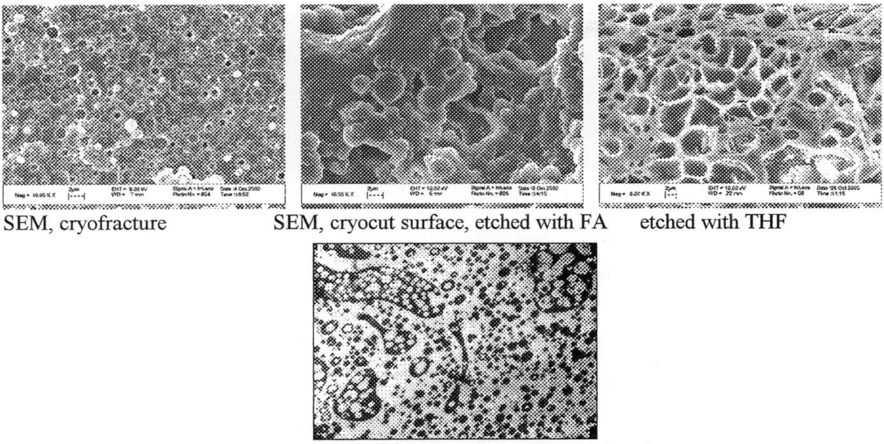

SEM, cryofracture SEM, cryocut surface, etched with FA etched with THF

TEM, PA6 stained (image size 34 x 24.5 μm)

Fig. 2. Morphology of PA6/SAN = 70/30 blend extruded at 40 rpm and injection molded.

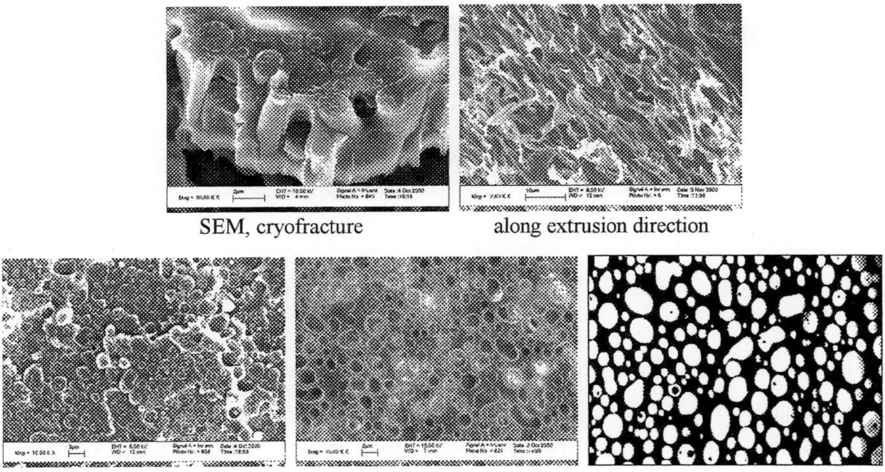

SEM, cryofracture along extrusion direction

SEM, cryofracture SEM, cryocut surface, etched with THF TEM, PA6 stained (image size 34 x 24.5 μm)

Fig. 3. Morphology of PA6/SAN = 70/30 blend extruded at 80 rpm.

Fractures made along the extrusion direction clearly show that both materials form a co-continuous structure with elongated interconnected fibrils of both components in extrusion direction. SAN forms the convex surface part with PA6 filling the spaces between the SAN

strands. However, standard morphological investigations, as used for the blends mixed at 40 rpm, are not able to represent the co-continuous structure unequivocal. Investigating cryofractured surfaces or selectively etched cut surfaces oriented perpendicular to the extrusion direction could give the impression of particles of SAN in a PA6 matrix. Also TEM carried out on samples cut parallel to the extrusion direction could lead to the conclusion that PA6 forms the matrix phase with SAN particles in it. This result indicates the importance of carefully performed morphology investigations.
The morphological findings could be confirmed by solubility tests. Figure 4 shows vials with the blend materials dissolved in THF and formic acid, respectively. The strand extruded at 40 rpm clearly shows solubility in THF and insolubility in formic acid indicating SAN to be the matrix component. The injection molded 40 rpm sample shows partial solubility in both solvents in that there is a undissolved part in formic acid and THF indicating a “mixed structure” morphology. The strand extruded at 80 rpm retains its shape in both solvents indicating a continuous structure of both components. Quantitative analysis of the soluble parts showed a blend fraction soluble in formic acid of 29.8 wt% (PA6 content 30 wt%) and of 64.5 wt% soluble in THF (SAN content 70 wt%) confirming the existence of a co-continuous structure.

strand 40 rpm | extruded at 40 rpm, injection molded | strand 80 rpm

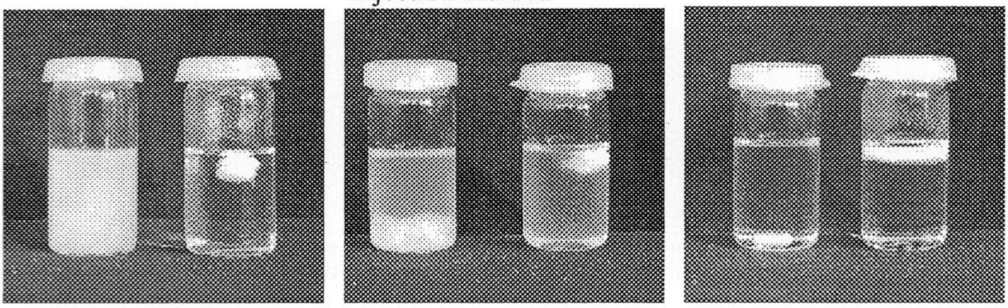

Fig. 4. Blends dissolved in THF (left vials) and formic acid (right vials).

Melt Rheological Behavior of the Blends

Melt rheological behavior is expected to show differences between the different types of morphology. Recently it was found by different authors that the elastic part of the complex viscosity, G', is very sensitive to the formation of co-continuity. Steinmann et al.[4] demonstrated that the value of G' at very low frequencies, representing the shape sensitive relaxation time range, is a suitable criterion for co-continuity. At the phase inversion composition, G' showed a significant maximum for all investigated PS/PMMA and SAN/PMMA blend systems having different viscosity ratios. This strong enhancement of the storage modulus was explained by extra stresses associated with the shape relaxation of the interconnective structures. On the other hand, Galloway and Macosko[5] found maxima in G' at the onset compositions of co-continuity for PEO/PS blends exhibiting a broad co-continuous range. They explained these G' maxima with the interfacial area maxima at these compositions.[6] Prochazka et al.[7] confirmed these G' maxima for PEO/PVDF blends investigated at low frequencies (0.01 and 0.1 rad/s).

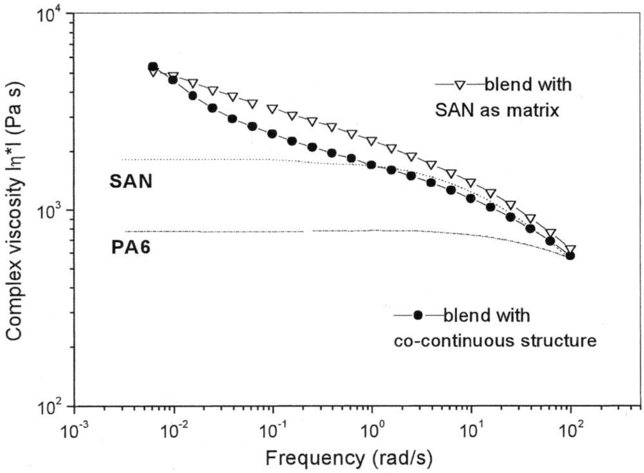

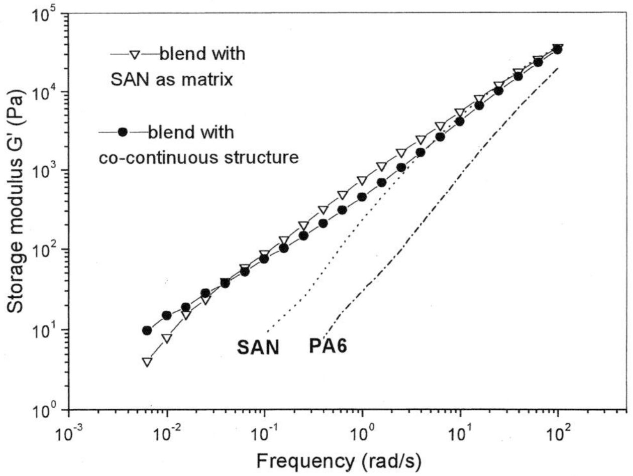

Fig. 5. Complex viscosity |η*| and storage modulus G' versus frequency for SAN/PA6 = 70/30 blends at 240°C.

Although all these considerations are made by changing the composition to switch from a dispersed to a co-continuous structure, we would like to determine if this criterion is also applicable to distinguish between a dispersed and a co-continuous structure having the same composition. Figure 5 shows the results of frequency sweeps performed at the mixing temperature.

At high frequencies, the blend based on the higher viscosity component as the continuous phase (SAN) and with PA6 as filler particles exhibits a higher blend viscosity than the co-continuous blend. In the co-continuous blend, both phases contribute more equally to the ultimate blend viscosity. However, the shapes of the viscosity versus frequency curves are different. At lower frequencies a significant increase in the blend viscosity of the co-continuos blend can be observed. This increase in viscosity seems to be characteristic for co-continuous blends and was also found by Steinmann et al.[4] by comparing dispersed and co-continuous PS/PMMA blends having different compositions. On the other hand, the G' behavior is complex. According to the higher G' value of SAN as compared to PA6 the blend based on SAN has a higher G' value at frequencies between 100 and 0.04 rad/s. Again the shape of the curve versus frequency is different from that of the co-continuous blend. By further decreasing the measurement frequency, G' becomes higher for the co-continuous blend than for the dispersed one. At 0.01 rad/s the dispersed blend exhibits a G' value of 7.9 Pa whereas the co-continuous one shows a G' value of 14.7 Pa. Even if the difference between the behavior of the co-continuous and the dispersed blend seems to be small, it was found to be characteristic and reproducible by measuring different samples. In addition, comparison of the first and third frequency sweep and the SEM micrographs before and after rheological measurements did indicate that there are no changes in the morphology type and structure size range caused by the rheological tests.

Dynamic-Mechanical Analysis (DMA) of the Blends

It is expected that co-continuous blends combine the moduli of the components in a much more effective way than dispersed blends where it is mostly dependent on the matrix behavior.[8,9] Gergen[10] presented an impressive example for SEBS/PC = 70/30 blends where in one case PC was the dispersed phase and in the other where co-continuous structure existed; for the co-continuous structure the storage modulus-temperature-dependency reflects a greater contribution of the rigid PC phase. Therefore, DMA should be a suitable method to distinguish between both structures.

Figure 6 shows the behavior of G' versus temperature.

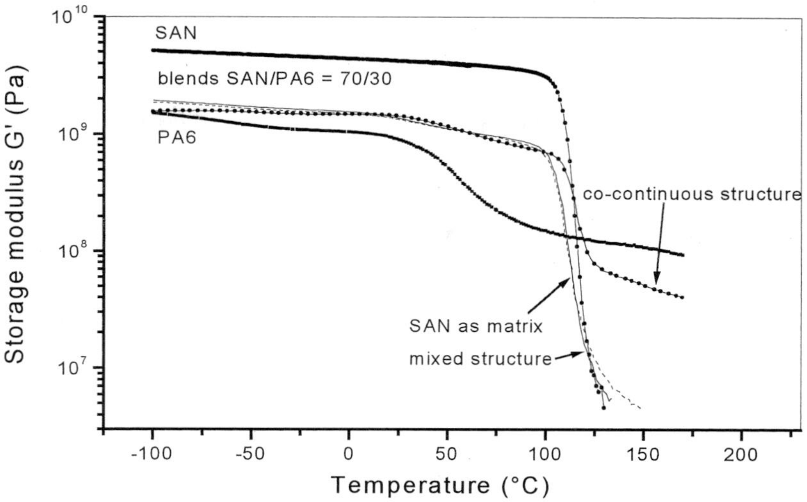

Fig. 6. Storage modulus G' versus temperature.

The differences between the G' values of the blends are only small in the temperature range up to the T_g of SAN. The co-continuous structure shows slightly lower values than the other blends. Above the T_g of PA6, the decrease in G' is slightly higher in the co-continuous blend but above the T_g of SAN, significant differences become evident. The co-continuous blend exhibits much higher G' values than the other blends. Whereas, the dispersed blend and the blend with the mixed structure show a decrease in G' comparable to that of SAN; the blend with the co-continuous structure remains a much higher G' values in the range of that of PA6. The differences between the dispersed blend and the blend with the mixed structure are very small indicating that in the mixed structure, SAN is the main matrix component.

Differential Scanning Calorimetry (DSC)

The system contains an amorphous phase (SAN) and a crystallizable phase (PA6). It is known from the literature that when PA6 forms the disperse phase in blend systems, a "fractionated" crystallization behavior may be seen. This means that in the case of a lower crystallization or glass transition temperature of the matrix material, PA6 does not crystallize at temperatures typical for the bulk crystallization but it crystallizes with the matrix phase.[11] Fractionated crystallization was shown by Frensch et al.[11] for PVDF/PA6 blends and by Pompe et al.[12] for PP/PA6 blends. For the PP/PA6 system Pompe et al.[12] calculated the critical PA6 particle size to be about 4 μm. Based on the particle sizes found in the blends under investigation here, fractionated crystallization behavior could be expected in the dispersed blend.

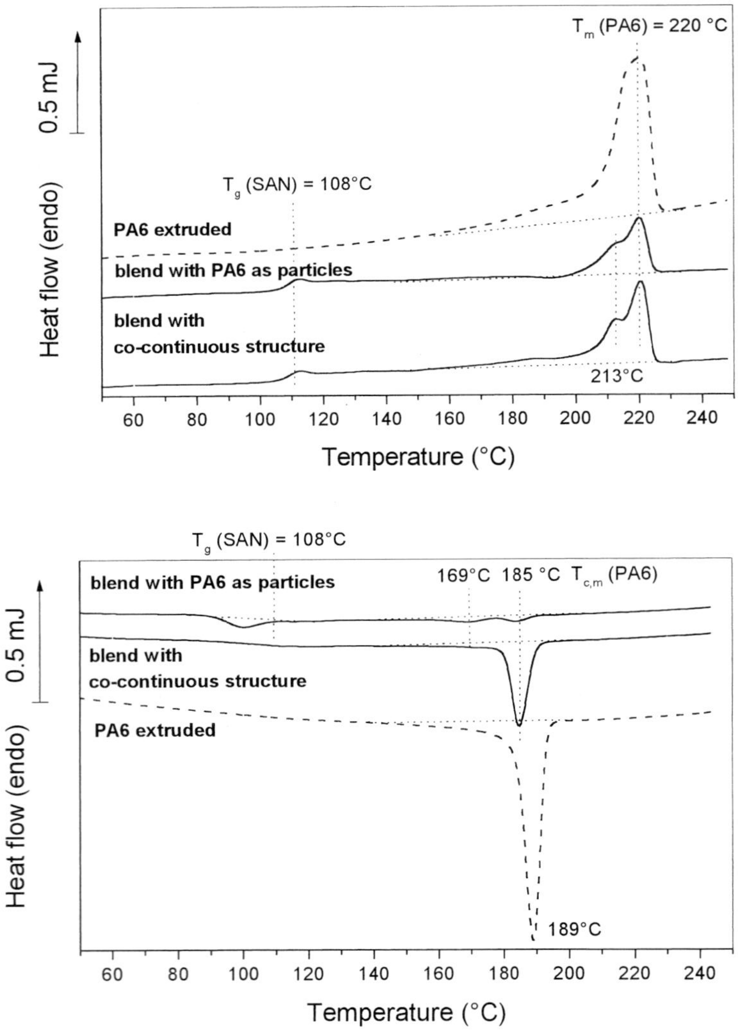

Fig. 7. Second heat (top) and cool (bottom) scans of SAN/PA6=70/30 blends and extruded PA6.

The results show that there really is a fractionated crystallization in the blend with SAN as matrix and PA6 as dispersed phase. In the co-continuous blend PA6 crystallizes in the temperature range typical for the extruded PA6; whereas, in the dispersed blend only a small amount crystallizes at this temperature. Some of the PA6 phase appears to crystallize below the T_g of the SAN phase. During the second heat, a cold crystallization at about 190°C is observed for the dispersed blend. However, the melting curves are very similar in both blends with a double peak (213°C and 220°C) indicating a transition from the γ to the α crystal type of PA6 during heating (Figure 7). Additional WAXD investigations indicate that the unprocessed as well as the extruded PA6 exhibits only α crystallites whereas, all the blends only show γ crystallites. Therefore, the differences in the crystallization behavior should not be due to different types of crystals. The crystallinity is clearly reduced in the dispersed blend which is evident by both DSC and WAXD results. Although different types of crystals are present in the PA6 component, the co-continuous blends exhibits about the same crystallinity as the processed PA6. Comparison of SEM micrographs on samples before and after DSC treatment did not show changes in the morphology type or phase size range.

Summary and Conclusion

In a PA6/SAN blend near to its phase inversion composition it is possible to create different types of morphology by changing the melt blending conditions. Under the conditions selected, a screw speed of 40 rpm led to PA6 dispersed in SAN. Injection molding of the dispersed structure resulted in a "mixed structure" with SAN as main matrix and PA6 particles; whereas, some of the PA6 particles act as a matrix for SAN particles. Increasing the screw speed to 80 rpm produced a co-continuous morphology with an elongated orientation in the extrusion direction.

The co-continuous type of structure is especially difficult to identify by microscopic methods. Only the combination of different methods (SEM of cryofractures perpendicular and along the extrusion direction, SEM of selectively etched surfaces, and TEM) finally gives a clear picture of the morphology.

Solubility investigations confirmed the morphologies observed in the morphological investigations. This is a suitable and easy way to unequivocally detect co-continuous structures if one or both components have specific solvents.

Oscillatory melt rheological investigations show a clear difference between the dispersed and the co-continuous structure. At high frequencies, the blend based on the higher viscosity component as the continuous phase (SAN) and with PA6 as filler particles exhibits a higher melt viscosity than the co-continuous blend in which the viscosities of both components contribute more equally to the blend viscosity. However, in the low frequency range the co-

continuous blend shows similar η^* and higher G' values than the dispersed one. It seems possible to deduce the type of morphology from the shape of the viscosity versus frequency and G' versus frequency curves when measured at low enough frequencies. A necessary condition (especially to get reliable values at low frequencies) is that the morphology has to be stable over the time frame of rheological investigation which was true for the blend system under investigation.

DMA indicated G' differences especially above the T_g of SAN. Blends with SAN as matrix show a sharp decrease in G' whereas the co-continuous blend retains higher moduli due to the greater contribution from the higher modulus of PA6 in this temperature range. Therefore, DMA is also a suitable method to detect co-continuous structures when comparing the G' curves to that of the blend components.

DSC investigations could be used to distinguish between PA6 as dispersed phases and as part of a co-continuous structure. If PA6 forms the dispersed phase, its crystallization is prevented at the typical PA6 bulk crystallization temperature and a fractionated crystallization occurs. If PA6 forms a continuous phase as in a co-continuous structure, PA6 shows normal crystallization behavior. However, this method is limited to systems where phases are able to show a fractionated crystallization.

Acknowledgement

The authors are grateful for the financial support granted by the Max-Kade-Foundation (New York) to P.P. during her stay at the University of Texas at Austin. We would like to thank Dr. Gisela Pompe (IPF Dresden) for assistance with the DSC measurements and result discussion and Dr. Dieter Jehnichen (IPF Dresden) for the WAXD measurements.

[1] N. Kitayama, H. Keskkula, D.R. Paul, *Polymer* **2000,** *41*, 8041.
[2] A. Galeski, A.S. Argon, R.E. Cohen, *Macromolecules* **1991,** *24*, 3945.
[3] C.W. Macosko, *Macrom. Symp.* **2000**, *149*,171.
[4] S. Steinmann, W. Gronski, Ch. Friedrich, *Rheol. Acta* **2002**, *41*, 77.
[5] J.A. Galloway, C.W. Macosko, *Conference Proceedings*, 3rd Pacific RIM Conference on Rheology, **2001**, July 08-13.
[6] J.A. Galloway, M.D. Montminy, C.W. Macosko, Polymer **2002**, *43*, 4715.
[7] F. Prochazka, C. Carrot, M. Castro, C. Celle, J.-C. Majesté, *Conference Proceedings,* PPS-18 Guimares Portugal, **2002**, June 16-20.
[8] L.A. Utracki, *Polymer Alloys and Blends: Thermodynamics and Rheology*; Hanser Publ.: München, **1989**, p. 178 ff.
[9] D.R Paul, S. Newman, *Polymer Blends, Volume 1*, Academic Press: New York, **1978**, 11-13.
[10] W.P. Gergen, R.G. Lutz, S. Davison, Hydrogenated Block Copolymers, Thermoplastic Elastomer Interpenetrating Polymer Networks. In: *Thermoplastic Elastomers*, 2nd ed.; G. Holden, N.R. Legge, R Quirk, H.E. Schroeder, Eds., Hanser Publ.: Munich, **1996,** 507.

[11] H. Frensch, H. Harnischfeger, B.-J. Jungnickel, Fractionated Crystallization in Incompatible Polymer Blends. In: *Multiphase Polymers: Blends and Ionomers,* L.A. Utracki, R.A. Weiss, Eds., ACS Symposium Series, **1989**, 101.
[12] G. Pompe, P. Pötschke, J. Pionteck, J. Appl. Polym. Sci. **2002,** *86*, 3445.

Interfacially Induced Additional Damping Peaks in Dynamic Mechanical Spectra – Micromechanical Transitions in Multipolymeric Materials

Didier Colombini, Frans H. J. Maurer*

Department of Polymer Science & Engineering, Lund University, Lund Institute of Technology – Box 124, SE – 22100 Lund, Sweden
E-mail: didier.colombini@polymer.lth.se

Summary: The viscoelastic properties of several hypothetical multiphase polymeric materials were investigated in relation to their phase-property dependencies and microstructures. Theoretical mechanical considerations based on the self-consistent interlayer model were performed to point out that specific geometrical arrangements into phases of a set of properties of the pure constituents can lead to interfacially induced damping peaks in dynamic mechanical spectra (DMS). Such additional contributions in DMS were referred to micromechanical transitions to distinguish them from ordinary molecular transitions.

Keywords: additional damping peak, interphase, micromechanical transition, polymer blend, viscoelastic properties

Introduction

The mechanical and viscoelastic properties of multiphase polymeric materials depend on molecular relaxation processes and morphology. Although these relaxations can usually be associated with each component, their appearance also depends on the chemical and physical interactions between the phases. These interactions may evolve in the existence of an interphase, which can lead to an additional damping peak in the dynamic mechanical spectra (DMS) of such multipolymeric systems [1-5]. However, an unexpected additional mechanical transition in DMS does not necessarily reflect an interfacial relaxation process.

Recently, much attention has been focused on predicting the viscoelastic properties of complex polymer systems in relation to their morphologies. As a result, numerous examples of *micromechanical transitions* (MMT) have been reported in the field of particulate multipolymeric systems [6-14]: the origin of such an additional damping peak has been explained by the change in the relative moduli values of the components in the matrix / interphase / particle structure of the multicomponent polymeric materials.

 DOI: 10.1002/masy.200350808

Basically, it is now well understood that the response of a multiphase polymeric material, to an external force field, is strongly affected by the viscoelastic properties of each component, the interactions between the different phases, the composition, the morphology, and the properties and volume fraction of the interphase. The present work deals with the use of self-consistent mechanical modelling to point out the importance of the microstructure (or geometrical arrangement of phases) and the phase-property-dependence on the viscoelastic damping properties of multipolymeric materials.

Theoretical Considerations – The Interlayer Model [15,16]

The self-consistent interlayer model used in this work, has been extensively described [7,14-16] and requires the definition of a representative volume element (RVE) in order to predict the dynamic mechanical shear and other physical properties of multipolymeric materials. An illustration of a RVE consisting of three concentric spheres embedded in an equivalent homogeneous medium, is depicted in Figure 1.

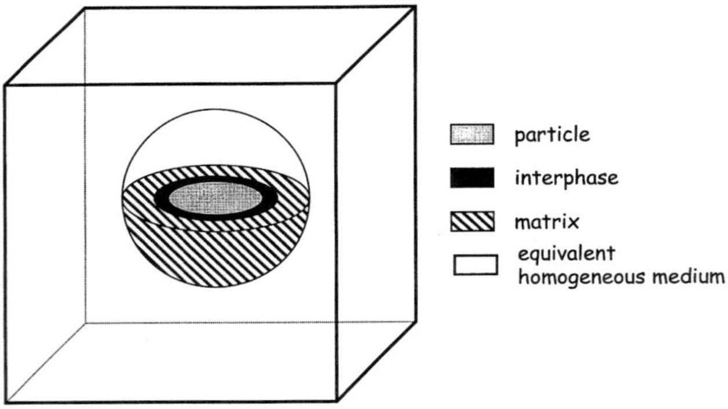

Figure 1: Illustration of a RVE used in the interlayer model

On the one hand, such a geometrical arrangement is representative of the morphology, the particle (phase 1) being surrounded by a shell of interphase (phase 2) which is covered by a shell of matrix (phase 3). On the other hand, the composition of the multicomponent material is also taken into account since the radii of the concentric spheres are chosen in accordance with the volume fractions of the different phases.

Hypothetical Materials

The viscoelastic properties of the hypothetical pure constituents (matrix, interphase, and particle) used as input data in the calculations are given in Figure 2 and Figure 3.

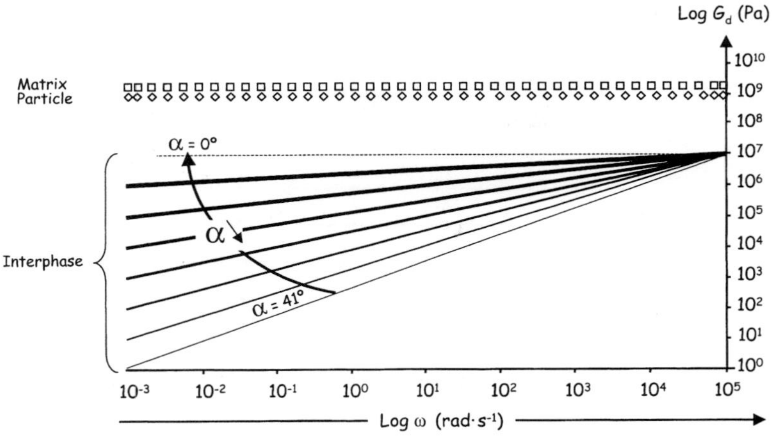

Figure 2: Viscoelastic properties of the hypothetical pure constituents used as input data in the numerical simulations: **dynamic shear modulus (G_d) *versus* frequency (ω)**

The use of isothermal data over a wide frequency range has been thought relevant since both the dynamic shear modulus, G_d, and the phase angle, δ, are frequency (ω) and temperature (T) dependent quantities, and directly related to each other by the following relationship [17]:

$$\delta\,(\omega, T) \approx \frac{\pi}{2} \left| \frac{\partial \text{ Log } G_d(\omega_i, T)}{\partial \text{ Log } \omega_i} \right|_{\omega_i}$$

In agreement with our recent extended investigation [14], both the continuous (matrix) and the dispersed (particle) phase were considered in the calculation to be in the glassy state with constant viscoelastic properties over the whole frequency range. On the same basis [14], their respective Poisson ratios ν_{matrix} and $\nu_{particle}$ were kept constant at 0.33.

To point out the interphase-property-dependence of the viscoelastic properties of multipolymeric systems, we decided to perform numerical simulations assuming a significant variation of the dynamic shear modulus of the interphase component. Basically, an angular shift, α, of $G_d^{interphase}$ has been considered. As illustrated in Figure 2, α is defined as the angle between the dashed line at $G_d^{interphase}$ equal to 10^7Pa (α=0°) and the variation of $G_d^{interphase}$ (from 10^7Pa to 10^x Pa with $x \in [0;6]$). Unless otherwise stated, $\nu_{interphase}$ was considered constant at 0.5 in calculation.

Worth noting is the absence of a damping peak of any of the pure constituents over the whole frequency range considered (Figure 3).

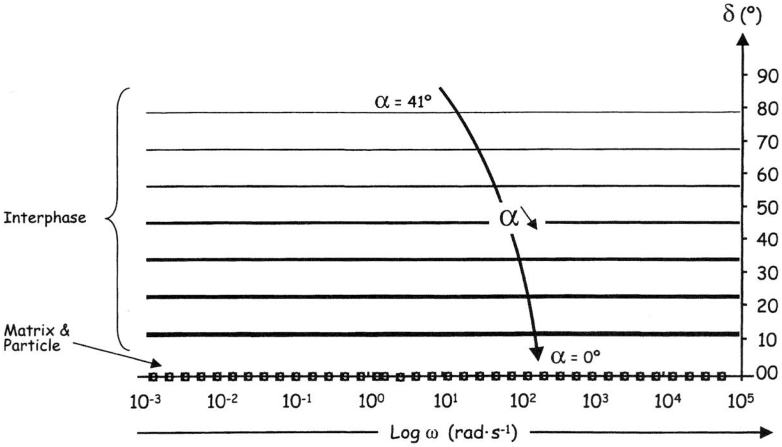

Figure 3: Viscoelastic properties of the hypothetical pure constituents used as input data in the numerical simulations: **phase angle (δ)** ***versus*** **frequency (ω)**
The evolution of α (defined in Figure 2) is highlighted by the arrow.

Results and Discussion

Figure 4 shows the theoretical results obtained for a hypothetical matrix / interphase / particle systems with the same blend composition of 70/5/30 volume parts, respectively. For the entire range of angular shift of $G_d^{interphase}$, the predicted values of both G_d^{blend} and $\tan\delta_{blend}$ are given versus the frequency. It can be seen in Figure 4 that all the theoretical evolutions of the loss factor present a damping peak.

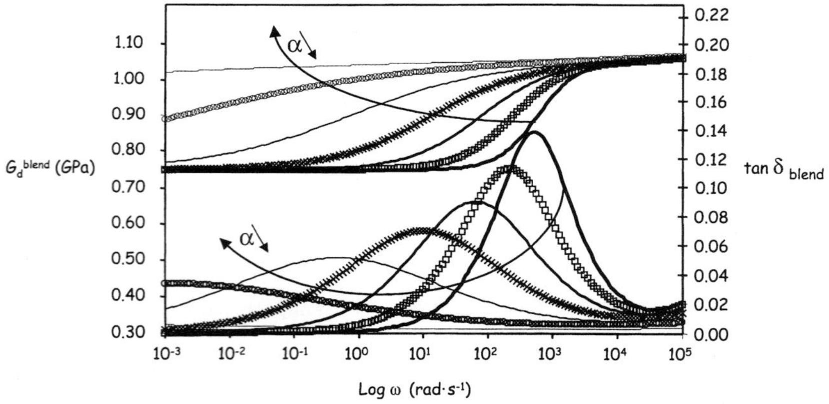

Figure 4: Theoretical viscoelastic properties (G_d and tan δ) of a matrix / interphase / particle (70/5/30) blend *versus* frequency as a function of the angular shift α of $G_d^{interphase}$
The arrow highlights the decrease of α from 41° to 0°.

At the same time as tan δ_{blend} reaches its maximum value, G_d^{blend} drops from a lower level to a higher. It can be noticed that when α decreases:

- the magnitude of the tan δ_{blend} peak decreases and the frequency location of its maximum value is significantly shifted toward lower frequencies,
- the width of the tan δ_{blend} peak increases but,
- the areas under the damping curves remain logically constant since the predicted drops of G_d^{blend} are almost similar.

As the phase angle δ was defined to be constant for each component in the hypothetical blends (see Figure 3), such predicted maxima of tan δ_{blend} have to be related to the change in the relative moduli of the components, since only $G_d^{interphase}$ was varied in our numerical simulations. In agreement with previous theoretical investigations [7,14], such interfacially induced additional damping peaks in DMS are referred to *micromechanical transitions* (MMT) to distinguish them from other ordinary molecular transitions.
It was then also deemed to be of interest to investigate how a change in either the volume fraction or the Poisson ratio of the interphase component could affect the curves shown in Figure 4. First of all, Figure 5 presents the results of numerical simulations based on matrix / interphase / particle blends with blend compositions of 70/x/30 (x=1,2,5,10 vol. parts) as a function of the dynamic shear modulus of the interphase component (α=41°). The calculated quotients of the dynamic shear modulus of the blend and either that of the matrix or that of the particle are given in Figure 5a. In addition, Figure 5b shows the corresponding values of tan δ_{blend} for the various 70/x/30 blend compositions. Same as in Figure 4, whereas the input data do not include any damping peak over the whole frequency domain considered in the simulations, all the theoretical predictions of the loss factor tan δ_{blend} show a maximum. Moreover, whatever the blend composition, both relative moduli jump from a lower to a higher level (as indicated by the vertical arrows in Figure 5). Similar observations based on the influence of the blend composition on the viscoelastic characteristics of MMT in DMS have been already discussed elsewhere [6-8,14]. On the basis of Figure 5, it can be recalled that:

- the position and magnitude of the jump in the moduli ratios (Figure 5a) as well as the maximum of tan δ_{blend} (Figure 5b), occur at higher $G_d^{interphase}$ values with increasing volume fraction of the interphase component,
- the more significant the slope in the change of the moduli ratios, the higher the value at the maximum of the tan δ_{blend} peak.

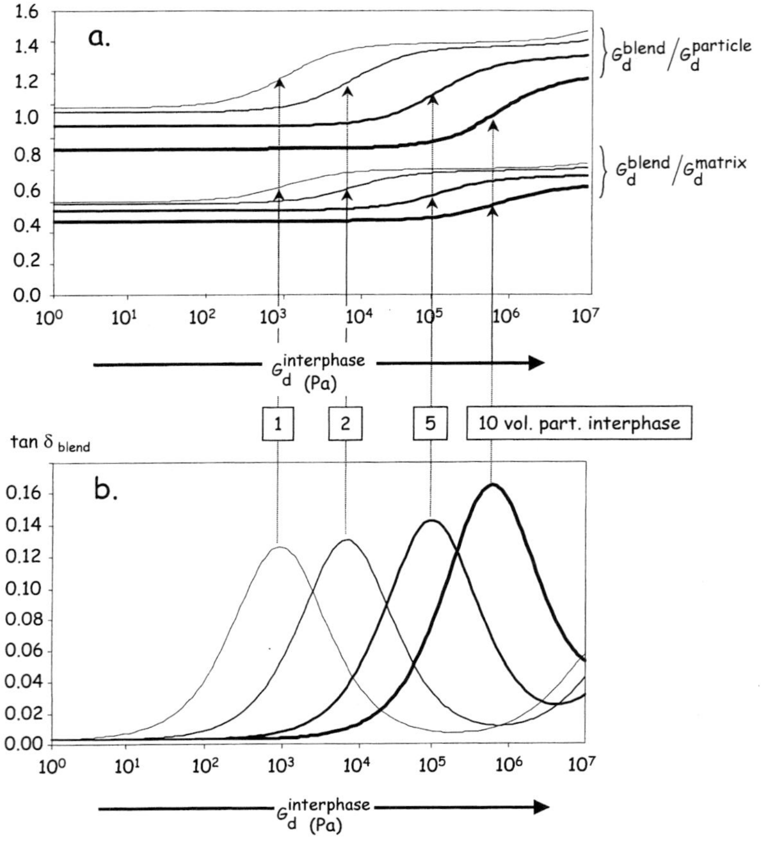

Figure 5: Modelling of the viscoelastic properties of hypothetical matrix / interphase / particle blends

a. ($G_d^{blend} / G_d^{matrix}$) and ($G_d^{blend} / G_d^{particle}$) versus $G_d^{interphase}$

b. loss factor tan δ_{blend} versus $G_d^{interphase}$

(70/x/30 with x = 1,2,5 or 10 vol. parts, $\nu_{matrix} = \nu_{particle} = 0.33$ *and* $\nu_{interphase} = 0.50$*)*

Finally, Figure 6 shows the results of numerical simulations based on two blend compositions (either 70/5/30 or 70/1/30) for ternary hypothetical matrix/interphase (α=41°)/particle. In order to investigate the influence of $\nu_{interphase}$ on the expected MMT (already observed for $\nu_{interphase}=0.5$, see Figure 5), the numerical calculations were performed by considering either $\nu_{interphase}=0.5$ or $\nu_{interphase}=0.499$. On the basis of previous investigations [7,14], ν_{matrix} and $\nu_{particle}$ were kept constant at 0.33. Basically, the way to present the predicted viscoelastic data in connection with the properties of the interphase component is similar in Figure 6 to that of Figure 5. It can be seen (Figure 6b) that the position of the maximum of tan δ_{blend} appears to be really sensitive to the changes of $\nu_{interphase}$ from 0.5 to 0.499. Nevertheless, such an amazing observation is significantly dependent on the volume fraction of the interphase, since the observed changes (Figures 6a and 6b) are much more significant for the 70/1/30 blend than for the 70/5/30 system. In any cases, such a noticeable influence of $\nu_{interphase}$ strongly indicates that the compressibility of the interfacial layer in a ternary polymer blend substantially affects the overall viscoelastic response of the multicomponent material.

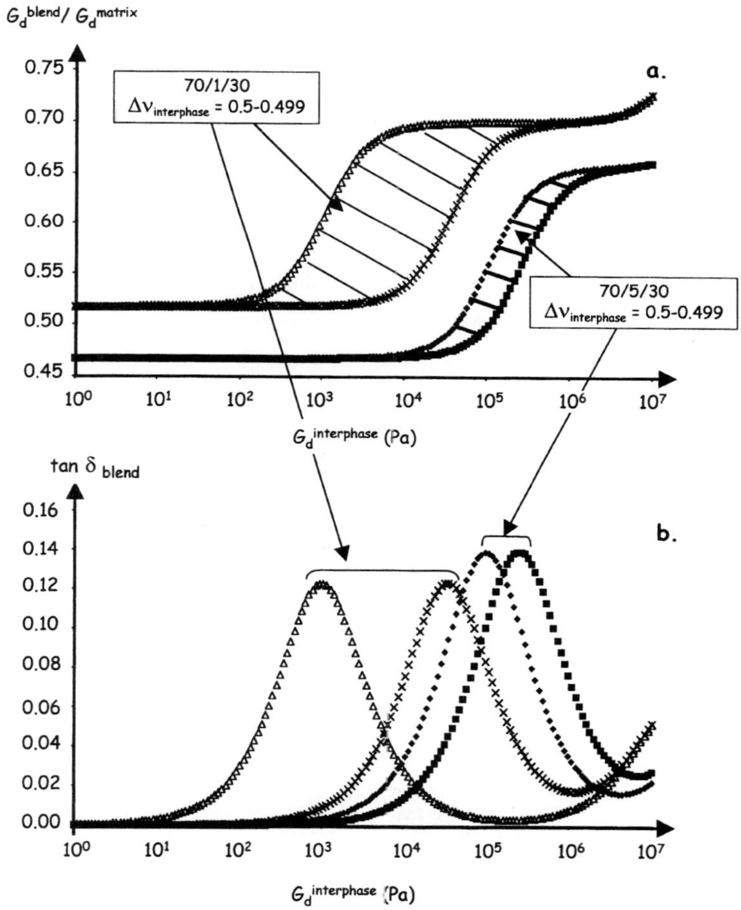

Figure 6: Modelling of the viscoelastic properties of hypothetical matrix / interphase / particle blends (70/5/30 or 70/1/30). **Composition dependence of the theoretical influence of $\nu_{interphase}$** ($\nu_{interphase}$ = 0.5 or 0.499)

a. (G_d^{blend} / G_d^{matrix}) versus $G_d^{interphase}$

b. tan δ_{blend} versus $G_d^{interphase}$

Conclusion

This work was mainly based on the use of the interlayer model to demonstrate the importance of the microstructure (or geometrical arrangement of the phases) and the phase-property-dependence on the overall viscoelastic damping properties of multipolymeric materials. Such a theoretical investigation based on self-consistent mechanical modelling can be reasonably considered as relevant arguments to justify the existence of "spurious" additional damping peaks in some experimental DMS reported in the literature for multipolymeric materials. Thus, it was emphasised that in some cases, a specific geometrical arrangement into phases of a set of properties of the pure constituents (none of them including any loss peak), can lead to an interfacially induced damping contribution in DMS. Such a phenomenon, so-called *micromechanical transition* definitively differs from classical molecular transitions, which solely reflect the molecular relaxation of macromolecules. Especially, it might be kept in mind that a MMT is an indicator of the manifestation of the interfacial area in a multicomponent polymeric system (the compressibility of which strongly influences the viscoelastic characteristics of the MMT in DMS), since this phenomenon requires the presence of a specific interphase, whose properties affect the overall relative moduli of the blended material and its components.

Acknowledgements

We gratefully acknowledge the Centre for Amphiphilic Polymers from Renewable Resources (CAP), Lund University, Lund – Sweden, for the financial support of this work.

[1] L.A. Utracki, "*Polymer Alloys and Blends – Thermodynamics and Rheology*", Hanser Verlag Publishers: Munich, Vienna, New York, **1989**
[2] S.W. Shalaby, "*Thermal Characterisation of polymeric materials*", Turi, E. A., Ed., Academic Press: London, **1981**
[3] Y.S. Lipatov, F.G. Fabulyak, V.V. Shifrin, *Polym. Sci. USSR* **1976**, 18, 866
[4] J.L. Thomason, "*Interfaces in Polymer, Ceramic and Metal Matrix Composites*", Ishida, H., Ed.; Elsevier Science: New York, **1988**
[5] L.T. Drzal, *Adv. Polym. Sci.* Ser. No 75, **1986**
[6] H. Eklind, S. Schantz, F.H.J. Maurer, P. Jannasch, B. Wesslen, *Macromolecules*, **1996**, 29, 984
[7] H. Eklind, F.H.J. Maurer, *Polymer*, **1996**, 37, 2641
[8] H. Eklind, F.H.J. Maurer, *Polym. Sci., Phys. Ed.*, **1996**, 34, 1569
[9] C.O. Hammer, F.H.J. Maurer, *Polymer*, **1998**, 39, 4243
[10] D. Colombini, J.J. Martinez-Vega, G. Merle, E. Girard-Reydet, J.P.P. Pascault, J.F. Gerard, *Polymer*, **1999**, 40, 935
[11] D. Colombini, G. Merle, N.D. Alberola, *J. Macromol. Sci. - Physics*, **1999**, B38 (5&6), 957
[12] D. Colombini, G. Merle, N.D. Alberola, *Macromol. Symp.*, **2001**, 169, 235
[13] D. Colombini, G. Merle, N.D. Alberola, *Macromolecules*, **2001**, 34, 5916
[14] D. Colombini, F.H.J. Maurer, *Macromolecules*, **2002**, 35, 5891
[15] F.H.J. Maurer, "*Polymer Composites*", Sedlacek, B., Ed.; W. de Gruyter & Co: Berlin, **1986**, 399
[16] F.H.J. Maurer, "*Controlled Interphases in Composite Materials*", Ishida, H., Ed.; Elsevier Science: New York, **1990**, 491
[17] H.C. Booij, G.P.J.M. Thoone, *Rheol. Acta*, **1982**, 21,15

Macromol. Symp. **2003**, *198*, 91-101

Blends of Isotactic Polypropylenes and a Plastomer: Crystallization and Viscoelastic Behavior

María L. Cerrada, Oscar Prieto, José M. Pereña, Rosario Benavente, Ernesto Pérez*

Instituto de Ciencia y Tecnología de Polímeros (CSIC). C/ Juan de la Cierva, 3. 28006 Madrid, Spain

Summary: Blends covering the entire range of compositions of a metallocenic ethylene-1-octene, CEO, copolymer and two conventional isotactic polypropylenes, iPP, of different molecular weights have been prepared, analyzing the effect of composition and molecular weight on the crystallization (studied by DSC and X-ray diffraction) and viscoelastic behavior (DMTA). It was found that those blends rich in the iPP component show a behavior practically coincident with the weighted addition of the two components. On the contrary, significant deviations were found for the blends where the CEO copolymer is the major component. These deviations are considerably more important in the case of the blends with the iPP of higher molecular weight. Moreover, both components are not miscible, exhibiting the glass transitions of the two neat components. The area under the loss tangent curves provides a preliminary information about how the toughness is enhanced using this type of impact modifier, though it provokes a significant reduction of stiffness.

Keywords: blends, crystallization behavior, ethylene-1-octene copolymer, impact resistance, isotactic polypropylene, relaxation, stiffness

Introduction

Isotactic polypropylene, iPP, exhibits valuable properties and a very competitive price. However, its poor low-temperature fracture toughness is an important drawback for several applications, so that it is often blended with elastomeric polymer materials which improve very effectively the impact resistance of iPP. The use of single-site highly-active metallocene catalysts has allowed the synthesis of ethylene copolymers with a very homogeneous comonomer distribution and narrow molecular weight distribution. Moreover, high comonomer contents are easily incorporated, in such a way that very low density polyethylenes can be obtained, exhibiting elastomeric properties.[1,2] Therefore, these metallocene polyethylenes are good candidates for their use as impact modifiers of iPP and the properties of the corresponding blends will depend very much on crystallinity, morphology and degree of dispersion,[3] which, in turn, seems to be very much

 DOI: 10.1002/masy.200350809

affected by the polymer miscibility in the molten state.[4,5] The aim of this work is to analyze the crystallization behavior and the corresponding structural studies on blends of iPP and an ethylene-1-octene copolymer and the further influence of the macroscopic morphology on the viscoelastic mechanisms exhibited.

Experimental

The two conventional commercially available iPP's, PP1 and PP2, respectively, analyzed in the current paper have been supplied by Repsol-YPF. A commercial metallocenic catalyzed CEO copolymer with a 9.3-mol % comonomer content, was utilized in the blends and has been supplied by Exxon Chemical. Table 1 shows the characteristics of the various plain polymers.

Table 1. Sample characteristics of the commercial materials analyzed.[a]

Sample	Isotacticity (% mmmm)	Mol % 1-octene	M_w	M_w/M_n	MFI
PP1	92.6	-	224000	4.2	35
PP2	91.8	-	349000	4.0	8.5
CEO	-	9.3	115000	2.5	1.1

[a]Data supplied by the manufacturers.

Blends with different content in CEO: 25, 50, 67, 75% in weight, labeled as PP1CEO75/25, PP1CEO50/50, PP1CEO33/67, PP1CEO25/75, PP2CEO75/25, PP2CEO50/50, PP2CEO33/67 and PP2CEO25/75, respectively, were prepared in a Haake Rheocord 9000 at 210°C and at 40 rpm for 10 min. After blending and homogenization of the two components, sheets specimens were obtained as films by compression molding in a Collin press between hot plates (200°C) at a pressure of 1.5 MPa for 8 min and subsequently quenched to room temperature.

Wide-angle X-ray diffraction patterns were recorded in the reflection mode at room temperature by using a Philips diffractometer with a Geiger counter connected to a computer. Ni-filtered CuK_α radiation was used. The diffraction scans were collected over a period of 20 minutes in the range of 2θ values from 3 to 43 degrees, using a sampling rate of 1 Hz. The goniometer was calibrated with a silicon standard.

The samples were also studied by either wide, WAXS, or small-angle X-ray scattering, SAXS, employing synchrotron radiation (λ= 0.150 nm) in the beamline A2 at HASYLAB (Hamburg,

Germany). Two linear position-sensitive detectors were used. The SAXS one, at a distance of 235 cm from the sample, was calibrated with the different orders of the long spacing of rat-tail cornea (L = 65 nm). It was found to cover a spacings range from 5 to 55 nm. The WAXS detector, covering a 2θ range from about 10 to 30 degrees, was calibrated with the different diffractions of crystalline PET. A heating rate of 8°C/min was used, acquiring data simultaneously in the two detectors in time frames of 15 s.

Calorimetric analyses were carried out in a Perkin-Elmer DSC7 calorimeter, connected to a cooling system and calibrated with different standards. The sample weights ranged from 5 to 7.5 mg. A temperature range from -70°C to 200°C has been studied and the used heating rate was 20 C min^{-1}. For crystallinity determinations, values of 209 and 290 J g^{-1} have been taken as the enthalpy of fusion of a perfectly crystalline material in iPP[6] and polyethylene, PE[7], respectively.

Viscoelastic properties were measured with a Polymer Laboratories MK II dynamic mechanical thermal analyzer working in the tensile mode. The real (EÑ) and imaginary (EÕ) components of the complex modulus and the loss tangent (tan δ) of each sample were determined at 1, 3, 10 and 30 Hz, over a temperature range from -150 to 150°C, at a heating rate of 1.5°C min^{-1}.

Results and Discussion

Figure 1 shows the X-ray diffractograms, acquired at room temperature, of the different pure components and blends. The two upper patterns in both figures, corresponding to the pure iPP components, exhibit the five main diffractions characteristic of the monoclinic α modification[8,9] of iPP (the samples are commercial polymers synthesized with traditional Ziegler-Natta catalysts, and no sign of the β or γ modifications[9-11] is detected). There is, however, a noticeable difference between the diffractograms of the two iPP samples. Thus, PP1 presents approximately equal intensities for the two first reflections (110 and 040), while the 110 peak is much more intense than the other in sample PP2. This feature is in relation to the fact that the film specimens have been obtained by compression molding, and the molecular weight differences between the two iPP samples seem to lead to different degrees of orientation on the final samples. This effect becomes less important as the CEO content increases in the blends.

Regarding the diffractogram corresponding to pure CEO, it can be observed that it presents a rather low crystallinity (as corresponds to its relatively high comonomer content) and the 110 and 200 reflections characteristic of the orthorhombic modification of PE are barely seen.[12]

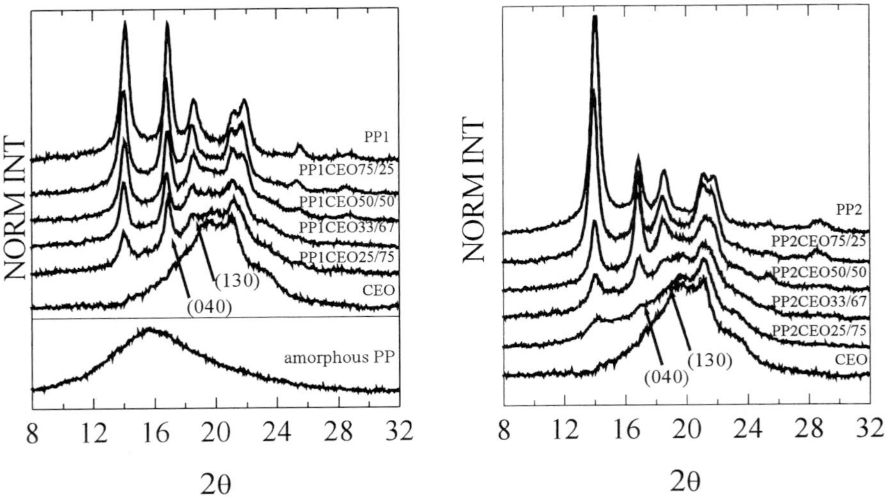

Fig. 1: X-ray diffraction patterns, at room temperature, of the different PP1/CEO (left plot) and PP2/CEO (right plot) blends. From top to bottom (left): PP1, PP1CEO75/25, PP1CEO50/50, PP1CEO33/67, PP1CEO25/75, CEO and totally amorphous PP sample (lower frame), respectively. And, (right): PP2, PP2CEO75/25, PP2CEO50/50, PP2CEO33/67, PP2CEO25/75 and CEO, respectively.

Focusing the attention on the blends, one important difference is observed between the two kinds of blends. The experimental diffractograms of the PP1 blends practically coincide with those that can be estimated by weighting the patterns corresponding to the two pure components. On the contrary, those from PP2 lead to iPP crystallinities for the high CEO contents considerably smaller than the expected ones. This fact is reflected, for instance, in the much smaller iPP crystallinity exhibited by PP2CEO25/75 blend when compared with PP1CEO25/75. For the smaller CEO contents, however, the experimental and expected crystallinities are about the same. The iPP crystallinity in the blends can be estimated by subtracting the corresponding amounts of the CEO profile and of the amorphous component of iPP, the latter represented by the lower diffractogram in the left plot in Figure 1 corresponding to an elastomeric, totally amorphous, PP sample.[13] After normalization to the iPP content in the blend, the normalized crystallinities are presented in Table 2 showing that crystallinity starts to diminish as CEO content increases, slightly in the blends of PP1 and rather steeply in the case of PP2.

Table 2. Normalized iPP crystallinities estimated from WAXS and DSC measurements and DSC crystallization temperatures, for the iPP component in PP1, PP2 and their corresponding blends.

Specimen	f_c^{WAXS}	f_c^{DSC}	T_c (°C)	Specimen	f_c^{WAXS}	f_c^{DSC}	T_c (°C)
PP1	0.64	0.49	115	PP2	0.61	0.49	109
PP1CEO75/25	0.64	0.48	114	PP2CEO75/25	0.61	0.46	110
PP1CEO50/50	0.64	0.48	113	PP2CEO50/50	0.61	0.44	106
PP1CEO33/67	0.61	0.61	117	PP2CEO33/67	0.43	0.35	102
PP1CEO25/75	0.57	0.41	114	PP2CEO25/75	0.20	0.14	56

Dynamic crystallization was performed by DSC, WAXS and SAXS to learn the cause of the lower crystallinity developed in PP2 blends with the highest CEO contents. The DSC curves in the left plot of Figure 2 show that the crystallization of either the PP1 or the CEO components occurs at approximately the same temperature for all compositions (around 115°C for PP1 and 56°C for CEO). On the contrary, the results for the blends of PP2 exhibit a characteristic behavior of the crystallization peak corresponding to the PP2 component. The 33% PP2 blend shows already an appreciable decrease of the crystallization temperature and that peak is absent in the case of PP2CEO25/75 (though a wide flat peak is observed with a rather small enthalpy). Moreover, it is important to note that now the exotherm corresponding, eventually, to the CEO crystallization is considerably more intense than in the case of PP1CEO25/75 (and the expectations considering the composition in CEO of that blend, i.e .75% of the enthalpy corresponding to pure CEO).

WAXS analysis at room temperature and dynamic DSC crystallization results suggest, consequently, that crystalization has been either diminished and delayed in PP2CEO33/67 and PP2CEO25/75. Then, crystllinity values are lower to those expected and crystallization occurred at lower temperature, respectively, as listed in Table 2 and Figure 2.

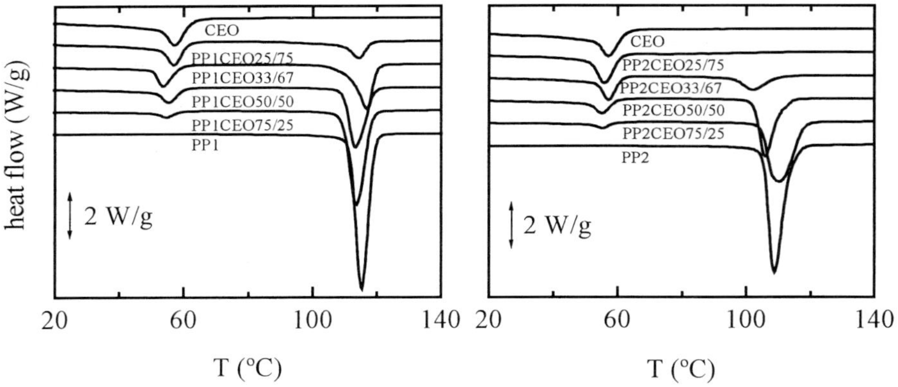

Fig. 2. DSC cooling curves corresponding to the PP1/CEO (left plot) and PP2/CEO (right plot) blends. From top to bottom (left): CEO, PP1CEO25/75, PP1CEO33/67, PP1CEO50/50, PP1CEO75/25 and PP1, and (right): CEO, PP2CEO25/75, PP2CEO33/67, PP2CEO50/50, PP2CEO75/25 and PP2, respectively.

The SAXS and WAXS results for dynamic crystallization display similar characteristics to those described for the DSC measurements in PP2CEO50/50, as depicted in Figure 3 for SAXS experiments. Each component crystallizes in their usual temperature range. However, crystallization again takes place at considerably lower temperatures in PP2CEO25/75. This feature is more clearly noticed in the analysis of the SAXS invariant,[14] as observed in Figure 4. These results reflect again that the crystallization of the iPP component in the PP2CEO25/75 blend takes place together with that for CEO. The corresponding derivative of the SAXS invariant (right plot of Figure 4) is in a very good agreement with the DSC results. The conclusion is, therefore, that the crystallization of the iPP component in the case of blend PP2CEO25/75 is considerably inhibited or, better, slowed down,[15] since it takes place at a much lower temperature, with an important proportion of iPP crystallizing simultaneously with the CEO component. An additional problem is the possibility of formation of the mesomorphic modification[14,16] of iPP instead of the regular α form. The fact that this mesomorphic modification is obtained under fast quenching conditions and that the crystallization rate of iPP in blend PP2CEO25/75 seems to be diminished considerably, opens up the possibility of formation of this mesomorphic phase.

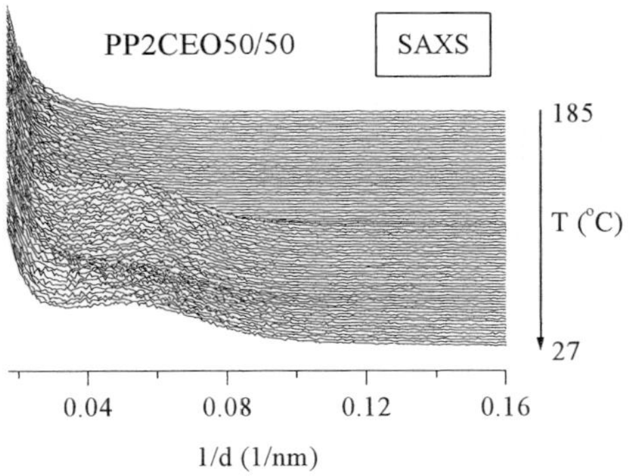

Fig. 3. Real-time SAXS profiles obtained with synchrotron radiation for PP2CEO50/50 and PP2CEO25/75 blends in a crystallization experiment at 8 °C/min.

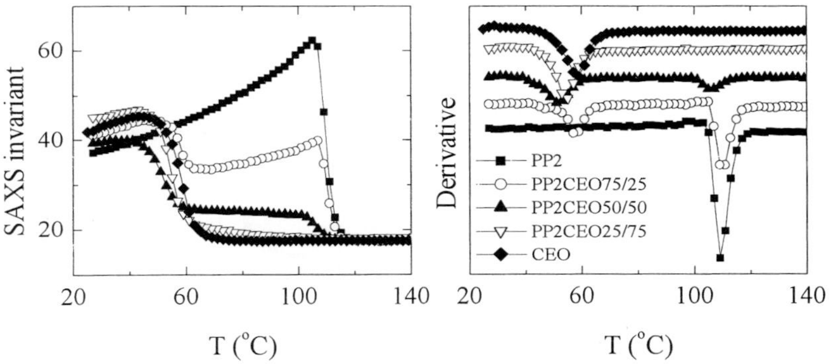

Fig. 4. Relative SAXS invariant (left plot) and derivative of the SAXS invariant (right plots) for PP2 and their respective blends with CEO.

Two different interpretations are possible to explain these results. The first one is related to the assumption that CEO is able to dissolve iPP in the molten state when the composition of iPP in the blend is low, so that the dilution effect will delay the crystallization of iPP, and only when the PE counterpart crystallizes, the iPP concentration in the non-crystalline regions will increase enough to be able to crystallize. The second interpretation is based on the supposition that the immiscible blend of these components will be subdivided into isolated regions (droplets) of small size which contain, on average, a significantly reduced number of heterogeneous nuclei where the iPP crystals are going to grow from. Looking at the behavior of the present blends, specially that of PP2CEO25/75 (see Figure 2), and taking into account the morphology observed in an optical microscope, part of the iPP in this blend crystallizes as very small crystallites along a broad exotherm at temperatures intermediate between the exotherms of pure iPP and CEO, and another part crystallizes simultaneously with CEO. Therefore, the two interpretations given above may be occurring in the present case. The important feature to be remarked is the great influence of the molecular weight of iPP. Considering these interpretations, this effect may arise from differences in the transport term which is more significant for PP2 blends since its higher molecular weight, accentuated by the dilution effect, or, alternatively, by changes in the particle size distribution of the dispersed phase.

Figure 5 shows the viscoelastic response of PP2, CEO and their corresponding blends, which qualitatively is quite similar to that presented by blends obtained from PP1. The different relaxation processes observed are analyzed separately as follows in order of increasing temperatures.

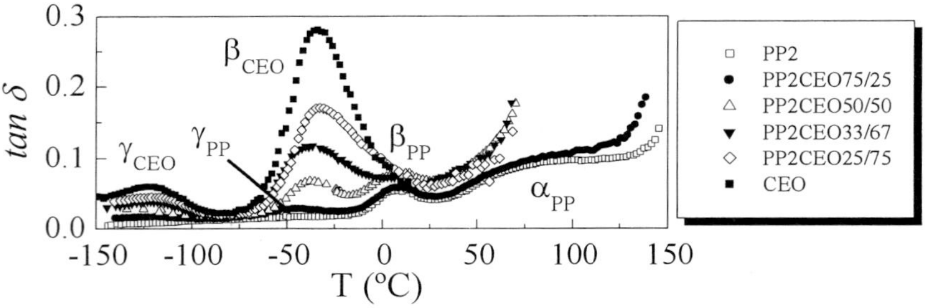

Fig. 5. Temperature dependence of the loss tangent for PP2, CEO and their blends.

The γ_{CEO} relaxation corresponds to the observed γ relaxation in polyethylene, which was firstly attributed to crankshaft movements of polymethylenic chains.[17,18] There is, however, a body of opinions which support one or more of the various model for restricted conformational transitions as kink formation, inversion and migration.[19-21] This type of motion requires chains containing sequences of three or more methylenic units. As *Figure 5* displays, the γ_{CEO} relaxation location is quite unaffected by the blend composition due to the immiscibility of the two polymers studied. In addition, the intensity of such a process decreases and the apparent activation energy associated to it slightly increases as the iPP component increases in the blend.

The γ_{PP} relaxation, appearing around -50°C, has been attributed to local motions in the amorphous region of iPP.[22] Figure 5 shows that this relaxation is being masked in the blends by the more prominent β_{CEO} process.

The β_{CEO} relaxation corresponds to that one, which has been universally detected in branched polyethylenes at temperatures around -20°C, but it, sometimes, appears, though weakly, in some samples of linear polyethylene. Experimental evidence has been reached by DSC and MDSC about the location of the glass transition at around -48°C in the CEO and its glass fiber composites.[23] Accordingly, the β_{CEO} relaxation in CEO and in its blends with either PP1 or PP2 is considered the relaxation process associated to the glassy-rubbery transition in this component. Its location is rather independent of the type of iPP utilized for the blending.

The β_{PP} process has been identified with the glass transition temperature of the amorphous regions in iPP.[22] In the blends, as seen in Figure 5, this relaxation appears as an unambiguous peak for the PP2CEO75/25, PP2CEO50/50 blends (also in PP1CEO75/25, PP1CEO50/50, respectively). As iPP content decreases in the blend, as occurred in PP2CEO33/67, the maximum becomes a shoulder that significantly overlaps with the β_{CEO} mechanism in the side of high temperatures. Therefore, and in spite of the considerable overlapping, the two glass transitions are observed in the different blends, indicating the immiscibility of the two components.

Finally, the α_{PP} relaxation is clearly observed in the two neat iPP's and in the blends with the highest content of this component. It is attributed to movements in the iPP crystalline regions.[22] As the CEO content increases in the blends under study, the mechanical strength at temperatures higher than room temperature considerably diminishes and this motion cannot be detected. Accordingly, the tan δ parameter starts to increase without showing the characteristic shoulder exhibited by this relaxation.

A rough estimation of the impact strength can be obtained from the area under the loss tangent curve.[24] Such estimation has been made for the present samples by integration from -150°C to 30°C. The corresponding results for the PP1 blends are shown in Figure 6 as a function of the storage modulus of the sample. It can be observed that the incorporation of the plastomer leads to a significant increase on the impact resistance, although it is accompanied by a rather pronounced diminishment on the storage modulus.

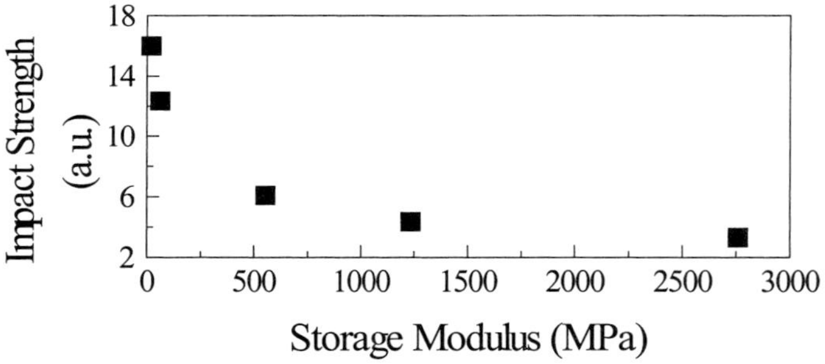

Fig. 6. Variation of the toughness (estimated by integration under the loss tangent curve) and storage modulus in iPP1, CEO and their blends.

Conclusions

The current study has established a diminishment in crystallinity and crystallization temperature corresponding to the iPP component in PP2CEO25/75 and PP2CEO33/67 blends compared with their homologous PP1CEO25/75 and PP1CEO33/67. These features point out an inhibition of the crystallization process of the iPP component in the case of the blends rich in CEO. This inhibition is considerably more pronounced for the PP2 blends. The existence of two independent glass transitions indicates immiscibility of the two components in the two kinds of blends. Moreover, the area under the loss tangent curve is considerably increased as CEO content does in the blend, pointing out an enlargement in the impact strength though it is accompanied by a clear decrease on the storage modulus. Consequently, depending upon the final requirements for a specific application, the balance between stiffness and toughness has to be found.

Acknowledgments

The financial support of CAM (Project 07G/0038/2000) and Ministerio de Ciencia y Tecnología (projects MAT2001-2321 and MAT2001-1731) and the supply of the samples by Repsol-YPF and EXXON Chemical are gratefully acknowledged. M.L. Cerrada is grateful to CSIC (I3P Programme) for her financial support. The synchrotron work (in the polymer line of Hasylab at DESY, Hamburg) was supported by the IHP Programme "Access to Research Infrastructures" of the European Commission (Contract HPRI-CT-1999-00040). We thank the collaboration of the Hasylab personnel, and specially Dr. S. Funari and Dr. A. Meyer, responsibles of the polymer beamline.

[1] S. Bensason, J. Minick, A. Moet, S. Chum, A. Hiltner, E. Baer, *J. Polym. Sci.:Polym. Phys.* **1996**, *34*, 301.
[2] S. Vanden Eynde, V. B. F. Mathot, M. H. J. Koch, H. Reynaers, *Polymer* **2000**, *41*, 4889.
[3] A. Galeski, Z. Bartzak, M. Pracella, *Polymer* **1984**, *25*, 1323
[4] M. Yamaguchi, K.-H. Nitta, H. Miyata, T. Masuda, *J. Appl. Polym. Sci.* **1997**, *63*, 467.
[5] J. H. Lee, J. K. Lee, K. H. Lee, C. H. Lee, *Polymer J.* **2000**, *32*, 321.
[6] J. R. Isasi, L. Mandelkern, M. J. Galante, R. G. Alamo, *J. Polym. Sci., Polym. Phys.* **1999**, *37*, 323.
[7] B. Wunderlich, in "*Macromolecular Physics*", Academic Press, New York, 1980.
[8] G. Natta, P. Corradini, *Nuovo Cimento Suppl.* **1960**, *15*, 40.
[9] A. Turner-Jones, J. M. Aizlewood, D. R. Becket, *Makromol. Chem.* **1964**, *75*, 134.
[10] S. V. Meille, D. R. Ferro, S. Brückner, *Macromol. Symp.* **1995**, *89*, 499.
[11] E. Pérez, D. Zucchi, M. C. Sacchi, F. Forlini, A. Bello, *Polymer* **1999**, *40*, 675.
[12] M. L. Cerrada, R. Benavente, E. Pérez, *J. Mater. Res.* **2001**, *16*, 1103.
[13] S. Mansel, E. Pérez, R. Benavente, J. M. Pereña, A. Bello, W. Röll, R. Kirsten, S. Beck, H.-H. Brintzinger, *Macromol. Chem. Phys.* **1999**, *200*, 1292
[14] W. J. O'Kane, R. J. Young, A. J. Ryan, W. Bras, G. E. Derbyshire, G. R. Mant, *Polymer* **1994**, *35*, 1352.
[15] J. Li, R. A. Shanks, R. H. Olley, G. R. Greenway, *Polymer* **2001**, *42*, 7685.
[16] R. Androsch, B. Wunderlich, *Macromolecules* **2001**, *34*, 5950.
[17] T. F. Schatzki, *J. Polym. Sci.* **1962**, *57*, 496.
[18] R. F. Boyer, *Rubb. Chem. Technol.* **1963**, *36*, 1303.
[19] R. H. Boyd, R. S. Breitling, *Macromolecules* **1974**, *7*, 855.
[20] R. H. Boyd, *J. Polym. Sci, Polym. Phys.* **1975**, *13*, 2345.
[21] N. J. Heaton, R. Benavente, E. Pérez, A. Bello, J. M. Pereña, *Polymer* **1996**, *37*, 3791.
[22] C. Jourdan, J. Y. Cavaille, J. Perez, *J. Polym. Sci.: Polym. Phys.* **1989**, *27*, 2361.
[23] M. L. Cerrada, R. Benavente, E. Pérez, J. Moniz-Santos, M. R. Ribeiro, *Polymer* **2001**, *42*, 7197.
[24] S. H. Jafari, A. K. Gupta, *J. Appl. Polym. Sci.* **2000**,*78,* 962.

Thermal and Mechanical Behavior of Amorphous and Semi-crystalline Poly(Vinylidene Fluoride)/Poly(Methyl Methacrylate) Blends

Jlidi Jarray,[1] *Fadhel Ben Cheikh Larbi,*[1] *Faustine Vanhulle,*[2] *André Dubault,*[1] *Jean Louis Halary**[2]

[1] Laboratoire de Physico-chimie des Matériaux Polymères, Institut Préparatoire aux Etudes Scientifiques et Techniques, BP51, 2070 La Marsa, Tunisia
[2] Laboratoire de Physico-chimie Structurale et Macromoléculaire (UMR 7615), Ecole Supérieure de Physique et Chimie Industrielles de la Ville de Paris, 10, rue Vauquelin, 75231 Paris cedex 05, France

Summary: Various PVDF/PMMA (poly(vinylidene fluoride)/poly(methyl methacrylate)) blends were selected for mechanical testing in compression. At low PVDF content (less than 50/50 w/w), the blends remain amorphous and PVDF and PMMA are fully miscible. In PVDF-richer blends, PVDF crystallizes in part, leading to a PMMA-enriched homogeneous amorphous phase. In this study, the degree of crystallinity was set at equilibrium by appropriate annealing of the samples before testing. Mechanical analysis was focused on the low deformation range, and especially on the yield region. Depending on the test temperature and blend composition, three types of response were identified, depending on whether plastic deformation is influenced: 1) by the PMMA secondary relaxation motions, 2) by the PVDF/PMMA glass transition motions, or 3) by the crystallite-constrained PVDF chains.

Keywords: blends, mechanical relaxation, plastic deformation, poly(methyl methacrylate), poly(vinylidene fluoride)

Introduction

In the 1970's, the PVDF/PMMA blends have been studied extensively, mainly in relation to PVDF piezoelectric properties.[1-7] Much attention has been paid to such problems as miscibility of the amorphous phase, crystallization of PVDF in various allotropic phases, and molecular origin of the PVDF/PMMA interactions. Moreover, blending with PMMA was described as an original way to force PVDF to crystallize into the piezoelectric phase, which is thermodynamically unstable in pure material.[8] More recently, the interest for the PVDF/PMMA was re-actualized, because of some potential applications as coatings and in building trade. In this new context, many studies was devoted to the characterization of the mechanical properties of the blends. The purpose of the present publication is to focus on the small deformations in compression and to analyze the yield region as a function of blend composition and, in turn, degree of crystallinity of the samples. As the strains are systematically lower than about 0.1, the PVDF crystallites, whenever they exist, are supposed not to deform. Therefore, the observed plastic deformation is expected just to depend on the

 DOI: 10.1002/masy.200350810

characteristics of the PVDF/PMMA amorphous phase. Emphasis will be put on chain mobility, according to the methodology already proposed for PMMA[9] and many other amorphous polymeric materials.[10-14]

Experimental

Materials. The PMMA used in this study was an amorphous polymer of high molecular weight, prepared by conventional free radical polymerization. Its glass transition temperature, as determined by DSC at heating rate of 10K/min, was Tg_A = 110°C. The grade of used PVDF presented a melt viscosity of 830 Pa.s at 230°C, a glass transition temperature Tg_F = -40°C, a melting temperature Tm_F = 165-172°C and a maximum degree of crystallinity of about 50%. Both materials were kindly provided by ATOFINA.

Blending procedure. Homogeneous dilute solutions were prepared by dissolving 5w% of polymeric materials in dimethylformamide, under continuous stirring at 50°C for 24 h. Then, PVDF and PMMA were co-precipitated, by pouring the solution drop by drop into a large excess of water. Finally, the polymeric powder was dried at 100°C for 15 h in a vacuum oven. Blends prepared in this way covered the composition range from 10/90 to 90/10 w/w.

Samples for mechanical testing. The dry powder was compression molded in the form of sheets (200*200*3 mm^3) under press, at 200°C for 30 mn. Then, small bars (6*3*3 mm^3) were cut from the sheets by using a diamond saw. As justified in the results and discussion section, the bars of composition likely to crystallize (i.e. containing at least 50 w% of PVDF) were subjected to an annealing step of 60 h at 150°C, prior to mechanical testing.

DSC experiments. Measurements were performed on a modulated differential scanning calorimeter (DSC universal v3.0G, from TA Instruments), by using a heating rate of 10K/min and a 1Hz sinusoidal modulation with an amplitude of ± 1°C. The characteristic temperatures Tg and Tm were taken at the onset of the heat capacity jump and at the extremum of the melting peak, respectively, both being relative to the in-phase signal.

Stress-strain curves. Mechanical measurements were performed on a MTS-810 testing system, equipped with an isothermal chamber and non-lubrificated plates. The experiments covered the temperature range from –50°C to Tg – 20 K. The strain rate was always equal to $2.10^{-3}s^{-1}$. As a first approximation, nominal stress was considered instead of the true stress, since the strains under inspection are quite low.

Results and Discussion

DSC Data

Glass transition temperature of the amorphous blends. As expected from earlier studies,[7] the blends containing up to 40w% PVDF are totally amorphous and their DSC traces just show a heat capacity jump at Tg. Dependence of Tg_{blend} on composition is shown in Figure 1. Its asymmetric character results from the occurrence of specific interactions between the blend components. It can be accounted for by using a modified Fox equation of the type :

$$\frac{w_A + kw_F}{Tg_{blend}} = \frac{w_A}{Tg_A} + k\frac{w_F}{Tg_F} \qquad (1)$$

where w_A and w_F are the weight fractions of PMMA and PVDF, respectively, and k is an empirical fit constant.

DSC traces of the blend PVDF/PMMA 50/50. For the sample rapidly quenched from the melt (without further annealing), the thermogram shows a Tg_{blend} around 40°C (in agreement with Figure 1), and then an exothermic peak around 140°C. This indicates that some PVDF crystallization occurs during the temperature scan. In order to set the degree of crystallinity at equilibrium, the samples were annealed at 150°C for various times until a plateau is reached. Reasonable estimates of PVDF crystalline weight fraction, y, were deduced from the melting peak areas. It turns out (Figure 2) that the critical annealing time was around 50 h, and that further DSC scan did not reveal the presence of any crystallization peak. In turn, Tg_{blend} was shown to increase with annealing time until a constant value is also reached for 50 h (Figure 3). This feature is indicative of the enrichment of the amorphous phase in PMMA, as the result of gradual PDVF crystallization.

DSC data for the other semi-crystalline PVDF/PMMA blends. As the blend 50/50 is the less crystallizable material of the series, it is obvious that the PVDF-richer blends get *a fortiori* the equilibrium degree of crystallinity after the annealing step (60 h at 150°C).

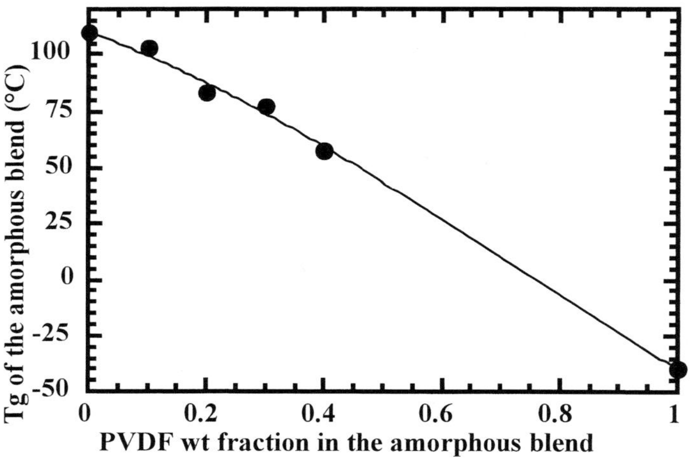

Fig. 1. Glass transition temperature versus composition for the amorphous blends (circles: experimental values; solid line : best fit from Equation (1) with k = 0.76).

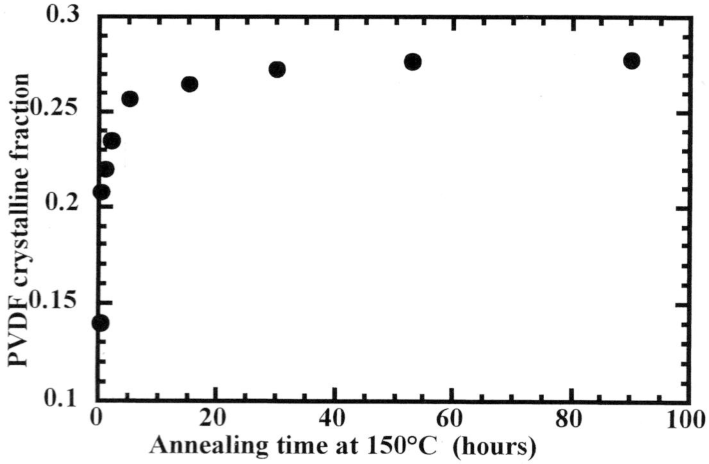

Fig. 2. PVDF crystalline fraction versus annealing time at 150°C for PVDF/PMMA 50/50.

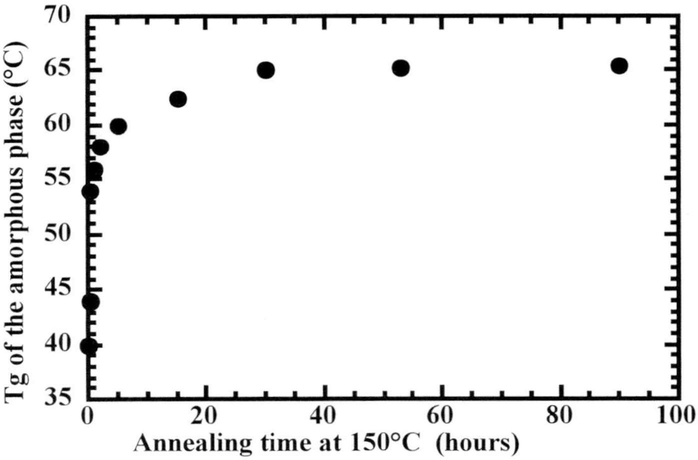

Fig. 3. PVDF/PMMA 50/50: effect of the annealing time (at 150°C) on the glass transition temperature of the amorphous phase.

In Figure 4, the PVDF crystalline fraction, y, (in weight) is plotted as a function of the overall percentage of PVDF in the blend. This quantity strongly increases with increasing PVDF amount in the blend. This behavior is consistent with the parallel increase in blend density. In addition, Tm also increases (Figure 5), suggesting that the crystallites grow in size and degree of perfection with increasing percentage of PVDF in the blend.[3] Consideration of the amorphous phase Tg (Figure 6) shows that this quantity levels off for the highest PVDF amounts. As shown by simple calculations, this result cannot be uniquely explained by the PMMA enrichment of the amorphous phase, which accompanies PVDF cristallization. To this end, let us first calculate the PVDF weight fraction, w_{Fa}, in the amorphous phase from the crystallinity data thanks to the relation :

$$W_{Fa} = \frac{W_F\,(1-y)}{1-yW_F} \qquad (2)$$

and then deduce the calculated value of Tg_{blend} from equation 1 (with k = 0.4842). As shown in Table 1, the values derived from the mixing law deviate from the experimental measurements, and the PVDF-richest blends present the largest deviations. These apparent discrepancies originate probably from the rigidification of those of the amorphous PVDF skeletal bonds which are located along the chain in the vicinity of bonds involved in the crystallites. The more crystalline the blend, the more numerous the amorphous bonds which are constrained by the crystalline phase, and the higher the Tg_{blend}.

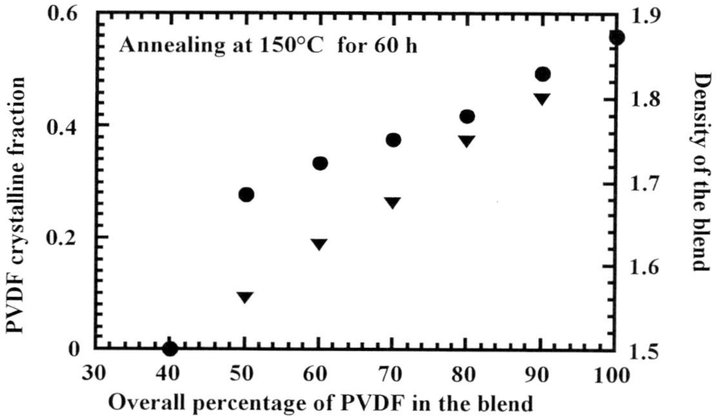

Fig. 4. PVDF crystalline fraction (circles) and density of the blend (triangles) versus blend composition.

Table 1. Thermal properties of semi-crystalline PVDF/PMMA blends.

PVDF/PMMA	w_F	w_{Fa}	Experimental Tg_{blend} °C	Calculated Tg_{blend} °C
50/50	0.50	0.42	66	58
60/40	0.60	0.50	54	45
70/30	0.70	0.60	45	27
80/20	0.80	0.70	54	12
90/10	0.90	0.82	53	-7

Analysis of the Stress-strain Curves

General features. Figure 7 shows the shape of typical stress-strain curves covering the entire range of blend composition, namely : a) the PMMA-richest blend 10/90, b) the PMMA-poorest amorphous blend 40/60, and the PVDF-richest blend 90/10. For the amorphous blends, the curves a) and b) have in common to present a usual profile, including an initial linear part, typical of the elastic response, then a curvature corresponding to the inelastic response before the yield point is attained at the maximum stress σ_y. They distinguish from each other by the stress changes after the yield point.

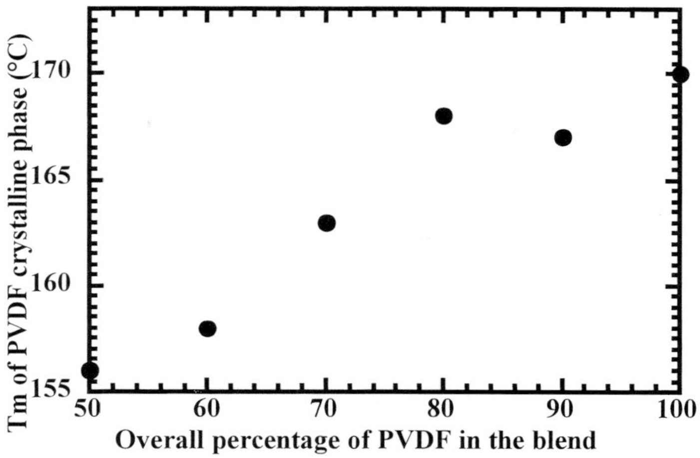

Fig. 5. Melting temperature of the PVDF crystalline phase versus blend composition for the semi-crystalline blends.

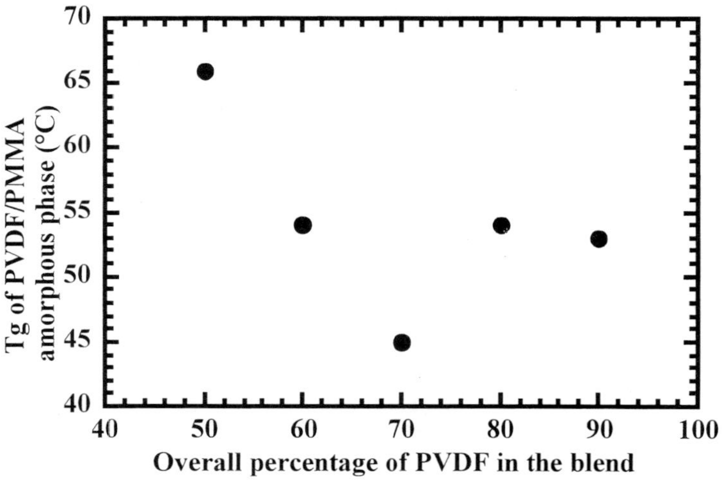

Fig. 6. Glass transition temperature of the amorphous phase versus blend composition for the semi-crystalline blends.

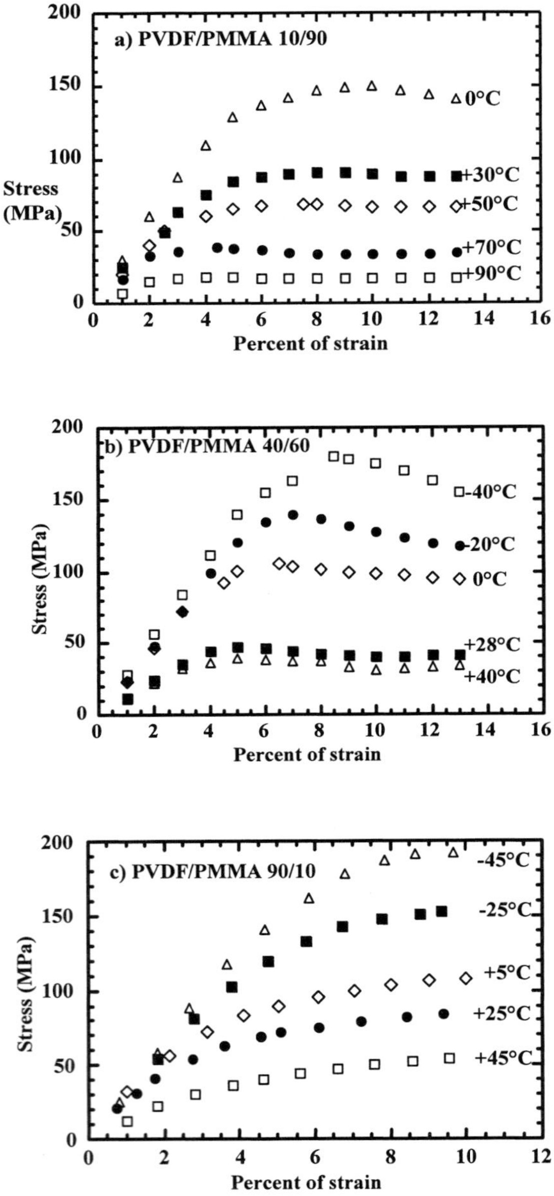

Fig. 7. Stress-strain curves for three typical blend compositions at various test temperatures.

A marked softening is observed on the curve b) until a plateau is reached, characterized by the plastic flow stress, σ_{pf}. On the other hand, almost no strain softening shows up on the curves a). For the semi-crystalline materials (curves c)), both yield point and strain softening are hidden by the occurrence of a strong strain hardening, which develops very rapidly at the end of the domain of elastic response. Conventionally, a 'pseudo-yield' can be defined, as sketched in Figure 8.

Influence of chain mobility on the observed behavior. In the case of amorphous materials, it is recognized that the plastic deformation events are closely related to the chain motions.[15-18] Recent reports illustrated the suitability of such connections for various materials, including methyl methacrylate copolymers,[9,10] semi-aromatic polyamides,[11,12] epoxy resins[13] and polystyrene/poly(phenylene oxide) compatible blends.[14]

A simple way to visualize the situation encountered for the PVDF/PMMA amorphous blends is to calculate first the mechanical energy, W_y, required to attain the yield point, which is approximately given by (see Figure 8):

$$W_y = \sigma_y \left(\varepsilon_y - \frac{\varepsilon_{ey}}{2} \right) \qquad (3)$$

and then to follow the decrease of W_y with increasing temperature.

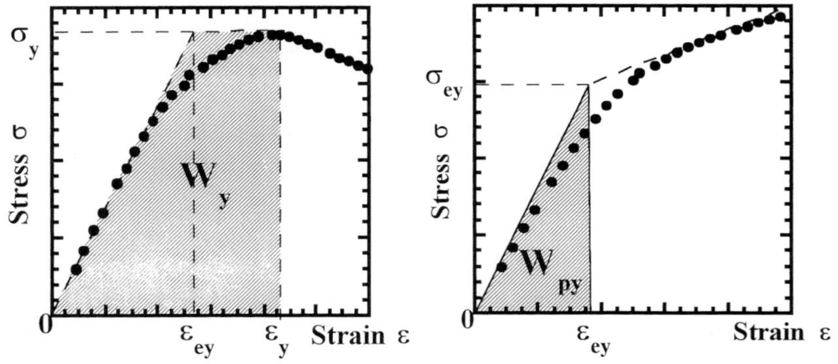

Fig. 8. Evolution of the mechanical energy to yield for the two types of stress-strain curves (left: with softening; right: with hardening).

As shown in Figure 9, the plots of W_y versus test temperature reveal the occurrence of two well-separated profiles. In the high temperature range, W_y extapolates linearly to zero at a temperature close to Tg_{blend}, just meaning that plastic flow becomes there a spontaneous, thermally activated process. Interestingly, in the lower temperature range, another linear profile is observed. In that case, W_y extrapolates to zero in the temperature range of the PMMA β relaxation,[9] putting emphasis on the influence of the β motions on the deformation process. These trends are summarized in Figure 10.

The same procedure holds for the semi-crystalline blends, provided W_y is replaced, in the absence of true yield point, by W_{py} calculated as (see again Figure 8):

$$W_{py} = \frac{\sigma_{ey}\varepsilon_{ey}}{2} \qquad (4)$$

A unique straight line accounts for the temperature dependence of W_{py}, as illustrated in Figure 11 on the example of PVDF/PMMA 90/10. In this case, a residual value of σ_{ey} would be observed between Tg ad Tm as the result of the semi-crystalline character of the materials. However, W_{py} vanishes as σ_{ey} is extrapolated to zero, at a temperature very close to the experimental Tg_{blend}. This result holds whatever the semi-crystalline blend under consideration (Figure 12). The fact that yield properties are governed below Tα by the glass transition motions only can be understood very simply. Blend composition is such that the MMA units are rather scarce in these PVDF-rich blends. Therefore, PMMA secondary relaxation motions are not sufficiently numerous to generate the mobility microdomains which are likely to initiate the plastic deformation.

Molecular interpretation of strain softening and strain hardening. The shape of the PVDF/PMMA 10/90 stress-strain curves shown in Figure 7a resembles very much that reported in the literature for pure PMMA.[9] As a first approximation, it can be interpreted in the same way by neglecting the perturbation provoked by the scarce PVDF chains. Strain softening is very small in PMMA (and in PVDF/PMMA 10/90, too), because the β molecular motions responsible for yielding (so-called α-precursor β motions) are of same nature as the α motions responsible for plastic flow. On the other hand, strain softening enlarges as long as the blend PVDF content increases (see Figure 7b) because the PVDF chain segments contribute significantly to the α process. Therefore, PMMA β motions are no longer the precursors of the blend α motions and, in turn, the stress to be applied is much higher for yielding than for plastic flowing, as previously reported for MMA-based random copolymers including stiff units.[9] In other words, while comparing amorphous samples in isostructural conditions, difference in strain softening amplitude (σ_y - σ_{pf}) reflects the difference in the nature of the motions occurring at yield and plastic flow.

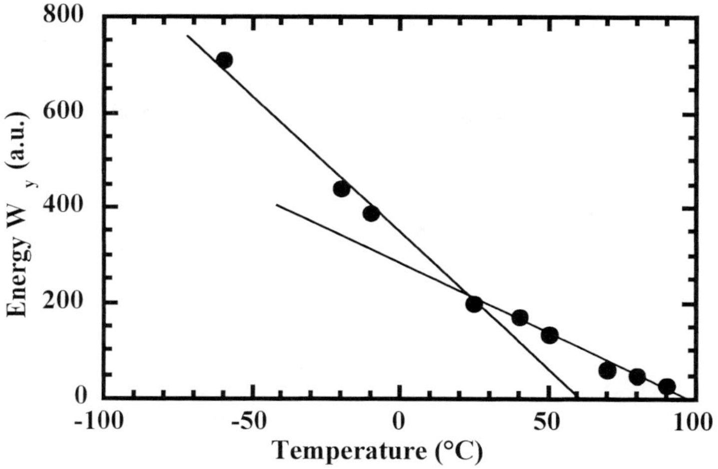

Fig. 9. Variation of the mechanical energy to yield, W_y, as a function of temperature for the amorphous 10/90 blend.

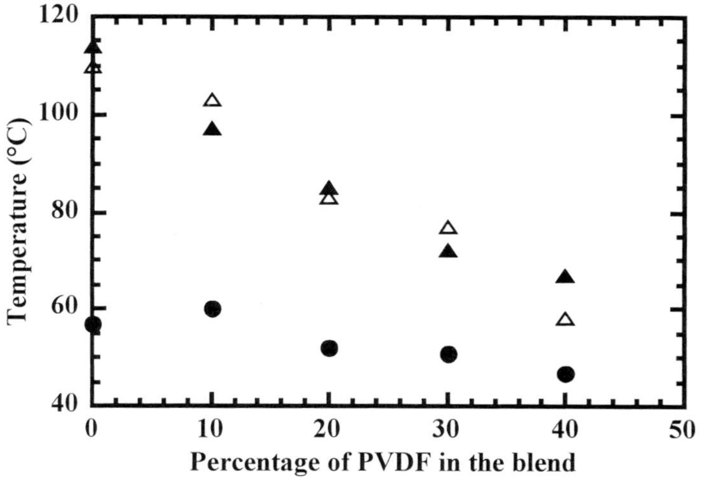

Fig. 10. Zero energy extrapolated temperatures as a function of amorphous blend composition (circles : low temperature regime; filled triangles : high temperature regime). The values of Tg_{blend} (open triangles) are also given for sake of comparison.

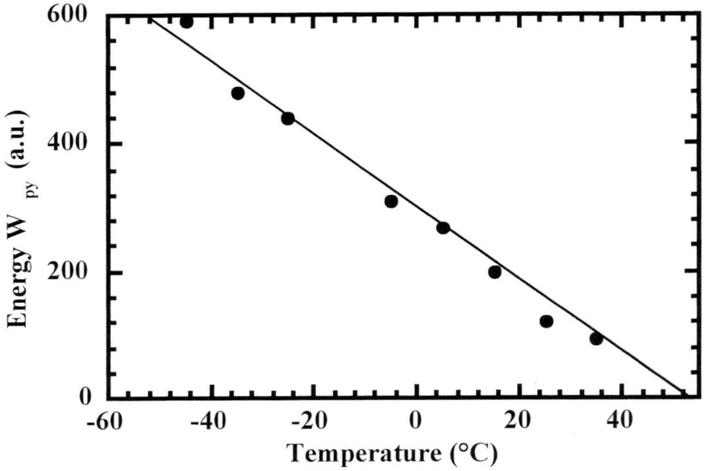

Fig. 11. Variation of the mechanical energy to yield, W_{py}, as a function of temperature for the semi-crystalline 90/10 blend.

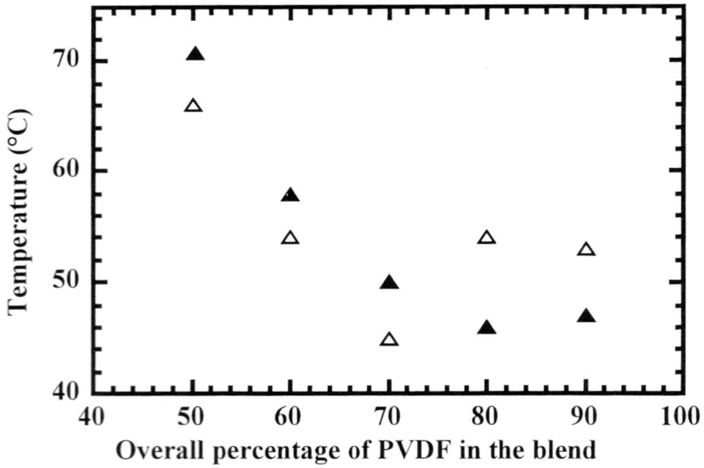

Fig. 12. Zero energy extrapolated temperatures (filled triangles) as a function of overall PVDF content for the semi-crystalline blends. The values of Tg_{blend} (open triangles) are also given for sake of comparison.

As far as the PVDF-rich semi-crystalline blends are concerned, the occurrence of the strain hardening phenomenon, which even conceals the true yield point (Figure 7c), is similar, from a phenomenological viewpoint, to what has been observed for nanosilica particle-filled PMMA.[19] In the latter case, a tentative explanation for the early hardening was that the PMMA chain deformation is limited by the vicinity of the undeformable silica particles. If so, some chains would reach very rapidly their limit of extensibility, and therefore an increased stress would be required to further deform the sample. On may suggest the same explanation for the PVDF amorphous segments constrained by the crystallites. Therefore, the higher the PVDF content in the blends, the more numerous the constrained segments and, in turn, the earlier the appearence of strain hardening. This finding is actually verified by comparing the behavior of the blends 50/50, 60/40, 70/30, 80/20 (stress-strain curves not shown here), and finally 90/10 (Figure 7c). It is supported by the above DSC data, which provide evidence for the presence of highly constrained PVDF chain segments in the amorphous phase.

Conclusions

For the amorphous PVDF/PMMA blends, a quantitative mixing law accounts for the variations of Tg as a function of blend composition, and the yield properties are influenced by the PMMA secondary relaxation motions in the lower temperature range and, at higher temperatures, by the blend α motions, which are the mechanical expression of the glass transition motions.

For the semi-crystalline PVDF/PMMA blends, the glass transition temperature of the amorphous phase, calculated from crystallinity data, deviates from the mixing law, as the expression of constrained chains. The higher the overall PVDF content, the larger the effect. Only a 'pseudo-yield' point is observed and the yield properties are governed by the Tg motions. As in PMMA nanohybrids, the crystallites reduce the limit extensibility of some PVDF segments in the amorphous phase, and a strong strain hardening effect is observed on the stress-strain curves.

WAXS and SAXS measurements are currently in progress with the aim of characterizing the PVDF crystalline phase in more details and getting further information on the constrained chains.

Acknowledgments

This study was supported by the 'Comité Mixte de Coopération Franco-Tunisienne' (CMCU) under contract No 01F1302. The authors would like to thank Dr François Beaume (ATOFINA, Serquigny, France) for kindly providing the PVDF and PMMA pure polymers and for his interest in this study. Thanks are also due to Mrs Amalia de Roffignac for her contribution to the experimental work.

[1] J.S. Noland, N.N.C. Hsu, R. Saxon, J.M. Schmitt, *Adv. Chem. Ser.* **1971**, *99*, 15.
[2] T. Nishi, T.T. Wang, *Macromolecules* **1975**, *8*, 909.
[3] T. Nishi, T.T. Wang, *Macromolecules* **1977**, *10*, 421.
[4] O. Faria, R.L. Moreira, *Polymer* **1999**, *40*, 4465.
[5] K. Saito, S. Miyata, T.T. Wang, Y.S. Jo, R. Chujo, *Macromolecules* **1986**, *19*, 2450.
[6] C. Léonard, J.L. Halary, L. Monnerie, D. Broussoux, B. Servet, F. Micheron, *Polymer, commun.* **1983**, *24*, 110.
[7] C. Léonard, J.L. Halary, L. Monnerie, F. Micheron, *Polym. Bull.* **1984**, *11*, 195.
[8] C. Léonard, J.L. Halary, L. Monnerie, *Macromolecules* **1988**, *21*, 2988.
[9] P. Tordjeman, L. Tézé, J.L. Halary, L. Monnerie, *Polym. Eng .Sci.* **1997**, *37*, 1631.
[10] L. Tézé, J.L. Halary, L. Monnerie, L. Canova, *Polymer* **1999**, *40*, 971.
[11] S. Choe, B. Brulé, L. Bisconti, J.L. Halary, L. Monnerie, *J. Polym. Sci., Polym. Phys. Ed.* **1999**, *37*, 1131.
[12] B. Brulé, L. Bisconti, J.L. Halary, L. Monnerie, *Polymer* **2001**, *42*, 9073.
[13] D. Rana, V. Sauvant, J.L. Halary, *J. Mater. Sci.* **2002**, in press.
[14] C. Creton, J.L. Halary, L. Monnerie, *Polymer* **1999**, *40*, 199.
[15] J.C. Bauwens, *J. Mater. Sci.* **1972**, *7*, 577.
[16] C.J. Bauwens-Crowet, *J. Mater. Sci.* **1973**, *8*, 968.
[17] J.M. Lefebvre, B. Escaig, *J. Mater. Sci.* **1985**, *20*, 438.
[18] J. Perez, "Physique et Mécanique des Polymères Amorphes", Tec. Doc. Lavoisier , Paris 1992.
[19] M. Mauger, Thesis of the University Pierre and Marie Curie (Paris VI), Dec. 2000.

Macromol. Symp. **2003**, *198*, 117-134

Morphology and Micromechanical Behaviour of Styrene/Butadiene Block Copolymers and Their Blends with Polystyrene

Rameshwar Adhikari,[1] *Goerg H. Michler,**[1] *Konrad Knoll*[2]

[1] Institute of Materials Science, Martin-Luther University Halle-Wittenberg, D-06099 Halle/Saale, Germany
[2] BASF Aktiengesellschaft, Polymers Laboratory ZKT/I-B1, D-67056 Ludwigshafen, Germany

Summary: The correlation between the morphology and the deformation mechanism in styrene/butadiene block copolymers having modified architecture and in blends with homopolymer polystyrene (hPS) was studied. It was demonstrated that the morphology formation in the block copolymers is highly coupled with their molecular architecture. In particular, the micromechanical behaviour of a star block copolymer and its blends with polystyrene was investigated by using electron microscopy and tensile testing. A homogeneous plastic flow of polystyrene lamellae (*thin layer yielding*) was observed if the lamella thickness was in the range of 20 nm. The deformation micromechanism switched to the formation of craze-like deformation zones when the average PS lamella thickness changed to about 30 nm and more.

Keywords: blends, electron microscopy, micromechanism, styrene/butadiene block copolymers, thin layer yielding

Introduction

The use of polymeric materials in everyday life is increasing day by day, where often a balance of mechanical and optical properties is desired. From the economical point of view, one may be interested to produce the plastics as cheap as possible. In order to be successfully competent in the market, these products should possess attractive price and interesting property profile. On the other hand, the concern of the recyclability of the plastics is also being a very important aspect from both ecological and economical points of view.

For the applications like food packaging films, beverage drinking cups etc., an excellent optical property is desired in addition to the optimum level of the mechanical parameters such as toughness, stiffness and strength.[1] These demands should be fulfilled by the plastics which are

 DOI: 10.1002/masy.200350811

easy to synthesise, and are processable by easy means. The goal may be achieved by the heterogenisation of commodity plastics like polystyrene. The heterogenisation of polystyrene, a brittle polymer in the bulk state, by the incorporation of a small amount of rubber (e.g., polybutadiene PB) may lead to a drastic increase in the toughness but the product is opaque.[2,3] A common example of toughened polystyrene is the high impact polystyrene (HIPS). Hence, one of the challenges in academia and industries in the field of polymer research is the development of tough and transparent heterogeneous polymers, e.g., toughened polystyrene.

It has been demonstrated that the polymers which break macroscopically in a brittle manner (e.g., poly(methyl methacrylate) PMMA, polystyrene PS) are intrinsically ductile.[2,4] The theoretical maximum natural draw ratio (λ_{max}) of polystyrene is about 4 and that of PMMA is about 3.[2] This means that these polymers can show an elongation to break of several hundred percent. Hence the main goal of the toughness modification of the brittle plastics like polystyrene is to transfer their intrinsic ductility to the macroscopic scale.

It is well known that joining the polystyrene chains with the polybutadiene chains with covalent bonds leads to the intramolecular phase separation of these chains and result in the formation of highly ordered structures called microphase separated structures (MSS) whose dimension lies in the range of gyration-radius of the molecules.[5] Through the variation of molecular weight, composition, chain architecture and the processing conditions, the dimension, nature and orientation of these structures can be adjusted.[5-10] This allows to produce the transparent nanostructured heterogeneous polymeric materials and enables a precise control over their mechanical properties.

Because of their higher production costs, the block copolymers are seldom used as pure materials. Styrene/butadiene block copolymers, for example, are often used in combination with homopolystyrene (e.g., general purpose polystyrene GPPS).[1,10] The compatibility of the added polystyrene with the block copolymer is limited by concentration and molecular weight amount of the homopolystyrene (hPS) chains relative to the corresponding block of the block copolymer.[5,9] The molecular weight of the polystyrene blocks is, however, practically restricted by the requirement of the rheological properties.

In our pervious studies, we reported about the toughening mechanism in the asymmetric styrene/butadiene star block copolymers and the influence of molecular architecture on their microphase morphology and micromechanical behaviour.[11,12] The aim of this study is to investigate the influence of the modified molecular architecture of styrene/butadiene block copolymers in a narrow composition range (ca. 70 volume % polystyrene) on their morphology and to analyse, in particular, the micromechanical behaviour of the blends consisting of a star block copolymer and general purpose polystyrene (GPPS).

Experimental Section

Materials and Sample Preparation

The important characteristics of the block copolymers and homopolystyrene studied are listed in Table 1. The details of the morphology of the block copolymer may be found in the references mentioned in the parentheses. In particular, the morphology formation in the linear triblock copolymers and their correlation with the tensile properties was studied by tapping mode scanning force microscopy.[13] In this work, we report the equilibrium morphology of the bulk materials investigated by means of transmission electron microscopy (TEM).

The block copolymers have nearly identical chemical composition (about 70 volume % of styrene). Except the sample LN1-S74, which is a linear symmetric SBS triblock copolymers, the remaining block copolymers possess non-classical molecular architecture (asymmetric structure, tapered transition, etc.). The samples were provided by the BASF Aktiengesellschaft, Ludwigshafen.

The block copolymer chosen to prepare blends with polystyrene was an asymmetric star block copolymer (ST2-S74, volume fraction of PS, $\Phi_{ST} = 0.74$; see Table 1). It has 4 asymmetric SBS arms in average, one of which is much longer than the others. The longest arm is styrene-rich while the shorter ones are butadiene-rich.[1,12]

Two different types of polystyrene homopolymer (hPS) samples were used: one of them has a lower molecular weight $M_w = 15{,}000$ g/mol and $M_w/M_n = 1.29$ (named as PS015) and the other one has a higher molecular weight $M_w = 190{,}000$ g/mol and $M_w/M_n = 2.3$ (named as PS190). The latter was a general purpose polystyrene (GPPS).

The samples were prepared by two different methods: solution casting and injection moulding (mass temperature 250°C and mould temperature 45°C). The mixtures were prepared in various weight ratios: 20, 40, 60 and 80 wt-% hPS. Solution casting was performed using a neutral solvent toluene allowing the solvent to evaporate from the sample solution (3 % weight/volume) over a period of about two weeks. The films, about 0.5 mm thick, were dried and annealed for 48 hours in a vacuum oven under reduced pressure.

Table 1. The characteristics of the block copolymers investigated, LN and ST stand for linear and star architectures, respectively.

samples	§M_w (g/mol)	§M_W/M_n	*Φ_{ST}	remarks
LN1-S74	87,700	1.07	0.74	symmetric SBS triblock, neat transition with the pure PB mid-block, linear architecture [13]
LN2-S74	105,100	1.13	0.74	asymmetric S_1BS_2 triblock copolymer, ($S_1 \neq S_2$), tapered transition [13]
LN4-S65	139,200	1.20	0.65	random PS-co-PB copolymer as rubbery mid-block, symmetric outer PS blocks, linear architecture [1,13]
ST2-S74	184,000	1.69	0.74	highly asymmetric star architecture, each arm with S_1BS_2 structure ($S_1 \neq S_2$), tapered transition [1,11,12]
PS015	15,200	1.29	100	polystyrene homopolymer
PS190	190,000	2.30	100	polystyrene homopolymer

* total styrene volume fraction determined by Wijs titration and procedures
§ number average (M_n) and weight average (M_w) molecular weight determined by gel permeation chromatography (GPC) using PS calibration

Tensile Testing

Macroscopic tensile tests using the injection moulded bars were carried out using universal tensile machine (Zwick 1425) at a cross head speed of 50 mm/min at room temperature (23°C) according to ISO 527.

Transmission Electron Microscopy (TEM)

Morphological details of undeformed as well as deformed samples were examined using transmission electron microscopy (TEM, Joel 200 kV). For the TEM investigations, a small block of specimen taken from the middle of a tensile bar approximately 0.10 mm below the surface was trimmed to a pyramidal shape and immersed in an osmium tetroxide (OsO_4) solution for several days at room temperature (23°C) to selectively stain the butadiene phase. Ultra-thin sections of the samples were cut from each block using a Leica Ultracut-C Ultramicrotome operated at room temperature (23°C). Sections from strained samples were taken from the location close to the fracture surface. Lamellar long period and lamellar thickness were quantified from the transmission electron micrographs using a special image processing program and Fourier transformation.

Results and Discussion

Influence of Block Copolymer Architecture on Their Morphology

Classically the nature and the dimension of the diblock copolymer morphologies are adjusted by the variation of composition and molecular weight. In the strong segregation limit, with increasing polystyrene (PS) content, body centred cubic spheres, hexagonal arranged cylinders or 3D 'gyroid' network of polystyrene (PS) domains dispersed in the matrix of polyisoprene were observed in poly(styrene-block-isoprene) SI diblock copolymers.[5,8] With further increase in the PS content, the alternating layers of PS and PB lamellae and the structures mentioned above in the reversed order were found (Figure 1, top). As demonstrated by our works, one may notice that the microphase separated morphology of the styrene/butadiene based block copolymers can be considerably changed also by the variation of the molecular architectures (Figure 1, bottom).

In consideration of the classical diblock copolymer phase diagram, hexagonal PB cylinders in the matrix of PS phase would be expected in the SBS triblock copolymers with 74% volume fraction of polystyrene without any modification of the molecular architecture (Figure 1a, bottom). The results of our investigations showed that the relative volume fraction of the hard and the soft components may be altered significantly by the modification of the molecular structure of the styrene/butadiene block copolymers: a 'two-component three-phase' morphology consisting of alternating layers of polystyrene and polybutadiene (where the PB lamellae contain scattered PS domains) appeared in the tapered asymmetric star block copolymer (Figure 1b, bottom). This characteristic 'two-component three-phase' morphology was observed in the copolymer having polystyrene chains in the molecule core.

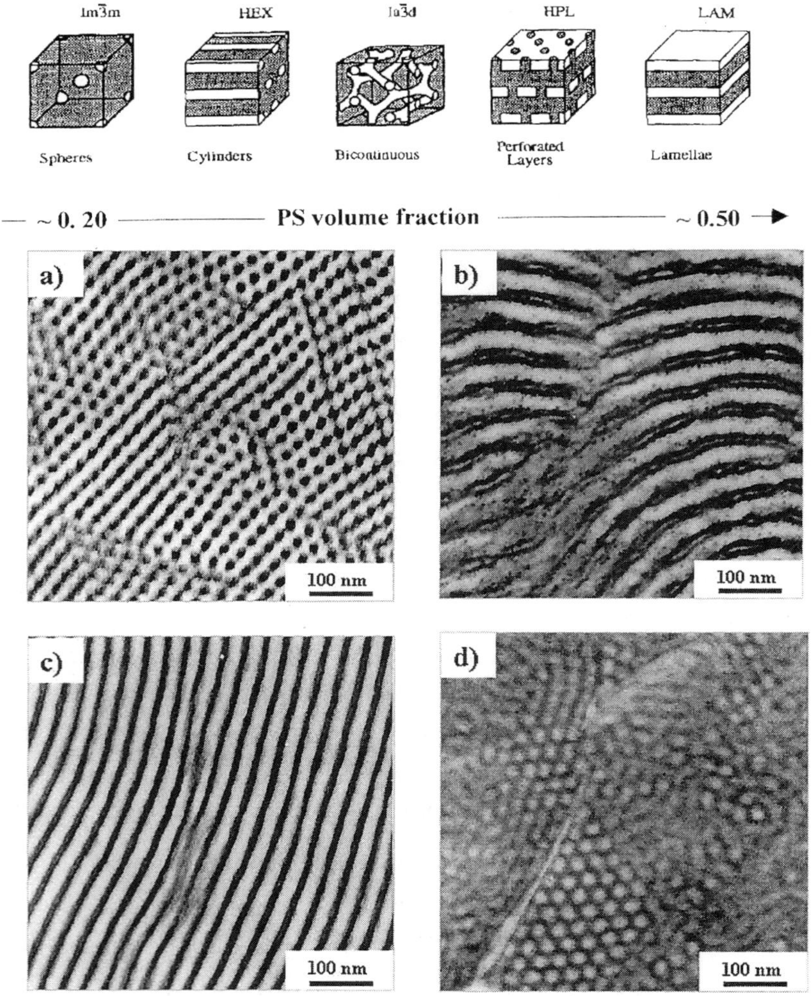

Fig. 1. Top: Scheme representing the morphology of styrene/butadiene diblock copolymer as a function of polystyrene content (the picture from ref. 5, p. 27); bottom: TEM micrographs showing the morphology of the investigated block copolymers – a) LN1-S74, b) ST2-S74, c) LN2-S74 and d) LN4-S65.

The PS core with a total molecular weight of about 20,000 g/mol, which is enough for the formation of the phase separated domains, makes it possible that the scattered cylinder-like domains are formed inside the PB lamellae in this star block copolymer. Such a morphology was reported in an ABC triblock copolymer by Shibayama et al.[14]

Keeping the overall chemical composition constant (Φ_{PS} = 0.74) and introducing quite a long tapered transition between the polybutadiene (PB) and the polystyrene (PS) blocks in an asymmetric SBS triblock copolymer (LN2-S74, see Figure 1c, bottom) further modifies the block copolymer morphology. In LN2-S74, an unambiguous lamellar morphology was observed. The formation of lamellae-like (as in ST2-S74) or clear lamellar (as in LN2-S74) morphology in the asymmetric block copolymers is also associated with a large shift in the glass transition temperature of polybutadiene phase ($T_{g\text{-}PB}$). The formation of lamellar morphology in ST2-S74 and LN2-S74 instead of the hexagonal PB cylinders (one similar to that of LN1-S74 presented in Figure 1a, bottom) is the result of following factors: i) asymmetric architecture (shift of phase diagram) and ii) presence of a part of polystyrene in the butadiene phase (increase in volume fraction of the soft phase).

Table 2. Glass transition temperature of the soft and the hard phases in the investigated block copolymers (data obtained by differential scanning calorimetry, DSC, heating rate 10°C/min).

Polymer	$T_{g\text{-}PB}$ (soft phase)/°C	$T_{g\text{-}PS}$ (hard phase)/°C
LN1-S74	-98	101
LN2-S74	-53	101
LN4-S65	-34	70
ST2-S74	-79	104

The presence of a fraction of PS chain segments in the PB phase is suggested by a significant increase in the $T_{g\text{-}PB}$ (see Table 2). Additionally, the presence of tapered interface might favour an increase in volume fraction of the soft phase.[1] In the literature, it was discussed that the tapered transition has a strong influence on the interface width; and it may even lead to a separate phase, which has its own glass transition temperature.[15] However, we observed only a strong shift of $T_{g\text{-}PB}$ towards higher temperature in the tapered SBS block copolymers relative to the neat ones.[16]

Randomising the rubbery block (e.g., by the incorporation of a styrene/butadiene random copolymer as the rubbery block) may shift the soft phase volume fraction to even higher magnitude enabling the formation of 'gyroid'-like morphology, the latter being expected in a neat diblock copolymer at a composition range of $\Phi_{PS} \sim 0.32$. In a star block copolymer whose structure is analogous to that of ST2-S74 (see Table 1 and ref. [11,12]), 'gyroid'-like morphology was observed.[17]

The shift of morphology scheme, discussed so far, is valid for the block copolymers independent of their molecular topology. As shown by our TEM investigations of a linear styrene/butadiene triblock copolymer having a random styrene-co-butadiene centre block and total 65 vol.-% styrene (Figure 1d, bottom), the morphology consists of the dispersed polystyrene domains. Dispersed PS domains in the matrix of styrene-co-butadiene is expected since the volume fraction of the PS blocks in this copolymer is about 0.32.

It should be mentioned that a modification in molecular architecture, which leads to an increase in the effective volume fraction of the soft component, must be associated with the intervention of the stiff polystyrene segments in the butadiene phase which comprises the flexible chains at room temperature. This hinders the mobility of the soft phase (here the PB phase) and hence causes an increase in the glass transition temperature ($T_{g\text{-}PB}$). In all the block copolymers with modified architectures an increase in the $T_{g\text{-}PB}$ was found (see Table 2).

In fact, besides the volume fraction of the phases, the thermodynamics at the interface is the key factor that plays a decisive role in determining the interfacial curvature, i.e., via the interaction parameter of the two species in contact.[5] Thus a change in the chemical nature of the blocks at the interfacial region (neat versus tapered or random) should affect the thermodynamic equilibrium and the resulting morphology.

The equilibrium morphologies of the block copolymers outlined above are stable ones. These were observed independent of preparation conditions. The processing history influences mainly the orientation and order of the microdomains and not their nature and dimension.[16] The results discussed present an excellent evidence of possibility of modifying the block copolymer phase diagram via a change in molecular architecture. Thus, the architectural modification of block copolymers may open an important way of controlling their morphology and mechanical properties.

Morphology Formation in Block Copolymer/PS Blends

It is well known that the length of the homopolymer chains relative to that of the corresponding block of the block copolymer plays a vital role in the phase behaviour of the block copolymer/homopolymer blends.[5,9] The molecular weight of the longest PS block in the star block copolymer ST2-S74 is in the range of 70,000 g/mol.

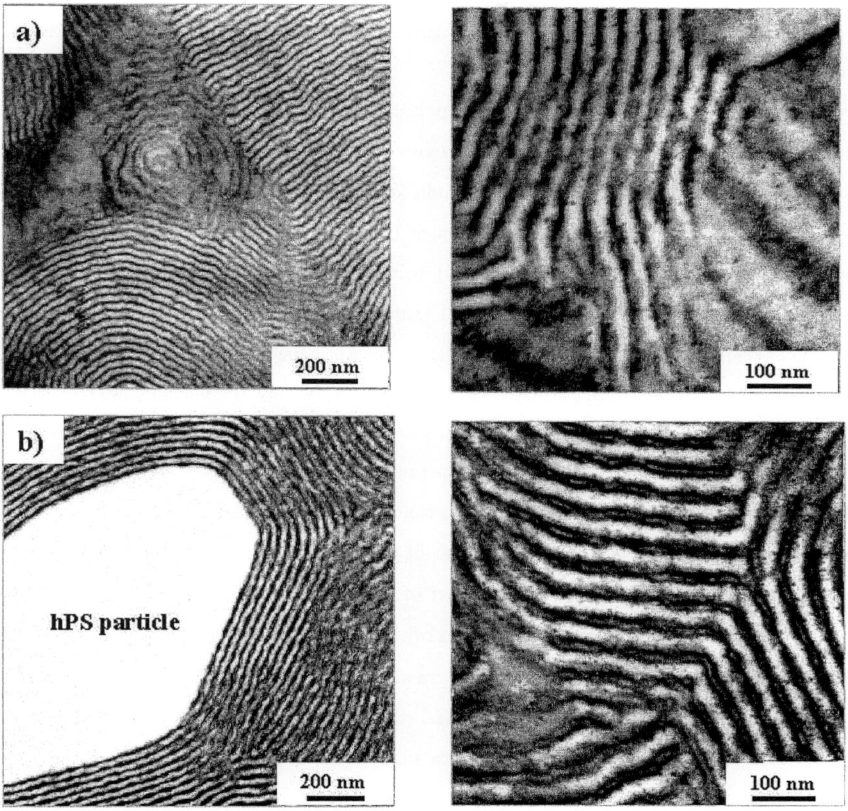

Fig. 2. Lower (left) and higher (right) magnifications of TEM micrographs of the star block copolymer/hPS blends as a function of hPS molecular weight M_{hPS}: a) M_{hPS} = 15,200 g/mol and b) M_{hPS} = 190,000 g/mol; hPS content 20 wt.-% in each case; solution cast films.

Figure 2 shows representative TEM micrographs of the two blends of ST2-S74 with 20 wt.-% hPS. The samples were prepared by solution casting. The polystyrene samples differ in the molecular weight. PS015 and PS190 have the weight average molecular weight (M_w) of 15,200 g/mol and 190,000 g/mol, respectively (Table 1). Details about the morphology and micromechanical behaviour of the blends may be found in another publication.[18]

The molecules of the sample PS015 are shorter than the longest polystyrene block of the parent star block copolymer. Hence these chains may diffuse into the PS blocks of the block copolymer and swell the PS lamellae (Figure 2a). This leads to the formation of a microphase separated morphology (absence of PS particles). The PS lamellae in this blend thicker (about 22 nm) than in the pure star block copolymer (about 20 nm).

Increasing the molecular weight of the added polystyrene (hPS) decreases the entropy of these chains and the hPS chains cannot diffuse deeply into the PS block of the block copolymer. These hPS chains tend to segregate towards the middle of the domains and finally, if the molecular weight is too high, these are even expelled out of the PS domains leading to the formation of macrophase separated hPS particles dispersed in the matrix of microphase separated block copolymer matrix at low hPS concentration. The molecular weight of the PS190 is obviously much higher than the corresponding block in the block copolymer. Hence the added polystyrene predominantly forms the hPS particles leaving the morphology of the surrounding matrix unchanged (Figure 2b). The lamellar thickness (20 nm) and the spacing (about 42 nm) in the macrophase separated blend is the same as in the pure block copolymer.

It should be mentioned that the polystyrene samples used in this work have a wide molecular weight distribution. Hence, a low molecular weight part of PS190 might still be added to the PS block of the block copolymer. Moreover, a part of PS015 may even be added to the butadiene phase of the pure block copolymer as suggested by recent works of Feng et al.[19]

The situation changes when the blends are mixed in extruder and processed by injection moulding. Due to the high speed processing and rapid cooling conditions during the moulding process, there is not enough time for the components to form the structures that may be expected under equilibrium conditions.

Figure 3 presents the TEM micrographs of the pure block copolymer and two blends prepared by injection moulding. Details on the morphology and micromechanical behaviour of these blends may be found in.[17] Since the thickness of the structures doesn't exceed 100 nm, the blends may be regarded as being microphase separated. This is the reason why the blends were nearly as transparent as the pure block copolymer itself. The microphase separated structures are oriented in the direction of shear force (i.e., along the injection direction) in the pure block copolymer as well as in the blends.

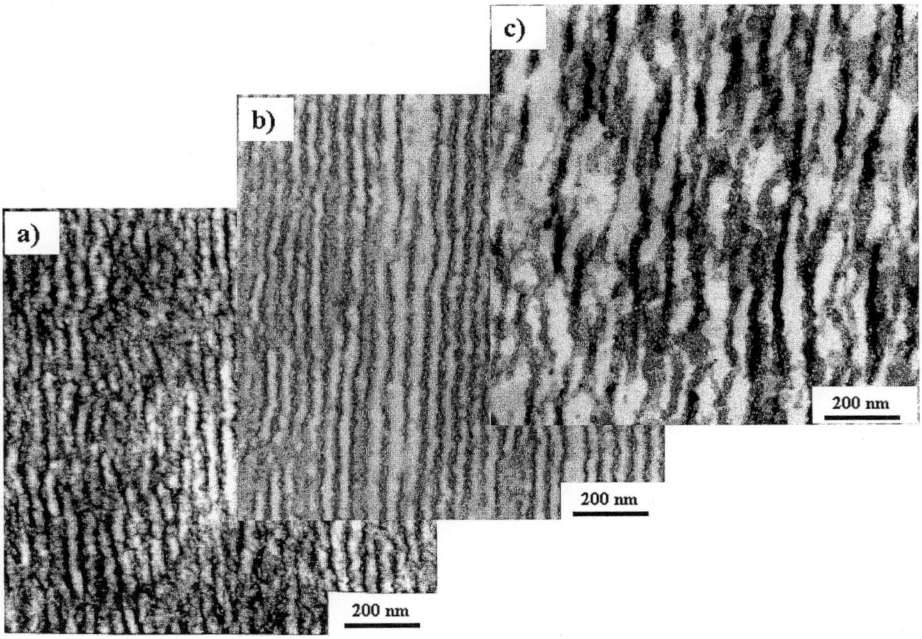

Fig. 3. Representative TEM images showing the morphology of injection moulded samples: a) pure ST2-S74, b) blend with 20 wt.-% PS190 and c) blend with 60 wt.-% PS190, injection direction vertical.

The thickness of the PS lamellae (or PS strands) increases continuously with increasing PS190 content in the blends. At the same time, the thickness of the PB lamellae doesn't change significantly. This suggests that the homopolystyrene chains are accommodated at the PS block lamellae of the pure copolymer. At higher PS content, the PB lamellae appear as elongated worms dispersed in the PS matrix Figure 3c.

To demonstrate the result of the added hPS to the PS block lamellae of the star block copolymer, the thickness distribution of the PS lamellae in pure ST2-S74 and the blend with 20 wt.-% PS190 is represented in Figure 4. One can see that the average thickness of the PS lamellae changes from ca. 20 nm (pure block copolymer) to about 30 nm (blend with 20 wt.-% PS). The average value of the PS lamella thickness continuously increases with increasing PS190 content in the blends.

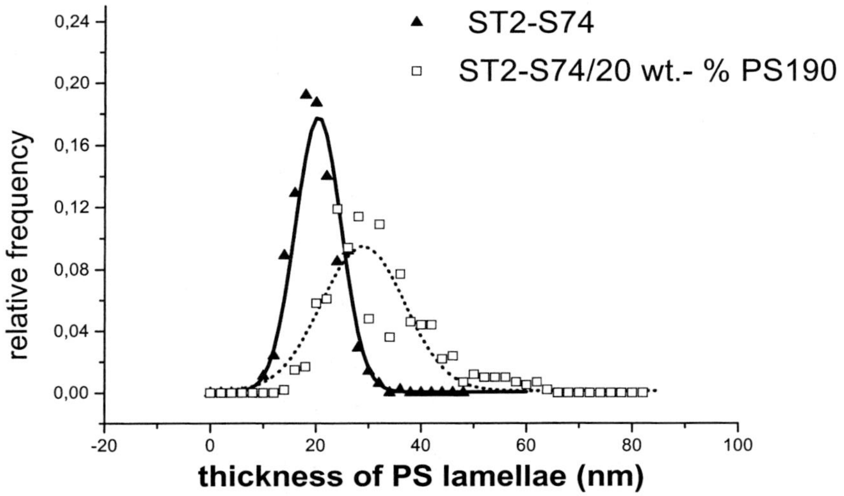

Fig. 4. Distribution of PS lamella thickness in ST2-S74 and a blend with 20 wt.-% PS190.

Micromechanical and Mechanical Behaviour

Micromechanical and mechanical behaviour of the phase separated block copolymer systems are generally controlled by two variables: molecular parameters and phase morphology. The molecular parameters include the molecular construction (diblock, triblock, star block, graft block structure, and the location of the rubbery block in the molecule etc.), molecular weight, chain topology (star block, graft block structure etc.).

Our previous works showed that the molecular structure of the block copolymers affects also the phase volume ratio (soft/hard volume ratio) and hence enables the formation of morphologies which would be expected in another composition range in a neat diblock copolymer.[13] The resulting morphology has the decisive impact on the micromechanical and mechanical behaviour of these block copolymer systems. Obviously, the block copolymers

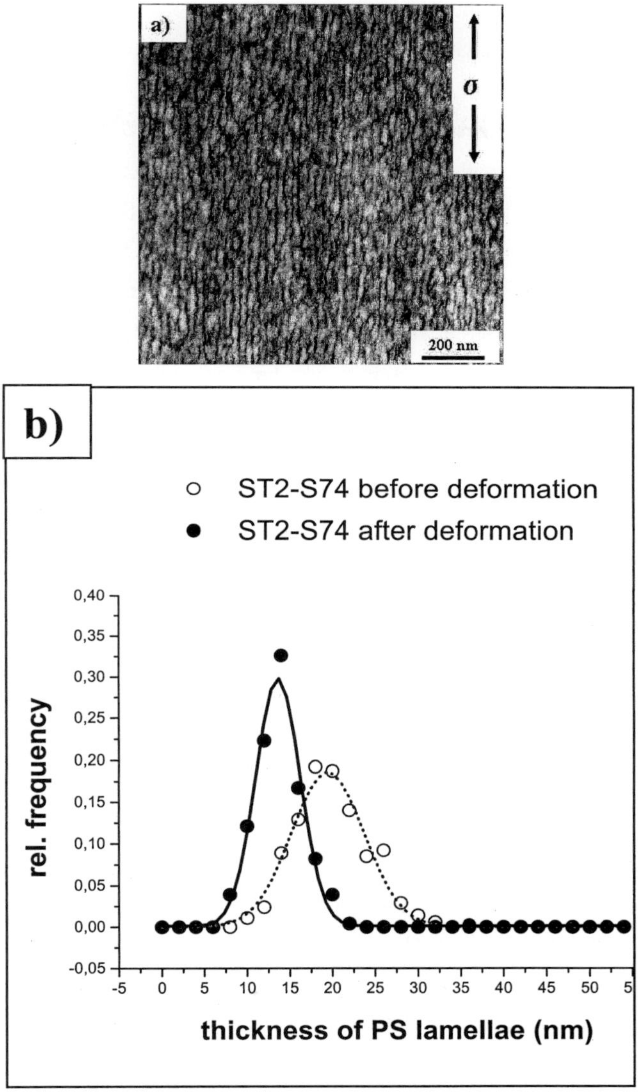

Fig. 5. a) TEM micrograph showing the morphology of the injection moulded ST2-S74 after deformation up to fracture, sample location close to the fracture surface (compare with Figure 3a), and b) the distribution of PS lamella thickness in ST2-S74 before and after deformation.

presented in Figure 1, due to the presence of different morphologies, are characterised by different micromechanical behaviour and result in entirely different kinds of macroscopic mechanical properties.

Figure 5a shows the morphology of the injection moulded star block copolymer ST2-S74 after tensile deformation (compare with Figure 3a). The sample was taken from the locations close to the fracture surface after tensile testing. Quantification of the PS lamella thickness in this block copolymer before and after deformation is presented in Figure 5b. It can be easily noticed that the average thickness of the PS lamellae has been reduced by from about 20 nm to about 12 nm, i.e., a reduction by about 50 %. Assuming the same degree of reduction in the lateral direction of the lamellae during deformation, a stretching ratio of $\lambda \sim 4$ (a local strain of about 300%) may be calculated. It is interesting to note that this calculated strain of the lamellae is in the same range as the macroscopically measured value of the strain at break of the bulk injection moulded specimens (Figure 6a). This value corresponds also to the maximum draw ratio of craze fibrils in bulk polystyrene samples.[3]

The fact, that the local strain of the PS lamellae approximately corresponds to the macroscopic strain at break of the compact star block copolymer, suggests that the ductility of the star block copolymer is a consequence of the homogeneous plastic deformation of the PS lamellae. Of course, the deformation of the rubbery phase (which may be regarded as being in the liquid state with the T_g of about -76°C) combined with the slipping of the glassy layers have contributed partly to the high elongation achieved by the compact star block copolymer (see Figure 6).

Since the PS lamellae in the investigated block copolymers are about 20 nm thick, it can be easily noticed that the drawing of the PS lamellae is quite similar to that of the craze fibrils in the polystyrene homopolymer. Both craze fibrils and PS lamellae have similar dimensions (thickness) and also show a similar extension ratio. Due to a very low glass transition temperature of the butadiene phase in the block copolymers, it may be regarded as a liquid and has analogous function as the microvoids in the PS craze. This means that the intrinsic ductility of the PS can be achieved by reducing the dimension of the PS layers down to a critical value (in the range of 20 nm in styrene/butadiene triblock copolymers). Based on the results obtained from a number of lamellar block copolymers, we formulate the mechanism of '*thin layer yielding*'. As the name already suggests, the PS layers are capable of undergoing a homogeneous plastic deformation

a)

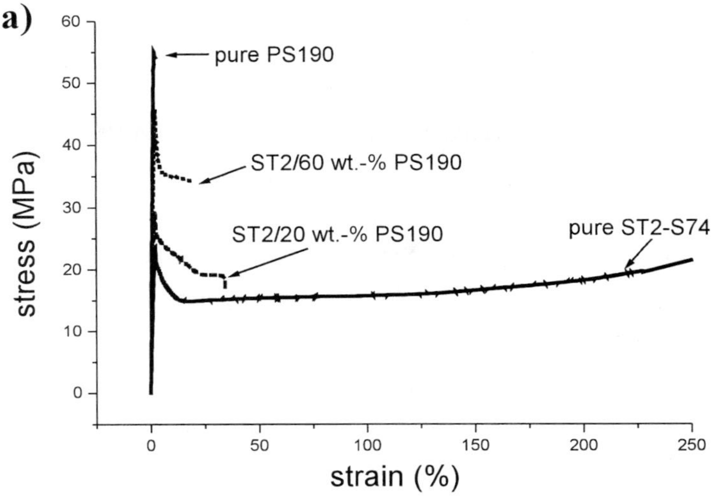

b)

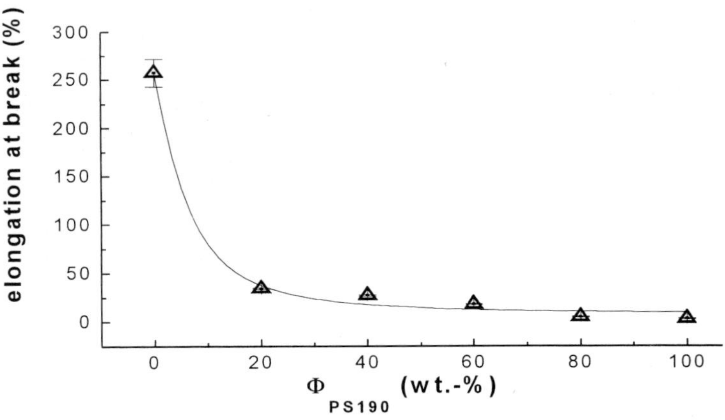

Fig. 6. a) Stress-strain curve of the star block copolymer ST2-S74 compared with that of pure polystyrene PS190 and blends containing 20 and 60 wt.-% PS190; b) elongation at break plotted against the PS190 content in the ST2-S74/PS190 blends.

under tension if the thickness of these layers lies below a critical value. As a result, the elongation to break which is a measure of ductility, drastically increases. Above this critical thickness the material behave brittle (see Figure 6b).[11]

The plastic deformation of glassy PS lamellae in the block copolymer studied is in line with earlier studies of Kawai et el. on unoriented films. Using TEM and SAXS, they showed that, during plastic deformation, shearing and kinking processes occurs prior to the disruption of glassy PS lamellae.[20,21] At higher strains, formation of ‘fish-bone structure' (chevron-morphology) characterized by a four-point SAXS pattern was observed.

Recently, Thomas and co-workers explored the deformation behavior of oriented lamellar SBS triblock.[22] During parallel and perpendicular loading, destruction of lamellar morphology and formation of chevron morphology, respectively were observed. The homogeneous plastic flow of glassy lamellae is a cause of destruction of lamellar morphology leading to a reduction in lamellar spacing during parallel deformation as reported by Thomas and co-workers.[22]

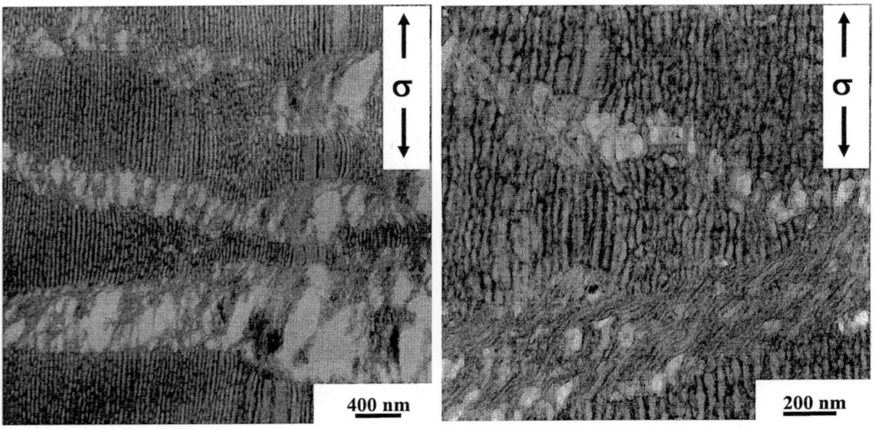

Fig. 7. Lower (left) and higher (right) magnification of TEM micrographs showing the morphology of injection moulded ST2-S74/20 wt.-% PS190 blend and crazes after deformation, deformation direction is shown by an arrow; samples stained by OsO_4.

To examine the influence of the PS layer thickness, we studied the deformation behaviour of blends of the star block copolymer ST2-S74 with polystyrene homopolymer PS190 prepared by injection moulding. With increasing homopolymer content, the PS lamellae thickness increases (see Figure 4). With increasing PS content, the Young's modulus and yield stress was found to increase linearly[18] while the elongation at break decreased from over 200% to about 34% (Figure 6b). This transition from ductile to brittle behaviour is associated with a transition in the deformation mechanism from homogeneous plastic flow to the formation of craze-like zones. The localization of deformation in these craze-like zones is shown by the TEM micrograph of the ST2-S74/20 wt.-% PS190 blend (Figure 7). The transition in deformation mechanism from ductile plastic yielding of PS lamellae to crazing takes place when the thickness of the PS lamellae reaches a value of ca. 30 nm.

The craze-like deformation zones in the blend with 20 wt.-% PS190 are characterised by a high degree of stretching of the lamellae accompanied by void formation. Locally, an elongation ratio of $\lambda > 4$ may be estimated. However, the macroscopically achievable elongation of this sample is relatively low (only about 34 %) due to the extreme localisation of the deformation zones. Hence, one can correlate the transition in the deformation mode from the homogeneous plastic yielding to the formation of local craze-like zones with the thickness of the polystyrene layers. These results are strong evidence of the proposed 'thin layer yielding' mechanism.

Conclusions and Future Works

The results may be summarised as follows:

a. Modified block copolymer architecture leads to a change in the nature of microphase separated morphologies to be formed in the solid state.

b. The lamellar systems having glassy/rubbery alternating layers deform via '*thin layer yielding*' mechanism, where the glassy layers act as craze-fibrils separated by rubbery layers instead of microvoids. This mechanism may be used as an alternative toughening mechanism in the brittle polymers.

In spite of the strong evidence in support of the '*thin layer yielding*' mechanism, the exact magnitude of the critical thickness and its dependence on temperature and loading rate is still unknown. The role of the interface between the polymer layers in this connection is also of interest. The investigation of these aspects are underway.

Acknowledgements

The research was supported by the Kultusministerium des Landes Sachsen-Anhalt (Project: 'Neue Funktionswerkstoffe auf der Grundlage Schwach entmischter Blockcopolymere') and the Deutsche Forschungsgemeinschaft (Project: 'Innovationskolleg – Neue Polymermaterialien'). We thank the BASF Aktiengesellschaft for co-operation and supply of materials and Ms. S. Goerlitz for the TEM investigations. Generous helps from Dr. R. Godehardt, Mr. W. Lebek, Dr. R. Lach, Dr. R. Weidisch and Dr. T. A. Huy are acknowledged.

[1] K. Knoll and N. Nießner, *Macromol. Symp.* **1998**, *132*, 231.
[2] C. B Bucknall, in "*The Physics of Glassy Polymers*", R. .N. Haward and R. J. Young Eds., chapter 8, p. 363, Chapman & Hall, London 1997.
[3] G. H. Michler, "*Kunststoff-Mikromechanik: Morphologie, Deformations-und Bruchmechanismen*", Carl Hanser, München 1992.
[4] R. J. M. Smit, W. A. M. Brekelmans, H. E. H. Meijer, *J. Matt. Sci.* **2000**, *35*, 2855.
[5] I. W. Hamley, "*The Physics of Block copolymers*", Oxford Science Publications, Oxford 1998.
[6] C. Lee, S. P. Gido, Y. Poulos, N. Hadjichristidis, N. B. Tan, S. F. Trevino and J. W. Mays, *J. Chem. Phys.* **1997**, *107*, 6460.
[7] S. T. Milner, *Macromolecules* **1994**, *27*, 2333.
[8] M. W. Matsen, *J. Chem. Phys.* **2000**, *113*, 5539.
[9] H. Hasegawa and T. Hashimoto, in "*Comprehensive Polymer Science, Suppl. 2*", S. L. Aggarwal and S. Russo Eds., p. 497, Pergamon, London 1996.
[10] G. Holden, "*Understanding Thermoplastic Elastomers*" p. 15, Carl Hanser Verlag, Munich 2000.
[11] G. H. Michler, R. Adhikari, W. Lebek, S. Goerlitz, R. Weidisch, K. Knoll, *J. Appl. Polym. Sci.* **2002**, *85*, 683.
[12] R. Adhikari, G. H. Michler, W. Lebek, S. Goerlitz, R. Weidisch, K. Knoll, *J. Appl. Polym. Sci.* **2002**, *85*, 701.
[13] R. Adhikari, R. Godehardt, W. Lebek, R. Weidisch, G. H. Michler, K. Knoll, (**2001**), *J. Macromol. Sci. B : Polym. Phys.* **2001**, *40*, 833.
[14] M. Shibayama, H. Hasegawa, T. Hashimoto and H. Kawai, *Macromolecules* **1982**, *15*, 274.
[15] F. Buehler, W. Gronski, *Makromol. Chem.* **1986**, *187*, 2019.
[16] R. Adhikari, "*Ph. D. Thesis*" Universität Halle-Wittenberg, Germany 2001.
[17] R. Adhikari, M. Buschnakowski, R. Godehardt, T. A. Huy, G.H. Michler, K. Geiger, K. Knoll, in preparation
[18] R. Adhikari, G. H. Michler, S. Henning, R. Godehardt, K. Knoll, *J. Appl. Polym. Sci.* submitted
[19] H. Feng, Z. Feng, H. Yuan and L. Shen, *Macromolecules* **1994**, 27, 7830.
[20] M. Fujimura, T. Hashimoto, H. Kawai, *Rubber Chem. Technol.* **1978**, *51*, 215.
[21] T. Hashimoto, M. Fujimura, K. Saijo, H. Kawai, J. Diamant, M. Shen, *ACS Advances in Chemistry Series – Multiphase Polymers,* **1979**, *176*, 257.
[22] Y. Cohen, R. J. Albalak, G. J. Dair, M. S. Capel, E. L. Thomas, *Macromolecules*, **2000**, *33*, 6502

Interfacial Chemistry and Morphology of In-Situ Compatibilised PA-6 and PBT Based Blends

Martin van Duin,[*1] *Marnix van Gurp,*[1] *Luc Leemans,*[1] *Monique Walet,*[1] *Marcel Aussems,*[1] *Philippe Martin,*[2] *Roger Legras,*[2] *Ana V.* Machado,[3] *Jose A. Covas*[3]

[1] DSM Research, P.O. Box 18, 6160 MD Geleen, The Netherlands
E-mail: martin.duin-van@dsm.com
[2] Laboratoire des Hauts Polymères, Université Catholique de Louvain, Belgium
[3] Department of Polymer Engineering, Universidade do Minho, Portugal

Summary: Reactive blending is the preferred route for obtaining rubber-toughened engineering plastics. In this study the chemistry and morphology of in-situ compatibilised PA-6/EPM-g-MA blends and PBT/E-MA-GMA blends are compared. Despite the different chemical structures formed (PA-6: cyclic imide; PBT: ester), the different side-reactions that take place (PA-6: chain scission; PBT: rubber crosslinking) and the different reaction rates (PA-6: very fast; PBT: rather slow), a general description of reactive blending is proposed. First, a very fast physics-controlled dispersion of the rubber phase into the polycondensate matrix occurs (from milimeter to micrometer scale). Then the interface, thus generated, is covered via graft-copolymer formation. This results in a stabilisation against coalescence and in a reduction of the interfacial tension, enabling further dispersion of the rubber phase. In the case of PA-6/EPM-g-MA blends the graft formation occurs so quickly that when the physics-controlled dispersion is completed, the final dispersion is achieved. In the case of PBT/E-MA-GMA blends the graft formation is much slower and the chemistry-controlled dispersion can hence be considered as a secondary process, yielding the final morphology.

Keywords: blend, graft copolymer, in-situ compatibilisation, morphology, reactive blending

Introduction

The main reasons for the commercial success of polyamide-6 (PA-6) and poly(butyleneterephthalate) (PBT) as engineering plastics are their high strength and stiffness and their high heat resistance, combined with their relatively low cost. PA-6 has the disadvantage of high water adsorption and, consequently, has a lower dimensional stability and poorer electrical properties than PBT. Since both thermoplastics are brittle when subjected to impact conditions, they are often compounded with glass fibres or rubber. In practice about 20 wt.% of a finely

 DOI: 10.1002/masy.200350812

dispersed (≤ 1 μm) rubber phase with good adhesion to the thermoplastic matrix has been shown to ensure sufficient rubber toughening.[1,2] To obtain such small rubber particles and good adhesion, pre-synthesised core-shell rubbers may be used. Alternatively, compatible blends may be produced via reactive blending, i.e. in-situ compatibilisation.[3,4] PA-6 is typically toughened with ethylene-propylene rubber (EPM) using maleic-anhydride-modified EPM (EPM-g-MA) as the reactive compatibiliser. In the case of PBT one often resorts to epoxy-containing rubbers, such as ethylene-(methyl acrylate)-(glycidyl methacrylate) terpolymers (E-MA-GMA).

In the present paper the results of our (partly published) studies on PA-6/EPM-g-MA blends and PBT/E-MA-GMA blends[5-14] will be combined. The similarities and the differences of these blends will be highlighted and conclusions with respect to reactive blending on a more general level will be drawn.

Experimental

For experimental details the reader is referred to the original studies.[5-14] Only a few general remarks will be made here. Experiments with representative low-molecular-weight model compounds have been carried out, since these allow relatively straightforward characterisation of the reaction products, in order to determine the chemical structures formed during the interfacial reactions between the polycondensate end groups and the rubber functionalities. Selective fractionation and hydrolysis schemes have been developed and applied, in order to identify the graft copolymers and to quantify the conversion of the reactive groups and/or the amount of copolymer formed. The morphology of the heterogeneous thermoplastic/rubber blends has been investigated using electron microscopy in combination with image analysis.

Besides blends produced in batch kneaders or extruders, the development of the various physico-chemical phenomena was also studied along the screw axis of a co-rotating twin-screw extruder by using special sampling devices.[15] As shown in Figure 1 (top), by turning a bolt (2) in the barrel wall, the cavity (4) can be positioned towards the opening in the wall of the extruder (1), allowing polymer melt to flow from the screw channel into the cavity. By turning the bolt 180°, the melt sample is accessible from the outside of the extruder. The sample is removed, quickly quenched and analysed off-line. Figure 1 (bottom) shows the extruder layout and typical locations of the sampling devices. They are positioned along the staggering kneading blocks and/or left elements, where inner pressure develops for forcing the melt sample into the cavity.

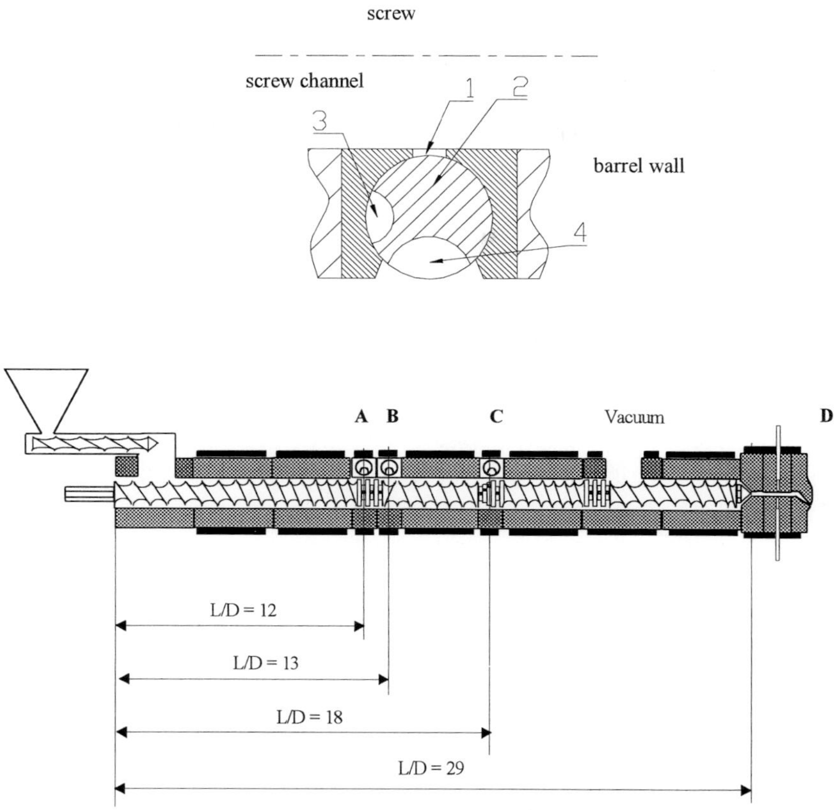

Fig. 1. Schematic drawing of sampling device (top) and set-up of extruder with positions of sampling devices along screw axis (bottom).

PA-6/EPM-g-MA Blends

In the case of PA-6/EPM-g-MA blends produced in the melt, the graft copolymer is formed upon reaction of the PA-6 amine end group (PA-NH_2) with the cyclic anhydride of EPM-g-MA, yielding a cyclic imide link under formation of water (Figure 2).[5,6,16-18] If the concentration of anhydride is larger than that of the amine groups, an indirect reaction between the PA-6 amide group (PA'-CONH-PA) and the anhydride will occur, again resulting in an imide link, but also in a new PA-6 carboxylic end group (PA'-COOH) due to scission of the PA'-6 chain. The reaction between anhydride and amine group is extremely fast, that between anhydride and amide is much slower.

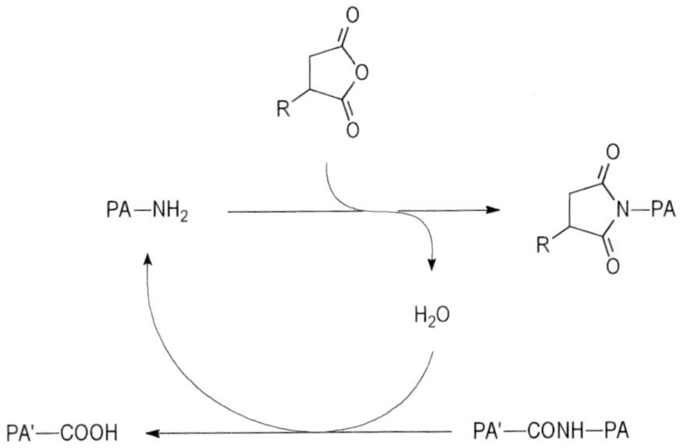

Fig. 2. Reaction scheme illustrating in-situ compatibilisation of PA-6/EPM-g-MA blends.

Selective extraction with formic acid at room temperature allows quantification of the PA-6 graft content in PA-6/EPM-g-MA blends. Hydrochloric acid treatment allows identification of the cyclic imide and quantification of the MA conversion. The ratio of the PA-6 graft content and the degree of anhydride conversion provides an estimate of the number-average molecular weight of the PA-6 grafts. Some typical results for 50/50 (w/w) blends of PA-6 and PA-6.6 with EPM-g-MA and poly(styrene-co-MA) (SMA)[5] and for 80/x/20-x (w/w/w) PA-6/EPM/EPM-g-MA blends[7] are presented in Tables 1 and 2, resp. The amount of PA-6 graft and the MA graft conversion depend clearly on the type of blend components used and the blend composition. The MA graft conversion is high and the molecular weight of the PA-6 grafts is low in comparison with that of the initial PA-6 polymer, both demonstrating the occurrence of PA-6 chain scission.

Table 1. Characteristics of 50/50 (w/w) blends of polyamides and MA-containing polymers produced on a co-rotating twin-screw extruder (240/255 °C, 200 rpm).[5]

blend	grafted PA (wt.%)	MA graft conversion (%)[a]	M_n free PA (kg/mol)[b]	M_n PA graft (kg/mol)[b]
PA-6/EPM-g-MA	14	100		2.0
PA-6/SMA	12	42		0.2
PA-6.6/EPM-g-MA	13	100	16	1.8
PA-6.6/SMA	10	44	8.5	0.16

a: MA content of original EPM-g-MA and SMA is 1.4 and 28 wt.%, resp.; b: M_n of original PA-6 and PA-6.6 is 13 and 20 kg/mol, resp.

Table 2. Characteristics of PA-6/EPM/EPM-g-MA blends produced on a co-rotating twin-screw extruder (220 °C, 200 rpm, 6 kg/hr).[7]

PA-6/EPM/EPM-g-MA blend (w/w/w/)	grafted PA (wt.%)	MA graft conversion (%)[a]	M_n PA graft (kg/mol)[b]	average particle size [range] (μm)
80/20/0	0	-	-	9.0 [2.0 – 31]
80/15/5	0.38	100	1.4	2.6 [1.5 – 23]
80/10/10	0.8	96	1.7	2.3 [1.5 – 8.0]
80/5/15	0.8	89	1.2	2.2 [1.5 – 5.2]
80/0/20	0.8	86	1.0	0.33 [0.1 – 0.7]

a: MA graft content of original EPM-g-MA is 0.49 wt.%; b: M_n of original PA-6 is 13 kg/mol.

The low molecular weight of PA grafts has also been explained by the high concentration of low-molecular-weight PA at and/or the fast diffusion to the interface.[17, 18] The morphology of the PA-6/EPM-g-MA blends consists of a (sub-)micrometer dispersion of rubber spheres. The rubber dispersion often contains PA-6 occlusions. When PA-6.6. is used instead of PA-6, crosslinking of the rubber phase via H_2N-PA-NH_2 oligomers (including 1,6-hexanediamine) occurs, yielding a coarse and irregular morphology,[5] which corroborates the results found by Paul et al.[19]

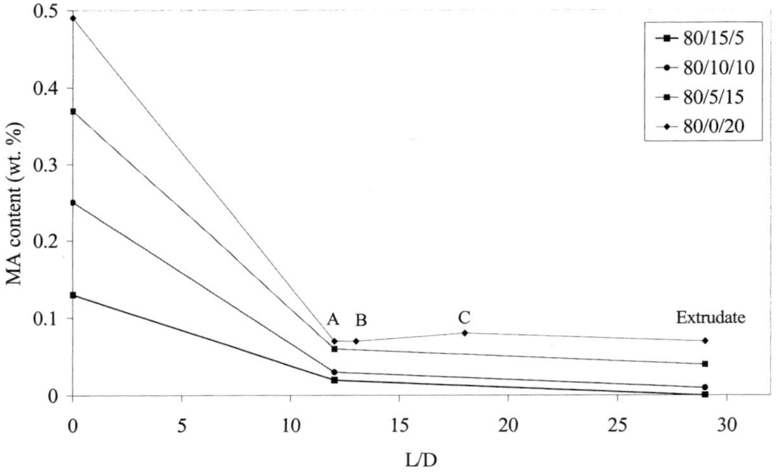

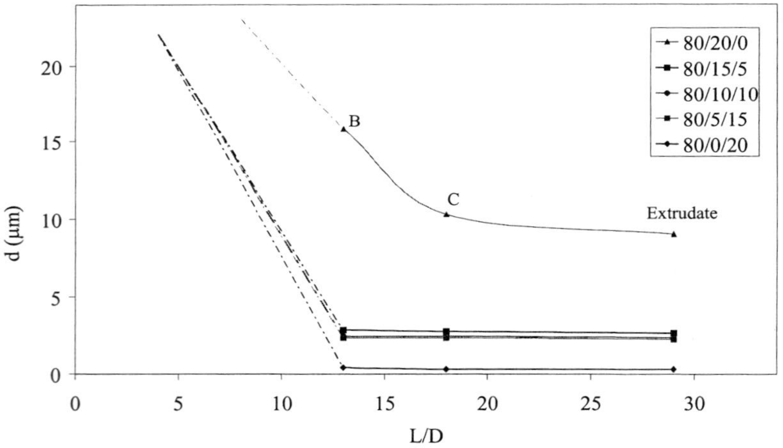

Fig. 3. MA graft content of EPM-g-MA (top) and of average rubber particle size (d) (bottom) along screw axis during production of 80/20-x/x (w/w/w) PA-6/EPM/EPM-g-MA blends.[7]

Samples were taken along the screw axis during the production of PA-6/EPM/EPM-g-MA blends.[7] For all compatibilised blends the reaction between PA-6 and EPM-g-MA occurs in the first kneading zone, resulting in a MA graft conversion of 80% (Figure 3; top). With the exception of the non-compatibilised blend, the dispersion of the rubber phase occurs also extremely fast, i.e. from milimeter granulate to sub-micrometer particles within the first kneading zone (Figure 3; bottom). Another extruder lay out with 6 sampling devices in series at the first kneading zone has shown that the PA granulate melts over an axial distance of only a few centimetres and both chemical and morphological development occur over the same distance. The actual dispersion process seems to be rather chaotic,[8, 9] which confirms the observations by Scott and Macosko:[20] very large domains, sheets, elongated threads, threads breaking up into droplets and small droplets are all observed in the same sample. Further along downstream, the PA-6 graft content, the residual MA content and the average rubber particle size do not change significantly. However, some last large rubber particles are finely dispersed along the screw axis, resulting in a decrease in the particle size distribution, and also degradation of the PA-6 matrix occurs.[10]

PBT/E-MA-GMA Blends

In the case of PBT/E-MA-GMA blends produced in the melt the main reaction is the opening of the epoxide rings of E-MA-GMA as a result of attack of PBT carboxylic end groups (PBT-COOH), resulting in ester linkages (Figure 4: top: first reaction).[11] This reaction is not that fast and the epoxide conversion is far from complete, but it still predominates over the similar reaction between PBT hydroxyl end groups (PBT-OH) with rubber epoxide[21] and over PBT-OH/E-MA transesterification (Figure 4: bottom). The latter reaction does occur in non-compatibilised PBT/E-MA blends though, but is very slow and the degree of conversion is too low to obtain a fine rubber dispersion.[12, 22] In addition to the graft reaction, also crosslinking of the E-MA-GMA rubber occurs (Figure 4: top: second reaction), initiated by the secondary hydroxyl group actually formed during graft formation.

Selective extraction with trifluoroacetic acid allows the determination of the non-grafted PBT content. Subsequent extraction with chloroform yields the amount of free E-MA-GMA. The remaining residue consists of both E-MA-GMA and PBT. Using m-cresol, a common solvent for PBT and E-MA-GMA, at 130 °C it was shown that this residue is crosslinked. Hydrolysis with t-butylammonium hydroxide shows that the crosslinking is not due to HOOC-PBT-COOH chains acting as a crosslinker. Table 3 gives some typical results obtained for PBT/E-MA/E-MA-GMA blends produced in an extruder.[12-14] Increasing the GMA content of E-MA-GMA clearly results in higher PBT graft contents, but also in increased crosslinking of the rubber phase.

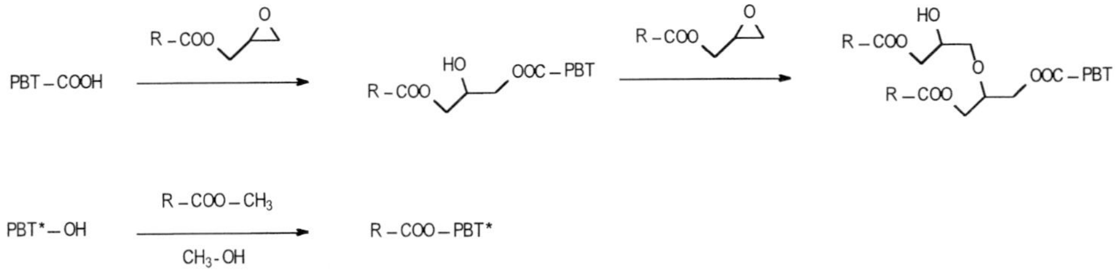

Fig. 4. Reaction scheme illustrating in-situ compatibilisation of PBT/E-MA-GMA blends.[11]

Table 3. Characteristics of 80/20 (w/w) PBT/E-MA-GMA blends produced on a co-rotating twin-screw extruder (280 °C, 200 rpm, 10 kg/hr) [12-14].

E-MA-GMA	free PBT (wt.%)	free rubber (wt.%)	molecular weight between crosslinks (kg/mol)[a]	average particle size (μm)	interfacial copolymer density (chain/nm^2)
0 wt.% GMA[b]	80	18	3.8	0.97	0
1 wt.% GMA	78	5.7	1.8	0.26	0.044
2 wt.% GMA	74	0.68		0.26	0.11
8 wt.% GMA	68	4.3	1.1	0.12	0.19
8 wt.% GMA (48% modified[c])	71	0.23	0.58	0.21	0.27
E-MA/E-MA-GMA (10/10; w/w)	72	5.0	1.7	0.17	0.21

a: of dispersed rubber phase determined with solid-state ^{1}H NMR relexation time measurements; b: E-MA copolymer; c: E-MA-GMA modified with p-*t*-butylbenzoic acid.

The morphology of PBT/E-MA-GMA blends is usually a fine sub-micrometer dispersion. However, the rubber particles are somewhat irregular in shape and do not have a smooth interface, which is probably due to the crosslinking of the E-MA-GMA phase.[11] Using Atomic Force

Microscopy in the electrostatic resonance mode it is shown that in the case of PBT/E-MA-GMA blends the crosslinking reaction is indeed initiated at the interface by the graft reaction and then proceeds towards the centre of the rubber particles.[14] In blends of PBT with acid-modified E-MA-GMA the crosslinking occurs homogeneously throughout the rubber particles, since the modified rubber already contains the secondary hydroxyl groups. Still, even when this modified E-MA-GMA is used, a sub-micrometer rubber dispersion is obtained.

Samples were taken along the screw axis during PBT/E-MA-GMA blend production.[14] The formation of PBT grafts proceeds relatively slowly and continues to the end of the extruder. The crosslinking of the rubber phase also continues to the end of the extruder. The dispersion of the rubber phase seems to follow a two-stage process, viz. first a fast dispersion to micrometer scale, followed by a final dispersion to sub-micrometer scale (Figure 5). Characterisation of samples obtained via batch kneader mixing at 250 °C as a function of time showed that during short

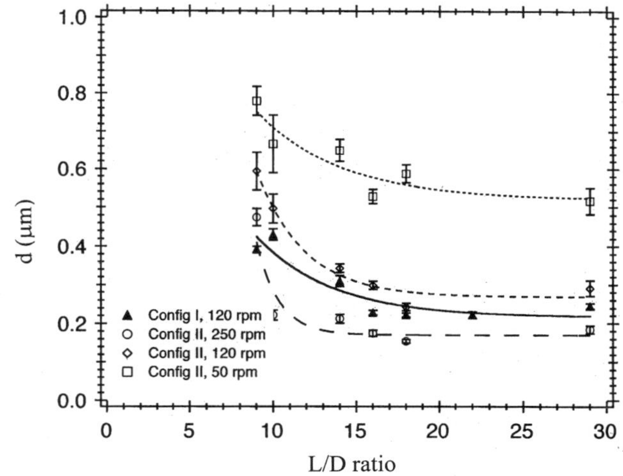

Fig. 5. Average rubber particle size (d) along screw axis during production of 80/20 (w/w) PBT/E-MA-GMA blends.[14]

mixing times (< 30 sec) a very fast dispersion occurs, but hardly any grafting.[12, 13] At intermediate mixing times the small rubber particles are grafted. At prolonged mixing times (> 15 min), rubber particles seem to be linked by HOOC-PBT-COOH chains. Increasing the screw speed enhances clearly the dispersion in the physical regime.

A plot of the average particle size versus the free PBT content shows a fair correlation for a large set of PBT blends produced in both a batch kneader and an extruder (Figure 6).[12-14] This suggests strongly that graft formation is limited by the available interface. The interfacial graft density can be calculated by combining the PBT graft content and the rubber particle size. This quantity varies by an order of magnitude and is dependent on all the experimental variables (PBT molecular weight and carboxylic end group content, rubber epoxide content, blend composition, processing temperature and screw speed).

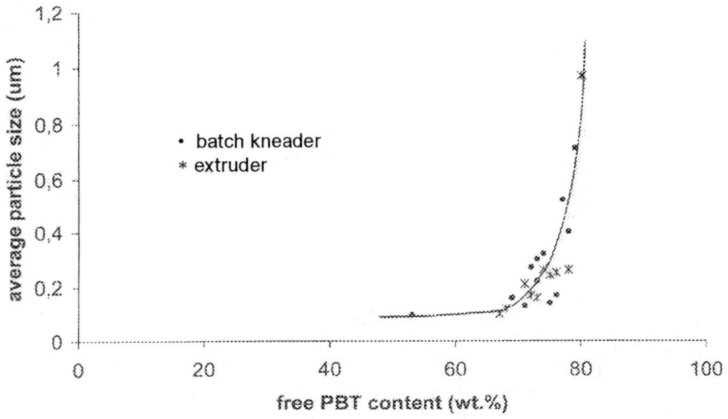

Fig. 6. Average rubber particle size as a function of free PBT content for PBT/E-MA/E-MA-GMA blends produced in a batch kneader and an extruder.[12-14]

Conclusions

Reactive blending of PA-6/EPM-g-MA and PBT/E-MA-GMA blends is obviously a complex process with continuous and reciprocal interactions between chemistry, morphology and rheology. In spite of the facts that the chemical structures formed in these two types of blends differ (PA-6: cyclic imide; PBT: ester), different side-reactions take place (PA-6: chain scission; PBT: rubber crosslinking) and the reaction rates are also quite different (PA-6: very fast; PBT: rather slow), general conclusions may nevertheless be drawn. First, the physical dispersion, governed by the viscosity (ratio) and the interfacial tension, results in the rapid formation of a fine rubber dispersion, i.e. from milimeter to micrometer scale within a few seconds. This results in a decrease of the diffusion distance between the polycondensate chains and the rubber interface by a factor of 10^3 and an increase of the available interface by a factor of 10^9, enhancing dramatically the rate of interfacial reactions. In the case of the PA-6/EPM-g-MA blends, the graft reaction is so fast that the interface is immediately covered with PA-6 chains. This results in a fast reduction of the interfacial tension and in prevention of coalescence and, hence, in an immediate further refinement of the dispersion. These phenomena occur so quickly that this sequence cannot be experimentally observed. In the case of the PBT blends the graft reaction is much slower and the subsequent morphology refinement takes place at a lower rate, which does allow the distinction between a physics- and a chemistry-controlled dispersion regime.

[1] L.A. Utracki (Ed.), *"Encyclopeadic Dictionary of Commercial Polymer Blends",* ChemTec Publishing, Toronto 1994.
[2] D.R. Paul and C.B. Bucknall, *"Polymer Blends",* John Wiley & Sons, New York 2000.
[3] S. Datta, D.J. Lohse, „*Polymeric Compaptibilisers*", Hanser Publishers, Munich 1996.
[4] C. Koning, M. van Duin, C. Pagnoulle, R. Jérome, *Progress Pol. Sci* .**1998**, *23*, 707.
[5] M. van Duin, M. Aussems, J.M.R. Borggreve, *J. Polym. Sci, Part A, Polym. Chem.* **1998**, *36*, 179.
[6] M. van Duin, R. Borggreve, "*Blends of Polyamides with Maleic Anhydride containing Polymers: Interfacial Chemistry and Properties*", in "*Reactive Modifiers for Polymers*", S. Al-Malaika (Ed.), Blackie Academic & Professional, London, 1997, ch.3.
[7] A.V. Machado, J.A. Covas, M. van Duin, *J. Polym. Sci., Polym. Chem.***1999**, *37*, 1311.
[8] A.V. Machado, J.A. Covas, M. van Duin, *J. Appl. Polym. Sci.* **2001**, *80*, 1535.
[9] M. Walet, M. van Duin, A.V. Machado, J.A. Covas, "*Morphology Development in PA-6/EPM-g-MA Blends*", poster on Polymer Processing Society 15, 's-Hertogenbosch 1999.
[10] J.A. Covas, A.V. Machado, M. van Duin, *Adv. Polym. Technol.* **2000**, *19,* 260.
[11] P. Martin, J. Devaux, R. Legras, M. van Gurp, M. van Duin, *Polymer* **2001**, *42*, 2463.
[12] P. Martin, J. Devaux, R. Legras, L. Leemans, M. van Gurp and M. van Duin, "*Reactive Compatibilisation of blends of PBT with Epoxide-containing Rubber. Part 1: The Influence of Processing Parameters*", submitted to *Polymer.*
[13] P. Martin, J. Devaux, R. Legras, L. Leemans, M. van Gurp and M. van Duin, *"Reactive Compatibilisation of blends of PBT with Epoxide-containing Rubber. Part 2: The Effect of Concentrations of Reactive Rubber*", submitted to *Polymer.*
[14] P. Martin, „*Chemistry, Morphology and Properties of PBT/Epoxide-containing Rubber Blends*", PhD thesis, Louvain-la-Neuve 2002.
[15] A.V. Machado, J.A. Covas, M. van Duin, *J. Appl. Polym. Sci.* **1999**, *71*, 135.
[16] M. Tessier, E. Maréchal, *J. Polym. Sci., Polym. Chem.* **1988**, *26A*, 2785.
[17] P. Maréchal, *"Comprehensive Study of Polyamide-6/Rubber Blends*", PhD thesis, Louvain-la-Neuve, 1994.
[18] B. de Roover, *« Etude des Mélanges constitués de Poly(m-xylylène adipamide) et de Polypropylene fonctionnalisé par l'Anhydride Maléique »,* PhD thesis, Louvain-la-Neuve, 1994.
[19] Y. Takeda, D.R. Paul, *J. Polym. Sci., Polym. Chem.* **1992**, *30,* 1273.
[20] C.E. Scott, C.W. Macosko, *Polym. Bulletin* **1991**, *26,* 341.
[21] Y.J. Sun, G.H. Hu, M. Lambla, H.K. Kotlar, *Polymer* **1996**, *37,* 4119.
[22] K.G. Gravalos, J.K. Kallitsis, N.K. Kalfoglou, *Polymer* **1995**, *36,* 1393.

Functionalization of Polyolefins by Reactive Processing: Influence of Starting Reagents on Content and Type of Grafted Groups

Elisa Passaglia,[1] *Serena Coiai,*[2] *Mauro Aglietto,*[3] *Giacomo Ruggeri,*[3] *Marco Rubertà,*[3] *Francesco Ciardelli**[1,3]

[1] CNR ICCOM Sezione di Pisa, Dip.Chimica e Chim. Ind.le, Università degli Studi di Pisa, Via Risorgimento, 35, 56126 Pisa, Italy
[2] Scuola Normale Superiore, Piazza dei Cavalieri, 7, 56100 Pisa, Italy
[3] Dipartimento di Chimica e Chimica Industriale, Università degli Studi di Pisa, Via Risorgimento, 35, 56126 Pisa, Italy

Summary: Previous studies carried out in our laboratory on the functionalization of ethylene polymers (homo and copolymers) through free radical processes, using mechanical mixers and/or extruders with diethylmaleate (DEM) and dicumyl peroxide (DCP), have shown that the presence of branched α-olefin blocks favours degradation versus functionalization and crosslinking whereas these last are more effective with linear ethylene blocks.
In this contest the present paper reports about attempts to extend the same processes to propylene polymers by benefitting of the indications of previous results. In particular the one step functionalization of EPM with two or more different groups was carried through the use of different monomers. Moreover the procedure was extended to polypropylene thus obtaining a significant functionalization degree flanked by a remarkable decrease of molecular weight.

Keywords: ethylene-propylene rubber (EPM), functionalization, FT-IR, melt processing, Poly(propylene) (PP)

Introduction

The free radical initiated functionalization of polyolefins (PO) is a method largely used from an industrial point of view to prepare modified materials used as chemical coupling agents to increase the polymer interfacial interactions with inorganic fillers and the miscibility with polar polymers as well as to produce polyolefins with improved adhesion and dyeability.
In the last two decades, thanks to the studies concerning the PO radical grafting,[1,2,3] mainly of maleic derivatives, a large number of results were reported and the commercial use of functionalized PO in polymer blends was substantially increased.
Modifying reactions promoted by free radicals, derived from thermal decomposition of initiator, and carried out on the molten material in extruders or mechanical mixers, are

 DOI: 10.1002/masy.200350813

versatile and easy to apply. They allow to obtain functionalized PO macromolecules by grafting to the PO backbone a limited amount (generally around 1% mol) of functional groups used for a growing number of industrial applications. On the other side the formation of extremely reactive not very selective free radical species brings to the contemporary occurrence of side reactions like degradation, chain extension and crosslinking. The modulation of the feed composition (monomer/radical initiator ratio)[4] was used in order to promote the desired reaction while suppressing the side reactions.

Also the functionalization degree (FD) depends both on the structure of the polyolefin and on the feed composition. The presence of isolated α-olefin units favours higher FD values due to the easier formation of macromolecular free radicals by hydrogen abstraction from tertiary C-atoms (LLDPE, VLDPE, EPR). Also the presence of double bonds increases FD due to a contribution of the "ene" reaction between these double bonds and the monomer. The polyolefins, which are degraded by chain scission, such as PP or SEBS, show the lowest FD.[4,5]

In previous papers[6] starting from the above mentioned results it was possible to find a proper feed composition range within which the FD was linearly dependent on functionalizing monomer concentration according to the equation:

$$FD = k_g \left(\frac{2 f k_D [DCP]_0}{k_t} \right)^{1/2} \cdot [DEM] \qquad (1)$$

where k_g = rate constant of the grafting reaction, k_D and k_t are the rate constant of initiator decomposition and termination reactions; [DCP] is the molar concentration dicumyl peroxide, the radical initiator, and [DEM] is the molar concentration of diethyl maleate, the functionalizing monomer. In proper composition range both degradation and crosslinking were negligible in respect to the grafting (functionalization) reaction. Thus the evolution of the process could be very well simulated by a mathematical kinetic model validated by DSC measurement of the heat evolution *vs* time.[7]

In order to further develop this very convenient process in the direction of producing functionalized propylene polymers, we have investigated in this paper the effect of monomer structure and feed composition on the reaction behaviour in the attempt to identify the best conditions for introducing different functional groups in one step and to extend the process also to propylene homopolymers.

Experimental Part

Isotactic Polypropylene (iPP) homopolymer having a melt flow rate (MFR) of 0.24 g/10 min was supplied by Borealis BE 50/E1271; poly(diethyl fumarate) (PDEF) was prepared by radical initiator (AIBN) as reported in the literature.[8]

Dicumyl peroxyde DCP (Sigma-Aldrich) was used as provided without further purification. Azobisisobutyronitrile AIBN (Sigma-Aldrich) and maleic anhydride MAH were recrystallized from acetone and benzene respectively. Diethylmaleate (DEM), Sigma-Aldrich product, was used after distillation under reduced pressure (0.33 mmHg, 53°C). The cetyl maleate monomer was a commercial product about 98% pure as checked by gas-chromatography.

Modified EPM samples were prepared using a Brabender Plastograph PL2100 mixer, torque and temperature data have been acquired by Brabender Mixing software WinMix ver.1.0. Two different temperatures: 180°C in the case of EPM1 run and 230°C in the case of EPM2 run were used and the reaction time was 10 minutes and rpm=30. The polymers recovered were extracted with boiling acetone (see Table 1).

The modified PP samples were prepared with the same procedure by melt-mixing the components at a temperature of 180°C and rotor speed of 50 rpm. 20 g of iPP were introduced into the Brabender chamber. After complete fusion, functionalizing molecules and peroxide were added to the molten bulk. The reaction was carried out for 25 min and then stopped. The recovered polymer was sequentially extracted with boiling acetone and toluene. In each case, the residues were dried until constant weight (see Table 2).

FT-Infrared spectra were performed with a Fourier Transform Spectrometer "Perkin Elmer FT-IR 1760-X" on films made by compression moulding. Proton magnetic resonance spectra were performed with a spectrometer "Varian Gemini 200 MHz"; the spectra of functionalized polymers were made in solution of 1,2,4-trichlorobenzene, using 1,1,2,2-tetrafluorodideuteroethane as internal standard, at the temperature of 90°C.

Results and Discussion

One Step Multifunctionalization of EPM

The contemporary use of DEM and MAH as monomers allowed to produce EPM macromolecules containing different functional groups;[9] however some interference were observed between the two monomers leading to a different composition in respect to the feed and low monomer conversion at least for DEM.[8] In order to attempt the preparation of a similar multifunctionalized EPM and avoid the problem of using a mixture of comonomers with different compatibility in the EPM melt, a monoester of maleic acid was used. The compatibility problem with the hydrophobic PO was solved by the use of a long alkyl chain (cetyl) which provided a substantial homogeneity of the reaction mixture.

Reaction samples were prepared starting from an ethylene-propylene copolymer (EPM) characterized by 22.7% propylene units with DCP as radical initiator (Table 1).

Table 1. Functionalization runs of EPM with cetyl maleate.

Runs	cetyl maleate monomer content %by mol	DCP peroxide content % by mol
EPM1[a)]	0.88	0.06
EPM2[b)]	0.88	0.06

[a)] the run has been carried out in the Brabender mixer at 180°C, rpm=30, t=10min
[b)] the run has been carried out in the Brabender mixer at 230°C, rpm=30, t=10min

The IR spectra have been acquired on the samples after the reaction and washed with boiling acetone. The presence (Figure 1) of four absorption peaks, partially overlapped, between 1700÷1900 cm^{-1}, related to carbonyl stretching, suggests that different groups are grafted to the polyolefin chain.

According to literature data the vibrations in the C=O stretching region can be assigned as follows: peak at 1735 cm^{-1}: ν_{co} ester group, peak at 1710 cm^{-1}: ν_{co} acid group, peaks at 1780 cm^{-1} and 1865 cm^{-1}: ν_{co} symmetrical and ν_{co} asymmetrical of the anhydride group.

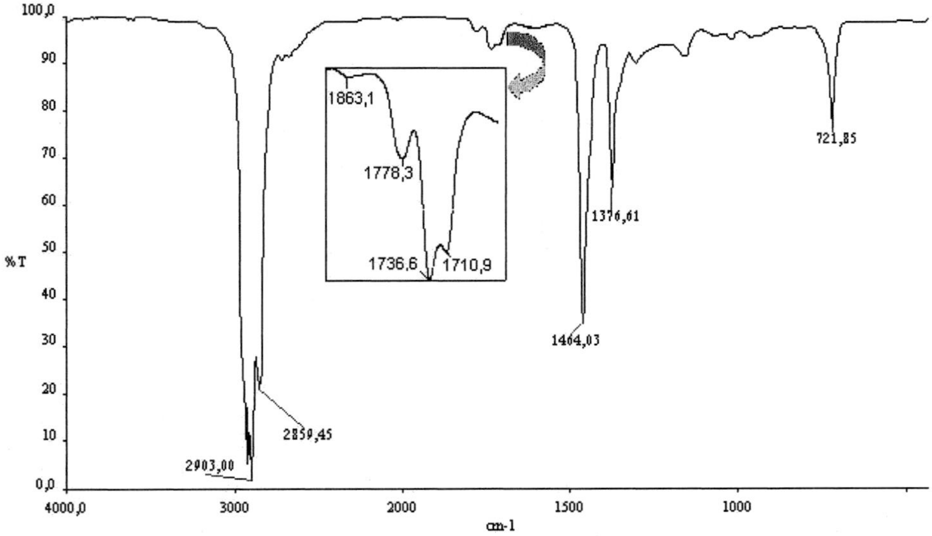

Fig. 1. IR spectrum of EPM2 run.

One can suppose that the three different types of groups reported in Figure 2 are grafted to the EPM chains. The presence of the anhydride group (B) suggests the possible intermediate formation of succinic acid groups (C) after the insertion which can undergo cyclization at high temperature. It is however not possible to exclude that maleic acid and maleic anhydride are formed from the maleic acid cetyl monoester before the grafting.

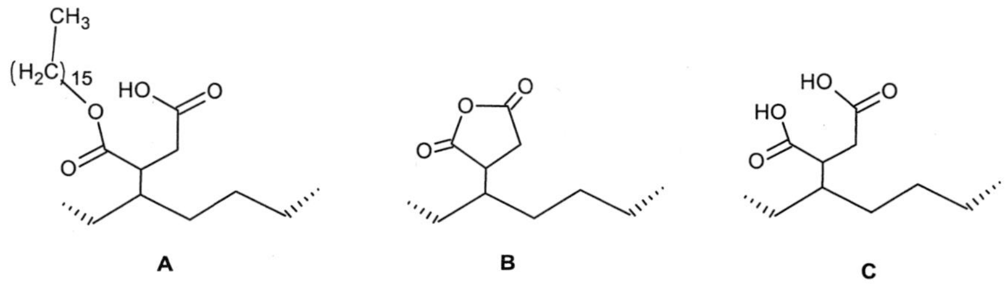

Fig. 2. Functional groups inserted onto polyolefin chain (EPM).

Indeed the thermogravimetric analysis (TGA) of the mono cetyl maleate carried out under an atmosphere of nitrogen consists of two steps. Starting from 100°C we observed a weight loss of 63% wt connected probably with the development of cetyl alcohol at T_{max} at 237°C. As this last temperature is close to that of reaction, the presence of the anhydride groups onto the polymer backbone can be associated with a similar mechanism.

The deconvolution of the EPM2 spectrum (Figure 3) has been performed to detect the functionalization degree and type of grafted groups.

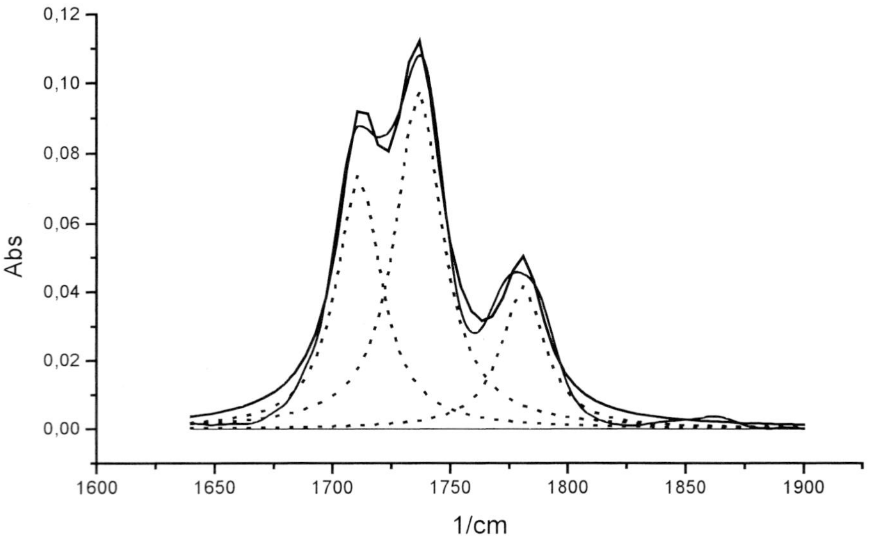

Fig. 3. Deconvolution of the carbonyl stretching region of the EPM2 IR spectrum.

The FD value referred to diethysuccinate groups were found to be $FD_{(SAH)} \cong 0.05\%$ mol. For the same sample ^{1}H-NMR signal at 4.15 ppm (-O-CH_2-) of the ester groups grafted onto the polymer chain, gives FD $\cong$ 0.2% mol. By considering the total amount of the primary radicals (0.12 to respect 100 monomeric units of the polymer) it is possible to calculate the value of the efficiency that is slightly larger than 2. Moreover, on the basis of the amount of the monomer in the feed, conversion of about 30% was calculated.

These results are quite in agreement with those obtained in the case of the functionalization reaction with diethylmaleate thus confirming preliminarily the same mechanism for both the functionalizing monomers.

One Step Multifunctionalization of PP

The functionalization reactions of iPP was carried out, above its melting temperature, with maleic acid derivatives and radical initiator under conditions granting low extent of side reactions for ethylene polymers.[6]

Two series of experiments have been performed. In the former (runs PP03-09, Table 2) diethyl maleate (DEM) and dicumyl peroxide (DCP), and in the latter, a mixture of DEM/maleic anhydride (MAH) 2:1 mol and DCP (runs PP10-11 Table 2) were used.

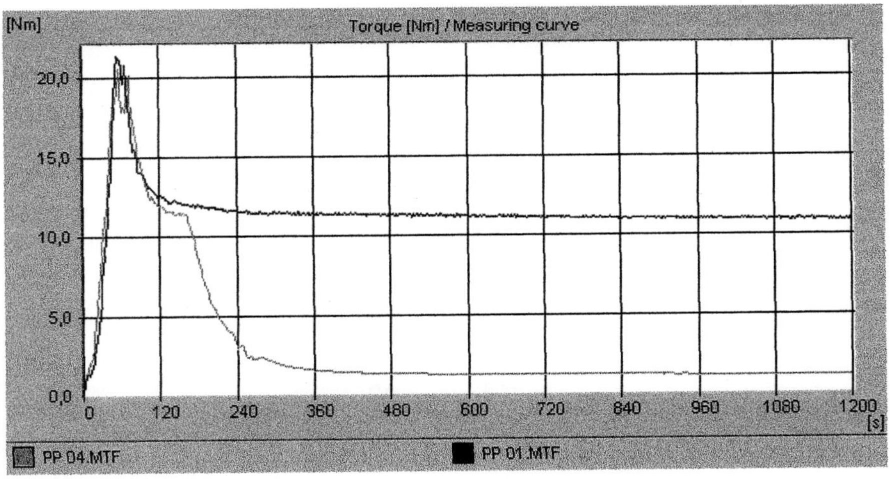

Fig. 4. Torque *vs* time for the runs PP01 (Polypropylene treated in Brabender at 180°C without peroxide and functionalizing monomers) and PP04 (PP with DCP and DEM).

The mixing torque graph (Figure 4) is similar for all the samples independently of the monomers/DCP fed ratio in the Brabender chamber with a sharp decrease of the torque values after addition of the monomer-DCP solution to the molten polymer. The described behaviour can be correlated to the viscosity decrease of the molten polymer caused by degradation reactions. In fact, in the case of isotactic polypropylene (iPP) the treatment with peroxide initiators in the melt gives to the polymer degradation through β-scission reactions.[10]

The samples were analysed by FT-IR after washing with boiling acetone before the functionalization degree (FD) determination. The standardization curve was obtained by preparing mixtures of known composition of iPP with poly(diethyl fumarate) (PDEF), according to the method usually adopted in our laboratory.[11]

Bands at 1736 cm^{-1} (A1) and at 1166 cm^{-1} (A2) have been taken into consideration. By plotting the peak areas ratio (A1/A2) versus the composition of the mixtures iPP/PDEF, ranging from 0.15% mol to 0.85% mol of PDEF, a good linear correlation has been obtained. This suggests that the A1/A2 band areas ratio can be safely used also for detecting FD of iPP samples functionalized by DEM grafting (PP-g-DES). The FD values range from 0.2% to 1% mol (Table 2) with DEM conversion values from 7 to 30%; these values are lower than those obtained in the case of polyethylenes (PE) functionalization probably due to the occurrence of secondary reactions with particular reference to the degradation process.

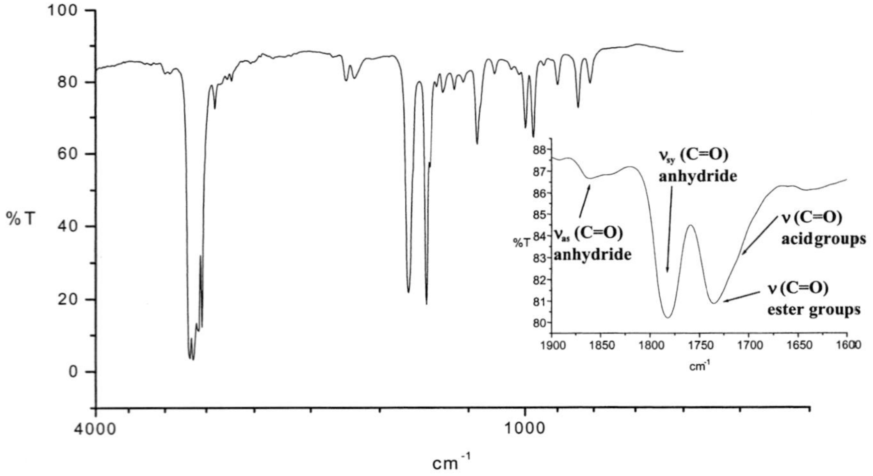

Fig. 5. FT-IR spectrum for PP10 sample.

With a DEM/MAH mixture FD determination is more complex as different types of groups are grafted on iPP chains. FT-IR spectra present superposition of diagnostic bands (ester, anhydride, acid obtained by hydrolysis of anhydride groups) (Figure 5).

The deconvolution of the spectra carried out by a non-linear fitting of experimental points allows to obtain separately the diagnostic band areas (Figure 6).

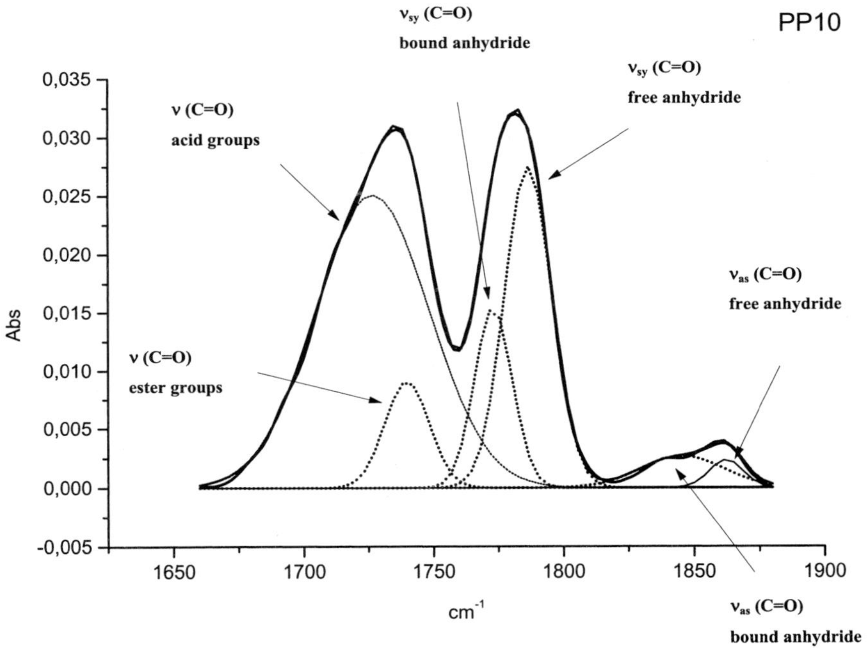

Fig. 6. Deconvolution of PP10 IR spectrum ranging from 1660 cm^{-1} to 1880 cm^{-1}.

The knowledge of the integral absorptivity (ε) for all the bands used, allows the evaluation of the FD values (Table 2) relative to DES, SAH and SA groups have been calculated:

$$FD_{(DES)} = \frac{\overline{\varepsilon_{1166}}}{\overline{\varepsilon_{DES}}} \cdot \frac{\overline{A_{DES}}}{A_{1166}} \qquad (2)$$

$$FD_{(SAH,SA)} = \frac{\overline{\varepsilon_{1166}}}{\varepsilon_{SA}} \cdot \frac{\overline{A_{SA}}}{A_{1166}} + \frac{\overline{\varepsilon_{1166}}}{\varepsilon_{SAH}} \cdot \frac{\sum_i \overline{A_{SAH,i}}}{A_{1166}} \qquad (3)$$

SAH = succinic anhydride grafted groups

SA = succinic acid grafted groups

Table 2. FD values of functionalized iPP samples.

Runs	DEM % mol	MAH % mol	DCP % mol	M/DCP[a) mol/mol	FD_{DEM} [b)]	FD_{MAH} [b)]	DEM conv.[c)]	MAH conv.[c)]
PP03	2.44	-	0.08	30.5	0.17	-	7.0	-
PP04	4.88	-	0.08	61.0	0.34	-	7.0	-
PP05	4.88	-	0.16	30.5	0.86	-	17.6	-
PP06	4.88	-	0.23	21.2	1.04	-	21.3	-
PP07	1.22	-	0.08	15.2	0.20	-	16.4	-
PP08	2.44	-	0.16	15.2	0.58	-	23.8	-
PP09	2.44	-	0.23	10.6	0.73	-	29.9	-
PP10	1.90	0.95	0.08	35.6	0.10	0.65	5.3	65.3
PP11	1.90	0.95	0.16	17.8	0.09	0.86	4.7	90.5

a) monomer/DCP molar ratio
b) number of inserted groups to respect 100 of monomeric units of the polymer
c) conversion (% by mol)

MAH seems more reactive than DEM; probably for its capability to give oligomer chains grafted onto iPP macromolecules.[9] Moreover, the mixing solid monomer MAH with liquid monomer DEM could favour its penetration into the molten polymer and promote the MAH grafting reactions. After the functionalizing reactions all samples are completely soluble in toluene (insoluble fractions <1% wt) and therefore it is possible to assume that changes of MW and structure of PP have taken place (Table 3). Indeed about 80% wt of starting iPP and also of iPP treated in Brabender mixer without additives is insoluble under the employed conditions.

These data show that the treatment of iPP in the melt with peroxide and maleate molecules yields good functionalization degrees, but at the same time the chains β-scission process, due to the presence of many tertiary carbon atoms, is responsible for extensive degradation and decrease of MW. The efficiency values of the peroxide for iPP is lower than in the case of polyethylene and EPM copolymer (Table 4) thus confirming the occurrence of secondary reactions reducing the macroradicals amount which reacts with the functionalizing monomers.

Table 3. Solubility and molecular weight of functionalized iPP samples.

Sample	DEM % mol	MAH % mol	DCP % mol	Residue toluene extraction % wt	[η] (dL/g)	MW_η
PP01[a)]	–	–	–	82.9	3.12	370000
PP02	–	–	0.08	1.1	1.12	100000
PP03	2.44	–	0.08	1.5	–	–
PP04	4.88	–	0.08	<1	1.20	110000
PP05	4.88	–	0.16	<1	–	–
PP06	4.88	–	0.23	<1	–	–
PP07	1.22	–	0.08	<1	–	–
PP08	2.44	–	0.16	<1	–	–
PP09	2.44	–	0.23	<1	–	–
PP10	1.90	0.95	0.08	2.5	1.25	120000
PP11	1.90	0.95	0.16	<1	–	–

a) Polypropylene treated in Brabender at 180°C without peroxide and functionalizing monomers

Table 4. Initiator efficiency in iPP, LLDPE and EPM polymer treatment with DEM and DCP as functionalizing agents.

Runs	DCP % mol	M/DCP[a)] mol/mol	Efficiency[b)]
PP04	0.08	61	2.1
PP05	0.16	30.5	2.7
PP06	0.23	21.2	2.3
E12[c)]	0.11	32.7	5.7
E16[c)]	0.22	16.4	3.9
EPR2-01[c)]	0.11	16.4	5.7
EPR2-08[c)]	0.05	36	9.7

a) monomer/DCP molar ratio
b) defined as the molar ratio between the functionalization degree and the amount of primary radicals[4]
c) data reported in the reference 4

Beside the grafting reaction, free radicals derived from the peroxide decomposition produces a remarkable degradation of PP chains under mixing conditions. Indeed the run PP02 shows that by treatment with the sole peroxide MWη goes from 370000 of the original polymer to 100000 of the processed PP.[6] The same effect occurs even in the presence of MAH or DEM (see runs PP04 and PP10) and is largely responsible of the increased solubility in toluene of the produced PP (Table 3). The average viscosimetric molecular weight (MWη) has been determinated by the Mark-Houwink equation.[12]

By successive self-nucleation and annealing DSC (SSA-DSC)[13,14] a distribution of melting points induced by SSA treatment has also been observed. PP01 sample has a melting peak centred at about 170°C. The PP02 sample and the functionalized samples showed lower values of Tm, around 165°C, and many peaks appear at lower temperatures probably related to the degradation by β-scission of PP macromolecules into shorter chains some of which having probably lower stereoregularity.

Final Remarks

The functionalization of different polyolefins in the melt in the presence of a free radical initiator and unsaturated monomers can be carried out with a significant control of the process by proper selection of the feed composition in relation to the starting PO structure.

In case of EPM the previously selected conditions to optimize the functionalization reactions against degradation and crosslinking can be successfully used to insert different functional groups either by starting with a mixture of monomers (DEM+MAH) or with a monoester of maleic acid. Under analogous feed composition polypropylene gives a significant functionalization but degradation cannot be avoided according to the fact that β-scission is not extensively hindered by the presence of maleate.

On the basis of these indications further study of the functionalization of PP is in due course with different reactive monomers.

Acknowledgement

Partial financial support by MIUR is gratefully acknowledged.

[1] G. Moad, *Prog. Polym. Sci.* **1999**, *24*, 81-142
[2] S. Al-Malaika, *"Reactive modifiers for polymers"*, Academic & Professional, London, UK, **1997**
[3] A.V. Machado, J.A. Covas, M. Van Duin, *Polymer* **2001**, *42*, 3649-3655
[4] E. Passaglia, M. Marrucci, G. Ruggeri, M. Aglietto, *Gazzetta Chimica Italiana* **1997**, *127*, 91-95
[5] E. Passaglia, S. Ghetti, F. Picchioni, G. Ruggeri, *Polymer* **2000**, *41*, 4389-4400
[6] F. Ciardelli, M. Aglietto, E. Passaglia, G. Ruggeri, *Macromol. Symp.* **1998**, *129*, 79-88
[7] E. Passaglia, P. Siciliano, F. Ciardelli, G. Maschio, *Polym Int.* **2000**, *49*, 949-952
[8] T. Otsu, O. Ito, N. Toyoda, S. Mori, *Makromol. Chem. Rapid Commun.* **1981**, *2*, 725-728
[9] E. Passaglia, L. Corsi, M. Aglietto, F. Ciardelli, M. Michelotti, G. Suffredini, *J.Appl.Polym.Sci.* **2002**, in press

[10] E: Borsig, M. Lazar, A. Fiedlerova, L. Hrčková, M. Rätzsch, A. Marcinčin, *Macromol Symp.* **2001**, *176*, 289-298
[11] M. Aglietto, R. Bertani, G. Ruggeri, F. Ciardelli, *Makromol Chem.* **1992**, *193*, 179-186
[12] J. Brandrup, E.H. Immergut, Editors "Polymer Handbook" Second Edition, IV-9, John Wiley & Sons New York, 1975
[13] M. L. Arnal, M. E. Matos, R. A. Morales, O. O. Santana, A. J. Muller, *Macromol. Chem. Phys.*, **199**, 2275 (1998)
[14] M. L. Arnal, V. Balsamo, G. Ronca, A. Sanchez, A. J. Muller, E. Canizales, C. Urbino de Navarro, *J. Therm. Anal. Cal.*, **59**, 451 (2000)

Reactive Compatibilization of Blends of PET and PP Modified by GMA Grafting

Mariano Pracella, Donatella Chionna*

Centro Studi sui Materiali Macromolecolari Polifasici e Biocompatibili, CNR,
Via Diotisalvi 2, Pisa 56126, Italy
E-mail: pracella@ing.unipi.it

Summary: The melt radical grafting of glycidyl methacrylate (GMA) onto isotactic polypropylene (PP) was carried out in Brabender internal mixer and the influence of reaction procedure, radical initiator concentration and addition of co-monomer (styrene) on the grafting efficiency was examined. The viscosity, the thermal behaviour and melt rheology of PP-*g*-GMA samples was then analysed as a function of grafted GMA content. Blends of poly(ethylene terephthalate) (PET) with PP and PP-*g*-GMA (5.2 wt% GMA), prepared in internal mixer, were characterised by SEM, DSC and melt viscosimetry. The morphological analysis of PET/PP-*g*-GMA blends (80/20, 50/50 w/w) pointed out a marked improvement of phase dispersion (with particle size of about 0.6 μm for 80/20 blend) and interfacial adhesion, as compared to non-compatibilized PET/PP blend. The results of mixing torque and thermal analysis supported the occurrence of in-situ compatibilization reaction between epoxy groups of GMA modified PP and carboxyl end-groups of PET in the melt.

Keywords: blends, functionalization, poly(ethylene terephthalate), polypropylene, reactive compatibilization

Introduction

In situ compatibilization of blends of polyolefins with polar polymers, as polyesters and polyamides, can be efficiently performed by suitable modification of polyolefin chains with reactive groups (i.e., carboxyl derivatives, epoxies, etc.) able to give rise, during melt blending, to chemical reactions with the functional groups of the polar component.[1] The formation of a graft copolymer between the polymer components causes a reduction of interfacial tension with an improvement of phase dispersion and promotes adhesion through interpenetration and entanglements at the polymer-polymer interface.

 DOI: 10.1002/masy.200350814

The free radical grafting of GMA onto polyolefins (PE, PP, EPR, etc) has been investigated by several authors in the recent years,[2-5] however only a few studies dealt with the compatibilization of blends of GMA modified polyolefins with Nylons or polyesters.[6-10] The use of GMA in reactive compatibilization presents several advantages as compared to other reactive monomers (acrylic acid, maleic anhydride, etc.),[10] due to the higher reactivity of epoxy group toward carboxyl, hydroxyl or amino groups, absence of water formation during the reaction and less environmental risk.

The present paper is aimed at examining the melt grafting procedures of GMA onto PP and the compatibilizing efficiency of the grafted polyolefin in blends with PET. The influence of reagents concentration and addition of a co-monomer (styrene) on the grafting degree and the properties of PP-*g*-GMA copolymers has been first analysed. Then, the morphology and the thermal and rheological behaviour of blends PET/PP-*g*-GMA with different composition have been studied and compared with those of unmodified PET/PP blends.

Experimental

Materials

Isotactic PP (Novolen 1100 N, Targor) with MFR =12 g/10 min (230 °C, 2.16 kg) and PET (SINCO Engineering) with I.V.= 0.74 dl/g were used throughout. Before use the materials were carefully dried under vacuum in order to reduce the moisture content: PET samples were heated up to 170 °C for 4 hours, PP was dried at 80°C. Glycidyl methacrylate (GMA, 97% purity) and styrene (STY, 99% purity) were purchased from Aldrich and used as received. Bis[1-(*tert*-butylperoxy)-1-methylethyl] benzene (BTP), dicumyl peroxide (DCP) and benzoyl peroxide (BPO), commercial products from Aldrich, were used as free radical initiators without further purification.

Grafting Procedures

The grafting was carried out at 190 °C in a Brabender Plasticorder internal mixer with sigma rotors rotating at 50 r.p.m., under nitrogen stream. The variation of torque moment during reaction was recorded against the time of mixing. Three different addition procedures were tested for the grafting experiments:[9] *A*) PP (pellets) was first introduced into the mixing chamber for 2 minutes in order to melt, then GMA and the initiator were added and the mixing continued for additional 5 minutes; *B*) PP (pellets), GMA and the initiator were pre-mixed at room temperature, then fed into the mixer at 190°C and mixed for 5 minutes; *C*) similar to the previous one, but using a limited amount of PP (10% of the total amount) ground into powder. To enhance the grafting efficiency of GMA, styrene was also used as a co-monomer, in molar ratios from 0.5 to 1.5 with respect to GMA.[3, 4]

The reaction products were purified by dissolution in refluxing xylene and precipitation into acetone at room temperature, under continuous stirring, in order to remove the unreacted monomers (GMA, STY), the homopolymerized GMA and GMA-STY copolymer (soluble in acetone). The precipitate, PP-*g*-GMA, was filtered, washed again with acetone and dried under a vacuum at 80°C overnight. The purified samples were then compression moulded into thin films and analysed by FTIR in order to determine the amount (in wt%) of GMA grafted onto PP. The solution viscosities of PP and grafted copolymers were measured at 135 ± 0.1 °C with an Ubbelhode viscosimeter in tetrahydronaphtalene. The molecular weight of the samples was calculated from intrinsic viscosity values using the relation reported by Danusso *et al.* for isotactic polypropylene.[11] The grafting conditions, the reagents concentration and the GMA content of PP-*g*-GMA copolymers examined are listed in Table 1.

Table 1. Grafting conditions, GMA content and molecular weight of PP-*g*-GMA copolymers.

Sample code	$[GMA]_g$ (wt%)	Grafting Procedure	BTP (wt%)	$[GMA]_i$ (wt%)	$[STY]_i/[GMA]_i$ (mol/mol)	M_v
PP	-	-	-	-	-	92000
PP-g-GMA1	0.4	B	0.2	5.7	-	41000
PP-g-GMA2	1.0	C	0.4	5.7	-	43000
PP-g-GMA3	2.7	B	0.4	5.7	1/1	32000
PP-g-GMA4	4.7	C	0.2	7.4	1/1	n.d.
PP-g-GMA5	5.2	C	0.4	7.4	1/1	25000

Blending and Characterization

Binary blends PET/PP and PET/PP-*g*-GMA5 with composition 80/20 and 50/50 w/w were prepared in the Brabender mixer (260 °C, 50 rpm, 7 min) under nitrogen flux, recording the torque moment until reaching of stationary conditions. Plain PET and PP were also processed in the mixer at 260 °C and 190 °C (50 rpm, 7 min.) respectively.

The morphology of the blends was examined by a Jeol T300 scanning electron microscope (SEM) on samples fractured in liquid N_2 and sputter-coated with Au. The size distribution of the dispersed phase in the blends was then analysed by measuring a consistent number of particles (200-300) for each sample by an image analysis programme (Scion Image). A polarizing optical microscope (Leitz Ortholux) equipped with hot stage was employed for analysing the phase structure in the blends. Thermal properties of the modified polyolefins and blends were examined with a Perkin Elmer DSC-II differential scanning calorimeter using a standard heating/cooling rate of 10 °C/min. Temperatures and heats of phase transitions were determined respectively from the maxima and areas of the crystallization and melting peaks. The degree of crystallinity was calculated from the values of melting enthalpy. Melt viscosities were measured at 180 °C by a CEAST capillary rheometer (L/D = 40) in the shear rate range 10^1–10^3 s^{-1}.

Results

GMA Grafting onto PP

The amount of grafted GMA in PP-*g*-GMA samples was determined from area ratio of FT-IR peaks at 1730 cm^{-1} (stretching of the carbonyl group of GMA) and at 2722 cm^{-1} (combination band), typical of the PP skeleton.[12] The latter peak was chosen as an internal reference. Quantitative measurements of grafted GMA were performed by using a calibration plot, as shown in Figure 1, obtained by the 1730/2722 peak ratios of PP-*g*-GMA samples with known GMA content, independently determined by a titration procedure in xylene solution.[13]

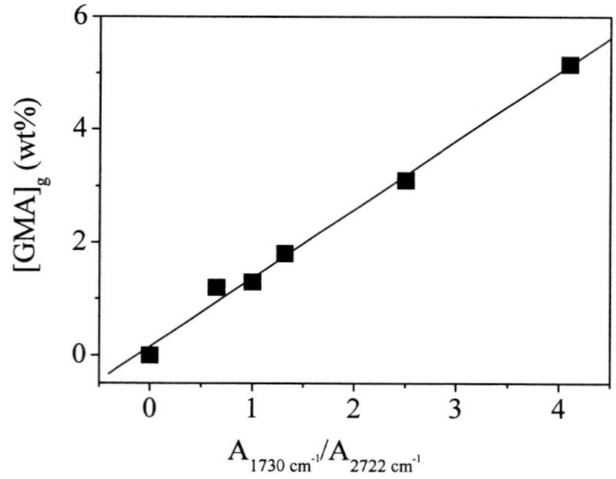

Fig. 1. Amount of grafted GMA vs. the ratio of FT-IR peak areas at 1730 and 2722 cm^{-1}.

The grafting degree of PP-*g*-GMA copolymers obtained by the various procedures *A*, *B* and *C*, with the same initial amount of GMA monomer ($[GMA]_i$= 5.7 wt%) and different content of peroxide (BTP), is reported in Table 2. In all cases the highest grafting degree was obtained by applying the procedure *C*. This can be ascribed to a better absorbency of monomer and initiator onto PP powder, giving rise to a more fast and complete reaction of polymer radicals with GMA. Among the three tested peroxides, BTP showed the highest grafting efficiency ($[GMA]_g$= 3.6 wt%) as compared to DCP ($[GMA]_g$= 2.4 wt%) and BPO ($[GMA]_g$= 1.0 wt%) in the same reaction conditions. The lower efficiency of BPO is likely due to its shorter half-lifetime.

Table 2. Amount of GMA grafted onto PP for different grafting procedures ($[GMA]_i$ = 5.7 wt%).

BTP content	Grafting Degree (wt%)		
(wt%)	A	B	C
0.2	0.1	0.4	0.6
0.4	0.3	1.5	2.3

The effect of amount of BTP on the grafting degree of copolymers is shown in Figure 2. The amount of grafted GMA increases with the peroxide concentration up to reach a maximum, corresponding to $[GMA]_g$= 4 wt%. For BTP contents > 0.5 wt%, degradation of PP and homopolymerisation reaction of GMA became prevailing, as indicated by the rapid decrease of torque, and thus of melt viscosity, following the grafting reaction. The grafting yield also increases with increasing the initial concentration of GMA and then level off at a value ($[GMA]_g$= 3-5 wt%) which depends on the amount of peroxide. Similar results have been reported for GMA melt grafted polypropylenes by using different radical initiators.[3,4]

According to literature reports [3-5], the addition of STY as co-monomer significantly promotes GMA grafting, owing to the higher reactivity of STY toward PP forming styryl-macroradicals, which in turn can easily react with GMA. As shown in Figure 3 for the grafting reaction in the presence of STY, with increasing the ratio STY/GMA (mol/mol) the grafting yield of GMA increases up to three-times as compared to the reaction without co-monomer. When the monomers ratio is higher than 0.5, the amount of the polymerised GMA exceeds that of grafted GMA. On the other hand, the addition of STY reduces the degradation of PP chains as evidenced (Figure 3) by the increase of mixing torque recorded at the end of the grafting process.[14] These samples were completely soluble in hot xylene indicating that no cross-linking takes place during the grafting reaction, owing to possible combination of PP-styryl macroradicals, as suggested by Wong and Baker.[12]

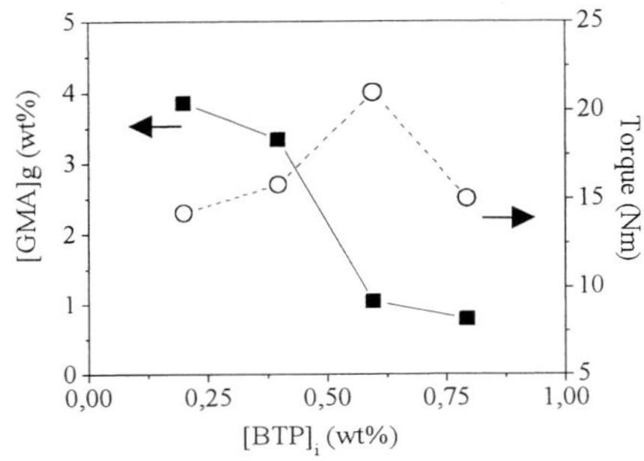

Fig. 2. Amount of grafted GMA and torque variation as a function of peroxide concentration [BTP] ([GMA]$_i$ = 5.7 wt%, procedure *B*).

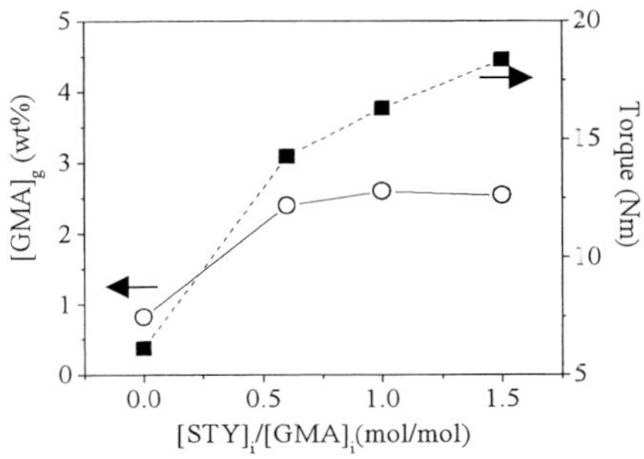

Fig. 3. Amount of grafted GMA and final mixing torque as a function of molar ratio [STY]/[GMA] ([GMA]$_i$ = 5.7 wt%, [BTP] = 0.2 wt%, procedure *B*).

Characterization of PP-g-GMA Copolymers

PP-*g*-GMA copolymers with different GMA content were analysed by viscosimetry, DSC and melt rheometry. Viscosity measurements of polymer solutions in tetrahydronaphtalene pointed out that the average molecular weight (M*v*) decreases with increasing the grafting degree from ca. $9.2*10^4$ for PP to $2.5*10^4$ for PP-*g*-GMA5 (Table 1). Samples with smaller GMA content (0.4-1.0 wt%) display a variation of M*v* comparable with that of samples with higher grafting degree, indicating that the decrease of molecular weight is mainly due to the presence of initiator at the reaction temperature, which induces large chain scission effects.[6]

Table 3. DSC data for PP-g-GMA samples on cooling and 2nd heating run (10°C/min).

Sample	T_c (°C)	ΔH_c $^{a)}$ (J/g)	T_m (°C)	ΔH_m $^{a)}$ (J/g)	X_{cr} $^{b)}$ (%)
PP	111.6	99.5	158.6, 165.5	104.5	57.5
PP-*g*-GMA1	112.1	99.5	160.0, 165.2	100.3	53.1
PP-*g*-GMA2	111.8	100.3	158.0, 165.1	105.3	55.8
PP-*g*-GMA3	119.5	102.4	160.6	108.7	57.5
PP-*g*-GMA4	120.1	95.3	162.8	102.8	54.5
PP-*g*-GMA5	119.1	87.4	160.6	86.9	46.0

a) Per gram of polymer component
b) $X_{cr} = \Delta H_m / \Delta H_m^o$ with ΔH_m^o = 188.9 J/g for PP

Crystallization temperatures (T_c), melting temperatures (T_m) and relevant heats of transition (ΔH_c , ΔH_m) of PP homopolymer and PP-*g*-GMA samples are summarized in Table 3.

Compared with pure PP, the crystallisation and melting behaviour of PP-*g*-GMA is markedly affected by the content of grafted GMA; in particular, with increasing the amount of grafted GMA T_c is shifted to higher temperature (from about 112 °C to 120 °C). These findings suggest that the nucleation process of PP crystals is enhanced in the GMA grafted polymers. Similar effects have also been observed for ethylene-propylene copolymers and PP modified with GMA[15,16] and they were ascribed to the nucleating capability of grafted groups. However, it must be taken into account that other factors, such as the presence of reaction impurities, molecular chain fractionation, etc., could significantly affect the crystallization of these copolymers.

A decrease of the overall crystallization (and melting) enthalpy is observed mainly for the sample with higher grafting degree (PP-*g*-GMA5), which implies a lower crystallinity degree due to the presence of GMA groups. The melting and reorganization phenomena of crystals are also influenced by the grafting degree: PP homopolymer and copolymers with lower GMA content display characteristic double melting peak (ca. 158-165 °C), whereas a single melting peak is observed for copolymers with GMA content > 1 wt%. For these latters an additional endothermal peak appears at lower temperature (ca. 147 °C), whose intensity increases with the amount of grafted GMA. Most likely, this peak could be ascribed to the presence of less perfect crystals of low molecular weight and/or GMA richest polypropylene chains.

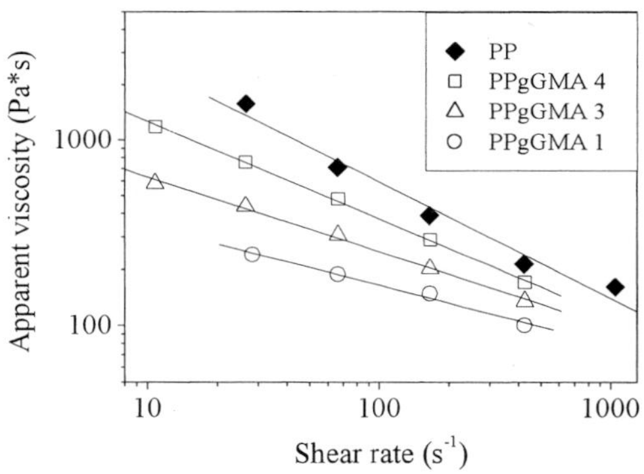

Fig. 4. Apparent melt viscosity of PP and GMA grafted PP samples vs. shear rate.

The rheological data, reported in Figure 4, show that the melt viscosity of the modified polylpropylenes is lower than that of homopolymer, as a consequence of chain scission phenomena during the grafting process.[8] However, the melt viscosity of PP-*g*-GMA samples increases with increasing the GMA grafted content and this could be ascribed to the effect of polar interactions between GMA groups in the copolymer chains, even if the occurrence of cross-linking between grafted chains can not be excluded at higher GMA contents.[12]

Blends of PET with PP and PP-g-GMA

SEM micrographs of PET/PP and PET/PP-*g*-GMA5 blends (80/20 and 50/50 w/w) are shown in Figure 5. Blends of PET and PP (Figures 5a-b) show the typical morphology of incompatible systems with a rough dispersion of the components and no adhesion between matrix (PET) and dispersed phase (PP). Otherwise, blends of PET and GMA modified PP (5,2 wt% GMA) show a more homogeneous morphology with finer dispersion of polyolefin phase and strong interfacial adhesion (Figures 5c-d).

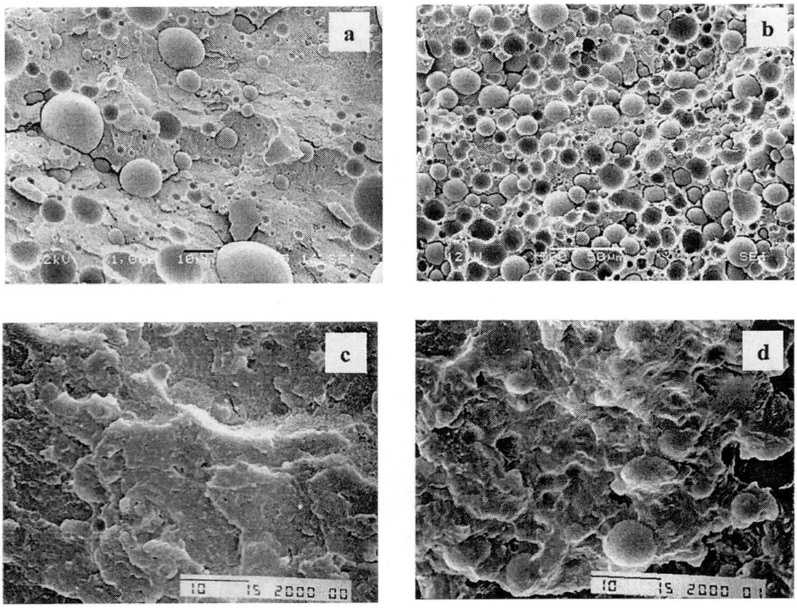

Fig. 5. SEM micrographs of fracture surfaces of blends PET/PP 80/20 (a) and 50/50 (b); PET/PP-*g*-GMA5 80/20 (c) and 50/50 (d) (bar length in figs. a, c, d: 10 μm; fig. d: 50 μm).

A wide particle size distribution is generally observed for the non-compatibilized blends, characterized by extended coalescence phenomena of dispersed phase due to the higher melt viscosity of PP as compared to PET. The number-average diameter, d_n, of dispersed particles is reported for all examined blends in the chart-bar of Figure 6. For blends with higher PET content (80/20) the average size of polyolefin particles decreases from about 2.6 μm for the non compatibilized blend to about 0.6 μm for that with PP-*g*-GMA5, while in the blends with composition 50/50, where PET is the continuous phase, the average size of PP dispersed phase changes from about 8 μm for PET/PP to 2.4 μm for PET/PP-*g*-GMA5. The compatibilizing effect resulting for the blends with modified PP is to be related with the interactions occurring at interface between epoxy groups of GMA and carboxyl end-groups of PET, leading to a drastic decrease of interfacial tension and suppression of coalescence.[17]

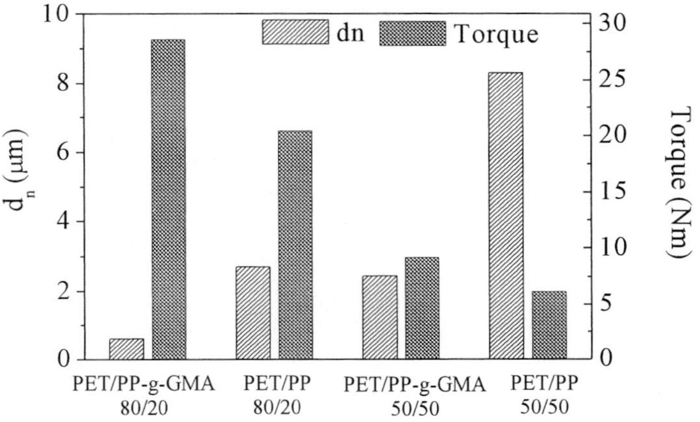

Fig. 6. Average diameter (d_n) of dispersed phase and final mixing torque for PET/PP and PET/PP-*g*-GMA5 blends.

The presence of intermolecular interactions in the modified blends is supported by the variation of mixing torque, as a consequence of changes of melt viscosity due to copolymerization reactions between the polymer components.[18, 19] As shown in Figure 6, a significant increase of torque has been found for PET/PP-*g*-GMA5 blends with respect to PET/PP blends at same composition. The higher torque value of the blend PET/PP-*g*-GMA5 80/20 can be related with the molar ratio between the reactive groups (mol COOH/mol GMA= 0.27) which is greater than that of PET/PP-*g*-GMA5 50/50 blend (mol COOH/mol GMA= 0.07). Further, it must be

noticed that, due to the bifunctionality of PET chain, additional reaction of hydroxyl end-groups of polyester with PP-g-GMA may occur giving rise to crosslinking phenomena.[10, 19]
The DSC analysis of the various blends showed a marked influence of the composition and GMA content on the crystallization behaviour of both polymer phases (Table 4). For PET/PP blends a shift of T_c of PP to higher temperature than that of pure PP (111.6 °C) was observed, indicating a nucleating effect of PET matrix (or likely its impurities) on the dispersed polyolefin. For PET/PP-*g*-GMA5 blends, a more pronounced increase of T_c of PP phase up to 122 °C was found, similar to that previously described for plain PP-*g*-GMA copolymers (see Table 3). Moreover, the melting parameters (T_m, ΔH_m) and the crystallinity of polymers in the blends with PP-*g*-GMA5 are generally lower than that of PET/PP blends (and pure components), in agreement with the enhanced phase dispersion of the formers. In particular, the lowest crystallinity value of PP (26.5%) is observed for PET/PP-*g*-GMA5 80/20 blend, which displays the lowest dispersed particle size.

Table 4. DSC data for PET/PP blends on cooling and 2nd heating run (10°C/min).

Sample	PET				PP			
	T_c (°C)	ΔH_c [a)] (J/g)	T_m (°C)	ΔH_m [a)] (J/g)	T_c (°C)	ΔH_c [a)] (J/g)	T_m (°C)	ΔH_m [a)] (J/g)
PET	182.0	-39.3	247.8	35.1				
PP					111.6	-100.0	165.5	104.5
PET/PP 80/20	185.9	-39.3	248.3	34.7	116.5	-96.1	162.5	92.8
PET/PP 50/50	179.7	-34.7	247.1	33.0	115.0	-100.3	163.5	101.1
PET/PPgGMA 80/20	189.8	-38.5	247.5	29.7	122.0	(-	160.5	(50.1)
PET/PPgGMA 50/50	181.2	(-13.4)	243.5	28.4	120.4	-96.1	163.7	77.7

a) Per gram of polymer component

Conclusions

The study of radical grafting of GMA onto PP in the melt showed that the grafting degree of the polyolefin can be enhanced by a suitable control of reactions parameters and procedures (monomer concentration, type and content of radical initiator, addition of co-monomer) which affect the grafting mechanism, as well as the chain degradation of PP and secondary reactions (GMA hopolymerization, crosslinking). The addition of styrene as co-monomer significantly improves the GMA grafting.

The molecular weight, the melt viscosity and phase behaviour of examined PP-*g*-GMA copolymers are strictly depending on the content of grafted GMA. With increasing the GMA content, a large change of crystallization and melting temperatures was found, as compared to PP homopolymer, suggesting the presence of PP crystals with different perfection and chain structure.

Blends of PET with PP-*g*-GMA5 displayed a neat improvement of phase dispersion and adhesion with respect to PET/PP blends. Mixing torque and DSC data supported the occurrence of reactions between GMA modified PP and PET in the melt, likely leading to the formation of a graft copolymer, acting as compatibilizer. The compatibilizing effect of GMA grafted PP in the examined blends (80/20 and 50/50) can be interpreted by taking into account the molar ratio between the epoxy groups and the functional end-groups of PET chains.

The improvement of phase morphology and physical-mechanical properties observed for PET/PP blends compatibilized with PP-g-GMA, as compared to those compatibilized with maleic anhydride grafted PP,[20] support the higher efficiency of GMA grafted polyolefin.

[1] S.B. Brown in "*Reactive Extrusion, Principles and Practise*", M. Xanthos Ed., Hanser Publ., Munich **1992**, chapt.4
[2] R. Gallucci, R.C. Going, *J. Appl. Polym. Sci.*, **1982**, *27*, 425.
[3] Y.J. Sun, G.H. Hu, M. Lambla, *Angew. Makromol. Chem.*, **1995,** *229*, 1.
[4] H. Cartier, G. H. Hu, *J. Polym. Sci., Part A: Polym. Chem.*, **1998**, *36*, 1053.
[5] N. Torres, J.J. Robin, B. Boutevin, *J. Appl. Polym. Sci.*, **2001**, *81*, 581.
[6] L. Chen, B. Wong, W. E. Baker, *Polym. Eng. Sci.*, **1996,** *36*, 1594.
[7] X. Zhang, G. Li, D. Wang, Z. Yin, J. Yin, J. Li, *Polymer*, **1998**, *39*, 15.
[8] M.F. Champagne, M. Huneault, C. Roux, W. Peyrel, *Polym. Eng. Sci.*, **1999**, *39*, 976
[9] F. Pazzagli, M. Pracella, *Macromol. Symp.*, **2000**, *149*, 225.
[10] Y.J. Sun, G.H. Hu, M. Lambla, H.K. Kotlar, *Polymer*, **1996**, *37*, 4119.
[11] F. Danusso, G. Moraglio, *Makromolekulare Chem.*, **1956**, *20*, 37.
[12] B. Wong, W.E. Baker, *Polymer*, **1997**, *38*, 2781.
[13] H. Huang, N. C. Liu, *J. Appl. Polym. Sci.*, **1998**, *67*, 1957.
[14] P. Chandranupap, S. N. Bhattacharya, *J. Appl. Polym. Sci.*, **2000**, *78*, 2405.
[15] X. Zhang, L. Li, Z. Yin, Y. Qi, J. Yin, *J. Appl. Polym. Sci.*, **1996**, *62*, 1933
[16] Z. Yin, X. Zhang, Y. Zhang, J. Yin, *J. Appl. Polym. Sci.*, **1997**, *63*, 1565.
[17] W. Hale, H. Keskkula, D.R. Paul, *Polymer*, **1999**, *40*, 365.
[18] M. Pracella, L. Rolla, D. Chionna, A. Galeski, *Macromol. Chem. Phys.*, **2002**, *203*, 1473.
[19] W. Loyens, G. Groeninckx, *Macromol. Chem. Phys.*, **2002**, *203*, 1702.
[20] M.Pracella, D. Chionna, B. Bulleri, 7th *European Symposium on Polymer Blends*, Lyon (France), May **2002**, Abs. C8.

Morphology and Mechanical Properties of Extruded Ribbons of LDPE/PA6 Blends Compatibilized with an Ethylene-Acrylic Acid Copolymer

*Francesco Paolo La Mantia,**[1] *Roberto Scaffaro,*[1] *Antonino Valenza,*[2] *Augusto Marchetti,*[3] *Sara Filippi*[3]

[1]Dipartimento di Ingegneria Chimica dei Processi e dei Materiali, Università di Palermo, Viale delle Scienze, 90128 Palermo, Italy
[2]Dipartimento di Chimica Industriale e Ingegneria dei Materiali, Università di Messina, Salita Sperone 31, 98166 Messina, Italy
[3]Dipartimento di Ingegneria Chimica, Chimica Industriale e Scienza dei Materiali, Università di Pisa, Via Diotisalvi 2, 56126 Pisa, Italy

Summary: Two grades of low density polyethylene (LDPE) were blended with polyamide-6 (PA) in the 75/25 and 25/75 wt/wt ratios and shaped into ribbons with a Brabender single screw extruder. An ethylene-acrylic acid copolymer (EAA) was used in the 2 phr concentration as a compatibilizer precursor (CP). The morphology of the ribbons and its evolution during high temperature annealing were investigated by scanning electron microscopy (SEM). The results confirmed that EAA does actually behave as a reactive compatibilizer for the LDPE/PA blends. In fact, in the presence of EAA, the interfacial adhesion is improved, the dispersion of the minor phase particles is enhanced and their tendency toward fibrillation is increased, especially for the blends with the higher molar mass LDPE grade. The mechanical properties of the latter blends were found to be considerably enhanced by the addition of EAA, whereas the improvement was relatively modest for the blends with the lower molar mass LDPE. The fracture properties of double end notched samples of the ribbons prepared with the blends containing the lower molar mass LDPE grade were also studied. It was shown that, despite of the increased interfacial adhesion caused by the presence of EAA, the latter plays a measurable positive effect on the fracture properties only for the blends with LDPE as the matrix.

Keywords: blends, impact resistance, low density polyethylene, mechanical properties, morphology, polyamide-6

Introduction

The preparation of compatibilized blends of polyolefins with condensation polymers has attracted considerable attention during the last few decades not only because of its scientific implications, but also in view of the interesting practical applications of these blends.[1-10] In particular, the compatibilization of blends of polyethylene (PE) with polyamide-6 (PA), carried out by the

 DOI: 10.1002/masy.200350815

addition of compatibilizer precursors (CPs) consisting of polyolefins containing maleic anhydride, acrylic acid, oxazoline, or other functional groups capable of reacting with PA to form PA-*g*-CP copolymers has been the object of a considerable number of scientific publications.[3-10]

In a previous work,[11] some of us made a detailed investigation of the effectiveness of ethylene-acrylic acid copolymers (EAA) as CPs for the reactive blending of low density polyethylene (LDPE) with PA. It was shown that EAA form immiscible, yet highly compatible, blends with LDPE, and give rise to relatively slow acidolysis reactions, when blended with PA, yielding EAA-g-PA copolymers. The scanning electron microscopy (SEM) investigation of the ternary LDPE/PA/EAA blends showed that the addition of 1-2 phr EAA enhances considerably the dispersion of the minor phase droplets, prevents their coalescence during subsequent annealing and/or reprocessing operations and increases interfacial adhesion. It was also found that no further improvement of these properties could be obtained by the use of higher proportions of these compatibilizers. The compatibilizing efficiency of the EAA copolymers was shown to be particularly prominent for the LDPE/PA blends containing an LDPE grade with a melt viscosity considerably higher than that of the PA.

In the present work, blends of PA with two grades of LDPE, with 75/25 and 25/75 wt/wt compositions, were shaped into ribbons with a Brabender single screw extruder. An EAA copolymer (Escor 5001, with 6.2 wt% acrylic acid) was used as a CP in the 2 phr concentration which was shown by previous morphological studies to produce optimum results. The morphology of the ribbons and its evolution during high temperature annealing were investigated by SEM. The ultimate mechanical properties of specimens cut out from the ribbons in the flow direction, were studied. The fracture resistance of double end notched (DEN) ribbon samples was also investigated. The results of the mechanical characterizations were discussed with reference to the morphology of the samples and to the effect of the EAA reactive compatibilizer.

Experimental

Two grades of LDPE, kindly provided by Polimeri Europa, were used: Riblene FF20 with a melt flow index (MFI) equal to 0.8 g/10 min and d = 921 Kg/m^3, referred to herein as LD08, and Riblene FC30 (LD03) with MFI = 0.27 g/10 min and d = 922 Kg/m^3. The polyamide-6 (PA) sample was kindly supplied by Snia Tecnopolimeri; its intrinsic viscosity in formic acid (85%) was

1.45 dL/g and the contents of amine and carboxyl end groups were 34 and 35 meq/Kg, respectively. The EAA sample, Escor 5001 by Exxon-Mobil Chemical Mediterranea, was an ethylene-acrylic acid copolymer with 6.2 wt % acrylic acid and MFI = 2.0 g/10 min.

Before use, all the polymers were accurately dried under vacuum for 12 h at 70°C (for the LDPE and the EAA samples) or 120°C (for the PA). The ribbons were extruded from a rectangular die (width 100 mm, thickness 1 mm) using a Brabender single screw extruder, at 240°C and 50 rpm. The residence time was about 2 min. The molten ribbons were taken on a device, consisting of a moving tape and an air cooled metal roll, that cooled the ribbon and drew it with a stretching ratio equal to 6 (calculated as the ratio of the die section to the section of the ribbon).

Scanning electron microscopy (SEM) observations of the surfaces produced by fracturing the ribbons under liquid nitrogen, either parallel and perpendicular to the flow direction, and coated with gold were made with JEOL JSM-5600 LV or Jeol T300 microscopes. Annealing experiments were made by heating ribbon samples in a nitrogen atmosphere at 240°C for different times.

Tensile tests were carried out with an Instron Mod. 4443 using samples of 10×100×0.6 mm cut out of the ribbons in the flow direction. All the samples were held in the laboratory atmosphere (T ≈ 25°C, r.u. ≈ 70%) for a week before testing. The barrel speed was held at 1 mm/min for 2 min and was then increased to 50 mm/min; the clamps distance was 30 mm.

The fracture properties were studied on DEN (double end notched) samples of the ribbons of 75/25 and 25/75 LD08/PA blends, without and with 2 phr EAA. The width of the samples was of ca. 45 mm and the ligament, i.e., the width between notches, was varied between 3 and 25 mm. The employed deformation speed was 500 mm/min.

Results and Discussion

Examples of the micrographs taken by SEM on the fracture surfaces of ribbons prepared from uncompatibilized 75/25 and 25/75 LD08/PA and LD03/PA blends are shown in Figure 1. It is clearly observed that fairly long fibrils, together with smaller, poorly elongated droplets are present in the blends containing LD08. The particles dimensions are somewhat smaller for the blends with PA as the minor phase (Figure 1a). For the blends with a LD03 matrix (Figure 1c), the PA particles are much larger and show practically no fibrillation. On the other hand, the 25/75 LD03/PA blend has distinctly lamellar morphology (Figure 1d). Notice that the viscosity of LD08 is close to that of PA, at the processing temperature, whereas that of LD03 is considerably higher. The micrographs in Figure 1 demonstrate that dispersion and fibrillation of the minor phase is only granted, under the employed extrusion conditions, if the viscosity ratio of the two phases is appropriate. However, practically no interfacial adhesion is obtained in the absence of a CP.

The SEM micrographs of the corresponding blends containing 2 phr EAA are shown in Figure 2. It is clearly observed that the morphology of the ribbons prepared from LD08 is only slightly improved, as far as the dimensions and the geometry of the minor phase particles are concerned, although the interfacial adhesion is enhanced. On the contrary, for the LD03 blends, the effect of the EAA addition is very strong: the size of the minor phase particles is reduced dramatically and their shape is also changed profoundly, as fibrillation is now evident. Clearly, the interactions brought about by the EAA-*g*-PA copolymers formed at the interface are such as to lower interfacial tension considerably and to allow fibrillation of the dispersed particles, despite of the difference of viscosity of the two phases. The latter effect is particularly evident for the 25/75 LD03/PA blend (cf. Figure 2d).

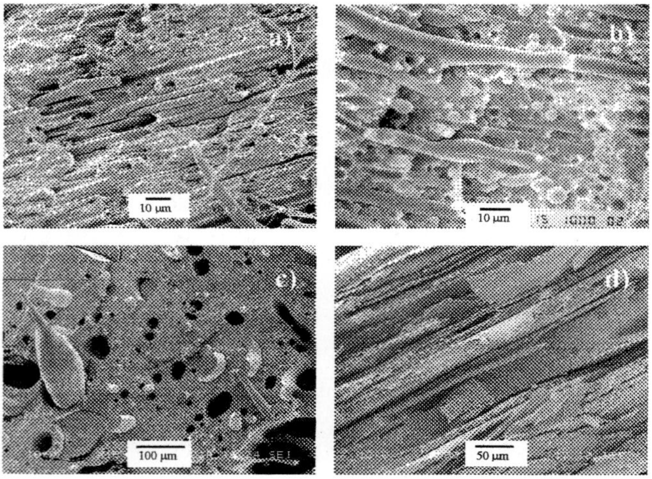

Fig. 1. SEM micrographs of the fracture surfaces of ribbons prepared from uncompatibilized blends: a) 75/25 LD08/PA; b) 25/75 LD08/PA; c) 75/25 LD03/PA; d) 25/75 LD03/PA.

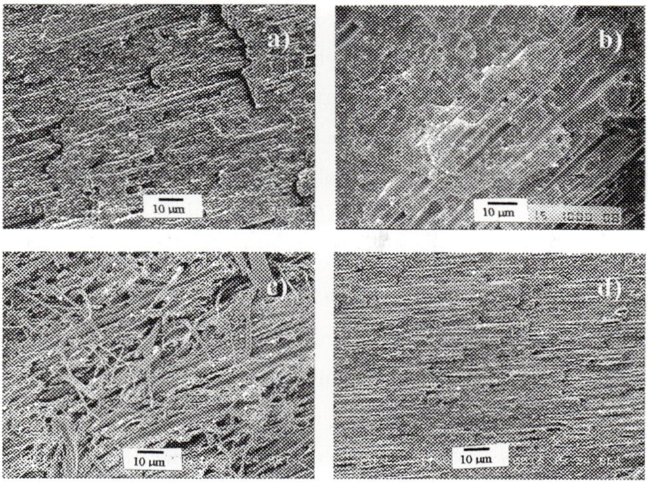

Fig. 2. SEM micrographs of the fracture surfaces of ribbons prepared from blends compatibilized with 2 phr EAA: a) 75/25 LD08/PA; b) 25/75 LD08/PA; c) 75/25 LD03/PA; d) 25/75 LD03/PA.

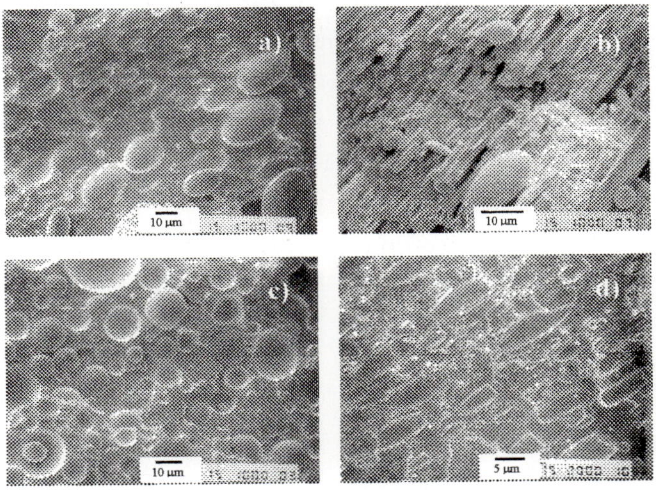

Fig. 3. SEM micrographs of the fracture surfaces of ribbons prepared from 75/25 (a, b) and 25/75 (c, d) LD08/PA blends, without (a, c) and with (b, d) 2 phr EAA, annealed at 235°C for 15 min (a, b, c) and 60 min (d).

Annealing experiments were carried out as described in the experimental section to investigate the effect of the presence of the EAA copolymer on the morphology changes due to loss of orientation and coalescence of the dispersed particles. The SEM micrographs shown in Figure 3 illustrate the behavior of the blends based on LD08. A comparison with the micrographs of unannealed ribbons shown in Figures 1 and 2 demonstrate that the speed of shape relaxation is considerably smaller for the compatibilized blends. In fact, nearly spherical droplets of the minor phase are already seen, for the blends without EAA, after 15 min annealing (Figures 3a and 3c), whereas only partial shape relaxation takes place for the compatibilized blends, even after 60 min. Also, the size of the dispersed particles remains smaller when the EAA copolymer is present. Similar annealing experiments were also carried out for the 25/75 LD03/PA blends, and the results are illustrated by the micrographs in Figure 4. It is clearly seen that very important morphological rearrangements take place in these blends upon annealing. Either the lamellar structure of the uncompatibilized blend (Figure 1d) and the fibrillar morphology of that containing EAA (Figure 2d) change gradually into those shown in Figure 4, which are typical for unoriented blends. The changes are much faster for the blend without EAA (Figure 4a), where LD03 spherical droplets are already seen after 25 min annealing, whereas much longer annealing times are needed to get the same result for the compatibilized blend (Figure 4b). Moreover, as it was expected on the basis of the results of previous studies,[11] a comparison of the two micrographs in Figure 4 demonstrates that the morphology is much coarser for the annealed blend without EAA, thus confirming that the addition of this CP at the mixing stage leads to a considerable reduction of interfacial tension in the LDPE/PA blends.

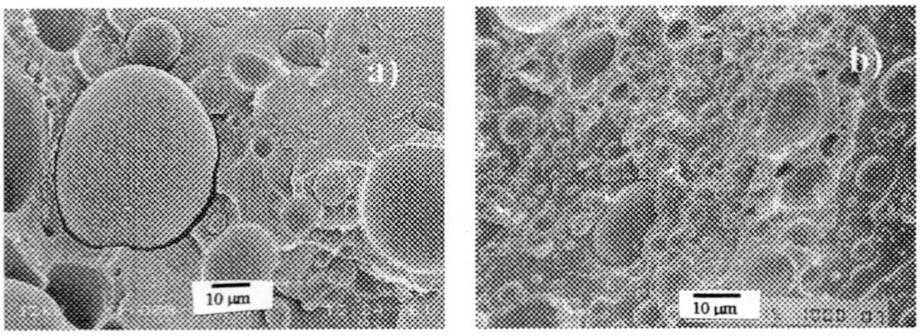

Fig. 4. SEM micrographs of the fracture surfaces of ribbons prepared from 25/75 LD03/PA blends, without (a) and with (b) 2 phr EAA, annealed at 235°C for 25 min (a) and 120 min (b).

The results of the tensile tests carried out on samples of the ribbons are shown in Table 1.

Table 1. Tensile mechanical properties of the ribbons.

Composition	E MPa	σ_b MPa	ε_b %
LD08/PA/EAA 75/25/0	193	21.9	370
75/25/2	190	20.0	395
25/75/0	295	49.4	310
25/75/2	292	51.1	330
LD03/PA/EAA 75/25/0	143	23.0	339
75/25/2	178	27.8	384
25/75/0	229	38.3	297
25/75/2	293	53.7	334

It can be seen that, in the absence of reactive compatibilizer, the elastic modulus E of the blends based on LD03 is lower than that of the blends containing LD08, for both compositions. This can probably be explained on the basis of the morphology of the former blends, shown in Figure 1, that are more irregular and practically void of oriented fibrils. The addition of 2 phr EAA brings about an improvement of the mechanical properties which is rather modest for the LD08/PA blends and appreciable for the LD03/PA blends. It is interesting that the properties of the compatibilized blends are nearly independent of the molar mass of the LDPE component, whereas they are critically influenced by the morphology. In fact, as it is clearly seen in Figure 2, the addition of EAA makes the LD03 blends of either composition attain a pronounced degree of orientation and fibrillation of the minor phase, so that their morphology becomes similar to that of the corresponding LD08 blends. This similarity is evidently paralleled by the similarity of the tensile properties. For the samples with fibrillar morphology, the improvement of mechanical properties, in particular the elongation at break, caused by the addition of EAA is generally small, as it is probably influenced scarcely by the increased interfacial adhesion and is mainly ascribable to the enhancement of dispersion and the reduction of size of the minor phase.

The technique normally used for measuring the essential work of fracture was employed to investigate the fracture behavior of samples of the LD08/PA ribbons, which were shown by SEM to posses fibrillar morphology even in the uncompatibilized state. The results are illustrated in Figure 5, where the specific work of fracture, measured as the ratio of the area under the stress-strain curve to the fractured area, is plotted as a function of the resistant width (ligament).

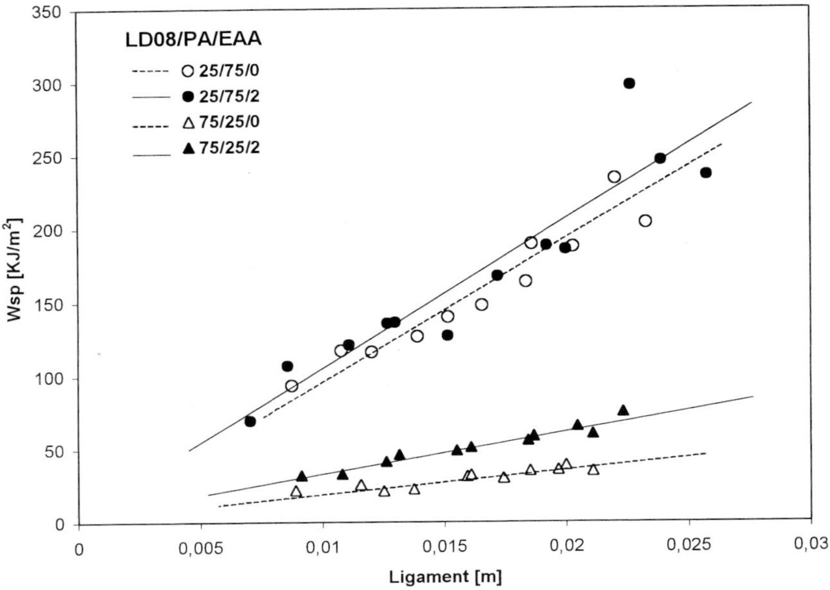

Fig. 5. Specific work of fracture of the LD08/PA/EAA ribbons as a function of the ligament width.

Although the employed measurement conditions (sample shape and deformation speed) were those commonly used for the determination of the essential work of fracture, as the intercept of the straight lines on the Y axis, the latter characteristic could not be determined reliably because the samples were very ductile and the specific work of fracture (Wsp) was mainly due to plastic deformation. However, the results shown in Figure 5 demonstrate that, for the blends with PA as the matrix, Wsp is hardly influenced by the presence of the EAA. Indeed, the experimental points shown in the Figure could practically be interpolated by the same straight line. On the other hand, compatibilization leads to a measurable increase of the work of fracture for the blends with LD08 as the main component, thus showing that the PA-*g*-EAA copolymers formed during blending act more efficiently as interfacial compatibilizers when LDPE forms the matrix. This result is in fair agreement with previous findings. In fact, the results of a former work,[11] in which the morphology of the same blends, in the undeformed state, was studied, revealed that the enhancement of interfacial adhesion caused by the presence of EAA is more pronounced in the blends with LDPE as the matrix.

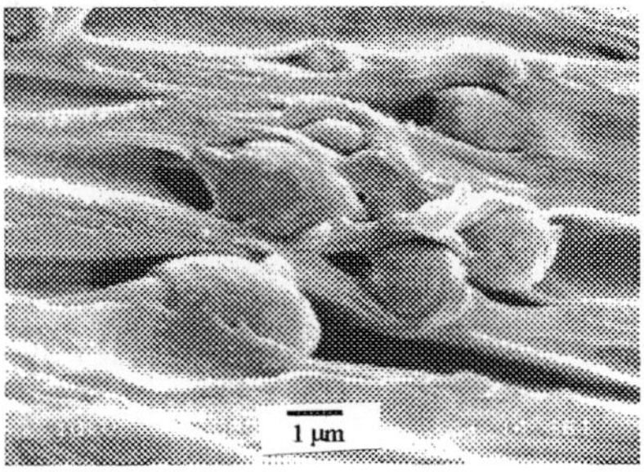

Fig. 6. SEM micrograph showing the morphology of a broken 75/25/2 LD08/PA/EAA ribbon, in the stretched region close to the fracture.

In order to get an experimental evidence of the interfacial adhesion between the LD08 matrix and the dispersed PA droplets, in blends compatibilized by the addition of EAA, the morphology of the stretched region of the broken ribbons used for the fracture experiments was investigated. The micrograph shown in Figure 6 was taken on a 75/25/2 LD08/PA/EAA ribbon. The small PA droplets embedded into the highly deformed LD08 matrix are seen to adhere quite tightly to the latter. No such behavior was displayed in the absence of the EAA reactive compatibilizer.

Conclusions

It has been shown that the morphology of extruded ribbons prepared from 75/25 and 27/75 LD08/PA and LD03/PA blends is appreciably improved if 2 phr of an EAA copolymer is added into them. Indeed, the dispersion of the minor phase droplets, their adhesion to the matrix and their deformation into fibrils were considerably enhanced. These effects have been shown to be particularly prominent for the blends containing the higher molar mass LDPE grade. The mechanical properties of the ribbons are also improved as a result of the addition of the EAA copolymer. The effect is modest for the LD08/PA ribbons, in which a fibrillar morphology is present even in the uncompatibilized state, but becomes appreciable for the LD03/PA samples whose morphology undergoes major changes when the compatibilizer is present. The determination of the work of fracture of the ribbons has confirmed that good interfacial adhesion between matrix and dispersed particles is obtained as a result of compatibilization, especially for the blends with a LDPE matrix.

Acknowledgment

This work was financially supported by the *Italian Ministry of University and Scientific and Technological Research* (MURST), contract n° MM09244737. Thanks are due to *Polimeri Europa, Exxon Chemical Mediterranea* and *Snia Tecnopolimeri* for providing samples of the polymers used in this work.

[1] W. Baker, C. Scott, G.-H. Hu, *Reactive Polymer Blending*, Hanser, Münich, 2001.
[2] F.P. La Mantia, *Adv. Polym. Techn.* **1993**, *12*, 47.
[3] F. Ide, A. Hasegawa, *J. Appl. Polym. Sci.* **1974**, *18*, 963.
[4] G. Serpe, J. Jarrin, F. Dawans, *Polym. Eng. Sci.* **1990**, *30*, 553.
[5] B.K. Kim, S.Y. Park, S.J. Park, *Eur. Polym. J.* **1991**, *27*, 349.
[6] A.R. Padwa, *Polym. Eng. Sci.* **1992**, *32*, 1703.
[7] R.A. Kudva, H. Keskkula, D.R. Paul, *Polymer* **1999**, *40*, 6003.
[8] A. Halldén, B. Ohlsson, B. Wesslén, *J. Appl. Polym. Sci.* **2000**, *78*, 2416.
[9] A. Valenza, A.M. Visco, D. Acierno, *Polym. Test.* **2002**, *21*, 101.
[10] C. Vocke, U. Anttila, J. Seppälä, *J. Appl. Polym. Sci.* **1999**, *72*, 1443.
[11] S. Filippi, V. Chiono, G. Polacco, M. Paci, L. Minkova, P. Magagnini, *Macromol. Chem. Phys.* **2002**, *203*, 1512.

Reactively and Physically Compatibilized Immiscible Polymer Blends: Stability of the Copolymer at the Interface

Charef Harrats,[1] Kristof Dedecker,[1] Gabriel Groeninckx,[1] Robert Jérôme[2]*

[1] Katholieke Universiteit Leuven, Laboratory of Macromolecular Structural Chemistry, Dept. of Chemistry, Celestijnenlaan 200 F, 3001 Heverlee-Leuven, Belgium
[2] Center for Education and Research on Macromolecules, Institute of Chemistry (B6), University of Liège, Sart-Tilman, 4000 Liège, Belgium

Summary: This paper reports on the interfacial behaviour of block and graft copolymers used as compatibilizers in immiscible polymer blends. A limited residence time of the copolymer at the interface has been shown in both reactive blending and blend compatibilization by preformed copolymers. Polystyrene (PS)/polyamide6 (PA6), polyphenylene oxide (PPO)/ PA6 and polymethylmethacrylate (PMMA)/PA6 blends have been reactively compatibilized by a styrene-maleic anhydride copolymer SMA. The extent of miscibility of SMA with PS, PPO and PMMA is a key criterion for the stability of the graft copolymer at the interface. For the first 10 to 15 minutes of mixing, the in situ formed copolymer is able to decrease the particle size of the dispersed phase and to prevent it from coalescencing. However, upon increasing mixing time, the copolymer leaves the interface which results in phase coalescence.
In PS/LDPE blends compatibilized by preformed PS / hydrogenated polybutadiene (hPB) block copolymers, a tapered diblock stabilizes efficiently a co-continuous two-phase morphology, in contrast to a triblock copolymer that was unable to prevent phase coarsening during annealing at 180°C for 150 minutes.

Keywords: compatibilization, graft and diblock copolymer, interfacial stability, reactive processing

Introduction

The location of the compatibilizer at the interface between the individual phases of a blend is a prime and necessary requirement for an efficient compatibilization and phase morphology stabilization upon further re-processing or thermal annealing. Indeed, a copolymer might be able to locate at the interface during the first minutes of the compounding process reducing

 DOI: 10.1002/masy.200350816

efficiently the size of the dispersed phase particles and preventing their coalescence. However, is the copolymer able to stay at the interface during long melt-mixing or melt-annealing operations? It is demonstrated in this work that either in reactively or physically compatibilized blends the compatibilizer can easily be rejected from the interface because of its energetically unbalanced design. The composition, molecular weight, and miscibility of the compatibilizer with the individual blend phases, its functionality when it is in-situ generated and its molecular architecture when it is physically added to the blend are the key parameters which affect its stable location at the interface.[1] A huge amount of literature clearly evidences the beneficial effect of various copolymers having various molecular characteristics on the compatibilization of immiscible polymer blends.

The stability of the copolymer at the interface between the blend phases has not been really focused on in literature. Few papers have reported on the stability of the phase morphology during annealing and ascribed the observed effect to the stability of the copolymer at the interface.[2, 3] The ability of the compatibilizer to stabilize a developed phase morphology can be used as an additional criterion for the evaluation of the compatibilization efficiency. E.g. a co-continuous phase morphology exhibits an intrinsic instability when subjected to further processing or thermal annealing.[4-7] We have previously reported that after only 3 minutes of annealing at 200 °C a 20LDPE/80PS blend having a fully co-continuous two-phase morphology exhibited a complete change to a droplet-in-matrix morphology.[2] Veenstra et al. have recently studied the extent of instability of a co-continuous phase morphology of a 50 / 50 polyether-ester / polystyrene blend as a function of temperature and annealing time.[8] In this case the phase morphology preserved its co-continuous character, but a pronouced phase coarsening was observed. Mekhilef et al. reported that the co-continuous phase morphology of 30/70 PS / PE blend also evolved to a droplet--in–matrix type morphology when annealed in the absence of shear.[9] After compatibilization using a SEBS triblock copolymer the stability of the co-continuous phase morphology was significantly improved. Nevertheless, the SEBS copolymer was not able to totally prevent the phase morphology from coarsening when the blend was annealed at 200 °C during 16 minutes.

A pure and a tapered diblock hPB-b-PS copolymers have shown substantial differences in their ability to stabilize a co-continuous two-phase morphology when PS rich (ca. 80 wt %) binary

LDPE / PS blends were annealed at 180 °C in the absence of shear.[2] Even at a 10 wt % concentration the pure diblock had a less efficient stabilizing activity than the tapered version. For instance, after only 3 minutes of quiescent annealing at 180 °C, the phase co-continuity was altered by the LDPE phase break-up and subsequent coarsening even in the presence of the pure diblock copolymer. When the tapered diblock was used, no phase morphology evolution was observed and, depending on the concentration, the co-continuous character was better preserved. This clear difference has prompted us to compare the emulsification activity of a triblock SEBS with a tapered diblock using the very unstable co-continuous phase morphology developed in a LDPE / PS 20 / 80 blend system. In this case and because the differences were expected to be very subtle, the blends were subjected to longer annealing times of 150 minutes.

Experimental

Table 1. Characteristics of the homopolymers and copolymers used.

PA6	3 samples from DSM: Mw_1: 18 000; Mw_2: 24000; Mw_3 : 44 000
PPO	Mw : 54000
PMMA	DIAKON MG 102 from Rhom & Haas, Mw: 69 000
PS1	Denka Styrol GP-1 Mw : 180 000
PS2	Polystyrol 158K BASF Mw: 100 000
SMAx	samples having different MA contents were used; x denotes the wt % of maleic anhydride reactive groups
LDPE	Shell 33 , Mw: 40 000
Triblock S-*b*-hB-*b*-S	Mw for the three sequences : 7500-35000-7500
Tapered diblock S-*b*-P(E-co-B)-*b*-hB	Mw can be given as : 23000-19000-28000

The blends were prepared using a conical twin screw mini-extruder (4 gr total mixing capacity) designed by DSM-Research.[10, 11] The mixing temperature was kept constant at 240 °C for the blend systems PA6 /(PMMA/SMA) and PA6 /(PS/SMA) and 270 °C for the blend systems PA6 /(PPO/SMA). When not specified, the screw speed was kept at 100 rpm. The miscible preblends (PS/SMA, PMMA/SMA and PPO/SMA) were also prepared using the mini-extruder. The preblends were cut into small pieces, and subsequently used in the final extrusion step with polyamide6. The polystyrene / LDPE blend system compatibilized using the triblock and the tapered S/hB copolymers was melt-mixed using a roll mill at 200 °C for 5 minutes at a roll speed of 30 rpm. The copolymer was first mixed with the minor phase; subsequently the major phase was progressively added to the mixture.

Discussions

In the reactive blends systems based on polyamide 6 and PMMA, PPO or PS, the SMA-g-PA6 compatibilizer is formed in-situ during mixing as a result of the reaction between the anhydride groups of SMA and the amino end groups of PA6. The imidation reaction involved is well documented and frequently employed in reactively designed blends based on polyamides.[11, 12-18]

Optimum Functionality of the SMA Compatibilizer

SMA compatibilizers having 14, 17, 20, 25, 28 and 33 weight % of maleic anhydride have been used in PA6 / PMMA blends. With respect to their compatibilizing efficiency, it is very important to take into account the miscibility behaviour between SMA and PMMA. PMMA/ SMA blends always show an LCST behaviour. At the extrusion temperature of 240 °C, SMA is miscible with PMMA for MA contents between 10 and 35 wt %. As the cloud point of the PMMA / SMA blends depends on the molecular weight of PMMA and SMA and on the blend composition, the miscibility behaviour was checked for the components used.

The compatibilized blends PA6 / (PMMA/SMA) under consideration have the composition 75/ (20/5). The miscibility behaviour was thus analyzed for the PMMA / SMA blends in a weight ratio of 80 / 20 ; they were prepared at 240 °C in a mini-extruder. The transparency of the blends was considered as a first indication of miscibility. The blends of PMMA with SMA14, SMA17, SMA20, SMA25 and SMA28 were transparent after 3 minutes of extrusion. The blend PMMA / SMA33 (80 / 20) was opaque at short extrusion times but became transparent after longer extrusion times (10 min). The blend PMMA/SMA8 remained opaque, even after longer extrusion times (10 min) indicating that this blend is not miscible at 240 °C. The miscibility of PMMA / SMA was evaluated using d.s.c. on blend samples quenched in liquid nitrogen. The results of the d.s.c. measurements are given in Figure 1 which clearly shows that SMA is more miscible with PMMA at the mid-concentration of about 20 wt % maleic anhydride. The more one deviates from this concentration in either directions, low or high MA content, the less the system is miscible.

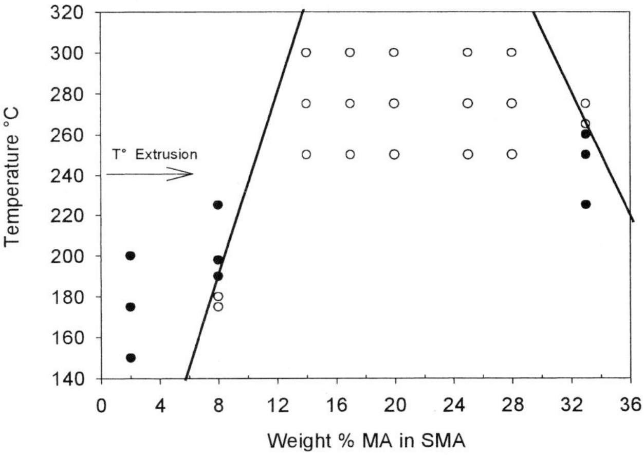

Fig. 1. Miscibility of PMMA / SMA 80/20 blends as a function of the MA content in SMA, (O) miscible; (•) immiscible.

The weight average particle diameter of PMMA dispersed phase in SMA compatibilized PA6 / PMMA blends is plotted in Figure 2 as a function of the extrusion time when SMA 20, SMA28 and SMA33 were used. Compared to the case of SMA20, a completely different behaviour is observed with SMA28 and SMA33 as compatibilizers. The dispersed phase particle size can still decrease after 3 minutes of extrusion for the blends containing SMA28 or SMA33, while this is not the case for SMA20. This behaviour can be directly correlated with the extent of miscibility of SMA and the minor phase PMMA. Indeed, in the beginning of the melt-extrusion process, SMA-g-PA6 graft copolymer is formed at the interface causing a significant particle size decrease. As the compounding process continues further, the SMA copolymer reacts more quantitatively so that an unbalanced graft copolymer, richer in PA6, is built-up. It has a strong tendency to be driven out from the interface into the PA6 phase where it is expected to be the most stable. This is made easier when the extent of miscibility of the SMA (28 and 33 wt % MA) with PMMA does not 'counter-balance' the unfavorable compositional effect (a graft copolymer richer in PA6 exhibits more interaction with the PA6 matrix). This correlation is also illustrated in Figure 3 with SMA copolymers covering a broader range of MA concentration. The weight average particle size of PMMA / PA6 blends is plotted as a function of the weight % of maleic anhydride of SMA copolymer after 20 minutes of extrusion time. The Figure clearly reveals that for the compatibilizers closer to the boundaries of the miscibility region of PMMA and SMA (refer to Fig. 1), a larger PMMA particle size is obtained.

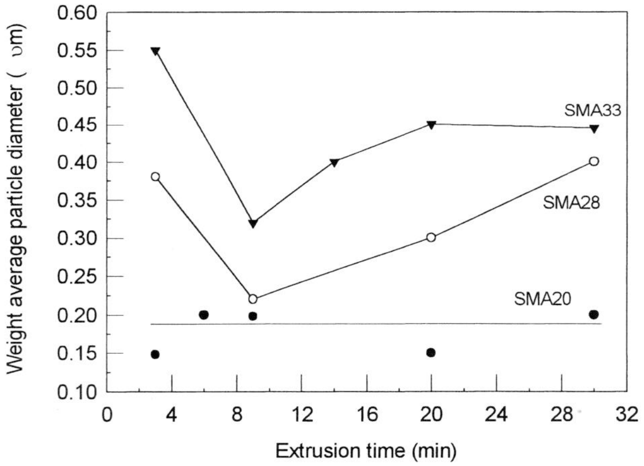

Fig. 2. Weight average particle diameter as a function of the extrusion time for the blend PA6 / (PMMA / SMA) 75 / (20 / 5) with different MA contents for SMA.

The unstability of the graft copolymer formed at the interface is also illustrated for melt-blended and SMA compatibilized PA6 /PPO blend system where the particle size of the PPO dispersed phase is plotted as a function of the extrusion time (Fig. 4). In this case SMA8 is less miscible with PPO than SMA2.[19] As can be seen, the particle size of SMA8 modified blend is smaller than that of SMA2 modified blend at short extrusion times. The kinetics of interfacial reaction is faster with SMA8 containing four times more MA reactive groups than SMA2. Beyond 15 minutes of extrusion time, the in-situ formed SMA-g-PA6 copolymer becomes reacher in PA6 grafts causing an unbalanced structural design which drives the copolymer out off the interface into the PA6 matrix, forming micelles. The moderate miscibility of SMA8 with PPO could no more 'counter-balance' the strong interactions the copolymer develops with the PA6 matrix. As a result of this 'decompatibilization' process, the PPO dispersed phase coarsens via a coalescence process.

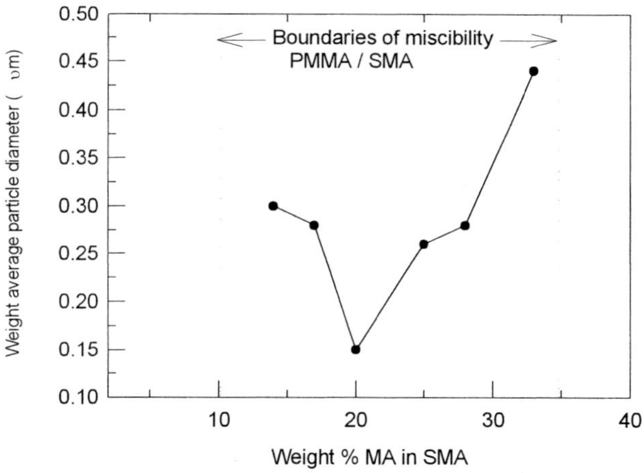

Fig. 3. Weight average particle diameter of the blend PA6 /(PMMA/SMA) 75 / (20 / 5) as a function of the MA content in SMA. All the blends contain 5 wt % of SMA and were compounded at 240 °C for 20 min.

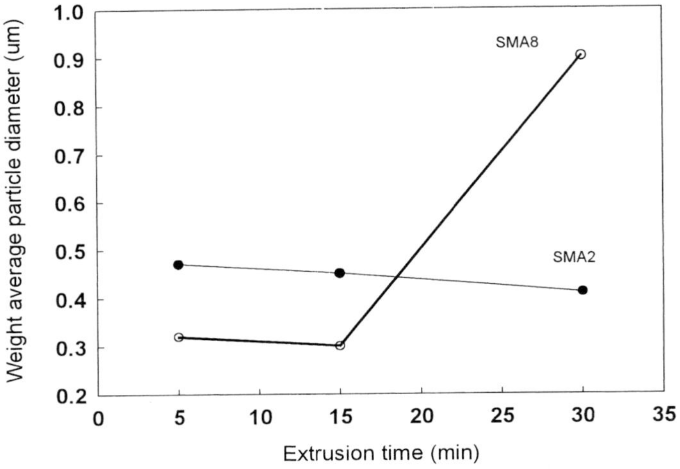

Fig. 4. Weight average particle diameter of PA6 / (PPO / SMA) blends in a weight ratio 75 / (20/5) as a function of the extrusion time. (●) SMA2 ; (O) SMA8.

A third example of copolymer expelling or pull-out from the interface between the individual phases of the blend is illustrated for the SMA reactively compatibilized PA6 / PS blend system. In this system the same SMA2 which is at the limit of miscibility with polystyrene homopolymer[20] results in the in-situ formation of SMA-g-PA 6 compatibilizer which is moderately stable at the interface. Indeed, we could show by doubling the PA6 matrix molecular weight and therefore its melt-viscosity that the formed SMA-g-PA6 copolymer was really driven out of the interface by the larger shearing forces exerted on it. The shearing forces developed by the PA6 matrix of which the molecular weight is 44000 were superior to the interactions the SMA has with the polystyrene minor phase. Contrary to the blend where the PA6 matrix has a molecular weight of 18000, the blend where the molecular weight of PA6 is 44000 exhibited a considerable particle coalescence process as a result of compatibilizer inefficiency. The graft copolymer is expelled from the interface forming micelles in the PA6 matrix, as can be seen from Figure 5a.[21] One would expect a dispersed phase size reduction (drop break-up process) when the matrix viscosity is higher (large MW) compared to a blend where the matrix viscosity is low, as is the case in Figure 5b. The opposite effect is observed which clearly demonstrated the 'just stable' location of the reactively formed SMA-g-PA6 copolymer at the interface. A larger PA6 matrix viscosity completely alleviated the compatibilization efficiency and caused the opposite effect.

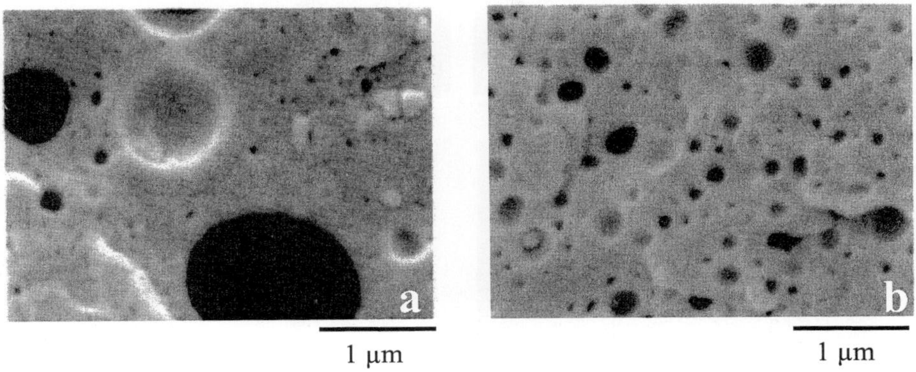

Fig. 5. SEM of cryo-smoothed and THF etched PA6 / (PS /SMA2) blend : 80 / (16 / 4); a) MW of PA6 = 44 000, b) MW of PA6 = 18 000.

The fourth example of this report illustrates a comparison between a tapered diblock copolymer S-*b*-P(E-co-B)-*b*-hB and a triblock S-*b*-hB-*b*-S used as physical compatibilizing agents in low density polyethylene / polystyrene blends. The two copolymers exhibit an excellent compatibilization activity in 80 LDPE/20PS blends where PS forms the dispersed phase. As illustrated in Figure 6, the unmodified blend exhibits, as expected, a coarse phase size (3 μm) as well as catastrophic mechanical properties (elongation at break : 30 %, tensile strength at break : 8.5 MPa). The addition to the blend of 10 wt % of the tapered diblock or the triblock copolymer results in a PS particle size reduction by a factor of 8. The copolymers were not only able to reduce the particle size of the PS phase, but efficiently improved the interfacial adhesion imparting to the modified blends a higher tensile strength and elongation at break (see table 2).

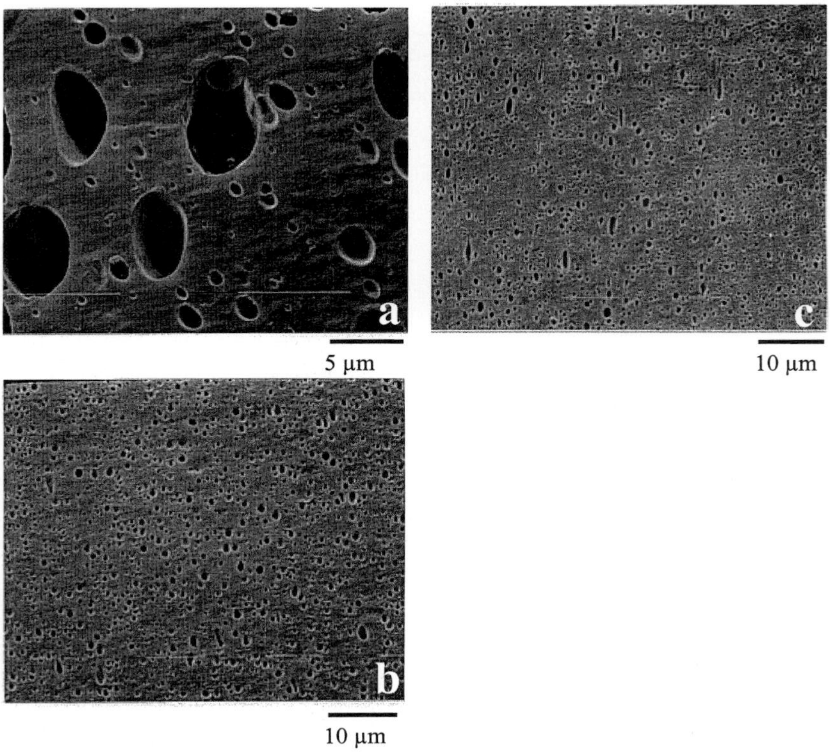

Fig. 6. SEM of cryo-smoothed and THF etched 80 LDPE/20 PS blends:
a) unmodified blend,
b) 10 wt % tapered diblock copolymer modified blend and
c) 10 wt % triblock copolymer modified blend.

In the symmetrically opposite composition of 20 LDPE / 80 PS where the phase morphology developed is co-continuous and not stable upon annealing, the two copolymers behaved differently. As shown in Figure 7 of the unmodified blend, the co-continuous phase morphology evolves to a dispersed phase morphology after as short as 3 minutes of quiescent thermal annealing at 180 °C. For the compatibilized blends a much longer annealing time is required to depict a phase morphology evolution. As shown in Figure 8 the triblock copolymer was not able to prevent phase coarsening to occur. The solvent extraction of PS phase using THF revealed that about 20 % of the phase continuity has been transformed into dispersion.

The blend modified with the tapered diblock copolymer was not only able to prevent the phase coarsening to occur but efficiently preserves the co-continuous character of the phase morphology (Fig. 9). The solvent extraction test carried out on this blend did not show any alteration of phase continuity.

Table 2. Tensile properties and PS phase particle size of unmodified and copolymer compatibilized 80 LDPE / 20 PS blends.

Blend	σ_b (MPa)	E_b (%)	PS Particle size (μm)
Unmodified blend	8.5	30	3
+ 10 wt % tapered copl.	17	350	0.4
+ 10 wt % triblock copl.	16	380	0.4

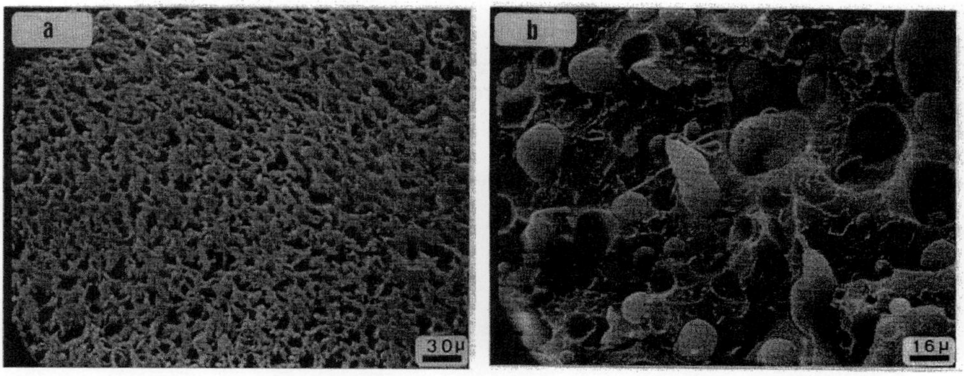

Fig. 7. SEM of unmodified 20LDPE /80PS blend (a) cryofracture and THF etched, non annealed sample, b) cryo-fracture, annealed for 3 minutes at 180 °C.

The clear superiority of a tapered diblock copolymer most likely results from the particular intermediate (tapered) sequence which consists of a mixture of styrene (A) and hydrogenated butadiene B monomeric units as : *-AAABBABABABAABBB-*. A similar sequence is expected to be neither miscible with polystyrene nor with polyethylene phases but has a strong tendency to locate at their interface forming an interphase. It can be considered as the driving sequence that forces the tapered diblock copolymer to locate efficiently and in a more stable manner at the

LDPE / PS interface than the triblock copolymer which is susceptible to partly distribute as own microdomain in either of the two phases, more preferably in the LDPE phase beause of its own phase morphology (PS domains in HPB matrix).

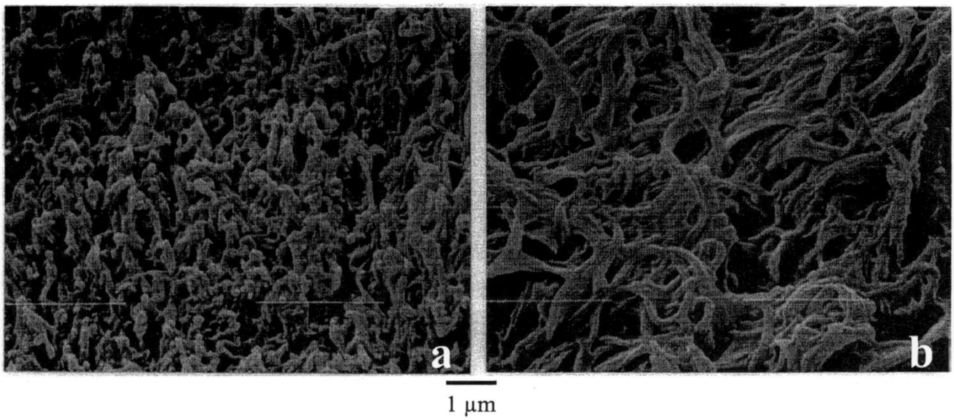

Fig. 8. SEM of room temperature fracture and THF etched 20 LDPE / 80 PS blend added with 10 wt % triblock copolymer : a) non annealed, b) annealed at 180 °C for 150 minutes.

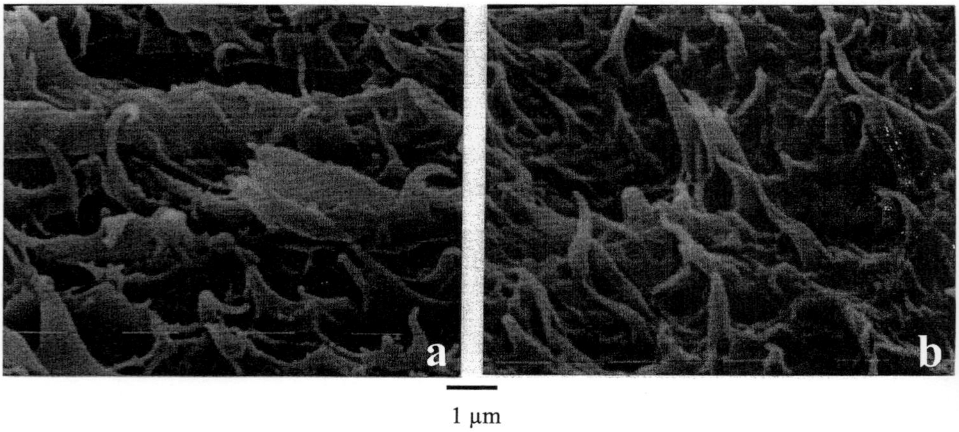

Fig. 9. SEM of room temperature fracture and THF etched 20 LDPE / 80 PS blends added with 10 wt % tapered diblock copolymer : a) non annealed, b) annealed at 180 °C for 150 minutes.

Conclusions

In this paper we reported examples from four different blend systems either reactive blends including PA6/PMMA, PA6/PPO and PA6/PS or physically compatibilized blends LDPE/PS, where the copolymer (graft or diblock) although efficient in developing a fine phase morphology was not able to locate permanently at the interface. A number of factors that cause its delocalization were illustrated. The stability of the copolymer at the interface may be altered because the balance of interaction between its own sequences and the individual phases of the blend is not suitable or the copolymer may organize into micelles in one of the phases. Reactive blending during longer times than usually experienced revealed a state of 'decompatibilization' that resulted in phase coarsening. Thermal annealing can also cause a phase morphology coarsening as a result of copolymer expelling from the interface.

The design of the copolymer compatibilizer must be based on the optimal interaction balance between its own sequences and the respective individual phases of the blend. Depending on the interaction balance, the compatibilizer might easily be rejected from the interface to the phase where it has the best thermodynamic affinity.

Acknowledgements

The authors K. Dedecker, C. Harrats. and G. Groeninckx are indebted to the Research Council of the KULeuven for financial support given to the MSC-Laboratory (GOA project 98 / 06).

[1] G. Groeninckx, C. Harrats and S. Thomas 'Reactive Blending with Immiscible Functional Polymers: Molecular, Morphological and Interfacial Aspects' In " Reactive Polymer Blending" W. E. Baker, C. Scott and G. H. Hua, Ed., Hanser Publishers, Munich, 2001.
[2] R. Fayt, C. Harrats, S. Blacher, R. Jérôme, Ph. Teyssié, J. Polym. Sci.: Physics Ed. 33, 801 (1995).
[3] C. Harrats, S. Blacher, R. Jérôme, J. Polym. Sci.: Physics Ed, (in press 2002).
[4] J. K. Lee and C. D. Han , Polymer, 40, 2521 (1999).
[5] D. Quintens, G. Groeninckx, M. Guest and L. Aerts, Polym. Eng. Sci., 30, 1474 (1990).
[6] D. Quintens, G. Groeninckx, M. Guest and L. Aerts, Polym. Eng. Sci., 30, 1484 (1990).
[7] H. Veenstra, J. V. Dam and A. Posthuma de Boer, Polymer, 41, 3037 (2000).
[8] H. Veenstra, J. V. Dam and A. Posthuma de Boer, Polymer, 40, 1119 (1999).
[9] N. Mekhilef, B. D. Favis and P. J. carreau, J. Polym. Sci. : Part B : polym. Phys. 35, 293 (1997).
[10] K. Dedecker and G. Groeninckx, Polymer 39, 4985 (1998).
[11] K. Dedecker and G. Groeninckx, J. Appl. Polym. Sci., 73, 889 (1999).
[12] M. Seadan, M. Lambla, M. Narkis, A. Siegmann and A. Tzur, Polym. Adv. Techn., 6, 679 (1995).
[13] GH. Hu, YJ. Sun and M. Lambla, Polym. Eng. Sci., 36, 676 (1996).
[14] E. Carone, U. Kopcak, MC. Goncalves and SP. Nunes, Polymer, 41, 5929 (2000).
[15] I. Kelnar, M. Stephan, L. Jakisch and I. Fortelny, Polym. Eng. Sci., 39, 985 (1999).

[16] B. Lu and TC. Chung, Macromolecules, 32, 2525 (1999).
[17] Y. Lee and K. Char, Macromolecules, 31, 7091 (1998).
[18] SC. Tjong, J. Mater. Sci. , 32, 4613 (1997).
[19] H. Wittler and G. Lieser, Makromol. Chem. Rapid. Commun., 14, 40 (1993).
[20] J. H. Kim, J. W. Barlow and D. R. Paul, J. Polym. Sci. Polym. Phys. Ed., 27, 223 (1989).
[21] L. Pan, T. Inoue, H. Hayami and S. Nishikawa, Polymer, 43, 337 (2002).

Dependence of the Interfacial Reaction and Morphology Development on the Functionality of the Reactive Precursors in Reactive Blending

*Z. Yin, C. Koulic, C. Pagnoulle, R. Jérôme**

Center for Education and Research on Macromolecules (CERM), University of Liège, Sart-Tilman B6a, 4000 Liege, Belgium

Summary: PMMA containing 50 wt% of anthracene-labeled PMMA chains end-capped by a phthalic anhydride group (anth-PMMA-anh) has been melt blended at 180°C with PS containing 33 wt% of chains end-capped by an aliphatic primary amine (PS-NH_2) and PS bearing 3.5 pendant amine groups (as an average) along the chains (PS-co-PSNH_2), respectively. The reactive chains have been synthesized by atom transfer radical polymerization. Conversion of anth-PMMA-anh into PS-b-PMMA and PS-g-PMMA copolymers has been monitored by SEC with a UV detector. The interfacial reaction mainly occurs in the initial melting and softening stage (<1.0 min.), although at a rate which strongly depends on the number of reactive groups attached to PS chains, the higher conversion being observed for the PS-co-PSNH_2 containing blends. The phase morphology depends on the architecture of the in-situ formed copolymer. Indeed, a coarser phase dispersion is observed in case of the graft copolymer compared to the diblock.

Keywords: anthracene labeling, polymer blend compatibilization, reactive blending

Introduction

Polymer blending is an efficient strategy to design new materials from commercially available polymers with an acceptable performances/cost balance.[1] Most polymers being immiscible, blends have to be compatibilized in order to decrease the interfacial tension, to restrict the phase coalescence and to improve the interfacial adhesion.[2-6] As a rule, compatibilizers are block or graft copolymers, which are either premade or formed in-situ in reactive blending. The molecular architecture of the preformed copolymers (block or graft copolymers) and, thus, the structure of the reactive chains used in reactive blending have an influence on the morphology development and the final physico-mechanical properties of the blends.[7] Fayt et al.[8-11] have

 DOI: 10.1002/masy.200350817

compared the efficiency of premade copolymers in improving the ultimate mechanical properties of polyblends. They concluded that a block copolymer is more efficient than a graft copolymer and a tapered diblock is superior to a pure diblock. In reactive blending, the molecular architecture of the in-situ formed copolymer has also an influence on its residence time at the interface. Inoue et al.[12,13] compared the in-situ formation of linear block copolymer and graft copolymer of the same molecular weight. Although the block copolymer left easily the interface, the graft copolymer with the trunk in the dispersed phase was stable at the interface. Nevertheless, mechanical properties were higher for the linear copolymer containing blends.[14] Moon et al.[15] showed the incidence of the reactive precursors structure on the reaction kinetics. End-functional poly(methyl methacrylate) reacts more rapidly than chains to which the reactive group is attached at mid-length. Pagnoulle et al.[16] investigated the influence of the architecture of the graft copolymer formed by the interfacial reaction of mutually reactive PP and SAN chains on the interfacial adhesion. Highly multi-graft copolymer formed at the SAN/PP interface results in relatively lower interfacial toughness than single graft copolymer.

In this work, extent of the interfacial reaction and development of the phase morphology of polystyrene/poly(methyl methacrylate) (PS/PMMA) blends have been investigated in relation to the architecture of the in-situ formed copolymer. For this purpose, PMMA chains end-capped by phthalic anhydride have been reacted, at 180°C, with primary amine terminated PS and randomly functional PS (3.5 NH_2 per chain), respectively. The end-functional PMMA chains have been labeled by anthracene in order to monitor the reaction progress in a straightforward and reliable way. Conversion of PMMA-anh into PS-b-PMMA and PS-g-PMMA copolymers has been monitored by size exclusion chromatography (SEC) with a UV detector. The phase morphology of the blends has been analyzed by transmission electron microscopy (TEM).

Experimental Section

Polymer Synthesis and Characterization

Molecular characteristics of the polymers used in this study are listed in Table 1. Molecular weight and molecular weight distribution were analyzed by size exclusion chromatography (SEC) in THF at 40°C, using a Hewlett-Packard 1037A refractive index and a UV detector.

PMMA and PS standards were used for calibration. The anthracene content of PMMA was determined by UV spectroscopy (Hitachi U-300).

Table 1. Characteristics of the polymers used in this study.

Polymers	Mn (g/mol)	Mw/Mn	Functionality (groups/chain)	Anthracene content (groups/chain)
PMMA	40000	1.2	-	-
PMMA	24000	1.1	-	-
anth-PMMA-anh	24000	1.2	0.84	2.0
$PS-NH_2$	27000	1.2	0.86	-
$PS-co-PSNH_2$	30000	1.25	3.5	-
PS	28000	1.2	-	-
PS	100000	1.2	-	-

Polystyrene end-capped by a primary amine ($PS-NH_2$) was prepared by atom transfer radical polymerization (ATRP) as detailed elsewhere.[17] The amine-functionality was estimated as the fraction of the PS chains that reacted in solution with PS end-capped by an isocyanate. This fraction was calculated from the SEC chromatograms before and after coupling, the elution peaks for the original and the coupled PS chains being completely separated one from each other.

Randomly functional polystyrene ($PS-co-PSNH_2$) was prepared by ATRP of a mixture of styrene and {1-methyl-1[3-(1-methylethylenyl)phenyl]ethyl} carbamic acid (TMITBC) without additional solvent. The carbamate groups were deprotected in a (2/1 v/v) dioxane-HCl (12M) mixture at room temperature. The reaction medium was then neutralized with ammonia and NaOH. The amine functionality was calculated from the content of TMITBC determined by FTIR as detailed elsewhere.[18]

Anthracene labeled poly(methyl methacrylate) end-capped by an anhydride (anth-PMMA-anh) was also synthesized by ATRP of a mixture of MMA and (9-anthracene)methyl 2-methyl-2-propenoate (anth-MMA, 2.6 wt%) in toluene at 85°C, using 2-hydroxyethyl 2'-methyl-2'-bromopropionate (HMB) as an initiator and CuBr/1,1,4,7,10,10-hexamethyltriethylene tetramine (HMTETA) as catalyst. This sample was contaminated by a very low fraction of higher molecular weight chains, as shown by the SEC-UV trace (Figure 1).

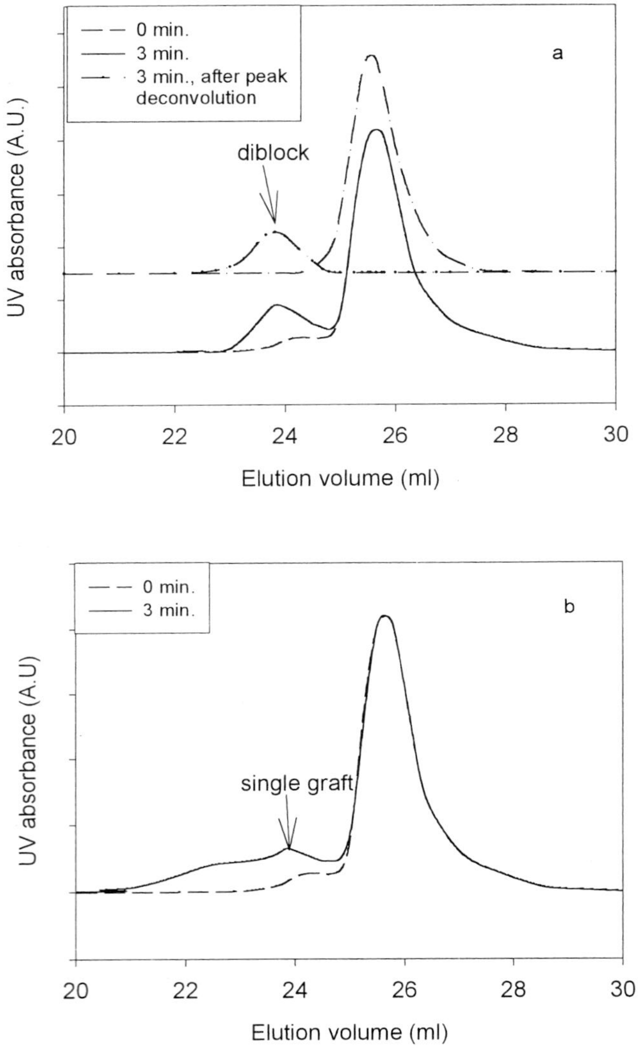

Fig. 1. SEC-UV (at 366 nm) traces for blends of anth-PMMA-anh with $PS-NH_2$ (a) and $PS-co-PSNH_2$ (b) after 3.0 min mixing.

The hydroxyl end-group of the PMMA chains was then reacted with trimellitic anhydride chloride in order to end-cap PMMA by an anhydride.[19] The anhydride content was estimated as the fraction of the anth-PMMA-anh chains that reacted with $PS-NH_2$ in solution. This fraction was extracted from the SEC-UV chromatograms (at 366nm, at which PS absorption is negligible) before and after coupling.

Blend Preparation

Blends were prepared in a twin-screw mini-extruder designed by DSM-Research (The Netherlands) at 180°C under a nitrogen flow. The capacity of the mixing chamber was 5 cm^3, and the material was recirculated for the requested period of time. 4 g of blends (25 wt% PMMA and 75 wt% PS) was melt blended at 150 rpm. The PS phase consisted of 33 wt% of PS-NH_2 (Mn= 27000 g/mol) or PS-co-PSNH_2 (Mn = 30000 g/mol) in non-reactive PS (Mn=100000 g/mol), and 50 wt% of anth-PMMA-anh (Mn = 24000 g/mol) was mixed with non-reactive PMMA (Mn= 40000 g/mol) in the PMMA phase. In non-reactive blends used as a reference, the PS phase consisted of 33 wt% PS of Mn = 28000g/mol and 67 wt% PS of Mn = 100000 g/mol. 50 wt% PMMA of Mn = 24000 g/mol was mixed with chains of Mn= 40000 g/mol in the PMMA phase. After blending, the extruded polymer was quenched into water in order to freeze in the phase morphology.

After reaction, the reactive blend was dissolved in a THF solution of phenyl isocyanate in order to block any unreacted PS-NH_2 or PS-co-PSNH_2 chains that might continue to react with anth-PMMA-anh. This solution was analyzed by SEC with a UV detector at 366 nm. In PS-NH_2 containing blends, the SEC-UV traces were fitted by two Gaussian curves, whose the relative area was used to calculate the conversion of anth-PMMA-anh into diblock copolymer.[19] In PS-co-PSNH_2 containing blends, the amount of anth-PMMA-anh chains converted into graft copolymer was quantified by substracting the contribution of the (anth-PMMA-anh) recorded before reaction from the SEC-UV trace for the reactive blend. The error was estimated at less than 10% in the two cases.

Phase Morphology

A Reichert-Jung Ultracut FC4 microtome equipped with a diamond knife was used to prepare ultrathin samples (ca. 50 nm), that were stained by RuO_4 vapor for 30 min, before being observed with a Philips CM 100 transmission electron microcope (TEM). The dark phase was PS as result of a high affinity for RuO_4. Size and size distribution of the PMMA dispersed phases were quantitatively analyzed with a computerized image analyzer (KS 100 Kontron imaging system). A total of 200-300 dispersed domains were considered per sample. The cross-sectional surface area of these particles was converted to an equivalent diameter by eqn. 1.

$$D_i = (4/\pi(\text{area}))^{0.5} \quad (1)$$

$$D_v = \sum n_i D_i^3 / \sum n_i D_i^2 \quad (2)$$

Results and Discussion

Figure 2 shows the dependence of the volume average diameter, D_v, on the mixing time for reactive and non-reactive blends. Within 1 min, the reactive blends yield a finer phase dispersion than the non-reactive one, which is evidence that a compatibilizer is formed at the interface, i.e. a diblock in case of the interfacial reaction of PS-NH_2 and a graft copolymer when PS-co-PSNH_2 is substituted for PS-NH_2.

The diblock appears to be superior to the graft copolymer in shifting the phase break-up/phase coalescence equilibrium in favor of a finer phase dispersion.

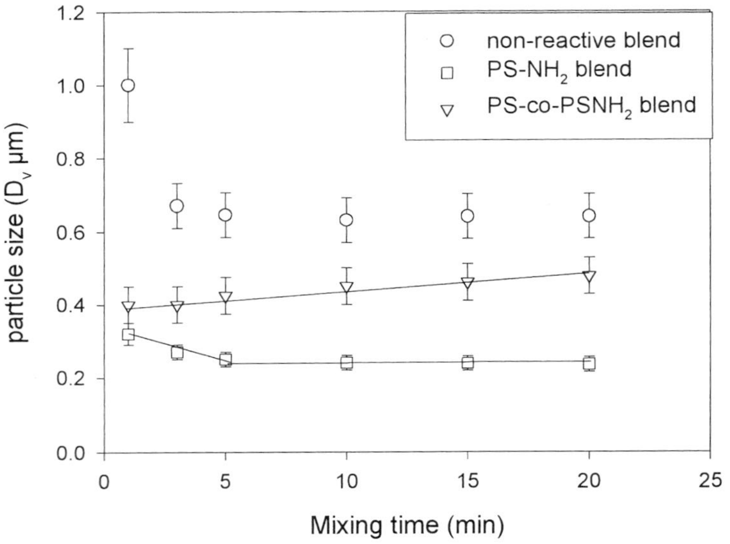

Fig. 2. Dependence of the volume average diameter, D_v, of PMMA/PS blends on the mixing time.

D_v is systematically smaller when a diblock rather than a graft copolymer results from the interfacial reaction. Moreover, Dv decreases during the first 5 min mixing and then levels off in case of diblock, in contrast to a constant although slow increase when the compatibilizer is a graft copolymer. This observation is in agreement with Fayt et al.,[8-11] who concluded that diblock copolymers are more efficient than graft copolymers. D_v is at least 2.5 times smaller after 1 min mixing for the reactive blends compared to the non-reactive system. Moreover, D_v goes on decreasing with increasing mixing time (up to 5 min) in the absence of compatibilization, which is not observed when the interfacial reaction occurs (at least in the time scale of Figure 2).

In order to account for the differences observed when monofunctional (amino)PS rather than polyfunctional PS chains react with monofunctional (anhydride) PMMA, the copolymer formation at the interface has been monitored with the mixing time.

Figure 1 shows the SEC-UV traces for PS-NH_2 (Figure 1a) and PS-co-PSNH_2 (Figure 1b) containing blends, respectively, after 3.0 min mixing. The signature of the diblock or graft copolymer is an elution peak for the converted anth-PMMA-anh chains observed at a lower elution volume than the original one. This additional elution peak is symmetric (at least after deconvolution) consistent with formation of a well-defined diblock in case of the PS-NH_2 containing blend. In contrast, a mixture of multi- and single-graft copolymers is formed in the PS-co-PSNH_2 containing blend. Conversion of anth-PMMA-anh into diblock copolymer was calculated from the area of the two deconvoluted peaks (Figure 1a).

A small amount (3.25%) of coupled anth-PMMA-anh chains at time zero was substracted from the SEC-UV traces of reactive blends. Conversion of anth-PMMA-anh into graft copolymer was estimated by substracting the anth-PMMA-anh SEC-UV peak before reaction from the trace recorded after 3 min mixing. It must be noted that PS-co-PSNH_2 (Mn : 35000 and 4.2 amines per chain) was previously reacted with PMMA-anh (Mn : 35000 and 0.85 anhydride per chain) at 170°C for 10 min (between the parallel plates of a rheometer) and that no significant formation of multigraft copolymer was observed by SEC-UV analysis.[18] Not only the mixing conditions were different, but all the chains were reactive which is not the case in this work. This difference is thought to originate from the interfacial surface area occupied by the copolymer at the interface, which is as small as the copolymer concentration is high.[18] In parallel, the probability that a second amine of the PS-co-PSNH_2 backbone of a single graft copolymer reacts, decreases with the surface area occupied at the interface. In this work, the reactive chains are diluted by non-reactive ones and the PS-co-PSNH_2 chains can adopt a flat conformation at the interface which is favorable to reaction of more than one amine per chain, consistent with the content of multi-graft copolymer which increases with the mixing time. Figure 3 illustrates the time dependence of the conversion of anth-PMMA-anh into diblock and graft copolymers, respectively.

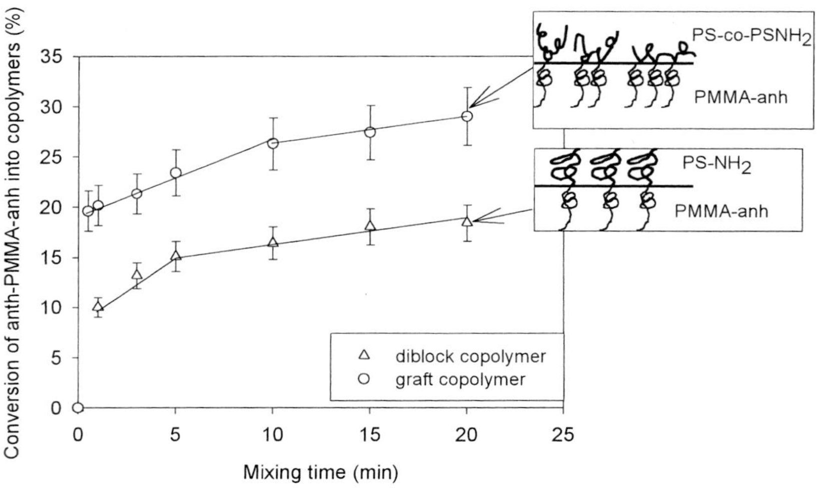

Fig. 3. Conversion of anth-PMMA-anh into diblock and graft copolymers in PS-NH_2 and PS-co-PSNH_2 containing blends, respectively.

Whatever the amine functionality of PS, the interfacial reaction mainly occurs extensively within a very short mixing time, consistent with the sharp breakdown in the progress of the morphology development (Figure 2) and with observation reported elsewhere.[20]

Figures 2 and 3 show an interesting parallelism in the time dependence of the diblock formation and the PMMA particle size in the PS-NH_2 containing blends. Indeed, after 1 min mixing the phase morphology is close to the stationary characteristic features observed after

5 min. Similarly, ca. half the PMMA chains converted at the stationary state, have reacted after 1 min, and the progress of the interfacial reaction is no longer substantial after 5 min.
The interfacial coverage by the diblock copolymer at the PMMA/PS interface has been estimated from eqn. 3, on the assumption that all the copolymer chains remain at the interface. This equation relies on two experimentally determined parameters; i.e., C_{cop} : the conversion of anth-PMMA-anh into diblock copolymer and the volume average particle diameter (Dv)

$$\Sigma = (\phi_w C_{cop} N_{av} \rho_{PMMA} D_v)/6Mn \qquad (3)$$

ϕ_w is the weight fraction of anth-PMMA-anh in the PMMA phase (0.5), N_{av} is the Aveogadro's number, ρ_{PMMA} is the PMMA density (1.15g/cm^3), Mn is the number average molecular weight of the anth-PMMA-anh chains. The maximum coverage of the interface (Σ_0) has also been extracted from the lamellar spacing of a symmetric PMMA-b-PS diblock copolymer.[21] Σ_0 = 0.158 chains/nm$_2$ for PMMA-b-PS copolymer of Mn = 51000g/mol. Figure 4 shows the interfacial coverage Σ (chains/nm^2) by the diblock copolymer in-situ formed in the PS-NH$_2$ containing blends.

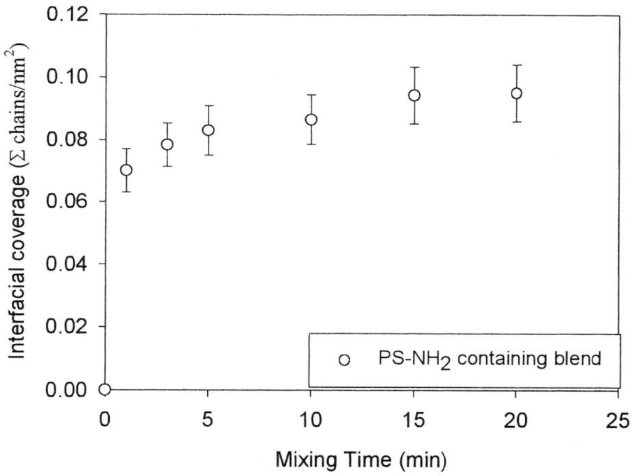

Fig. 4. Interfacial coverage Σ (chains/nm^2) by the diblock copolymer at the PMMA/PS interface in PS-NH$_2$ containing blend.

After 1.0 min mixing, Σ/Σ_0 is ca. 44%, and it increases up to 52% after 5 min, which is a coverage dense enough to stabilize the phase morphology, thus to decrease the interfacial tension and to protect the dispersed phases against coalescence. Because, at constant interfacial surface area (after 5 min mixing), the copolymer formation is dramatically slown down, it must be concluded that the residence time of the symmetric diblock copolymer of Mn = 51,000 is high which prevents not yet reacted amine and anhydride groups from colliding effectively at the interface and the interfacial reaction from progressing. The same situation has been previously reported in case of formation of symmetric diblock of high molecular weight (ca. 55000) at a flat interface.[22]

When PS-co-PSNH$_2$ is substituted for PS-NH$_2$, there is no more similarity in the way the phase morphology and the copolymer formation change with time (figs. 2 and 3). After 1 min mixing, the diameter of the PMMA phases increases continuously although very slowly, which indicates that the interfacial surface area tends to decrease consistent with a stabilization problem in line with D_v ca. 2 times larger than in the PS-NH$_2$ containing blends. Although not shown, the distribution of the PMMA domains is bimodal, which is indicative of phase coalescence. This observation may be surprizing because ca. 2 times more PMMA chains are converted into graft copolymer compared to the formation of diblock (Figure 3). Two pieces of information can be deduced. First, the multifunctionality of PS-co-PSNH$_2$ is favorable to the progress of the interfacial reaction. The 3.7 times higher concentration of amine in the PS phase when PS-co-PSNH$_2$ is used rather that PS-NH$_2$ is a reason for the progress of the interfacial reaction to be 2 times more important after 1 min. Moreover, when a single graft copolymer is at the interface, other amines of the same PS backbone are in a favorable position for reacting with PMMA-anh, at least as long as this backbone is lying flat on the interface. This conformation is reasonable when only part of the chains are reactive at least until the coverage density is not too high, thus apparently until 10 min mixing. This is consistent with a population of multigraft copolymer which coexists with the single graft copolymer and increases proportionally with the mixing time in the first 10 min. This issue cannot be discussed further, because the interfacial coverage by a mixture of single and multi-graft copolymers cannot be estimated properly. Secondly, the non complete stabilization of the interface in spite of a more extended interfacial reaction suggests that the graft copolymer is less efficient than the diblock and/or that part of it leaves the interface. In this respect, multigraft copolymer chains contain more PMMA than PS, as schematized in the insert of Figure 3, which may trigger diffusion away from the interface to the PMMA phase. When PS-NH_2 reacts with PMMA-anh, a symmetric diblock copolymer is formed at the interface, with individual blocks longer than the critical molecular weight for entanglement (Me = 13000 for PS[23] and 10000 for PMMA). This diblock resides at the interface for a long time and protects the PMMA domains against coalescence. When the interfacial formation of a graft copolymer is envisioned, the multifunctionality of the reactive PS chains diluted by unreactive PS leads to multigraft copolymer chains as extensively (compared to the single graft) as the mixing time is long, eventhough the amine groups are in molar excess with respect to the PMMA anhydride end-groups. The PS backbone of these chains cannot penetrate deeply the PS matrix, which explains that they are poorly efficient in preventing the PMMA domains from coalescing. The asymmetric composition of the multigraft copolymer is also an incentive for it to leave the interface.

Conclusion

Anthracene labeled PMMA chains end-capped by phthalic anhydride (anth-PMMA-anh) have been melt blended with PS-NH_2 and PS-co-PSNH_2 , respectively, in a mini-extruder. Only part of the chains in each phase is reactive. Progress of the PMMA-anh/PS-NH_2 and PMMA-anh/PS-co-PSNH_2 interfacial reaction has been monitored by SEC with a UV detector. Conversion of anth-PMMA-anh into diblock and graft copolymers has been determined from the relative area of the elution peaks for the unreacted PMMA and PMMA which is part of the copolymer. The interfacial reaction extensively occurs in the very first step of softening and melting of each constitutive component (<1.0 min.). Because of a higher content of amine and the polyfunctionality of the chains, PS-co-PSNH_2 reacts more rapidly with PMMA-anh than PS-NH_2. As a rule, after the initial step of mixing, the interfacial reaction is rapidly slown down.

In case of the diblock copolymer, as soon as the phase morphology remains constant (after 5 min), the progress of the interfacial reaction tends to level off. Thus, the copolymer stabilizes the interface, which remains occupied by the diblock chains and which is hardly accessible by the reactive chains. A coarser morphology is formed in the graft copolymer containing blends, although more PMMA chains are converted into graft than into diblock. Because a single-graft copolymer is not selectively formed, the coalescence barrier around the PMMA particles is less efficient and the residence time of the asymmetric copolymer chains at the interface may be shortened.

Acknowledgment

The authors are grateful to the "Services Fédéraux des Affaires Scientifiques Techniques et Culturelles" for general support to CERM and for a fellowship to Z.Y. in the frame of the "PAI-4/11:Supramolecular Chemistry and Supramolecular Catalysis". C.K. is "Aspirant" by the "Fonds National de la Recherche Scientifique" (F.N.R.S.), to which he is grateful for a research fellowship.

[1] C. Pagnoulle, C. Koning, M. Van Duin, R. Jérôme, *Progress in Polymer Science*, **1998**, *23*, 707.
[2] L. Leibler, *Macromol. Chem., Macromol. Symp.*, **1988**, *16*, 1.
[3] J. Noolandi, K.M. Hong, *Macromolecules*, **1982**, *15*, 482.
[4] U. Sundararaj, C.W. Macosko, *Macromolecules*, **1995**, *28*, 2647.
[5] N.C. Beck Tan, S.K. Tai, R.M. Briber, *Polymer*, **1996**, *37*, 3509.
[6] L.J. Norton, V. Smigolova, M.U. Pralle, A. Hubenko, K.H. Dai, E.J. Kramer, S. Hahn, C. Berglund, B. Dekoven, *Macromolecules*, **1995**, *28*, 1999.
[7] C. Pagnoulle, R. Jérôme, "Chapter 4: Key role of structural features of compatibilizing polymer additives" in *"Reactive Polymer Blending"*, W. Baker, C. Scott, G.H. Hu Eds., Hanser Gardner publications Inc., **2001**.
[8] R. Fayt, R. Jérôme, Ph. Teyssié, *J.Polym.Sci., Polym.Lett.Ed.*, **1981**, *19*, 79.
[9] R. Fayt, R. Jérôme, Ph. Teyssié, Ph., *J.Polym.Sci., Polym.Lett.Ed.*, **1982**, *20*, 2209.
[10] R. Fayt, R. Jérôme, Ph. Teyssié, *J.Polym.Sci., Polym.Lett.Ed.*, **1989**, *27*, 775.
[11] C. Harrats, R. Fayt, R. Jérôme, *Polymer*, **2002**, *43*, 863.
[12] P. Charoensirisombon, T. Chiba, S.I. Solomko, T. Inoue, M. Weber, *Polymer*, **1999**, *40*, 6803.
[13] P.Charoensirisombon, T. Inoue, M. Weber, *Polymer*, **2000**, *41*, 4483.
[14] P. Charoensirisombon, T. Chiba, K. Torikai, H. Saito, T. Ougizawa, T. Inoue, M. Weber, *Polymer*, **1999**, *40*, 6965.
[15] B. Moon, T.R. Hoye, C.W. Macosko, *Macromolecules*, **2001**, *34*, 7941.
[16] C. Pagnoulle, R. Jérôme, *Macromol.Symp.*, **2000**, *149*, 207-217
[17] K. Matyjaszewski, Y. Nakagawa, *Macromol.Rapid Commun.*, **1997**, *18*, 1057.
[18] Z. Yin, C. Koulic, C. Pagnoulle, R. Jérôme, *Macromolecules*, **2001**, *34*, 5132.
[19] Z. Yin, C. Koulic, C. Pagnoulle, R. Jérôme, *Can. J. Chem.Eng.*, Submitted.
[20] G.H. Hu, I. Kadri, *J. Polym. Sci. : Part B : Polym. Physics*, **1998**, *36*, 2153.
[21] C.W. Macosko, Ph. Guégan, A.K. Khandpur, A. Nakayama, Ph. Maréchal, T. Inoue, *Macromolecules*, **1996**, *29*, 5590.
[22] (a) Z. Yin, C. Koulic, C. Pagnoulle, R. Jérôme, *Macromol. Chem. Phys.*, **2002**, *203*, 2021; (b) Z. Yin, C. Koulic, C. Pagnoulle, R. Jérôme, *Langmuir*, in press.
[23] L.J. Fetters, D.L. Lohse, D. Richter, T.A. Witten, A. Zirkel, *Macromolecules*, **1994**, *27*, 4649.

Morphology of Reactive PP/PS Blends with Hyperbranched Polymers

Jürgen Pionteck, Petra Pötschke, Norbert Proske, Hanying Zhao, Hauke Malz,*[a] *Detlev Beyerlein, Ulrich Schulze, Brigitte Voit*

Institute of Polymer Research Dresden, Hohe Straße 6, D-01069 Dresden, Germany; E-mail: pionteck@ipfdd.de
[a] present address: Elastogran GmbH, KFE, P.O.Box 1140, D-49440 Lemförde, Germany

Summary: To study the efficiency of different mechanism for reactive compatibilization of polypropylene/polystyrene (PP/PS) blends main chain or terminal functionalized PP and terminal functionalized PS have been synthesized by different methods. While the in-situ block and graft copolymer formation results in finer phase morphologies compared to the corresponding non-reactive blends, the morphology development in the ternary blend system PP/PS + HBP (hyperbranched polymer) is a very complex process. HBP with carboxylic acid endgroups reacts preferably with the reactive sites of the oxazoline functionalized PS (PS-Ox) and locates mainly within the dispersed PS-Ox phase. A bimodal size distribution of the PS-Ox particles within the oxazoline modified PP (PP-Ox) matrix phase is observed with big PS-Ox particles (containing the HBP as dispersed phase) and small PS-Ox particles similar in size like the unimodal distributed particles in the non-reactive PP-Ox/PS-Ox blends. Factors influencing the morphology are discussed.

Keywords: blends, hyperbranched polymers, morphology, reactive compatibilization

Introduction

The compatibilization of immiscible blends can be done by the addition of emulsifiers (so called compatibilizers) as third component or by the use of reactive polymers, which can create block, graft, or star-like copolymers at the interface during the melt mixing process. While the first two mechanisms are under investigation for past several years, the in-situ formation of mixed star copolymers at the interphase of two immiscible polymers due to the addition of highly functional hyperbranched polyesters (HBP) is a rather new approach for compatibilization. Jannerfeldt et al. [1-3] investigated the grafting of aliphatic HBP onto a polypropylene (PP) chain. In this way the number of reactive sites per molecule is increased. The effect of this grafted PP in blends with polyamide 6 on interfacial properties and blend morphology was reported.

 DOI: 10.1002/masy.200350818

One of the goals of our work is the comparison of the different reactive compatibilization mechanisms, which are shown in Scheme 1. For this it is necessary to prepare proper reactive blend components. We used for this study the immiscible blend system PP/polystyrene (PS) since there are a variety of methods to synthesize or modify both components with different functionalities and PP and PS are stable during processing. As reactive compatibilization reaction the conversion of 2-substituted 1,3-oxazolines with carboxylic acids to mixed ester amides was chosen.

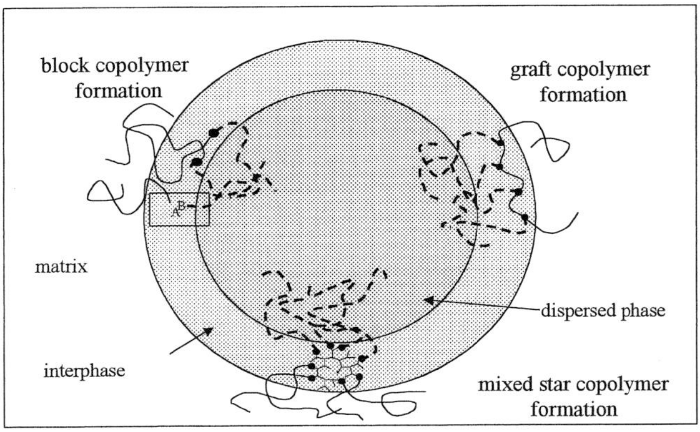

Scheme 1. Formation of block, graft, or star copolymers at the interface between immiscible polymers during melt mixing; A, B: different functionalities; black spots: coupling sites.

In previous studies the efficiency of the in-situ graft and block copolymer formation resulting in finer phase morphologies of reactive blends compared to the corresponding non-reactive blends was shown.[7,8] In this work we study the compatibilizing effect of the addition of polyfunctional carboxylic terminated HBP (which can react with 1,3-oxazoline terminated PP (PP-Ox) as matrix component as well as with 1,3-oxazoline terminated PS (PS-Ox) as dispersed phase component of the PP-Ox/PS-Ox blend) and compare the result with the in-situ compatibilization by the formation of PP-PS block- and graft-copolymers. The reason for the different behaviour will be discussed.

Experimental

Materials: Non-reactive isotactic PP was obtained by metallocene polymerization.[4] The molecular weight determined by GPC with PP standard was M_n = 21000 g/mol with a polydispersity (PD) of 2.05. The PP contains exactly one unsaturated endgroup per molecule, which can be converted into PP-Ox according Scheme 2.

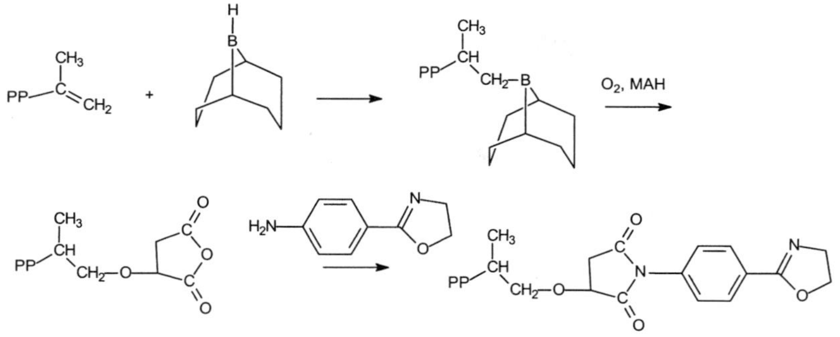

Scheme 2. Synthesis of 1,3-oxazoline terminated PP (PP-Ox).

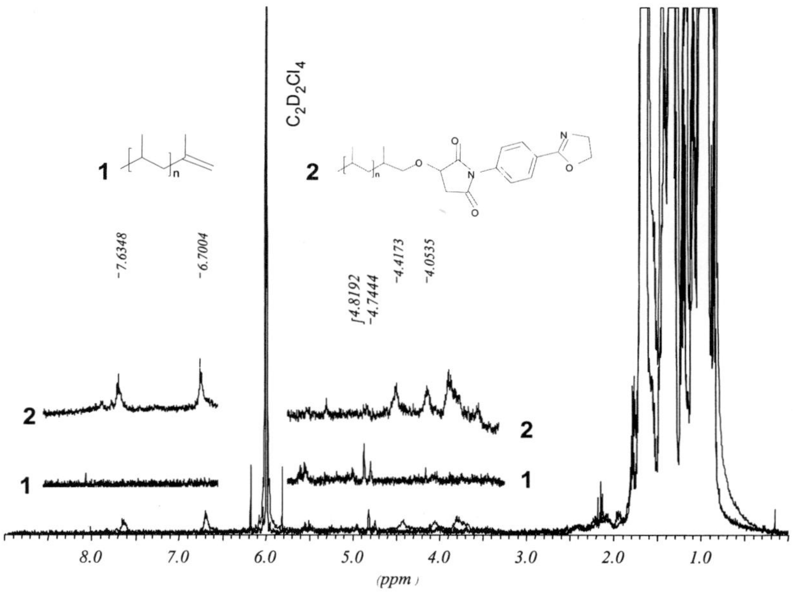

Fig. 1. ^{1}H-NMR spectra of the PP (1) and the oxazoline terminated PP-Ox (2) in $C_2D_2Cl_4$, 120 °C. The signals at 4.05 ppm and 4.42 ppm indicate the methylene protons of the oxazoline; the signals at 4.82 ppm, 4.74 ppm indicate the vinylidene-endgroup of the PP.

The hydroboration product is added to a solution of maleic anhydride (MAH) in THF. When small amounts of oxygen are added, MAH terminated PP as an intermediate product was formed. This can be converted into the PP-Ox by reaction with 2-(4-aminophenyl)-1,3-

oxazoline under the same conditions as described in.[5] The hydroboration and the conversion of PP-MAH to PP-Ox were done in dispersion. To improve the accessibility of the unsaturated sites in the dispersed PP particles small amounts of benzene were added to the tetrahydrofurane (THF) used as solvent. During the reactions no changes in the molecular weight could be observed. However, the conversion is not complete.

From comparison of the ^{1}H-NMR spectra (Fig. 1) of the unsaturated PP with that of the PP-Ox we calculated a functionality of 35 %, that means about every 3rd PP chain contains one 1,3-oxazoline endgroup.

Oxazoline-terminated PS was prepared by "living" radical polymerization using the TEMPO-method under analogous conditions like used for the synthesis of carboxylic terminated PS-COOH.[6-8] The initiator for the synthesis of PS-Ox was prepared by conversion of 4,4'-azo-bis(4-cyanopentanoic acid) (ABCPA) into the bischloride (Scheme 3). The product was dissolved in N,N-dimethylacetamide (DMAc) and slowly added to an equimolar mixture of 2-(4-hydroxyphenyl)-1,3-oxazoline and triethylamine (small excess) at 0 °C, stirred at room temperature for few hours and than carefully added to a saturated ice cooled sodium chloride solution. The precipitate was filtered off and washed with ice water. The functionality of the PS-Ox is about 75 %. The molecular weight is M_n = 28000 g/mol (PD = 1.2). Before using the PS in the blends the TEMPO group was removed by oxidation with m-chloroperbenzoic acid.[9] For comparison a non-reactive PS with Mn = 30000 g/mol (PD = 1.25) was synthesized using AIBN as initiator under similar conditions.

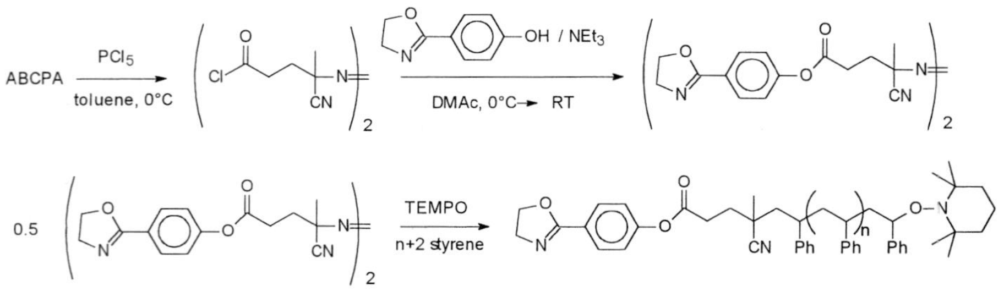

Scheme 3. Syntheses of PS-Ox by the TEMPO-method.

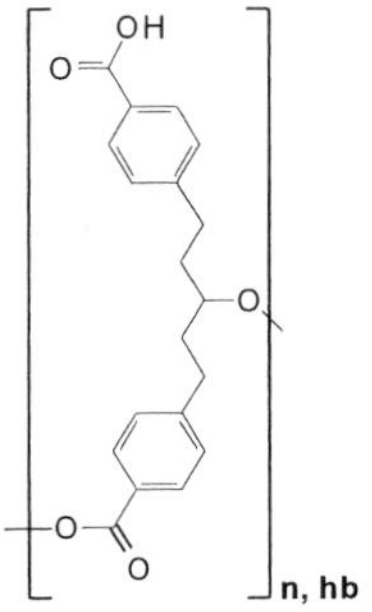

Scheme 4. Chemical structure of the HBP-COOH.

The carboxylic terminated hyperbranched polymer HBP-COOH (Scheme 4) with M_n = 26000 g/mol (GPC, PS standard) was obtained by condensation of 1,5-bis-(4'-carboxyphenyl)pentan-3-ol in presence of dibutyltindiacetate as catalyst. After 90 min stirring the product in argon atmosphere at 210 to 250 °C a vacuum of 0.06 mbar was applied for additional 4.5 h until no bubble development was observed. Details and the synthesis of the monomers are described in.[10] Even considering the uncertainties of the molecular weight determination (PS calibration, hyperbranched nature of the product) we have to expect a very high functionality of the HBP molecules in the range of almost 100 COOH groups per molecule.

Methods: The blends in the composition PP/PS = 2/1 (by weight) plus 5 wt-% HBP were prepared by melt mixing using a Micro Compounder (DACA Instruments; two conical co-rotating screws with a bypass allowing the material to circulate for defined periods; capacity of 4.5 cm^3) at 200 °C with 100 rpm for 5 min, if not otherwise mentioned. The strands coming out of the Micro Compounder were cryo cut by a microtome to analyse the morphology by means of an SEM LEO 435 VP (Leo Elektronenmikroskopie). The dispersed phases were etched under different conditions. Etching with THF for 4 h at room temperature dissolves the PS as well as the HBP phase. Etching with NaOH/water for 4 h dissolves only the HBP phase. The etched samples were sputtered with gold and analysed with an acceleration voltage of 10 kV. The viscosity and the viscosity changes during annealing were measured with an ARES rheometer (Rheometrics) using a parallel plate geometry (1.2 - 1.8 mm gap) under nitrogen atmosphere. Frequency sweeps (0.1 - 100 rad/s, 5 % strain) and time sweeps (60 min, 10 rad/s, 20 % strain) were performed at 200 °C. Infrared spectroscopy was performed using a IFS 66 (Bruker) in transmission mode on thin films prepared by solution casting from THF on KBr plates.

Results

The HBP is insoluble in both PP and PS, independent of the reactivity of the thermoplastics. However, the distribution of the HBP in PS is much finer than in PP (Fig. 2). This can be explained by the interactions between the polymers and the different viscosity ratios between the dispersed phase (HBP) and the matrix components. Due to the aromatic interactions between the HBP and the PS a better compatibility as compared to HBP/PP is achieved. In addition, the surface tension of HBP which is expected to be high because of the extremely high content of functional groups is more similar to that of PS (28.0 mN/m at 200°C [8]) than to that of PP (19.7 mN/m at 200°C[8]). Also the viscosity ratio which is next to the interfacial tension dominating the particle size when mixing non-reactive fluids[11] is different for both systems (Fig. 3). Assuming the validity of the Cox-Merz-rule the viscosity ratio was calculated for a shear rate of 100 s^{-1} that is, even if we do not know the mean shear rate in the Micro Compounder, a value typical for extrusion processing. The viscosity ratio λ changes from about 1 for HBP in PS and PS-Ox to about 3.5 for HBP in PP-Ox, and about 6.4 for HBP in PP. In the blend of PP-Ox with HBP a significant reduction in particle size is seen compared to the nonreactive mixture. whereas only marginal changes occur in the blend of PS-Ox with HBP. In addition to the rheological changes, a reduced interfacial tension due to the functionalization and the desired reaction may cause this reduction.

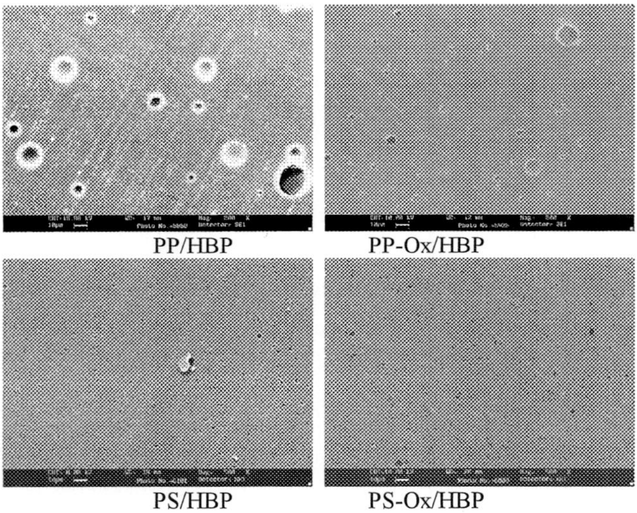

Fig. 2. Distribution of 5 wt% HBP in PP, PP-Ox, PS, and PS-Ox (frame size 240x180 μm).

Surprisingly, the distribution of the PS-Ox in the PP-Ox/PS-Ox blends is much finer than that of PS in the PP/PS blends (Fig. 4, left micrographs) though polymers with very similar molecular weights have been used and a change in the interfacial tension or a reaction between the two oxazoline terminated polymers is not expected. The reason is the strong influence of the functional groups on the viscosity of the PP. Fig. 3 shows that the viscosity of PP-Ox is higher than that of PP, whereas the viscosity of both PS is similar. The viscosity ratio between the dispersed phase (PS, PS-Ox) and matrix component (PP, PP-Ox), again calculated for 100 rad/s, changes from about 6.8 for the non-functionalized blend components to about 3.0 for the functionalized blend components.

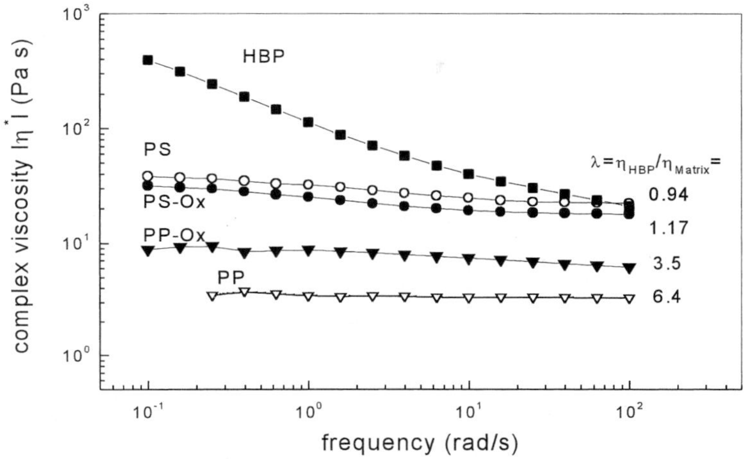

Fig. 3. Viscosity at 200 °C and viscosity ratio at 100 rad/s of the blend components.

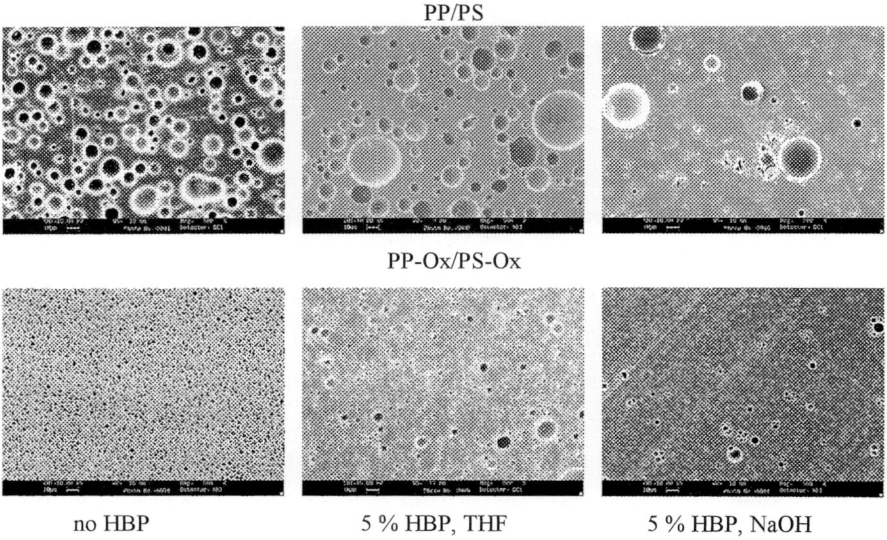

Fig. 4. Morphology of PP/PS=2/1 blends with and without addition of 5 wt% HBP (frame size 240x180 μm).

The strong influence of the viscosity ratio between the reactive blend components on the morphology development was already found for the formation of graft copolymers using oxazoline grafted PP and carboxylic acid terminated PS.[8] In case of the in-situ block copolymer formation during mixing of carboxylic terminated PP and oxazoline terminated PS we observed a similar strong effect of the viscosity ratio on the morphology (Fig. 5).[7] At unfavourable viscosity ratios, λ higher than 1, the effect of the reactivity is clearly visible while at favourable viscosity ratios, λ near 1, the reactive system has only a slight improved phase morphology compared to the non-reactive system. The favourable viscosity ratio results in even finer PS distribution than in the reactive system with more different viscosities.

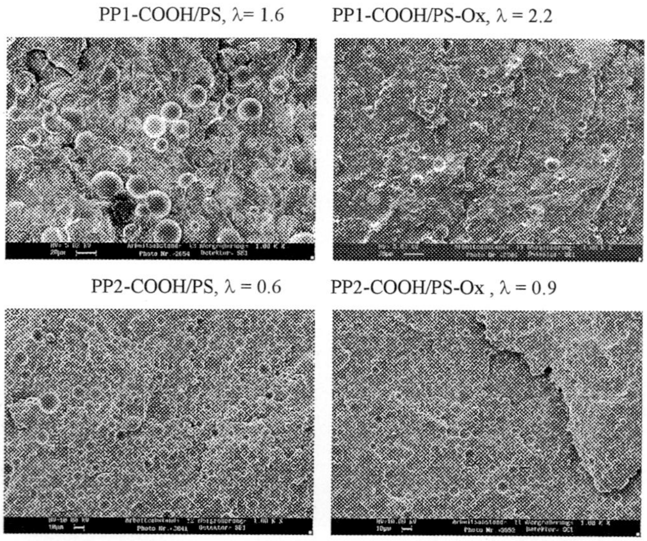

Fig. 5. Influence of the reactivity and viscosity on the morphology of PP/PS = 70/30 blends, SEM of cryofractures [7] (frame size 310x220 μm).

When 5 % HBP are added to the PP/PS blends (Fig. 4), the HBP distributes in a different manner as compared to PP-Ox/PS-Ox blends. In the samples etched with THF a bimodal distribution of the hole sizes resulting from the etching process is visible, in the reactive as well as in the non-reactive blend. When dissolving only the HBP phase by etching with aqueous NaOH it is clearly visible that the big holes in the PP/PS blend are formed from the HBP phase. However, even for the very big HBP particles there seems to be a PS layer around the HBP, visible as a bright shadow. Some other PS particles contain small drops of HBP but there are also unfilled PS drops. There is no HBP particle observable, which is in direct contact to PP. In PP-Ox/PS-Ox blends the HBP locates also mainly within the PS-Ox particles resulting in bimodal size distribution of the PS-Ox phase: small particles without entrapped HBP and large PS-Ox particles containing rather fine distributed HBP. The difference to the nonreactive blend consists in the existence of more than one (up to 15) very finely dispersed HBP drops in one PS-Ox particle, whereas the nonreactive blend always contains only one HBP particle in PS. Therefore, in both blends, only a very small part of the HBP is located at the interphase between the two main components as it would be necessary for acting as interfacial active compatibilizer.

The tendency of the HBP to locate within the PS-Ox particles increases with increasing mixing time. Fig. 6 gives examples of big PS-Ox particles containing a large number of HBP particles. The size of the PS-Ox particles filled with the HBP increases while the size of the unfilled PS-Ox particles (visible only as shadow) remains very fine. In order to proof if it is possible to locate the HBP at the PP-Ox/PS-Ox interphase, a mixture of HBP in PP-Ox was prepared initially by melt mixing for 30 min. After that the PS-Ox was added and the blend was melt mixed for further 5 min. In this case the "phase in phase" morphology is not visible. The HBP is distributed in the PP-Ox matrix as big particles and the PS-Ox exists as finely distributed particle.

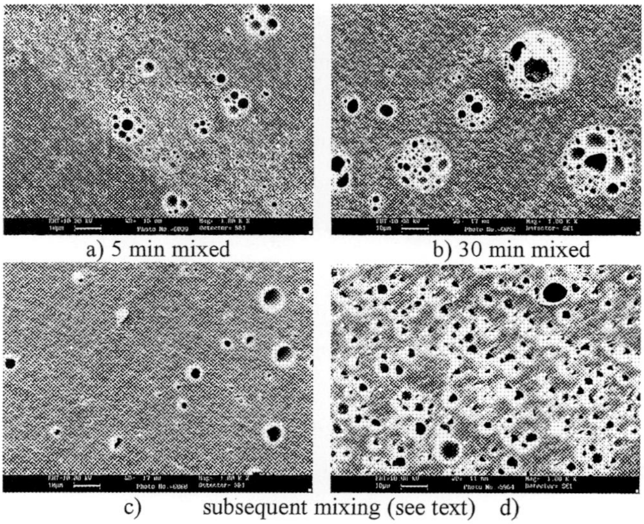

Fig. 6. Morphology of ternary systems PP-Ox/PS-Ox + HBP (2/1 + 5 wt%) in dependence on the preparation conditions, a) and b) simultaneous mixing, c) and d) subsequent mixing; a)-c)surfaces etched with NaOH, d) surface etched with THF (frame size 116x86 μm).

Discussion

The morphology of ternary PP/PS + HBP blends depends strongly on the preparation conditions and the reactivity of the system. Due to the polar character of the HBP it is more compatible with PS and therefore tends to migrate preferably into the PS phase. Because of the favourable viscosity ratio between HBP and PS this results in a fine distribution. When a reaction between the PS-Ox and HBP occurs, the HBP phase is fixed in the PS-Ox phase and finally the HBP is embedded as stabilized dispersed particle inside the PS-Ox particles. The growth of the HBP filled PS-Ox particles with mixing time can be explained by coalescence of the PS-Ox particles in the low viscous PP-Ox matrix because the interface is not stabilized. In addition, due to the reaction of the HBP with PS-Ox the viscosity of the HBP filled PS-Ox droplets increases (see below) resulting in worse viscosity ratios for fine droplet formation. However, inside the PS-Ox particles, the HBP particles do not coalesce but remain their small size. In case of PP/PS blends, the PS particles also coalesce but the HBP entrapped inside seems to coalesce within the PS particles, thus leading to the very big HBP particles. When mixing HBP with PS-Ox the viscosity raised with time detectable as an increase in the torque necessary for remaining a constant speed of the screws and as a continous increase in the pressure measured at the bottom of the Micro compounder. This effect could be also checked by measuring the complex viscosity with time of blends melt mixed for 5 min (Fig. 7). Only in case of PS-Ox/HBP the viscosity increased with time while in case of PP-Ox/HBP and PP-Ox/PS-Ox + HBP the viscosity remained constant over 60 min at 200°C. This indicates a reaction between HBP and PS-Ox but no significant reaction of the HBP with PP-Ox. Reasons for this may be the lower degree of functionalization of the PP-Ox and the relatively small PP-Ox/HBP interfacial area in these blends.

Parallel to that we could prove the chemical reaction between PS-Ox and HBP by FTIR. A decrease in the oxazoline absorbance ($\nu_{C=N}$=1654 cm^{-1}), an increase in the amid absorbance (ν_{CONH}=1680 cm^{-1}), and a shift in the region of carbonyl bands (ν_{CO}=1700-1730 cm^{-1}) were found when comparing a PS-Ox/HBP = 95/5 mixture with a blend melt mixed for 30 min at 200°C (Fig. 8).[12]

When mixing HBP first with PP-Ox, the HBP phase becomes a dispersed phase in PP-Ox exhibiting a large particle size caused by the unfavourable viscosity ratio. In addition, the HBP phase is immobilized by the interaction to PP-Ox. This hinders a finer distribution of the HBP particles as well as the thermodynamically preferred transport of the HBP towards the

PP-Ox/PS-Ox-interface and into the PS-Ox particles when the PP-Ox/HBP preblend is finally melt mixed with PS-Ox.

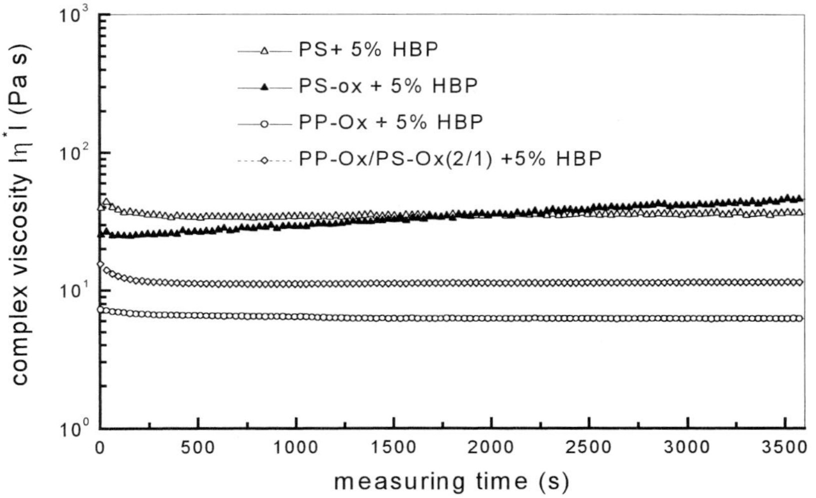

Fig. 7. Time dependence of the complex viscosity at 200°C of blends mixed for 5 minutes in the Micro compounder.

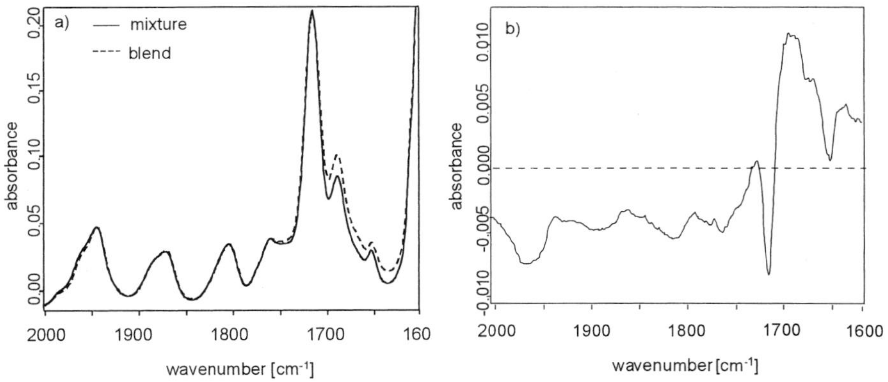

Fig. 8. Transmission IR spectra a) and difference spectrum b) of a PS-Ox/HBP=95/5 mixture and a blend melt mixed for 30 min at 200 °C.

Based on the discussions, it can be concluded that the COOH bearing HBP is not found as a suitable compatibilizer in the studied system of PP-Ox/PS-Ox. By simultaneous mixing of all components in both systems (reactive as well as nonreactive), HBP is not located at the interface. Instead, it resides entirely in the PS or PS-Ox phase (as one particle or many small particles, respectively). However, the mechanism observed offers a new possibility to create phase in phase morphologies that may have interesting properties which has to be studied in future work.

Acknowledgement:

The authors are grateful for the support of this work by the Deutsche Forschungsgemeinschaft (Collaborative Research Centre 287 – A11).

[1] G. Jannerfeldt, L. Boogh, J.-A.E. Manson, *J. Polym. Sci.: Polym. Phys.* **1999**, *37*, 2069.
[2] G. Jannerfeldt, L. Boogh, J.-A.E. Manson, *Polymer* **2000**, *41*, 7627.
[3] G. Jannerfeldt, L. Boogh, J.-A.E. Manson, *Polym. Eng. Sci.* **2000**, *41*, 293.
[4] A. Kaya, L. Jakisch, H. Komber, G. Pompe, J. Pionteck, B. Voit, U. Schulze, *Macromol. Rapid Commun.* **2000**, *21*, 1267.
[5] H. Zhao, J. Pionteck, Ch. Taesler, P. Pötschke, *Macromol. Chem. Phys.*, **2001,** *202*, 313.
[6] M. Baumert, R. Mülhaupt, *Macromol. Rapid Commun.* **1997**, *18*, 787.
[7] H. Malz: Synthesis of End-functionalized Polystyrenes and Polypropylenes (thesis, Dresden Technical University, Der Andere Verlag, Bad Iburg 1999.
[8] P. Pötschke, H.. Malz, J. Pionteck, *Macromol. Symp.* **2000**, *149*, 231.
[9] H. Malz, H. Komber, D. Voigt, J. Pionteck, *Macromol. Chem. Phys.*, **1998**, *199*, 583.
[10] D. Schmaljohann: Funktionalisierung von hochverzweigten Polyestern für den Einsatz als Beschichtungs- und Blendmaterial (thesis, Dresden Technical University), Herbert Utz Verlag Wissenschaft, München 1998.
[11] B. D. Favis in: "Polymer Blends, Volume 1: formulation", Eds. D.R. Paul, C.B. Bucknall, John Wiley &Sons, Inc., New York, 2000, Chapter 16 "Factors Influencing the Morphology of Immiscible Polymer Blends in Melt Processing"
[12] D. Beyerlein: Hochverzweigte Polyester in dünnen Schichten - Charakterisierung der Material- und Grenzflächeneigenschaften (thesis, Dresden Technical University), TENEA Verlag für Medien, Berlin 2002

Compatibilized Blends of Thermoplastic Polyurethane (TPU) and Polypropylene

Qi-Wei Lu,[1] *Christopher W. Macosko,**[1] *Jacques Horrion*[2]

[1]Department of Chemical Engineering and Materials Science, University of Minnesota, Minneapolis, MN 55455, U.S.A.
[2]Huntsman Polyurethanes, Everslaan 45, 3078 Everberg, Belgium

Summary: Compatibilized blends of thermoplastic polyurethane (TPU) and polypropylene (PP) were developed using amine (primary or secondary) functionalized PP's (PP-g-NH_2 or PP-g-NHR). The strategy of reactive compatibilization is based on fast reactions between amine functional groups and urethane linkages or traces of free isocyanates released by thermal degradation of TPU. Excellent compatibilization between TPU and PP was confirmed by rheological, morphological, and mechanical properties. Much finer domain size, higher interfacial adhesion, and more stable morphologies were clearly observed by scanning electron microscopy. Significant improvements in the overall mechanical properties (tensile, tear, abrasion) imply significantly more reaction between TPU and PP phases in the two TPU/PP blends containing PP-g-NH_2 or PP-g-NHR than a TPU/PP blend using PP-g-MA as a compatibilizing agent.

Keywords: amine functionalized polypropylene; blends; maleated polypropylene; reactive compatibilization; thermoplastic polyurethane (TPU)

Introduction

Thermoplastic polyurethane (TPU) and polyolefin (PO) blends have been investigated for their academic interest and commercial importance.[1-5] On one hand, PO is added into TPU to reduce TPU's cost and improve TPU's thermal stability, mechanical properties (e.g. modulus, strength, hardness), and processing performance. On the other hand, TPU is blended with PO to improve PO's properties (e.g. impact strength, adhesion, paintability). However, finding suitable compatibilizers for such an extremely immiscible system has been a great challenge. PO's functionalized (graft or random copolymers) with common functionalities including anhydride,[2,3,6] acid,[6,7] and acrylate[8,9] have been blended with TPU. While morphology and mechanical properties were reported, negligible reactions are expected between these functional groups and the urethane linkage (carbamate -NHCOO-). Also, reactions are quite limited between those functional groups and traces of free isocyanate released from thermal degradation of TPU.[10] Stutz[6] found negligible reactive coupling in TPU blends with acrylic acid and

 DOI: 10.1002/masy.200350819

maleic anhydride functional polymers (polyethylene and polystyrene). He attributed the negligible reaction to interfacial segregation, i.e. preference of the less polar soft segment of polyurethane for the interface. However, sufficiently fast interfacial reactions between highly reactive functional groups may overcome interfacial segregation. In a previous study of model urethane reactions,[10] both primary and secondary amine functionalities were found to be highly reactive toward urethane linkages at melt processing temperatures. Recently we have found a simple method to prepare primary and secondary amine functional polypropylenes (PP).[11]

In this work, compatible TPU/PP blends were developed using amine functional PP's as compatibilizers. Their compatibilization effects on the final blends were investigated by rheological, morphological, and mechanical measurements. The results were compared with two control blends: (1) a TPU/PP blend without any compatibilizer; (2) a TPU/PP blend containing a maleated PP (PP-g-MA) as the compatibilizing agent.

Experimental

Both TPU (Avalon 70AE, Shore hardness: 70A) and PP (13M11, random copolymer) were provided by Huntsman. The maleated PP (PP-g-MA, Fusabond MZ-109D) was obtained from DuPont. The anhydride graft content is 0.55 wt.% according to the supplier. The primary and secondary amine functional PP's (PP-g-NH_2: PP-g-MA/hexamethylenediamine, 1:1.5 by molar ratio of MA to diamine; PP-g-NHR, PP-g-MA/N-hexylethylenediamine, 1:1 by molar ratio of MA to diamine) were prepared according to the method we have developed.[11] All the raw materials were dried at 80 °C in a vacuum oven overnight before processing.

Melt blending of TPU, PP, and the functional PP's (PP-g-X, X = MA, NHR, or NH_2) was performed on a 16-mm co-rotating twin-screw extruder (Prism 16TSC, L/D = 25) at 200 °C and 70 rpm. The screw configuration contains two kneading sections and has been shown effective for melt blending.[12,13] Pellets of the three components were dry-blended before loading. The melt was continuously extruded into a strand through a 10 mm long and 2 mm diameter circular die. Blend compositions and torque and pressure readings during extrusion are listed in Table 1.

Dynamic shear viscosity of each of the blends was obtained using a dynamic stress rheometer (SR-200, Rheometric Scientific) with 25-mm parallel-plates and a 1-mm gap over a frequency range of 500 to 0.1 rad·s^{-1}. The temperature was controlled by forced convection of nitrogen to avoid oxidative degradation. Samples for the dynamic tests were compression-molded at 170 °C under a pressure of 0.55 MPa to round disks. Steady shear viscosity of each of the TPU blends was acquired on a capillary rheometer (RH2200, Rosand) with a circular capillary die (L/D = 30, 180° entrance angle). No corrections were applied to the data and only apparent viscosities were reported.

The TPU blends were cryo-fractured in liquid nitrogen, stained in RuO_4 vapor for 20 min, and coated with 20 Å platinum. Both secondary electron images for topology and backscattered electron images for phase contrast were taken on a scanning electron microscope (S-800, Hitachi) with an accelerating voltage of 10 kV. The interfacial area per unit volume was estimated from backscattered micrographs using a 2D digital image processing method,[14] which calculates interface perimeter/area from the number of pixels along the interface. Only one single image was analyzed, thus the interfacial area per volume values have uncertainties of about 15% particularly those with low interfacial area since cryo-fracturing may expose non-random surfaces.

Tensile specimen (ASTM D412 Type 1) were prepared by injection molding at 170 °C. The mold temperature was 40 °C. The tensile test (5500R, Instron) was run at room temperature with a loading speed of 200 mm/min. The testing standard used for PP is ASTM D638 (Type IV) with a speed of 50 mm/min. Abrasion resistance was evaluated according to DIN 53516 on a Zwick abrasion tester. Tear strength was measured according to DIN 53515 on an Instron 5500R tester at a speed of 500 mm/min. Mechanical test results are based on 3 specimens.

Results and Discussion

The torque and die pressure drop during extrusion are recorded (in Table 1). The values of TPU/PP and TPU/PP/PP-g-MA are close to each other because addition of PP-g-MA has two counterbalancing effects: lower blend viscosity; higher polarity in the PP phase. When the amine functional PP (PP-g-NH_2 or PP-g-NHR) is added into TPU/PP blends, both blending torque and pressure at the die entrance are noticeably higher, which gives a clear sign of interfacial reaction between amine (primary or secondary) functional groups on the PP and urethane linkages or traces of free isocyanates in the TPU.

Table 1. Torque and pressure during extrusion and melt viscosity.

Blend component	Weight ratio	Torque (N-m)	Pressure at die (psi)	Viscosity (Pa-s), 190°C	
				0.05 rad/s	10^3 s^{-1}
TPU/PP	70/30	8.4	288	1650	133
TPU/PP/PP-g-MA	70/25/5	8.7	265	2120	178
TPU/PP/PP-g-NHR	70/25/5	9.5	414	2200	231
TPU/PP/PP-g-NH_2	70/25/5	10.8	464	3240	221

Figure 1 shows shear viscosities of each of the blends at two temperatures (170 °C and 190 °C) as a function of steady shear rate or dynamic frequency. The values at 0.05 rad/s and 10^3 s^{-1} and 190 °C are listed in Table 1. The dynamic and steady shear viscosities of each blend follow the Cox-Merz rule. The viscosity ranking in general agrees with that of torques on the extruder: TPU/PP/PP-g-NH_2 > TPU/PP/PP-g-NHR > TPU/PP/PP-g-MA > TPU/PP. However, viscosity differences are relatively small. It's interesting to observe that the uncompatibilized TPU/PP blend has significantly lower viscosity than others at 190 °C and high shear rates. The abrupt viscosity drop is probably caused by interfacial slip between TPU and PP phases in this highly immiscible blend.[15] The interfacial tension between the soft segments of a poly(ester urethane) and PP was reported to be 8.2 mN/m and that between the hard segments of polyurethane and PP was 13.9 mN/m.[16] The TPU/PP/PP-g-NH_2 blend shows viscosity rising at low dynamic frequency indicating a yield stress, which is also seen in PP-g-NH_2 itself due to a lightly crosslinked structure caused by the primary amine functional groups.[11]

Scanning electron micrographs of each of the blends upon extrusion and after injection molding into a mold plaque are illustrated in Figure 2 and Figure 3, respectively. Since topology is represented by secondary electron images and phase contrast can be best viewed in backscattered electron images, combination of the two enables a more comprehensive understanding of the blend morphology. The unsaturated phenyl rings of the TPU can be selectively stained with RuO_4. Therefore, TPU phase displays in white and PP in black in the backscattered electron images. Amount of interface in each of the blends, in terms of perimeter of the interface divided by the area of the SEM image (i.e. interfacial area per unit volume for a random cut) is listed in Table 2.

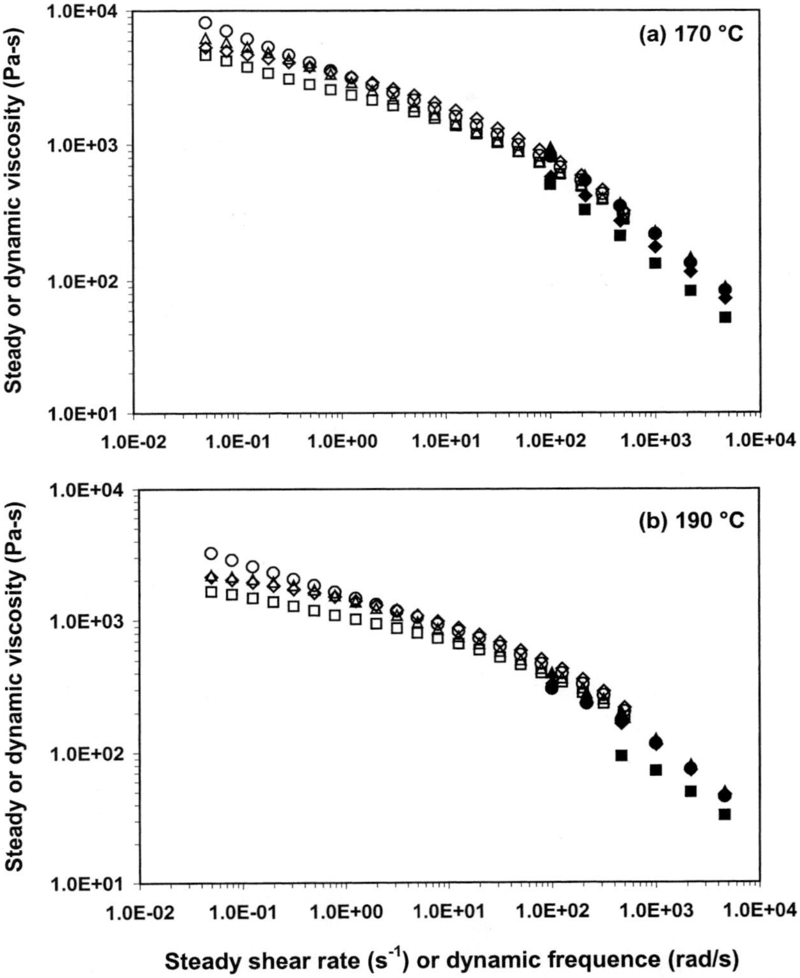

Figure 1. Master curve of viscosity vs. shear rate for different TPU/PP blends at 170 °C (a) and 190 °C (b): TPU/PP: steady (■), dynamic (□); TPU/PP/PP-g-MA: steady (◆), dynamic (◇); TPU/PP/PP-g-NHR: steady (▲), dynamic (△); TPU/PP/PP-g-NH_2: steady (●), dynamic (○). The data from capillary measurement were apparent values without corrections.

In the extrudate of the uncompatibilized TPU/PP blend (Figure 2a), both TPU and PP form long fibers that are well aligned along the flow direction with an aspect ratio greater than 20. The interface is very sharp due to the high interfacial tension. Interfacial failure during the cryo-fracturing (e.g. debonding, pull-out) is obvious. In addition, the morphology is unstable against processing and annealing. After the injection molding, fibers of the same material coalesce to form much larger laminar domains (Figure 3a). The amount of interface (in Table 2) is reduced 75% to a very low level (0.1 μm^{-1}) during coalescence.

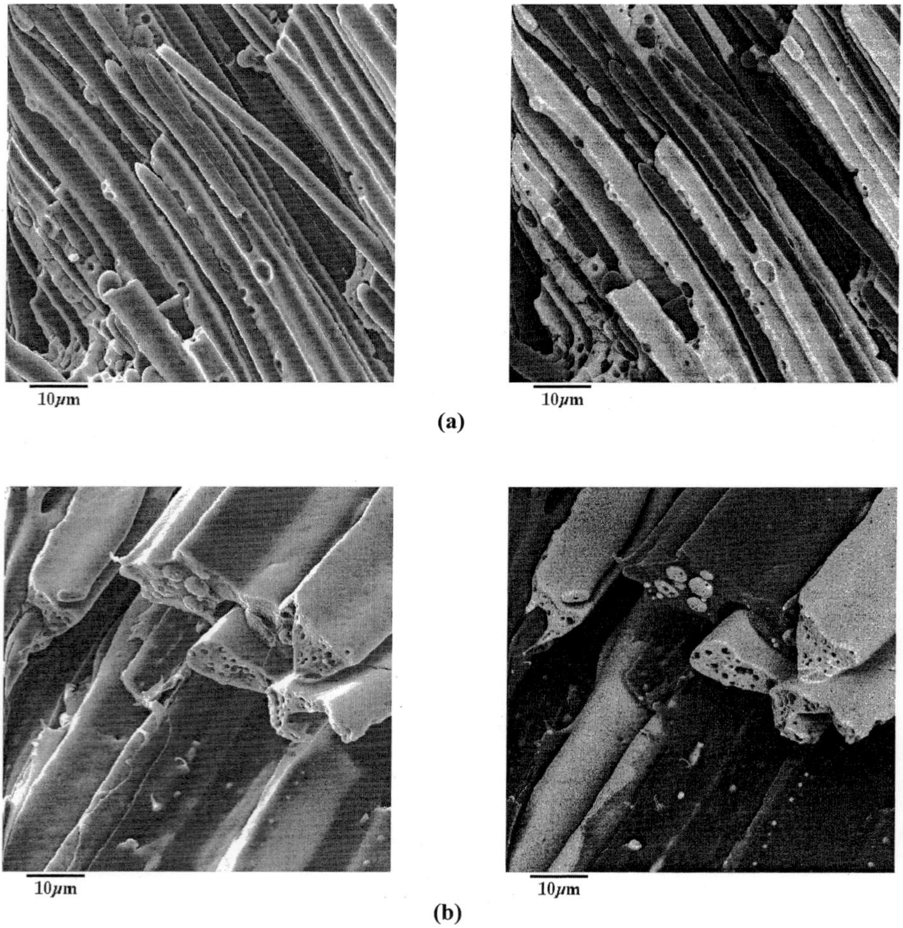

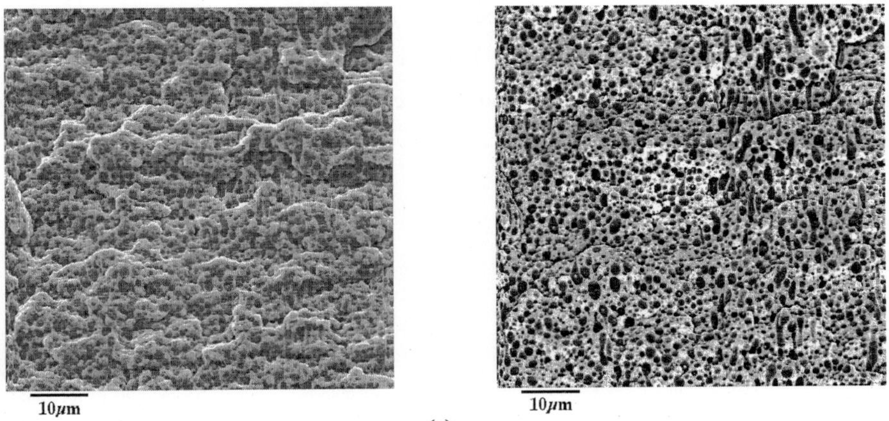

(c)

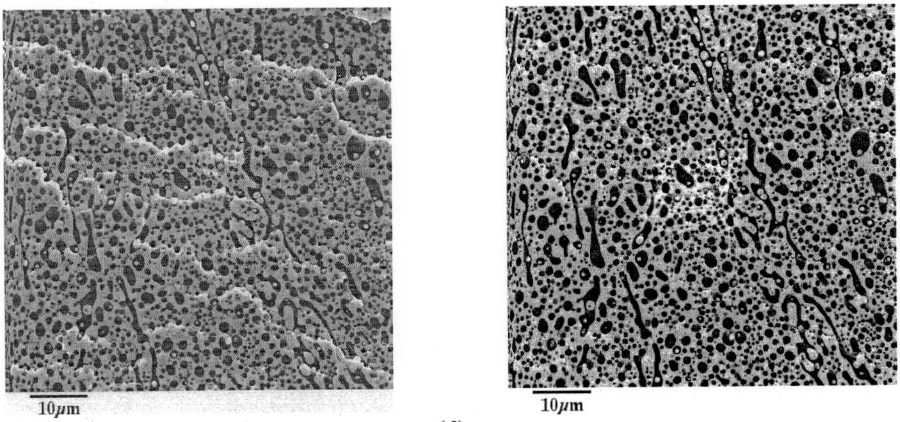

(d)

Figure 2. Scanning electron micrographs of different TPU/PP blends upon extrusion. Both secondary electron images for topology (left) and backscattered electron images for phase contrast (right; note: TPU, white; PP, black) are presented. (a) TPU/PP; (b) TPU/PP/PP-g-MA; (c) TPU/PP/PP-g-NHR; (d) TPU/PP/PP-g-NH_2.

Replacing 5 wt.% nonfunctional PP with PP-g-MA, the morphology upon extrusion becomes more laminar (Figure 2b). The morphology change from the uncompatibilized TPU/PP blend is mainly attributed to the reduced viscosity by adding the low-molecular-weight maleated PP. The interface is still sharp and delamination between layers can be clearly observed. Moreover, the domain size is still very large, as reflected in the interfacial area per unit volume (0.1 μm^{-1}). Nonetheless, the blend maintains a similar structure (Figure 3b) after the injection molding and the domain size decreases a little with more interface (0.2 μm^{-1}) probably generated by the high shear.

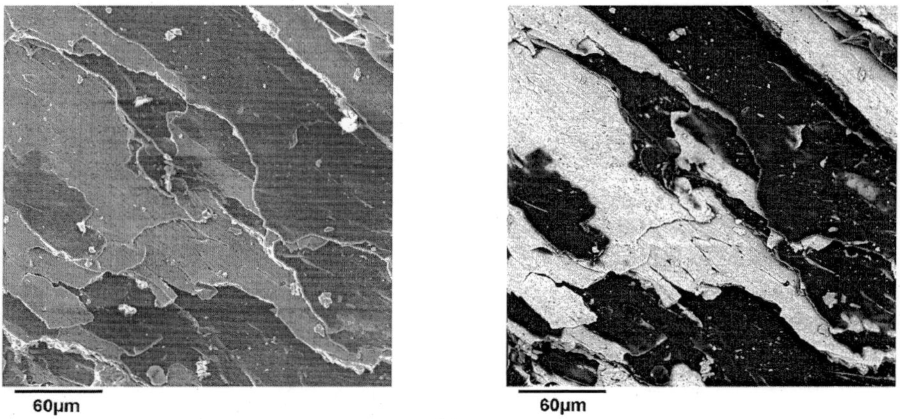

(a)

(b)

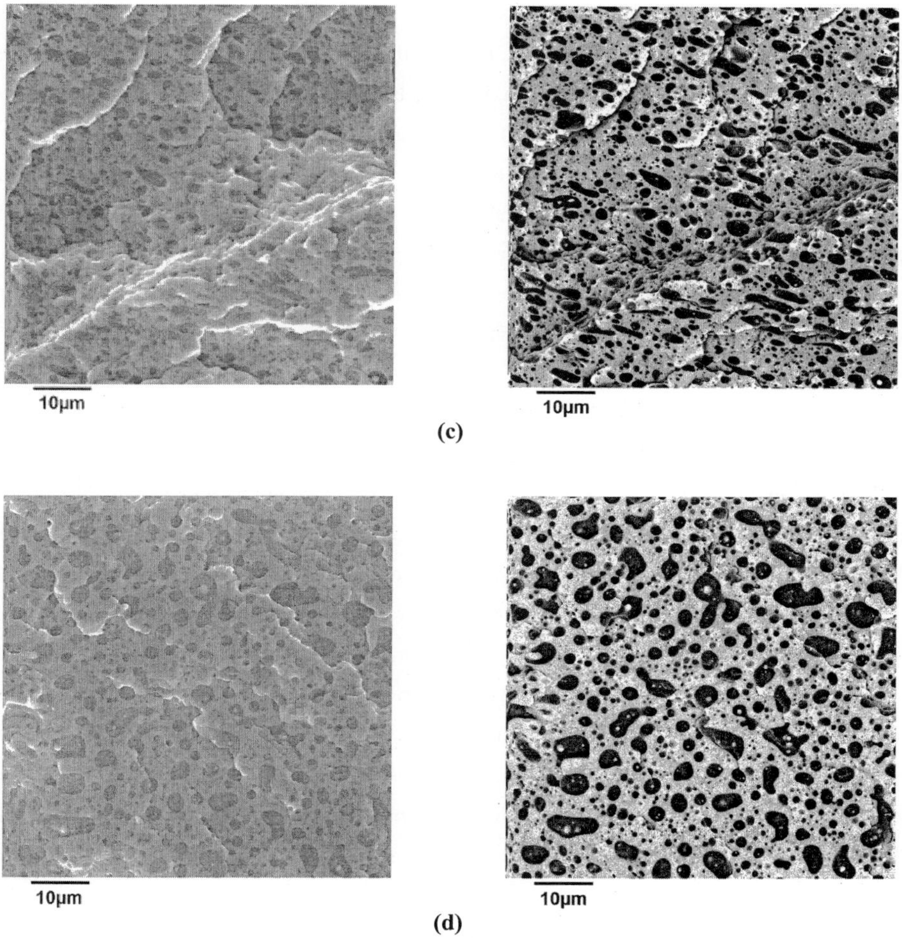

Figure 3. Scanning electron micrographs of different TPU/PP blends after injection molding. Both secondary electron images for topology (left) and backscattered electron images for phase contrast (right; note: TPU, white; PP, black) are presented. (a) TPU/PP; (b) TPU/PP/PP-g-MA; (c) TPU/PP/PP-g-NHR; (d) TPU/PP/PP-g-NH_2.

When 5 wt.% amine (primary and secondary) functional PP is added into the TPU/PP blend, the PP phase becomes particulate (Figures 2c and 2d). The interfacial areas are accordingly much larger (cf. Table 2) compared with the two previous blends. This is probably due to reduced interfacial tension and inhibition of coalescence caused by the reactively generated compatibilizers, i.e. TPU-b-PP block copolymers. All the PP particles on the cryo-fractured surface were broken trans-granularly, i.e. through the particle and no pull-out can be found, indicating the interfacial adhesion between the two phases is very strong. Both amine functional PP blends exhibit excellent morphological stability under a high shear rate during the injection molding. The particles coarsen slightly, i.e. less amount of interface (cf. Table 2). Particle size is somewhat smaller and morphological stability appears to be higher in the TPU/PP/PP-g-NHR blend than in the TPU/PP/PP-g-NH_2 in the micrographs both upon extrusion and after injection molding (Figures 3c and 3d, Table 2). Nonetheless, morphology differences are relatively small between the two amine functional PP blends.

Figure 4 illustrates the stress-stain curves of different TPU/PP blends, as well as TPU and PP homopolymers. The results of all the mechanical (tensile, tear, and abrasion) tests are summarized in Table 2. It's clearly seen that both elastic modulus and tensile strength are significantly higher after adding 30% PP into TPU. However, the uncompatibilized TPU/PP blend yields at 12% and breaks at a very low elongation. The tensile strength is only 5MPa. The TPU/PP/PP-g-MA blend shows considerable improvement with modulus, tensile strength, and elongation at break increased by 66%, 100%, and 220%, respectively. However, other mechanical properties such as tear strength and abrasion resistance are worse. Both TPU/PP/PP-g-NHR and TPU/PP/PP-g-NH_2 blends display higher moduli and much greater toughness. They show quite similar stress-strain behavior. The tensile strengths, close to those of TPU and PP homopolymers, are four times that of the uncompatibilized TPU/PP and twice that of the TPU/PP/PP-g-MA blend. The ultimate elongations, close to that of TPU homopolymer as well, have similar percentage increases. Meanwhile, the tear and abrasion resistance are also enhanced with higher tear strength and less mass loss upon abrasion. These significant improvements in the overall mechanical properties are due to finer morphology, and stronger interfacial adhesion between TPU and PP phases, all produced by the compatibilizers generated by interfacial reactions.

It is important to note that the amine modified PP's were used in relatively small amounts, 16% of the total PP. We have also found that half of that amount still works well.

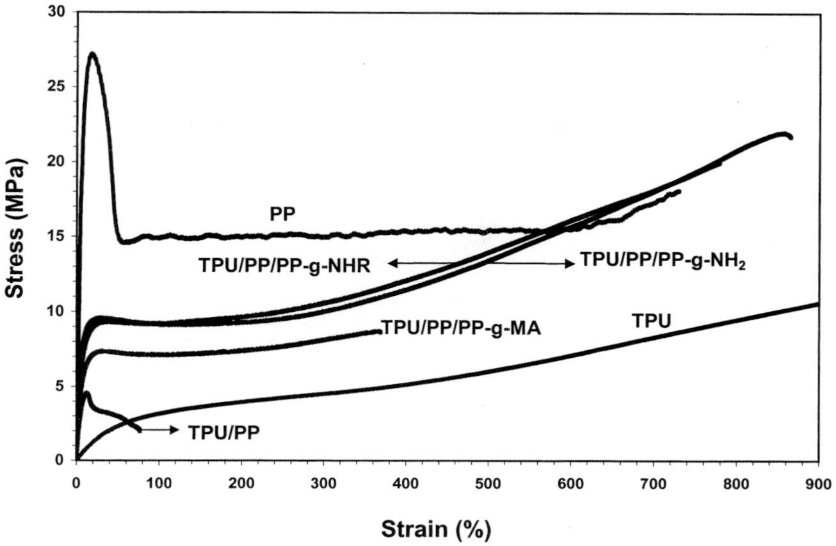

Figure 4. Tensile stress-strain results for different TPU/PP blends and TPU and PP homopolymers. The testing standard for PP is ASTM D638 (Type IV).

Table 2. Morphological and mechanical properties of different TPU/PP blends and TPU and PP homopolymers.

Blend component	Modulus (MPa)	Tensile strength (MPa)	Elongation at break (%)	Tear strength (N/mm)	Abrasion (mg loss)	Amount of interface (μm^{-1})	
						Upon extrusion	After injection molding
TPU*	(6.7)	23	825	65	42	-	
PP	490**	28**	730**	-	-	-	
TPU/PP	82	5	150	33	276	0.4	0.1
TPU/PP/PP-g-MA	136	10	480	25	318	0.1	0.2
TPU/PP/PP-g-NHR	172	20	800	38	173	1.6	1.5
TPU/PP/PP-g-NH_2	225	21	860	39	202	1.5	1.1

* Data provided by the supplier
** ASTM D638

Conclusion

Compatibilized TPU/PP blends can be produced by adding relatively small amount (~10%) of amine (primary and secondary) functional PP's (PP-g-NH_2 and PP-g-NHR). This compatibilization is the result of fast interfacial reactions between amine functional groups and urethane linkages or trace of free isocyanates released by thermal degradation of TPU. Much finer PP domains, better interfacial adhesion, and greater morphological stability were clearly observed. Mechanical properties (tensile, tear, abrasion) wee significantly improved in the two TPU/PP blends containing PP-g-NH_2 or PP-g-NHR compared to the uncompatibilized one, as well as the TPU/PP blend using PP-g-MA as a compatibilizing agent.

Acknowledgement

The authors are indebted to Jeffrey A. Galloway for help with the image analysis. This research was supported by Huntsman Polyurethanes, an international business of Huntsman International LLC.

[1] EP 347,794 (1989), Dow Chemical Co., inv.: V. S. Zabrocki.
[2] T. Tang, X. Jing, B. Huang, *J Makromol Sci – Phys* **1994**, *B33*, 287.
[3] K. Wallheinke, P. Pötschke, C. W. Macosko, H. Stutz, *Polym Eng Sci* **1999**, *39*, 1022.
[4] U.S. 6,174,959 (2001), Bayer Corp USA, invs.: J. F. Ciebien, R. L. Archey, H. L. Noble.
[5] WO 01/36535 (2001), Dow Chemical Co., invs.: H. Farah, F. Jr. Lerma.
[6] H. Stutz, P. Pötschke, U. Mierau, *Macromol Symp* **1996**, *112*, 151.
[7] K. Wallheinke, P. Pötschke, H. Stutz, *J Appl Polym Sci* **1997**, *65*, 2217.
[8] C. P. Papadopoulou, N. K. Kalfoglou, *Polymer* **1998**, *26*, 7015.
[9] C. P. Papadopoulou, N. K. Kalfoglou, *Polymer* **1999**, *40*, 905.
[10] Q. W. Lu, C. W. Macosko, T. R. Hoye, *J Polym Sci: Part A: Polym Chem* **2002**, *40*, 2310.
[11] Q. W. Lu, C. W. Macosko, J. Horrion, "Melt Amination of Polypropylenes", in preparation.
[12] M. Maric: Compatibilization of Thermoplastic/Silicone Blends, Ph.D. Thesis, Department of Chemical Engineering and Materials Science, University of Minnesota **1999**.
[13] M. Maric, C. W. Macosko, *Polym Eng Sci* **2001**, 41, 631.
[14] J. A. Galloway, M. D. Montminy, C. W. Macosko, *Polymer* **2002**, *43*, 4715.
[15] R. Zhao, C. W. Macosko, *J Rheol* **2002**, *46*, 145.
[16] H. Stutz, W. Heckmann, P. Pötschke, K. Wallheinke, *J Appl Polym Sci* **2002**, *83*, 2901.

Biodegradable Compositions by Reactive Processing of Aliphatic Polyester/Polysaccharide Blends

Philippe Dubois[1,*] *and Ramani Narayan*[2]

1)Laboratory of Polymeric and Composite Materials, University of Mons-Hainaut, Place du Parc 20, B-7000 Mons, Belgium
2)Department of Chemical Engineering, Michigan State University, East Lansing, MI 48823, USA

SUMMARY: Commercially available biodegradable aliphatic polyesters, i.e., high molecular weight poly(ε-caprolactone) (PCL) and polylactide (PLA), were melt blended with a well-known natural and biodegradable polysaccharide: starch either as corn starch granules or as thermoplastic corn starch after plasticization with glycerol. Conventional melt blending yielded compositions with poor mechanical performances as a result of lack of interfacial adhesion between the rather hydrophobic polyester matrix and the highly hydrophilic and moisture sensitive starch phase. Interface compatibilization was achieved via two different strategies depending on the nature of the polyester chains. In case of PLA/starch compositions, PLA chains were grafted with maleic anhydride through a free radical reaction conducted by reactive extrusion. The maleic anhydride-grafted PLA chains (MAG-PLA) allowed for reinforcing the interfacial adhesion with granular starch as attested by TEM of cryofracture surface. As far as PCL/ starch blends were concerned, the compatibilization was achieved via the interfacial localization of amphiphilic graft copolymers formed by grafting of PCL chains onto a polysaccharide backbone such as dextran. The PCL-grafted polysaccharide copolymers were synthesized by controlled ring-opening polymerization of ε-caprolactone proceeding via a coordination-insertion mechanism. These compatibilized PCL/starch compositions displayed much improved mechanical properties as determined by tensile testing as well as a much more rapid biodegradation as measured by composting testing.
Keywords: biodegradable, polyesters

Introduction

During the last two decades, considerable effort has gone into the development of biodegradable polymers, polymer blends and composites. Aliphatic polyesters such as polylactide (PLA) and poly(ε-caprolactone) PCL, represent important biodegradable polymers which are now finding commercial applications as single-use disposable items, in addition to the established applications as medical implants, sutures, and drug-delivery

 DOI: 10.1002/masy.200350820

systems.[1,2] Both polyesters are actually produced at the industrial scale : PCL by Solvay Interox in UK, Dow Union Carbide in US and Daicel in Japan, whereas PLA is supplied by, e.g., Cargill Dow in US, Mitsui Toatsu and Shimadzu in Japan. While ε-caprolactone is a petrochemical monomer, lactic acid and thus lactide are interestingly produced from non fossil renewable natural resources by fermentation of polysaccharide or sugar, e.g., from corn and beet, therefore allowing the biological cycle to be closed owing to the PLA biodegradation as well as the photosynthesis process.

Combining these synthetic polyesters with natural materials, e.g., polysaccharides such as starch, provides way of cost reduction and combined properties. Unfortunately, simple polyester composites and blends have poor properties because of the lack of interfacial adhesion.[3-5] Indeed it has been reported that the mechanical properties of the aliphatic polyester/starch blends made from various types of starch generally become poorer with increasing starch content. This can be attributed to the incompatibility between the hydrophobic polyester and the hydrophilic starch. Thus, it is expected that the properties of the polyester/starch blends could be improved by proper interface compatibilization.[6-12)]

This paper aims at reviewing two strategic pathways to improve the compatibility within polyester/granular starch blends. Aliphatic polyester/starch compositions have been prepared by mechanical blending of starch (granular or plasticized corn starch with a relative amylose/amylopectine content of 25/75) either with a maleic anhydride-modified polyester, i.e., maleic anhydride-grafted polylactide, or with poly(ε-caprolactone) compatibilized by amphiphilic PCL-grafted polysaccharide copolymers specifically localized at the PCL/starch interface.

Compatibilization of polylactide/starch compositions

In this first procedure, the PLA chains have been functionalized by maleation reaction in order to get reactive functions pending all along the polyester chains, and able to strongly interact with the hydroxyl groups of the polysaccharide phase. The grafting reaction of maleic anhydride (MA) onto the PLA backbone has been performed by reactive extrusion through a free radical process. 2,5-dimethyl-2,5-di-(tert-butylperoxy)hexane (Lupersol L101, ElfAtochem) was used as the free radical initiator.[13] Accordingly commercial PLA from Cargill Dow, containing 96% of L-lactic acid units, has been reacted in the molten state with maleic anhydride and in presence of Lupersol L101. Lupersol L101 is FDA approved as food additive and interestingly is characterized by an half life below 1 min at the extrusion temperature range, thus around 180°C. The maleation reaction has been conducted in a co-rotating intermeshing twin-screw extruder with a residence time ranging from 4 to 5 min. at a temperature ranging from 160 to 190°C. The relative content in maleic anhydride has been maintained at 2wt% on PLA basis. PLA being highly moisture sensitive, it was important to check first the effect of the extrusion conditions and more particularly of the extrusion temperature on integrity of the PLA chains. In absence of maleic anhydride and L101, the simple extrusion of PLA triggers a drastic degradation of the polyester chains as evidenced by the decrease of both molecular weight (Mn from ca. 120,000 down to ca. 80,000), solution viscosity, and melt viscosity (Table 1). This degradation occurs whatever the extrusion temperature studied, and more likely results from thermo-hydrolysis and back-biting side-reactions.[14]

Table 1 : Extrusion of PLA chains : effect of the extrusion temperature and L101 relative content on the PLA molecular weight (Mn), polydispersity index (Mw/Mn), intrinsic viscosity [η] and melt flow index under 2.16 kg load at 190°C (MI_2)

Extrusion Temp. (°C)	L101 (wt%)	Mn (GPC)	Mw/Mn	[η] (dL/g)	MI_2 (g/10min.)
-[a]	-	121,600	1.41	1.04	12.76
160	0.00	88,800	1.30	0.82	123.30
160	0.14	124,600	1.81	1.14	4.87
160	0.26	130,400	2.54	1.17	4.08
180	0.00	81,000	1.30	0.83	31.92
180	0.10	104,700	1.33	0.98	13.93
180	0.25	129,000	1.48	1.01	6.99
190	0.00	82,900	1.30	0.79	82.67
190	0.10	114,500	1.33	1.02	19.65
190	0.50	153,600	520[b]	1.72	ND

a) Not extruded sample
b) Highly branched PLA chains (microgel formation)

It is very interesting to point out that the simple addition of L101 to the extruded PLA allows to reduce the lost of molecular weight by free radical self-branching reaction. This is nicely evidenced by the extrusion of PLA conducted at 180°C in the presence of 0.25wt% L101 leading to PLA chains with molecular weight, molecular weight distribution and intrinsic viscosity values very similar to the values measured for the pristine (not extruded) PLA chains. It is however worth noting that at higher temperature and free radical initiator content (Table 1, last entry), highly branched PLA chains (even with microgel formation) are formed as shown by SEC with triple detection which displays a much larger radius of gyration, a large fraction of chains with apparent molecular weight (MW) above 1,000,000 daltons and a Mark-Houwink parameter "a" below 0.5, attesting for the highly branched structure and even the presence of microgels of limited solubility in THF at 25°C.[15] Actually the free radical chain branching of PLA more likely occurs by hydrogen radical abstraction in α position of carbonyl groups, followed by radical addition onto the carbon-carbon double bond of the enolate forms in equilibrium within the polyester chains. At larger extent, cross-linking and microgel formation may take place. Furthermore, free-radical chain scissions might occur and participate in the chain branching process as well.

The maleation reaction has been conducted in the presence of 2wt% maleic anhydride at 180°C. Clearly, by increasing the free radical initiator content from 0 to 0.50wt%, the content of maleic anhydride moieties grafted onto PLA chains does increase as checked by back titration of an excess of morpholine with HCl (Table 2).[13] The MW remain quite constant at a value around 85,000, which is more likely explained by a competition between the MW increase through chain branching and MW decrease by β-chain scission triggered by grafted maleic anhydride as shown hereafter (Scheme 1).

Table 2 : Maleation reaction of PLA chains carried out by reactive extrusion in the presence of 2wt% maleic anhydride at 180°C. Effect of L101 relative content on the PLA molecular weight (Mn), polydispersity index (Mw/Mn), and the content of grafted maleic anhydride

L101 (wt %)	Mn (GPC)	Mw/Mn	Maleation (wt %)[a]
0.00 [b]	81,000	1.30	-
0.00	81,200	1.25	0.066
0.10	95,300	1.27	0.227
0.25	89,000	1.30	0.475
0.50	85,100	1.31	0.653

a) Determined by back titration of excess morpholine with HCl
b) No maleic anhydride added to the reaction

Maleic anhydride-grafted PLA (MAG-PLA, here with 0.65wt% maleic anhydride) has been melt blended with granular corn starch in an internal mixer. Cryofracture surface as observed by SEM, displays improved PLA/starch interfacial interactions by using maleic anhydride-grafted polyester chains (Figure 1). Compared to non compatibilized starch/ PLA blends, surface dewetting is no longer observed around the starch particles. This likely results from the strong interactions between carboxylic anhydrides and the hydroxyl functions of the polysaccharide chains. Quantitative characterization of the MAG-PLA/starch compositions by tensile testing but also melt blending with plasticized starch still need to be performed and will be reported elsewhere.

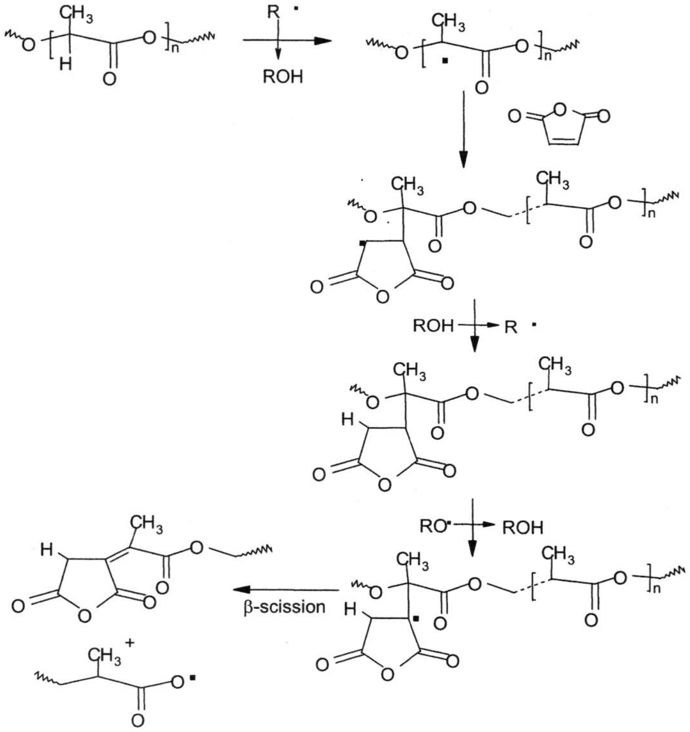

Scheme 1

Figure 1 : Scanning electron microscopy of cryofracture surface of a) corn starch/PLA conventional melt blend and b) corn starch/MAG-PLA (with 0.65wt% maleic anhydride)

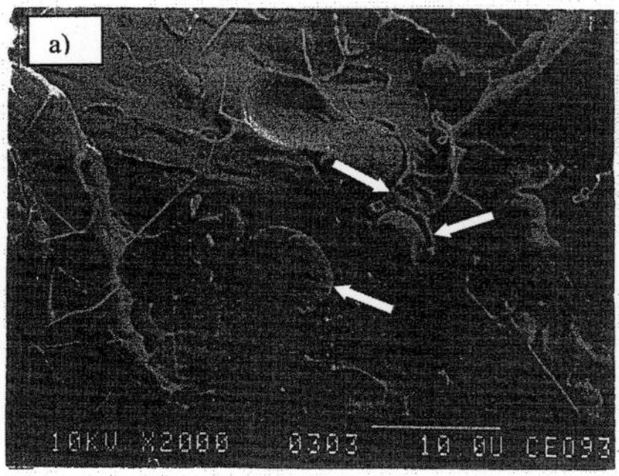

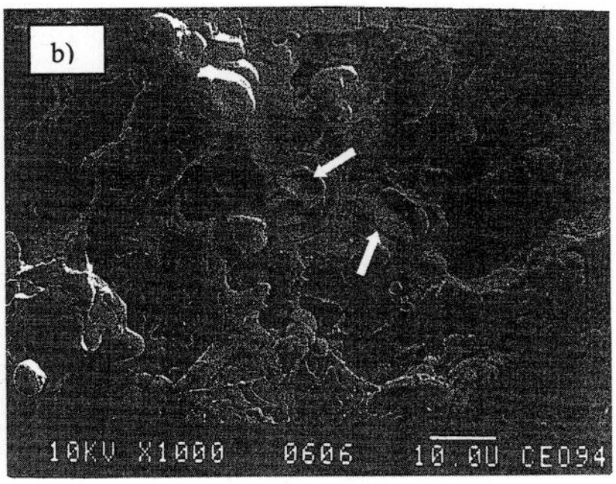

Compatibilization of poly(ε-caprolactone)/starch compositions

The second strategy relies upon the PCL grafting reaction onto a polysaccharide backbone. It means that the interface compatibilization has been achieved by using either preformed amphiphilic graft copolymers, actually PCL-grafted dextran (an exo-1,6-glucan polysaccharide) or by generating the graft copolymers via in situ ROP of ε-caprolactone directly initiated from the starch phase.

Thus, in a first series of experiments, PCL-grafted dextran copolymers have been prepared by a 3-step synthesis, involving successively a first partial and reversible protection of the glucosidic hydroxyl functions by silylation (using hexamethyldisilazane as the silylating agent). Then the recovered partially trimethylsilylated polysaccharide chains, now solubilized in organic solvents such as toluene or THF, have been used as multifunctional macroinitiators for the polymerization of ε-CL, the lactone ROP being initiated from the remaining free hydroxyl functions, by activation with either triethylaluminum or tin octoate. The grafting reaction and the ring opening polymerization are both under perfect control and proceed through the so-called coordination-insertion mechanism.[16] The third and last step relies upon the hydrolytic deactivation of the growing species and deprotection of the alcohol groups along the polysaccharide backbone, leading to the expected amphiphilic non ionic PCL-grafted dextran copolymers. This original three-step synthesis as well as the amphiphilic character of the so-obtained graft copolymers have been extensively reported elsewhere.[17] Compatibilization of PCL (CAPA 650 from Solvay Interox with Mn = 50,000)/granular starch blends with 40wt% of granular corn starch has been attempted by adding 5wt% of such amphiphilic graft copolymers. For instance, a copolymer with 60wt% of grafted PCL has been accordingly precipitated onto the starch particles before the melt blending step.[18] One clearly observed improved mechanical properties owing to this interface compatibilization : increase of both material rigidity (Young modulus increases from 265 to 309 MPa), stress at break (from 12.1 to 13.6 MPa) and ultimate strain (from 356 to 417 %). Stress at the limit of elasticity is enhanced as well with an increase from to 8.9 to 10.6 MPa. However, it must be stressed that in addition to the rather time-consuming 3-step synthesis of such amphiphilic copolymers, improved mechanical performances have been obtained only at the condition that the copolymers were first precipitated onto the starch granules assuring their interfacial location upon melt blending PCL with starch at 130°C. This is the reason why it has been decided to directly graft the PCL chains onto starch (either granular or plasticized) by in situ catalyzed ring opening polymerization of ε-caprolactone. Actually, this reaction has been performed in bulk in the presence of $AlEt_3$. Alkyl aluminums are known to be inactive towards the polymerization of lactones but when they react with the OH groups from starch, they yield aluminum alkoxide species able to initiate and further propagate the lactone polymerization.[16] Interestingly, a covalent ester bond is formed directly between the PCL grafts and the starch phase.[19]

First granular starch has been studied and dispersed in liquid ε-CL monomer under vigorous agitation. As a typical example, for an Al content as low as 0.1 wt% relative to the overall 50/50 wt% ε-CL/starch composition, almost quantitative monomer conversion (99 %) could be reached at 90°C within less than 5 min. A grafting efficiency as high as 90 % was determined by tentative selective extraction of non grafted PCL chains in a good solvent such as toluene. By SEM, one can observe that the recovered starch particles are individually and homogeneously embedded/coated within the covalently surface-grafted PCL chains leading to the formation of a “cocoon-like” structure. These surface-grafted PCL chains are long enough to allow crystallization from occurring. A melting temperature at ca. 55°C is measured by DSC. Furthermore, the mean particle size evolution along with the growth of the grafted PCL chains has been checked by laser light diffraction granulometry. Clearly a continuous increase in the mean diameter of the coated starch granules was detected all along the polymerization reaction.[20]

The in situ polymerization of ε-CL and concomitant PCL grafting have proved to be highly efficient with thermoplastic starch, i.e., starch previously plasticized with 25 wt% of glycerol by extrusion.[21] Again fast polymerizations were observed with monomer conversion close to 99 % reached within less than 10 min. and grafting efficiency ranging between 50 and 70 %.

PCL-grafted starch granules have been melt blended with CAPA 650 at 130°C. Typically PCL-grafted starch granules, actually grafted with 24 wt% of in situ polymerized PCL chains, have been investigated (Table 3). At same filler content (i.e., 40 wt%), the PCL covalent bonding allowed for improving both the elastic modulus and yield stress (σ_y) while keeping rather high ultimate properties (stress at break (σ_b) and elongation at break (ε_b)), at least at such high starch relative content. Such a positive mechanical behavior has been attested by markedly enhanced starch/PCL interfacial adhesion as observed by SEM on cryofracture surface. There is no longer surface dewetting/debonding of the PCL matrix all around the hydrophilic starch granules as was initially observed for simple melt blends without any interfacial compatibilization.[21]

Table 3 : tensile properties of PCL (CAPA 650)/granular starch compositions (with 40 wt% starch) melt blended at 130°C. Effect of the covalent PCL grafting onto starch surface (with 24 wt% of grafted PCL relative to starch phase)

PCL/starch (60:40)	σ_b (MPa)	ε_b (%)	Young modulus (MPa)	σ_y (MPa)
PCL/starch	12.1	360	265	9.0
PCL/PCL-grafted starch	14.5	350	320	16.1

Finally, the biodegradability of so-obtained PCL/starch blends has been investigated by composting.[22] For doing so, thin films (ca. 100µm thick) were buried in aerated composting bin for 120 days at 25-30°C (so called natural composting conditions), then followed by 20 days more at a higher temperature of 35-40°C (accelerated final composting). The film weight loss has been recorded along with the composting degradation time. Clearly the film weight loss increases with the starch content, here from 0 to 50 wt% polysaccharide (Table 4). The degradation started first within the starch phase and then occurred within the polyester matrix as observed by SEM.

Table 4 : time dependence weight loss of the PCL/starch films along with the composting degradation time (conditions : see text)

PCL/starch (wt/wt)	Weight loss (%)			
	50 days	90 days	120 days	140 days
100/0	0.45	0.86	1.11	21.18
90/10	1.20	1.63	2.20	40.31
80/20	1.54	1.85	2.30	42.41
70/30	1.73	2.80	3.50	47.21
60/40	6.30	11.10	11.30	53.36
50/50	6.40	11.71	13.36	67.21

It is very interesting to point out the effect of the interfacial compatibilization through either the precipitation of amphiphilic PCL-grafted dextran copolymers onto starch granules (PGD) or PCL-grafted starch (PGS) with only 4 wt% (relative to starch content) of surface-grafted PCL chains which were in situ polymerized as aforementioned. Surprisingly, both compatibilization strategies (PGD and PGS) favor the degradation of the PCL matrix (Table 5). For instance with 40 wt% of starch, the surface of which grafted by 4 wt% of PCL chains, 73 wt% weight loss were recorded after 140 days of composting. This is further evidenced by following the evolution of the PCL intrinsic viscosity along with the composting time (Figure 2). While almost no evolution is observed for the simple starch/PCL melt blend, both PGD and PGS compatibilized samples display a much more significant decrease of the PCL viscosity attesting for a much faster degradation. Such a behavior might find some explanation in the finer and more homogeneous polysaccharide dispersion within the PCL matrix and therefore a larger starch surface contact for the microorganisms.

Table 5 : time dependence weight loss of the PCL/starch films (with 40 wt% starch) along with the composting degradation time. Effect of the interface compatibilization : PGD and PGS (conditions : see text)

PCL/starch (wt/wt)	Weight loss (%)			
	50 days	90 days	120 days	140 days
60/40	6.30	11.10	11.30	53.36
60/40 PGD	4.85	9.06	11.19	67.00
60/40 PGS	12.92	18.96	21.76	73.00

Figure 2 : Evolution of the intrinsic viscosity of the PCL matrix with the composting degradation time

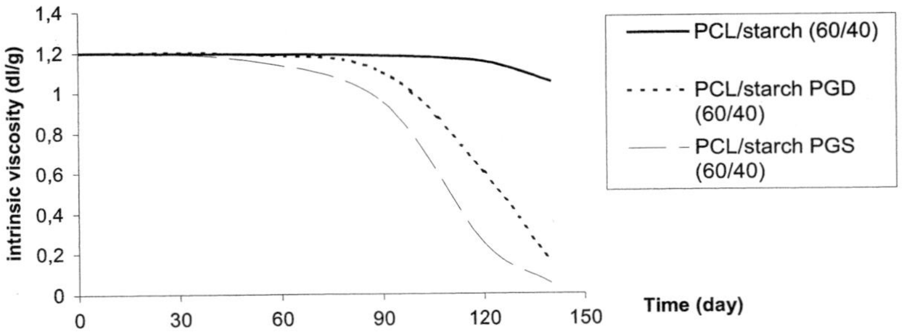

Conclusions

This study has emphasized that biodegradable compositions can be effectively reached by melt blending of granular or thermoplastic corn starch together with commercially available PCL or PLA after adequate compatibilization using reactive melt processing. Accordingly, high polysaccharide relative contents can be targeted leading to improved ultimate mechanical properties and much faster biodegradability as evidenced by composting testing.

Acknowledgments

This work was supported by the *Région Wallonne* and the *Fonds Social Européen* in the frame of *Objectif 1-Hainaut: Materia Nova.* Authors want to warmly thank all their co-workers who actively participated in this research (their name can be found in the references session) as well as their collaborators, namely Prof. E. Dellacherie from ENSIC in Nancy (France) and Dr R.P. Singh from NCL in Pune (India) and their co-workers, for their key-contributions and efficient cooperation. Ph. D. thanks *the « Services Fédéraux des Affaires Scientifiques, Techniques et Culturelles »* for general support (PAI V/03).

1. R. Narayan "Biodegradable Plastics" In Opportunities for Innovation: Biotechnology, NIST GCR 93-633. Gaithersburg, MD. National Institute for Standards and Technology, September 1993, p. 135.
2. M. Vert, J. Feijen, A.-C. Albertsson, G. Scott, E. Chiellini, editors. Biodegradable Polymers and Plastics; Royal Society Chemistry, Ed.; Redwood Press: Melksham, Wiltshire, 1992.
3. G.J.L. Griffin, *Am Chem. Soc. Div. Org. Coat. Plast. Chem.* **33(2)**, 88, (1973)
4. P.L. Nayak, *J. Macromol. Sc.- Rev. Macromol. Chem. Phys.* **C39(3)**, 481, (1999)
5. M.F. Koening, S.J. Huang, *Polymer* **36(9)**, 1877, (1995)
6. M.F. Koening, S.J. Huang, *Polymer* **36(9)**, 1877, (1995)
7. H. Pranamuda, Y. Tokiwa, H. Tanaka, *J. Environ. Polym. Degrad.* **4**, 1 (1996)
8. L. Avérous, L. Moro, P. Dole, C. Fringant, *Polymer* **41**, 4157 (2000)
9. E-J. Choi, C-H. Kim, J_K. Park, *J. Polym. Sc. Part B: Polym. Phys.* **37**, 2430 (1999)
10. E-J. Choi, C-H. Kim, J-K. Park, *Macromolecules* **32**, 7402 (1999)
11. M. Avella, M.E. Erico, P. Laurienzo E. Martuscelli, M. Raimo, R. Rimedio, *Polymer* **41**, 3875 (2000)
12. R. Mani, J. Tang, M. Battacharya, *Macromol. Rapid. Commun.* **19**, 283 (1998)
13. D. Carlson, Ph. Dubois, L. Nie, R. Narayan, *J. Appl. Polym. Sci.*, **72**, 477 (1999)
14. S. Gogolewski, M. Jovanovic, S.M. Perren, J.G. Dillon, M.K. Hughes, *Polym. Degrad. Stab.*, **40**, 313 (1993)
15. D. Carlson, L. Nie, R. Narayan, Ph. Dubois, *Polym. Eng. Sci.*, **38**, 311 (1998)
16. D. Mecerreyes, R. Jérôme, Ph. Dubois, *Adv. Polym. Sci.* **147**, 1 (1999)
17. I. Ydens, D. Rutot, Ph. Degée, J-L. Six, E. Dellacherie, Ph. Dubois, *Macromolecules*, 33, 6713 (2000)
18. E. Duquesne, D. Rutot, I. Ydens, Ph. Degée, Ph. Dubois, *Polym. Degrad. Stab.*, **73**, 561 (2001)
19. E. Duquesne, D. Rutot, Ph. Degée, Ph. Dubois, *Macromol. Symp.*, **175**, 33 (2001)
20. D. Rutot, Ph. Degée, R. Narayan, Ph. Dubois, *Composites Interfaces,* **7**, 215 (2000)
21. Ph. Dubois, M. Krishnan, R. Narayan, *Polymer* **40**, 3091 (1999)
22. R.P. Singh, J.K. Pandey, D. Rutot, Ph. Degée, Ph. Dubois, *Biomacromolecules*, submitted for publication.

Macromol. Symp. **2003**, *198*, 245-257

Phase Separation by Continuous Quenching: Similarities between Cooling Experiments in Polymer Blends and Reaction-Induced Phase Separation in Modified Thermosets

Ingo Alig, Maximilian Rüllmann, Marco Holst, Jianjun Xu*

Deutsches Kunststoff-Institut, Schloßgartenstraße 6, D-64289 Darmstadt, Germany

Summary: Small angle light scattering (SALS) has been applied to study the phase separation kinetics in a binary polymer mixture of poly(ethyl methyl siloxane) (PEMS) and poly(dimethyl siloxane) (PDMS). The phase separation was induced by cooling an initially homogeneous mixture with well defined cooling rates. The results have been compared to time resolved SALS and microscopy in the course of reaction-induced phase separation in mixtures of an epoxy resin and polysulfone (PSU).

For the critical PEMS/PDMS mixture with an upper critical point it was found in a continuous quenching experiment that the time evolution of the scattered light intensity $I(q,t)$ scales with the cooling rate. The similarity to the scaling behavior of $I(q,t)$ in isothermal experiments after fast quenches (scaled by the quench depth) is discussed. A secondary phase separation was found and has been explained by the competition between the growth of the two phase structure during cooling and the mutual diffusion without the assumption of gelation or vitrification.

For the epoxy/PSU mixture with 15% PSU, after the appearance of a bicontinuous structure a secondary phase separation was observed. Mixtures with higher PSU-contents formed epoxy-rich droplets in the PSU-rich matrix by nucleation and growth mechanism. The frustration of the structure growth can be explained by approaching vitrification of one or both phases.

The similarity between continuous cooling experiments in blends and the reaction-induced phase separation have been discussed in the generalized χN vs. composition phase diagram (N: degree of polymerization, χ: Flory-Huggins interaction parameter).

Keywords: continuous quenching, phase separation, polymer blends, thermosets

Introduction

The conventional experiment to investigate the phase separation kinetics in binary polymer mixtures is to transfer an initially homogeneous mixture by (almost) instantaneous changes of

 DOI: 10.1002/masy.200350821

temperature or pressure into the two phase region of the phase diagram and to detect the isothermal coarsening of the structure by time resolved microscopy or scattering experiments [1]. Due to the finite thermal conductivity it is practically impossible to realize an instantaneous transfer of a homogeneous mixture into the two phase region. An alternative experiment is to transfer a homogeneous mixture continuously into the two phase region. As schematically shown in the χN-phase diagram ($\chi(T, p)$: Flory-Huggins interaction parameter, N: degree of polymerization) in Scheme I a binary mixture can be continuously transferred from the one phase region into the two phase region by changing temperature, T, or pressure, p. This "χ-induced decomposition" can be realized by cooling (upper critical solution temperature) or heating (lower critical solution temperature) of the mixture. The cooling or heating rate becomes an additional parameter in this type of experiments. Alternatively, continuous pressure changes can be applied. Continuous quenching experiments are of practical importance for processing of polymer blends (e.g., mould injection).

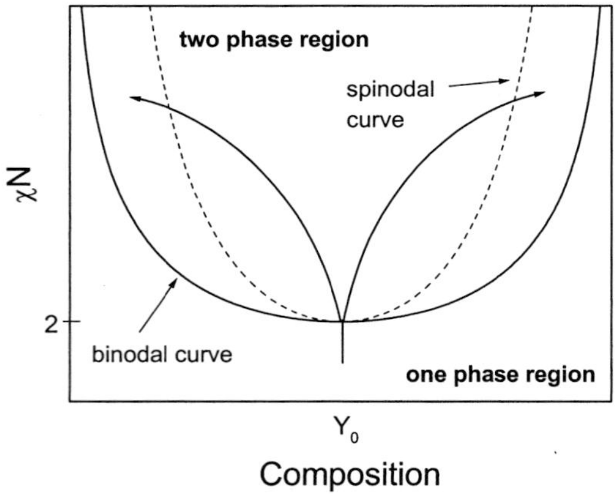

Scheme I. Continuous χ- and N-induced decomposition in the χN -phase diagram.

Two decades ago, Carmesin et al. [2] considered in a theoretical study the early stage of demixing of polymer blends during continuous quenching. More recently, Kukushkin and Slyonzov [3-5]] discussed in several theoretical papers the Oswald ripening under non-isothermal conditions. However, to our knowledge, no systematic experimental investigations of this type have been reported so far for polymer blends. Hashimoto et al. [6] and Hayashi et al. [7]

investigated the phase separation kinetics of polymer mixtures after a two step temperature change.

As obvious in Scheme I, a homogeneous mixture can also be transferred into the two phase region by increasing the degree of polymerization (reaction-induced or "*N*-induced" phase separation). A typical example is the reaction-induced phase separation of thermosets containing a chemically inert second component during isothermal curing experiments (see [8] and references therein).

It is the aim of this paper to discuss the similarities between continuous cooling experiments in polymer mixtures and the reaction-induced phase separation in quasi-binary thermoset blends. Therefore we applied time resolved small-angle light scattering (SALS) experiments and optical microscopy (OM) to mixtures of poly(ethyl methyl siloxane) and poly(dimethyl siloxane) while the mixtures were cooled with well defined cooling rates and to mixtures of an epoxy resin and polysulfone during isothermal curing.

Method

Recently we reported the details of a light scattering setup for investigation of phase separation kinetics in polymer mixtures after pressure jumps [9]. In this study the pressure cell was replaced by an electrically heated brass block, into which a sample holder was inserted. The sample thickness is 1 mm for the blend and 0.025 mm for the thermoset/thermoplastic mixtures. The accuracy of the temperature control of the sample was ± 5 mK for isothermal experiments and the cooling rates could be chosen between ± 10 mK/min and ± 0.3 K/min. The scattering vector q is given by $q = 4\pi n \sin(\theta/2)/\lambda_0$, where n is the refractive index, θ is the scattering angle, and λ_0 (= 632.8 nm) the wave length.

Samples

The polymer blend (System I) considered in this study is a binary mixture of poly(dimethyl siloxane) (PDMS) and poly(ethyl methyl siloxane) (PEMS), with molar masses of M_w = 18500 g/mol (M_w/M_n = 1.055) and M_w = 23927 g/mol (M_w/M_n = 1.049), respectively. The cloud point curve in Figure 1 (left part) shows an upper critical solution temperature. The spinodal (dashed) and the binodal curve (solid) are calculated using the Flory-Huggins theory.

In order to study the reaction-induced phase separation, blends of epoxy resin with polysulfone (PSU) have been prepared (for details see [10, 11]). The epoxy resin is diglycidyl ether of bisphenol-A (DGEBA, MY 790-1 from Ciba with M_n = 173.8 g/mol) cured with hexahydrophthalic anhydride (HHPA), using 2-ethyl-4-methylimidazole (EMI) as the catalyst. The mass ratio of DGEBA/HHPA/EMI was 120:100:2.4. PSU with M_n = 4900 g/mol and M_w = 11100 g/mol, was supplied by BASF with 99.9 % Cl-end group and 0.1 % OH-end group. The "phase diagram" in the coordinates of conversion versus volume fraction Φ_{PSU} of PSU is shown in Figure 1 (right part). The conversion was measured by differential scanning calorimetry and fitted by a modified Horie equation [10, 11]. The phase separation was investigated by OM, SALS and/or rheological measurements as described in [9, 11]. The gelation of the epoxy-rich phase ($\Phi_{PSU} < 0.15$) and the vitrification of the PSU-rich phase were determined from rheological experiments. The final conversion was estimated from calorimetric experiments. The vitrification curve, $T_g(\Phi_{PSU})$, was calculated from quasi-binary model using the Fox-equation and the DiBenedetto equation [11].

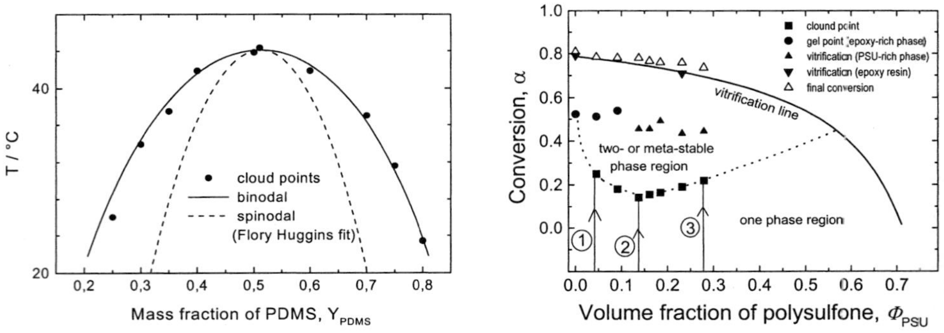

Figure 1. Phase diagram of the PDMS/PEMS mixture (left) and conversion versus composition diagram for the epoxy/PSU mixtures (right) cured at 80°C.

Continuous quenching of a binary blend

Figure 2 shows the time evolution of the scattered light intensity versus q for the critical mixture (Y_{PDMS} = 0.51) of PEMS/PDMS during a continuous cooling experiment (cooling rate: κ = 50 mK/min). After crossing the cloud point curve (t=0) a maximum in the scattering intensity (first peak) appears in the experimental accessible q-range. The maximum scattering intensity of the main peak I_{m1} increases and the related scattering vector q_{m1} decrease with time. This feature is typical for coarsening of a two phase structure (spinodal decomposition). At later times a second shoulder or maximum becomes visible. The maximum scattering intensity of the second peak/shoulder I_{m2} also increases and the related scattering vector q_{m2} decreases with time, which is an indication for the development of a secondary structure inside the domains of the bicontinuous primary structure.

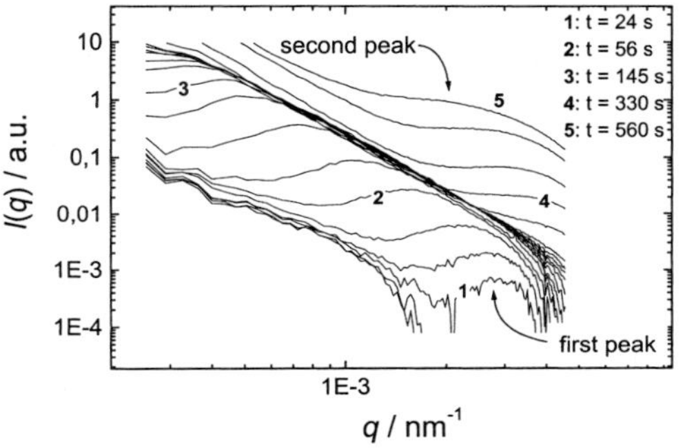

Figure 2. Time evolution of the scattering intensity during cooling of the critical PEMS/PDMS mixture ($Y_{PDMS} = 0.51$, cooling rate: $\kappa = 50$ mK/min).

The time evolutions of q_{m1} and q_{m2} are plotted in the left part of Figure 3 for three selected cooling rates κ. The corresponding values for I_{m1} and I_{m2} are shown in the inset. The time t is rescaled for both plots by the undercooling $\Delta T = \kappa \cdot t$. The values of $q_m(t)$ and $I_m(t)$ for different cooling rates can be shifted onto a master curve as shown in right part of Figure 3 for three selected cooling rates. This *cooling rate-scaling* is analogous to *quench depth-scaling* for instantaneous quenches and isothermal phase separation (see e.g. [1]). The reduced variables in Figure 3 are defined by: $q_{red} = q_m / q_{m,0}$ and $\Delta T_{red} = \Delta T / \Delta T_c = t_{red} = t / t_c$. $q_{m,0}$ is the q-vector at short times corresponding to a characteristic length l_c .

The inverse shift factors for the horizontal and vertical shifts are the characteristic undercooling ΔT_c (or the characteristic time t_c) and the characteristic scattering vector $q_{m,0} = q_m(0)$. In Figure 4 the values for ΔT_c (left part) and $q_{m,0}$ (right part) versus cooling rate κ are shown. The lines are fits with power laws ($\Delta T_c(\kappa) = \Delta\hat{T} \cdot \kappa^{\vartheta}$ and $q_{m,0}(\kappa) = \hat{q} \cdot \kappa^{\zeta}$) to the data from the first peak for a set of experiments with five different cooling rates. The fitting values are $\Delta\hat{T} = 1.71\,\mathrm{K}^{0.71}\mathrm{s}^{0.29}$ and $\vartheta = 0.29$ for ΔT_c and $\hat{q} = 8.77 \cdot 10^{-3}\ \mathrm{nm}^{-1}\mathrm{s}^{0.18}\mathrm{K}^{-0.18}$ and $\zeta = 0.18$ for $q_{m,0}$. In [12,13] it is shown that the exponents $\vartheta = 0.29$ and $\zeta = 0.18$ are related to well known critical exponents.

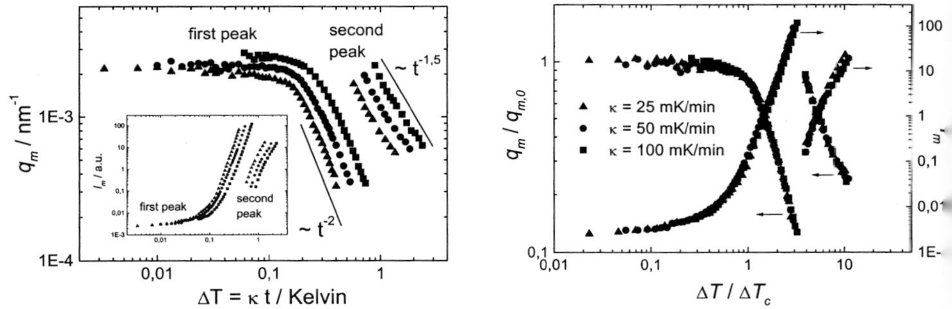

Figure 3. Evolution of scattering vectors (left part) for the first ($q_{m,1}$) and second peak ($q_{m,2}$) for the critical PEMS/PDMS mixture for different cooling rates κ. The time t is replaced by the undercooling: $\Delta T = \kappa\, t$. The corresponding intensities $I_{m,1}$ and $I_{m,2}$ are shown the inset. The right part shows the reduced plots for scattering vectors ($q_{m,1}$, $q_{m,2}$) and the intensities ($I_{m,1}$, $I_{m,2}$) versus the reduced undercooling, $\Delta T_{red} = \Delta T / \Delta T_c$.

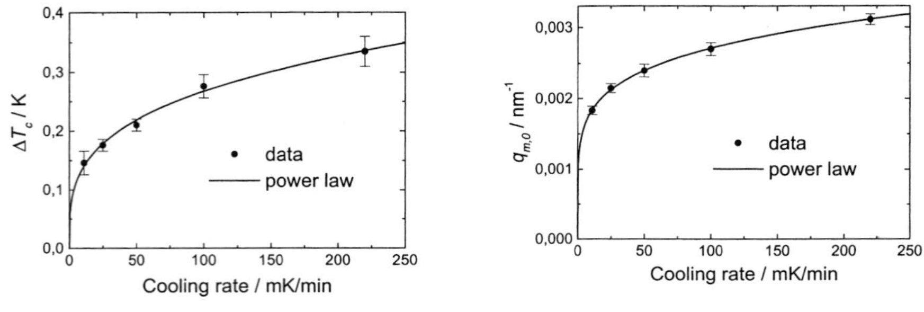

Figure 4. Scaling variables $q_{m,o}$ and ΔT_c (inverse shift factors from Figure 3) versus cooling rate κ for the critical PEMS/PDMS mixture.

As indicated in Figure 4, ΔT_c is the characteristic undercooling where both $q_{m,1}(\Delta T)$ and $I_{m,1}(\Delta T)$, show a crossover to a power law behavior. Based on the knowledge of structure growth in isothermal experiments, such a power law behavior in cooling experiments is expected for the late stage of demixing ($t > t_c$), by assuming a self-similar coarsening and thermodynamic equilibrium of the coexisting phases [12, 13]. Details of the expected power law behavior are discussed below.

As schematically shown in the phase diagram (Scheme II) the characteristic undercooling ΔT_c corresponds to the time t_c where for a given cooling rate κ the compositions of the two phases become identical to the corresponding binodal compositions. This follows from a thorough analysis of the relation between $q_{m,1}$ and $I_{m,1}$ [12,13]. For $t < t_c$ the trajectory of a cooling experiment is always in the two phase region and the coexisting phases are not in equilibrium. This situation corresponds to the early and intermediate stage of isothermal phase formation. In the early stage of demixing the characteristic structure length Λ is expected to decrease in cooling experiments with the undercooling by $\Lambda \propto \Delta T^{-\nu}$, where ν is critical exponent of the correlation length (Ising model: $\nu = 0.63$). This stage was too short to be detected in our SALS experiments. The late stage of demixing is usually characterized by a thermodynamic equilibrium of the coexisting phases (which are still growing) and a composition of the two phases which is identical to the binodal composition. In the continuous cooling experiment for $t > t_c$ the difference between the compositions of the phases is increasing with further undercooling $\Delta T(t)$ and the characteristic structure length $\Lambda(t)$ is expected to increase by a hydrodynamic coarsening mechanism [1] with $d\Lambda / dt \propto \Delta T^{z}$ [12, 13]. The exponent z is given by $z = \mu + \nu \cdot x_{\eta}$, where μ and x_η are the critical exponents of surface tension and viscosity respectively. The mean field value is $z = 1.5$ and the 3D-Ising model yields $z = 1.30$ [14]. A detailed analysis [12, 13] of the $q_{m,1}(\Delta T)$-curves yields $z = 1.68$, which is in the right order of magnitude. The $q_{m,1}$ ($= 2\pi/\Lambda$) vs. ΔT curves in Figures 3 indicate the expected power law behavior for $\Delta T > \Delta T_c$. Following these arguments in late stage of continuous cooling experiments a competition between the mass transfer due to diffusion and the growth of the two phase structure has to be expected. When the growth rate of the characteristic structure length $v_{str} \propto d\Lambda/dt$ becomes equal to or larger than the rate of the mean squared spatial displacement of the molecules ("diffusion velocity")

$v_{diff} \propto d\sqrt{D(t)t}\,/dt$ (*D*: mutual diffusion coefficient) the thermodynamic equilibrium of the coexisting phases is not guaranteed and a secondary phase separation inside the domains can occur (see Scheme II). The characteristic undercooling ΔT_s for this secondary phase separation is expected to depend on the cooling rate by $\Delta T_S \propto \kappa^{1/(2z+1-\nu^*)}$ [12, 13], where ν^* is the critical exponent for the diffusion coefficient (Ising model: $\nu^* = 0.63$, mean field theory: $\nu^* = 0.5$). For the reaction-induced phase separation the formation of a secondary structure inside the domains is usually explained by the increase of the viscosity due to gelation or vitrification. However, for PEMS and PDMS the viscosity does not change significantly in the temperature interval of the experiments and both glass transition temperatures are very low. Therefore the secondary phase separation in binary mixtures can be explained by kinetic arguments, without the assumption of a gelation or vitrification.

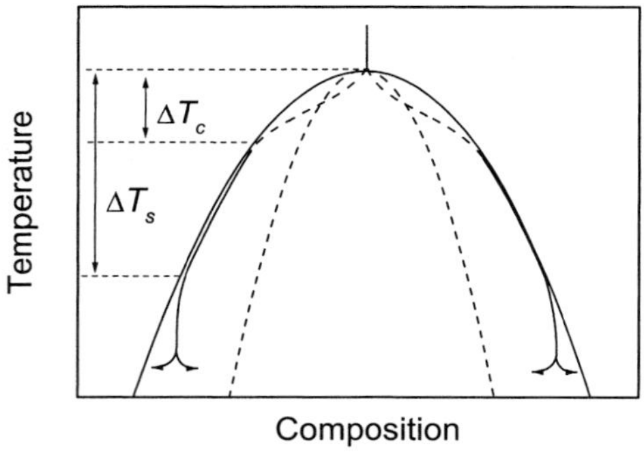

Scheme II. Cooling trajectory in the phase diagram of a binary mixture.

Reaction-Induced Phase Separation

In Figure 5 the time evolution of the scattering intensity during isothermal reaction of an epoxy mixture with 15 wt% PSU cured at 80°C is shown (see ② in Figure 1). Similar to evolution of the scattering intensity in the course of cooling of the PEMS/PDMS blend in Figure 2, two maximum in the scattering intensity become visible after the phase separation. The maximum intensity of the first peak, I_{m1}, increases and the related scattering vector, q_{m1}, decrease with time. The final morphology (see micrograph in Figure 6) is bicontinuous and the major scattering maximum can be related to reaction-induced spinodal decomposition. The open points q-values are calculated by $q_{m,1} = 2\pi/\Lambda_{OM}$, where the characteristic structure length Λ_{OM} was estimated from time resolved OM. The appearance of a second maximum in the scattering intensity at later times can be related to the secondary phase separation in the epoxy- and the PSU-rich phase. The micrograph in Figure 6 shows secondary structures in both phases. The maximum scattering intensity of the second peak, I_{m2}, is assumed to be an average of the secondary structures inside both phases. As expected I_{m2} increases with reaction time and the related q-value decreases (left graph in Figure 6).

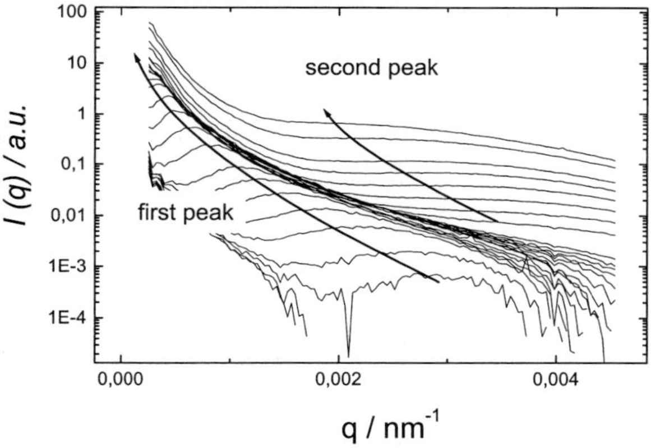

Figure 5. Time evolution of the scattering intensity *I(q)* during isothermal reaction of an epoxy mixture with 15 wt% PSU cured at 80°C.

The main characteristic of the epoxy mixture with 15 wt% PSU is the 'pinning' of the primary phase structure (corresponding to the first peak) as observed with OM occurs at about 17 min after the phase separation (leveling-off of $q_{m,1}$ in Figure 6). A deflection for the evolution of $q_{m,2}$ is also observed at this time. This 'pinning' was attributed to the viscosity increase due to approaching vitrification of PSU-rich phase.

The first appearance of the secondary structure is found to occur considerably before the gelation and vitrification of either phase [11]. It is explained by the competition between mutual diffusion and structure growth as for the polymer blends (see above). The corresponding trajectory for the "critical" thermoset/thermoplastic mixture is shown in a schematic conversion versus composition diagram (left figure in Scheme III). A secondary phase separation has been found inside both primary phases considerably before the gelation and vitrification of one of the phases.

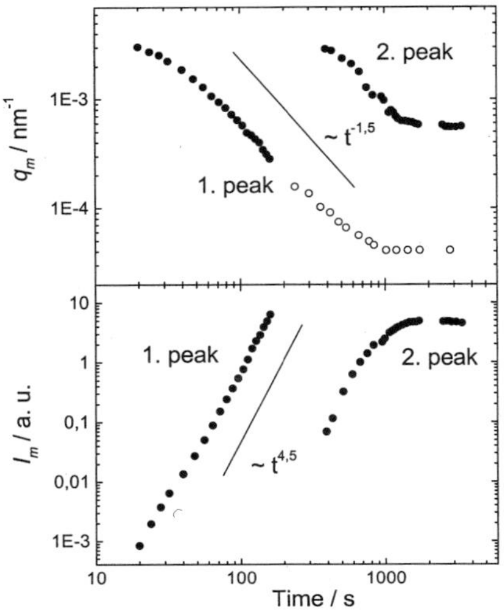

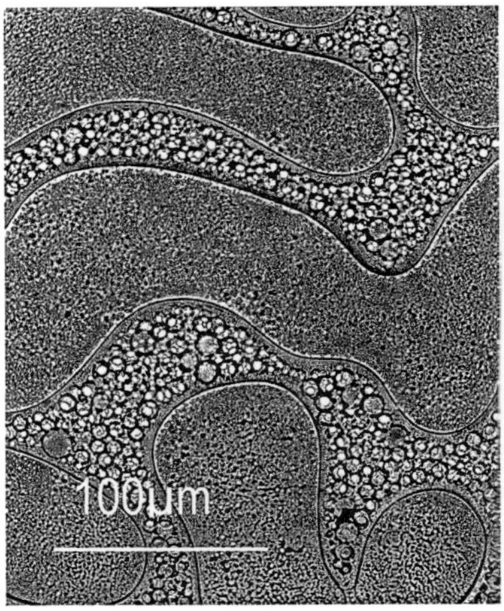

Figure 6. Time evolution of q_m (upper part) and I_m (lower part) for an epoxy / PSU mixture with 15 wt% PSU cured at 80°C. The open symbols are estimated from optical microscopic images. The right picture represents the 'pinned' domain structure 17 minutes after phase separation.

In Figure 7 the time evolution of the scattering vector at maximum intensity (upper part), the maximum scattering intensity (middle part) and the conversion (lower part)) for an epoxy/PSU mixture with 25 wt% PSU cured at 70°C, 80°C and 90°C are shown (see also ③ in Figure 1). As indicated in the electron micrograph in Figure 7 the mixture with 25 wt% PSU forms droplets of the epoxy-rich phase embedded in the PSU-rich matrix. Therefore the scattering maximum has to be related to scattering on droplets with a mean interspherical distance $\Lambda_s = 2\pi / q_{m,s}$. For the evolution of these epoxy-rich droplets a nucleation and growth mechanism can be assumed. The appearance of a maximum in *I(q)* in blends with a nucleation and growth mechanism is well known. It can be caused by coherent scattering of the droplet ensemble or a by depletion zones around the droplets [12,15,16]. Here we assume the first to be dominant because of a high droplet density. This can also explain higher order scattering peaks. The maximum has also been found for non-critical PEMS/PDMS mixtures and was simulated by the Rayleigh-Gans theory [12].

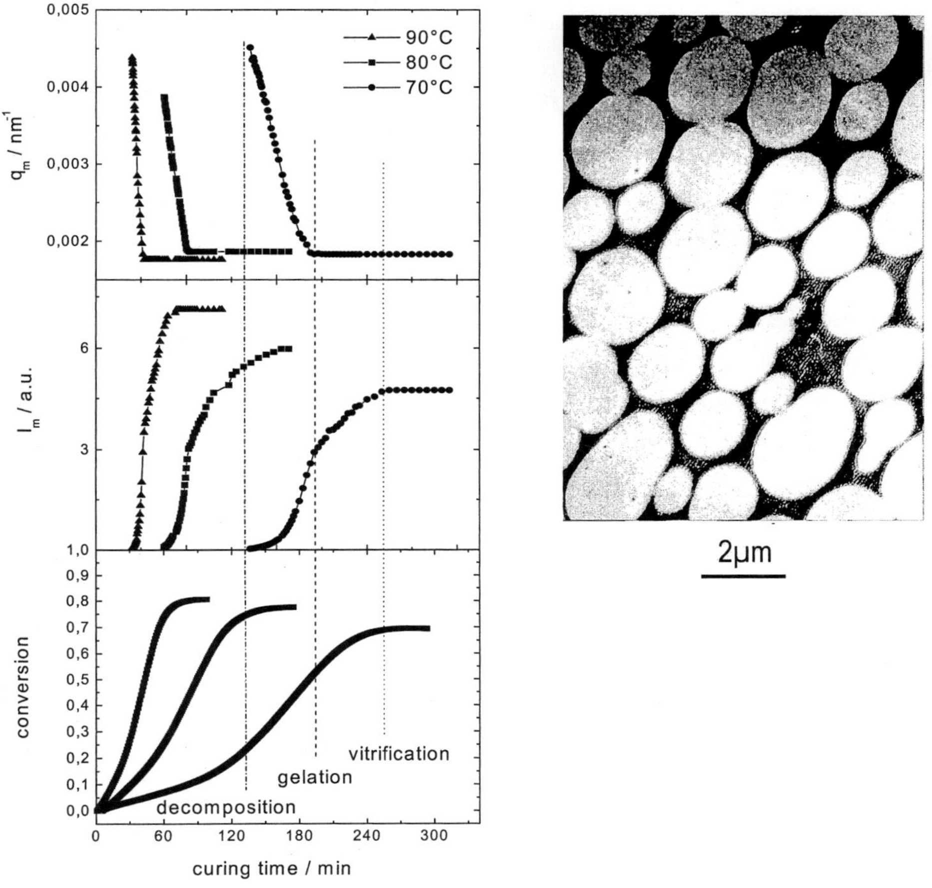

Figure 7. Time evolution of q_m (upper part), I_m (middle part) and conversion (lower part) for an epoxy/PSU mixture with 25 wt% PSU cured at different temperatures. The domain structure after 400 minutes curing at 80°C is shown in a transmission electron micrograph (right part).

It is obvious that the curves of q_m and I_m versus time for the three different curing temperatures have the same shape. The curves can be subdivided into four regions: (i) no scattering signal, (ii) strong decrease of $q_m(t_{cure})$ and strong increase of $I_m(t_{cure})$, (iii) leveling-off of $q_m(t_{cure})$ and lower slope in $I_m(t_{cure})$, and (iv) leveling-off of $I_m(t_{cure})$. The appearance of the maximum of scattered light intensity in the time interval (ii) is an indication for the phase separation. The epoxy reaction is assumed to proceed in both phases. As we have found by temperature-modulated calorimetry (isothermal experiments) the PSU-rich matrix vitrifies ($T_{cure} \cong T_g(t_{cure})$) at the beginning of the interval (iii), which can explain that structure growth stops. However, a considerable amount of reaction in the epoxy-rich phase is still occurring. Finally the reaction stops also in the epoxy-rich droplets by vitrification (interval iv). The characteristics for the "non-critical" thermoset/thermoplastic mixture, shown in the right figure of Scheme III, are the large differences in the conversions of the two phases which are necessary for vitrification, and that no secondary phase separation could be detected.

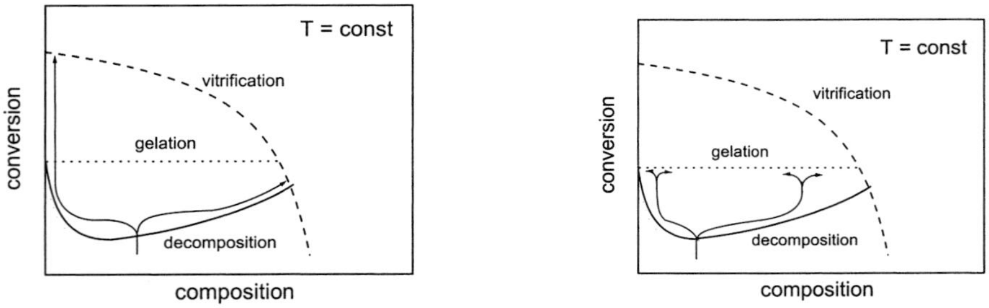

Scheme III. Conversion vs. composition diagrams with reaction trajectories for a "critical" (left) and a "non-critical" thermoset/thermoplastic mixture (right).

Similar to the *cooling rate scaling* for the binary blend one would expect here a *reaction rate scaling*. The similar shapes of the $q_m(t_{cure})$ and $I_m(t_{cure})$ curves for different temperatures in Figure 7 are an indication for that. To construct a master curve for reaction-induced phase separation the reaction time has to be replaced by the conversion and the dependence of the (local) viscosity on the conversion has to be taken into consideration.

Conclusion

Continuous quenches into the two phase region (Scheme I) can be realized by increase of χ (cooling or pressure ramps) or N (polymerization reaction). It has been shown that the continous χ-induced decomposition realized by cooling ramps in polymer mixtures and the N-induced decomposition in a reaction-induced phase separation show similar features in time resolved SALS experiments. For both cooling-induced and reaction-induced phase separation the appearance of a fine structure due to secondary phase separation was detected. It has been shown that secondary (and probably ternary and higher order) phase separation can be explained by the competition between structure growth and mutual diffusion.

Universality of structure growth for different cooling rates, *cooling rate-scaling*, was observed for the binary polymer blend and can be explained by the critical behavior of correlation length, diffusion coefficient, viscosity and surface tension. Similar behavior, *"reaction rate-scaling"*, can be expected for the polymerization-induced phase separation, where the viscosity increase during reaction has to be taken into account.

The financial support from the Bundesministerium für Wirtschaft und Technologie through the AiF (11540N) and the Deutsche Forschungsgemeinschaft (Al 396/3-2) is gratefully acknowledged.

[1] T. Hashimoto, in: *Materials Science and Technology* , E. L. Thomas, Ed., VCH, Weinheim 1993, Vol.12, Ch.6.
[2] H. O. Carmesin, D. W. Heermann, K. Binder, *Z. Phys.* B **1986**, *65*, 89.
[3] S. A. Kukushkin, V. V. Slyozov, *J. Phys. Chem. Solids* **1995**, *56*, 1259.
[4] S. A. Kukushkin, V. V. Slyozov, *J. Phys. Chem. Solids* **1996**, *57*, 195.
[5] S. A. Kukushkin, V. V. Slyozov, *J. Phys. Chem. Solids* **1996**, *57*, 601.
[6] T. Hashimoto, M. Hayashi, H. Jinnai, *J. Chem. Phys.* **2000**, *112*, 6886.
[7] M. Hayashi, H. Jinnai, T. Hashimoto, *J. Chem. Phys.* **2000**, *112*, 6897.
[8] J. P. Pascault, H. Sauterau, J. Verdu, R. J. J. Williams, *Thermosetting Polymers* , Marcel Dekker Inc., New York 2002.
[9] B. Steinhoff, M. Rüllmann, L. Kühne, I. Alig, *J. Chem. Phys.* **1997**, *107*, 5217.
[10] I. Alig, W. Jenninger, J.E.K. Schawe, *Polymer* **2000**, *41*, 1577.
[11] M. Holst, J. Xu, M. Rüllmann, M. Wenzel, I. Alig, *Macromolecules* (in preparation).
[12] M. Rüllmann, *PhD Thesis*, Darmstadt 2002.
[13] M. Rüllmann, I. Alig, *J. Chem .Phys.* (in preparation).
[14] A Stammer, B. A. Wolff, *Polymer* **1998,** *39*, 2065; W. Theobald, G. Meier, *Phys. Rev. E* **1995**, *51*, 5776
[15] G. E. Elicabe, H. A. Larrondo, R. J. J. Williams, *Macromolecules* **1997**, *30*, 6550.
[16] G. E. Elicabe, H. A. Larrondo, R. J. J. Williams, *Macromolecules* **1998**, *31*, 8173.

Solvent Activity Changes and Phase Separation During Crosslinking of Coating Films

*Miroslava Dušková-Smrčková,**[1,3] *Karel Dušek,*[1] *Petr Vlasák*[2]

[1]Institute of Macromolecular Chemistry, Academy of Sciences of the Czech Republic, 162 06 Prague, Czech Republic
E-mail: m.duskova@imc.cas.cz
[2]SYNPO, Ltd., 532 07 Pardubice, Czech Republic
[3]Department of Macromolecular Physics, Faculty of Mathematics and Physics, Charles University, Prague, Czech Republic

Summary: Changes in solvent activity and thermodynamic stability of a system undergoing crosslinking in the presence of a solvent are modeled by combination of branching theory with thermodynamics of polymer solutions and swelling of a crosslinked polymer. The system is considered as a quasiternary one in which the solvent, polydisperse soluble polymer (sol) and crosslinked network, respectively, represent the components. Crosslinking brings about growth of polymer structures but also changes in polymer-solvent interactions. The results are compared with onset of phase separation and changes in evaporation rates from films formed from a hydroxy-functional star oligomer and triisocyanate.

Keywords: crosslinking, phase separation, polymer-solvent interaction, solvent activity, swelling

Introduction

Solvent-based coating films are formed by simultaneous crosslinking of polymer network precursors and solvent evaporation. Usually, two-component systems are used: oligomeric or polymeric component carrying functional groups and a crosslinker of lower functionality (2-4) and lower molecular weight. Film formation by simultaneous crosslinking and solvent evaporation is a complex process in which the crosslinking rate and network structure development depend on the content and nature of the solvent and solvent evaporation depends not only on solvent volatility but also on the structure of the crosslinking system. Moreover, in some systems these changes induce a transition of the system from rubbery into glassy state and all important processes become controlled by segmental mobility.[1,2] In a drying film, diffusion is driven by gradients of chemical potentials. Evaporation of the solvent is controlled by solvent vapor pressure and this is determined by the solvent activity in the film. The solvent activity decreases with decreasing solvent concentration and usually increases with advancing reaction because the molecular weight of polymeric components or their crosslink density increase. Sometimes, the system becomes thermodynamically

 DOI: 10.1002/masy.200350822

unstable and phase separation sets in. In a crosslinked system, this happens when the content of solvent is equal or exceeds the maximum (equilibrium) degree of swelling of the network.[3-5] This is general case of reaction-induced phase separation widely studied.[6]

In this contribution, the changes of solvent activity and possible onset of phase separation resulting from crosslinking reaction in the presence of various amounts of solvent is studied theoretically and compared with some experimental data obtained for crosslinking system of a hydroxy-functional oligomer (tetrafunctional star) crosslinked with a triisocyanate. Before the gel point, the crosslinking system is considered as pseudobinary and, beyond the gel point, as pseudoternary; solvent, polydisperse branched polymer (sol), and crosslinked gel being considered as components. For modeling of the changes, the network formation theory is combined with the theory of polymer solutions/swelling of crosslinked gels. In this first approximation, the interactions are described by simple Flory-Huggins approach with concentration independent interaction parameters and the build-up of crosslinked structures using the statistical theory of branching processes in its ring-free version.

Experimental

The hydroxy-functional precursor was a star-shaped polyester with 4 primary hydroxyls of hydroxyl number 235.2 mg KOH/g and molecular weight 1031 g/mol was prepared by hyroxyl-anhydride-acid-epoxide reaction[7]. The polyisocyanate used was Desmodur 3300 (Bayer) - a trimer of 1,6-diisocyanatohexane of molecular weight 690 g/mol (VPO), number-average functionality 3.5 and second-moment average functionality 3.65.

The progress of reaction was monitored by transmission FTIR using the absorption by NCO groups at 2270 cm^{-1} or by chemical method in which unreacted isocyanate groups are reacted with excess of n-dibutylamine and the unreacted amine titrated by hydrochloric acid. Also, attenuated total reflection (ATR) mode of FTIR was used for monitoring NCO group conversion in the surface layer.

Evaporating films were cast on a metallic substrate using an blade applicator. Their typical wet thickness was 200 μm. Solvent evaporation was monitored by an automatic balance placed in an environmental chamber. Non-evaporating films, where the solvent concentration remained constant during crosslinking, were prepared between two glass plates. All experiments were carried out at temperature 25±2°C.

Phase separation occurred in the form of macrosyneresis beyond the gel point and was observed visually as liquid separated out of the gel or by ATR FTIR when the signals assigned to the polymer suddenly disappeared. The volume of the separated liquid was not determined.

Theoretical

Thermodynamics of Mixing and Swelling: The solvent vapor pressure, p_1, is proportional p_1^0 to the vapor pressure of the pure solvent and its activity in the film, a_1

$$p_1 = p_1^0 a_1 = p_1^0 c_1 \gamma_1 \tag{1}$$

where γ_1 is the activity coefficient and c_1 is the solvent concentration. The solvent activity is dependent on (a) solvent concentration, (b) thermodynamic interactions between segments of component units, and (c) structure of the polymer: molecular weights of sol, sol and gel fractions, and crosslink density that are functions of conversion of reactive groups. During film formation, the solvent concentration decreases by evaporation (a negative contribution to a_1 and the polymer structure develops (a positive contribution to a_1). Moreover, the thermodynamic interactions between solvent molecules and polymer segments change because reactive groups are transformed into bonds. In adhering films, stresses in plane develop. The resulting anisotropy of the swollen film was shown to have an effect on solvent activity[8]; this effect, however, will not be considered here.

The thermodynamics of the system is described by the Flory-Huggins-Staverman (FHS) theory of polymer solutions and Flory-Erman theory of rubber elasticity.[9]-[13] A similar Gibbs energy function as formulated in Ref. 14 was used by Frisch and Binder[15] for theoretical treatment of phase separation in interpenetrating polymer networks. These systems were considered as binary weakly crosslinked non-reactive systems with no solvent and no sol fraction.

The change in the Gibbs energy, ΔG, associated with mixing solvent with polymer component is assumed to be composed of mixing and network terms (beyond the gel point).

$$\Delta G = \Delta G_{\text{mix}} + \Delta G_{\text{net}} \tag{2}$$

The FHS mixing term for a ternary system (component 1 is solvent, component 2 soluble polymer, component 3 network polymer) is given by the equation[12],[13]

$$\frac{\Delta G_{\text{mix}}}{NkT} = \frac{\phi_1}{m_1}\ln\phi_1 + \frac{\phi_2}{m_2}\ln\phi_2 + \frac{\phi_3}{m_3}\ln\varphi_3 + \chi_{12}\phi_1\phi_2 + \chi_{13}\phi_1\phi_3 + \chi_{23}\phi_2\phi_3 \tag{3}$$

where N is the total number of lattice sites (occupied by polymer segments and solvent molecules), ϕ_1, ϕ_2, and ϕ_3 are volume fractions of the respective components and χ_{ij} are (concentration independent) binary interaction parameters. The quantities m_i are the ratios of volumes of component i to the volume of the lattice site; often a polymer segments is considered to occupy one lattice site and $m_i = V_i/V_s$, for polymers with identical segments but different degree of polymerization, P_i $m_i \propto P_i$. Sometimes the polymer segment is chosen such that $V_s = V_1$; then $m_2 = P_2 V_m$ (V_m is molar volume of polymer segment) and $m_1 = 1$ and $1/m_3 = 0$, because component 3 (network has infinite degree of polymerization).

The term ΔG_{net} is associated with deformation of the elastically active network chains (EANC) and other deformation dependent entropic changes originating in fluctuation of junctions (crosslinks). For Gaussian chains

$$\frac{\Delta G_{\text{net}}}{kT} = A n_e \left(\frac{L_x^2}{L_{0x}^2} + \frac{L_y^2}{L_{0y}^2} + \frac{L_z^2}{L_{0z}^2} - 3 \right) - B n_e \ln \frac{L_x L_y L_z}{L_{0x} L_{0y} L_{0z}} \tag{4}$$

where L_k is the length of the deformed sample in the direction of the k axis; L_{0k} is the corresponding value in reference state at which the chains are at their state of ease; n_e is the number of elastically active network chains; A is the front factor which, within the framework of the Flory-Erman junction-fluctuation rubber-elasticity theory, varies between $(f_e - 2)/f_e$ and 1 for phantom and affine models, respectively; f_e is the average functionality of an elastically active crosslink; B varies between 0 (phantom network) and $2/f_e$ (affine network).

The deformation ratios with respect to the reference state are defined by $\Lambda_k = L_k / L_{0k}$ and for an isotropic system $\Lambda_k(k{=}x,y,z) \equiv \Lambda$; Λ is related to volume fraction of the network polymer ϕ_3, and reference state at network formation, given by the volume fraction of the polymer at network formation ϕ_0, as $\Lambda = (\phi_3)^{-1/3}(\phi_0)^{1/3}$, so that

$$\frac{\Delta G_{\text{net}}}{kT} = 3n_e\left(A\Lambda^2 - B\ln\Lambda^3\right) = 3n_e\left(A\phi_3^{-2/3}\phi_0^{2/3} - B\ln(\phi_0/\phi_3)\right) \tag{5}$$

By passing to moles of EANC's and differentiating ΔG with respect to the number of moles of component I, the change of chemical potential, $\Delta\mu_I$ is obtained. The expression for the change of chemical potential of the solvent $\Delta\mu_1$ contains structural parameters of the crosslinked system (molecular weight averages, sol and gel fractions and their composition and concentration of EANC's) which are a function of conversion. They will be derived first.

Structural Changes Resulting from Crosslinking: The development of the network structure is described theoretically by the statistical Theory of Branching Processes (TBP). (cf., e.g., Refs.[16],[17]). The components A and B, are considered tetrafunctional and trifunctional, respectively. The component A carries f_A groups of type A, and the component B f_B groups B. All functional groups A and all functional groups B, respectively, have the same reactivity that is independent of conversion; intramolecular reactions are not considered at present. Due to limited space of this contribution, only basic results of the theoretical derivation are summarized below.

The basic probability generation function (pgf), $F_{0n}(z_A,z_B)$, for the distribution of reaction states of building units with respect to the number and type of formed bonds reads

$$F_{0n}(z_A,z_B) = n_A(1-\alpha_A+\alpha_A z_B)^{f_A} + n_B(1-\alpha_B+\alpha_B z_A)^{f_B} \quad (6)$$

where z_A, z_B are variables of the pgf, α_A and α_B are conversions of A and B groups, and n_A and n_B are molar fractions of components A and B, respectively. The critical conversion at the gel point for this system is described by the well known Stockmayer equation

$$(\alpha_A\alpha_B)_{crit} = \frac{1}{(f_A-1)(f_B-1)} \quad (7)$$

The extinction probabilities for bonds B→A and A→B, v_A and v_B, respectively, are given by

$$v_A = (1-\alpha_A+\alpha_A v_B)^{f_A-1} \qquad v_B = (1-\alpha_B+\alpha_B v_A)^{f_B-1} \quad (8)$$

The parameters needed in the thermodynamic equations are the weight or volume fractions of sol (w_s, φ_s) and gel (w_g, φ_g), number-average molecular weight or molar volume of sol ($(M_n)_s$, $(V_n)_s$), and conversions of A and B groups in sol and gel ($(\alpha_A)_s$, $(\alpha_B)_s$, $(\alpha_A)_g$, $(\alpha_A)_g$):

$$w_s = w_A(1-\alpha_A+\alpha_A v_B)^{f_A} + w_B(1-\alpha_B+\alpha_B v_A)^{f_B} \qquad w_g = 1-w_s \quad (9)$$

$$\varphi_s = \varphi_A(1-\alpha_A+\alpha_A v_B)^{f_A} + \varphi_B(1-\alpha_B+\alpha_B v_A)^{f_B} \qquad \varphi_g = 1-\varphi_s \quad (10)$$

$$(M_s)_n = \frac{n_A M_A(1-\alpha_A+\alpha_A v_B)^{f_A} + n_B M_B(1-\alpha_B+\alpha_B v_A)^{f_B}}{n_A(1-\alpha_A+\alpha_A v_B)^{f_A} + n_B(1-\alpha_B+\alpha_B v_A)^{f_B} - n_A f_A \alpha_A v_B(1-\alpha_A+\alpha_A v_B)^{f_A-1}} \quad (11)$$

$$(V_s)_n = \frac{n_A V_A(1-\alpha_A+\alpha_A v_B)^{f_A} + n_B V_B(1-\alpha_B+\alpha_B v_A)^{f_B}}{n_A(1-\alpha_A+\alpha_A v_B)^{f_A} + n_B(1-\alpha_B+\alpha_B v_A)^{f_B} - n_A f_A \alpha_A v_B(1-\alpha_A+\alpha_A v_B)^{f_A-1}} \quad (12)$$

$$(\alpha_A)_s = \frac{\alpha_A v_B}{(1-\alpha_A+\alpha_A v_B)} \qquad (\alpha_B)_s = \frac{\alpha_B v_A}{(1-\alpha_B+\alpha_B v_A)} \quad (13)$$

$$(\alpha_A)_g = \frac{1-v_B(1-\alpha_A+\alpha_A v_B)^{f_A-1}}{1-v_B(1-\alpha_A+\alpha_A v_B)^{f_A}} \qquad (\alpha_B)_g = \frac{1-v_A(1-\alpha_B+\alpha_B v_A)^{f_B-1}}{1-v_A(1-\alpha_B+\alpha_B v_A)^{f_B}} \quad (14)$$

w_A and w_B are mass fractions of components A and B, M_A and M_B the respective molecular weights, and V_A and V_B molar volumes. The concentration of elastically active network chains (EANC) is derived from the number of EANC's per building unit, N_e,

$$N_e = (n_A/2)\left[(f_A\alpha_A(1-v_B))(1-(1-\alpha_A+\alpha_A v_B)^{f_A-1}) - f_A(f_A-1)(\alpha_A(1-v_B))^2(1-\alpha_A+\alpha_A v_B)^{f_A-2}\right]$$
$$+(n_B/2)\left[(f_B\alpha_B(1-v_A))(1-(1-\alpha_B+\alpha_B v_A)^{f_B-1}) - f_B(f_B-1)(\alpha_B(1-v_A))^2(1-\alpha_B+\alpha_B v_A)^{f_B-2}\right] \quad (15)$$

The concentrations of EANC's in volume unit of sol+gel, ν_e, and in sol+gel, ν_{eg},

$$\nu_e = N_e \rho / M_{0n} \qquad \nu_{eg} = N_e \rho / M_{0n} w_g \tag{16}$$

where ρ is specific gravity of the polymer and M_{0n} is number-average molecular weight of component unit.

Thermodynamic Stability and Solvent Activity: The thermodynamic stability of the quasibinary (before the gel point) and quasiternary (beyond the gel point) systems will be examined in the future by calculating the spinodal and conditions for the critical point because they are a function of degree-of-polymerization averages only[12]. If phase separation occurred in system where the sol fraction is not negligible, fractionation of the sol molecules between liquid and gel phases would take place. This situation will not be examined here at present. Experiments give evidence that the system described here is one-phase until sol fraction is small; the separated liquid phase is almost a pure solvent. Therefore, solvent activity can be calculated until a_1 reaches unity which is the condition of phase separation for a binary gel-solvent system. From equations (2)-(5) and substituting $m_1 = 1$, $1/m_3 = 0$ and $m_2 = (V_2)_n/V_1$ ($(V_2)_n \equiv (V_s)_n$), one obtains the expression for the chemical potential of the solvent

$$\frac{\Delta\mu_1}{RT} = \ln\phi_1 + (1-m_2)\phi_2 + \phi_3 + \chi_{12}\phi_2^2 + \chi_{13}\phi_3^2 + (\chi_{12} + \chi_{13} - \frac{1}{m_2}\chi_{23})\phi_2\phi_3 + \nu_{eg}V_1[A\phi_3^{1/3}\phi_0^{2/3} - B\phi_3] \tag{17}$$

At constant ϕ_1, the volume fractions ϕ_2 and ϕ_3 change with volume fraction of sol and gel are dependent on conversion as

$$\phi_2 = (1-\phi_1)\varphi_s \qquad \phi_3 = (1-\phi_1)\varphi_g$$

This corresponds to situation in closed systems. Also, the equality $\phi_3 = \phi_0$ is valid. If ϕ_1 changes during network build-up (solvent evaporates), ϕ_0 is an integral value over ϕ_0-values at which EANC's were formed as shown in Ref.[1, 8]

As the composition of the polymer changes with conversion, and compositions of the sol and gel are different, the interaction parameters χ_{IJ} also change. For the present system, the change is not known, but can be estimated. What is known is the interaction parameter χ_{13} close to full conversion of functional groups (this value of χ_{13} is obtained by combination of equilibrium swelling and modulus of the swollen network). Also, we have found that in ketones χ_{13} decreases with increasing concentration of OH groups (results on off-stoichiometric systems). To get a rough estimate of the effect of conversion on the interactions, a linear dependence of interaction parameters on conversion was assumed. Thus, before the gel point

$$\chi_{12} = \chi_0 + (\chi_{13\text{fin}} - \chi_0)\alpha_A \tag{18}$$

Beyond the gel point (cf., Eqs. (13) and (14))

$$\chi_{12} = \chi_0 + (\chi_{13\text{fin}} - \chi_0)(\alpha_A)_s \qquad \chi_{13} = \chi_0 + (\chi_{13\text{fin}} - \chi_0)(\alpha_A)_g \tag{19}$$

χ_{23} is assumed to be proportional to the difference in concentration of OH groups in sol and gel

$$\chi_{23} = \chi_{13} - \chi_{12} = (\chi_{13\text{fin}} - \chi_0)[(\alpha_A)_g - (\alpha_A)_s] \tag{20}$$

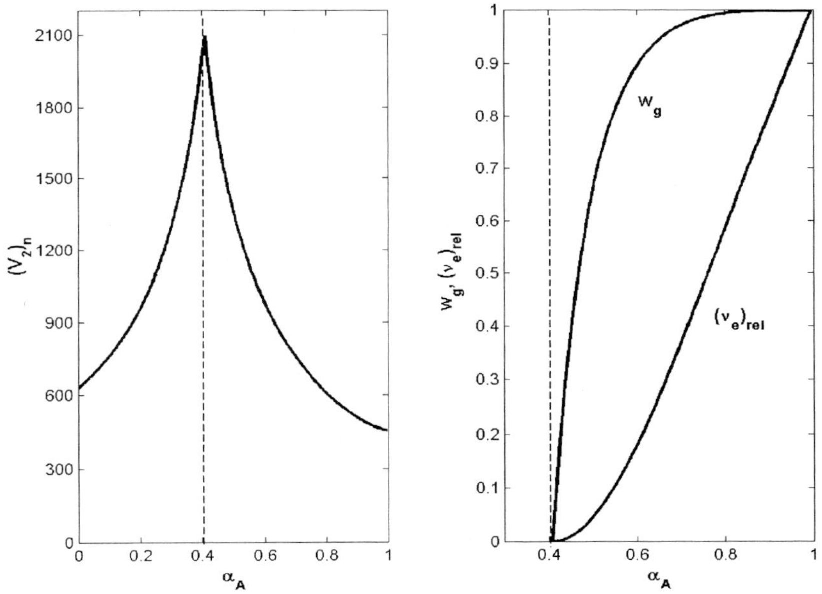

Fig. 1. Calculated dependence of number-average molar volume of sol $(V_2)_n$ [cm^3/mol], gel fraction w_g, and relative concentration of elastically active network chains [$(\nu_e)_{rel} = \nu_e(\alpha)/\nu_e(\alpha{=}1)$; ($\nu_e(\alpha{=}1) = 3.12\times10^{-3}$ mol/cm^3) for the system described in Experimental section with $f_A = 4$ and $f_B = 3$.

Results and Discussion

For $f_A = 4$ and $f_B = 3$, molecular weights and molar volumes specified in the Experimental section, a number of structural parameters were calculated. The changes in number-average molar volume of the sol, $(V_2)_n$, gel fraction, w_g, and concentration of EANC's, ν_e, are shown in Fig. 1. It is seen that the conversions of OH groups in sol, $(\alpha_A)_s$, and gel, $(\alpha_A)_g$, are different already at the gel point (Fig. 2). The number-average molar volume of sol passes through a maximum at the gel point. Close to 100%, the sol fraction is composed of initial components but their fraction is negligible (Fig. 1).

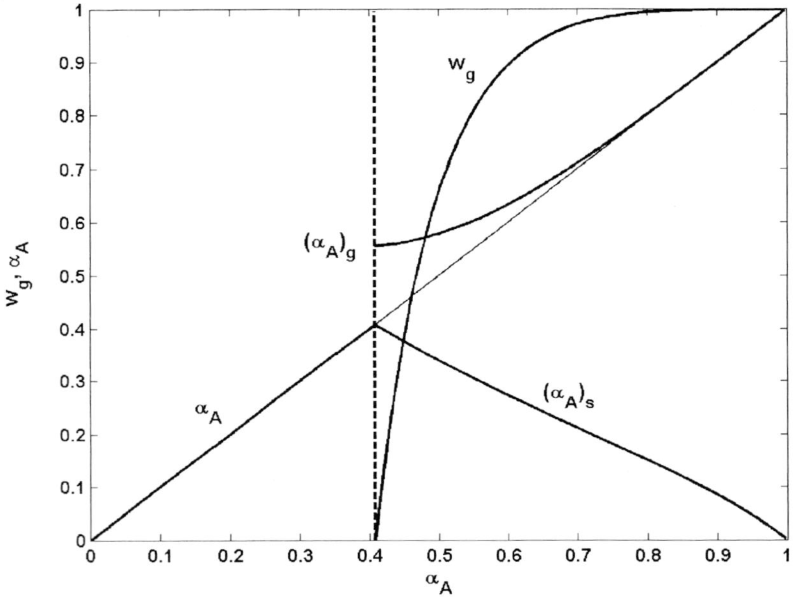

Fig. 2. Dependence of conversion of A (hydroxyl) groups in sol and gel and the gel fraction $(\alpha_A)_s$, and $(\alpha_A)_g$, respectively, on conversion for system described in Experimental section with $f_A = 4$ and $f_B = 3$.

These differences are reflected in differences in the interaction parameters χ_{12} and χ_{13} as well as non-zero value of χ_{23} (Fig. 3) estimated for methyl amyl ketone (MAK) (2-heptanone). The value of χ_0 was estimated to be 0.30, and the value of $\chi_{13\mathrm{final}} = 1.03$ (25°C) was determined experimentally by combined swelling and stress-strain measurements.

Figure 4 shows calculated dependencies of solvent activity on conversion for different concentrations of MAK valid for one-phase systems. The thermodynamic stability of the sol-gel-solvent system is not analyzed in this contribution and attention is paid mainly to the onset of solvent-gel phase separation well beyond the gel point. In this region, the condition for phase separation is given by $a_1 = 1$. The presence of the sol fraction is important just after the gel point when its molecular weight is high.

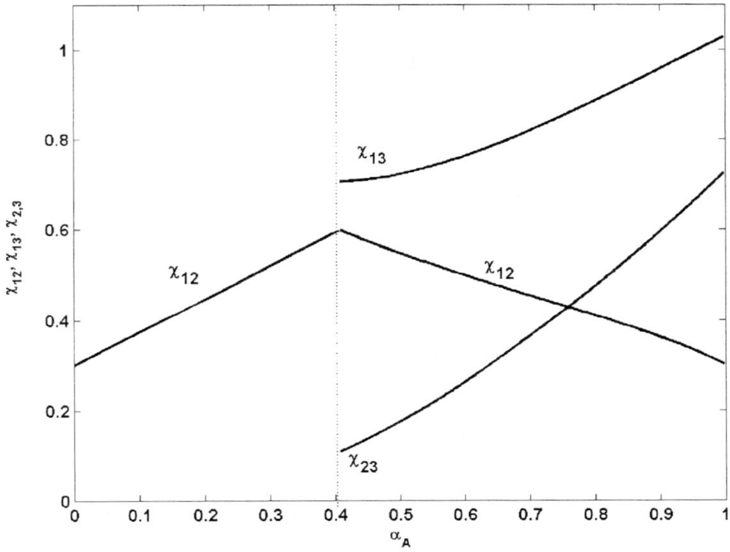

Fig. 3. Estimated change of the interaction parameters with conversion of functional groups for system described in Experimental section for interaction with methyl amyl ketone (MAK) with $f_A = 4$ and $f_B = 3$.

It can be seen that at constant solvent concentration the solvent activity increases with increasing conversion as a result of increase in both the molecular weight of the polymer before the gel point (or in crosslink density beyond the gel point) and interaction parameters χ_{12} and χ_{13}. Results of direct measurements of vapor pressure over crosslinking polymer-solvent systems are not available at present. However considering vapor pressure we have indications that the increase in the activity coefficient of the solvent can compensate for the decrease in its concentration during solvent evaporation coupled with crosslinking. Thus, the solvent activity is more or less constant or even slightly increases. This tendency is seen by examining the evaporation curves for the "slow" solvent MAK expressed as increase in content of solids (or weight fraction of polymeric species) (Fig. 5).

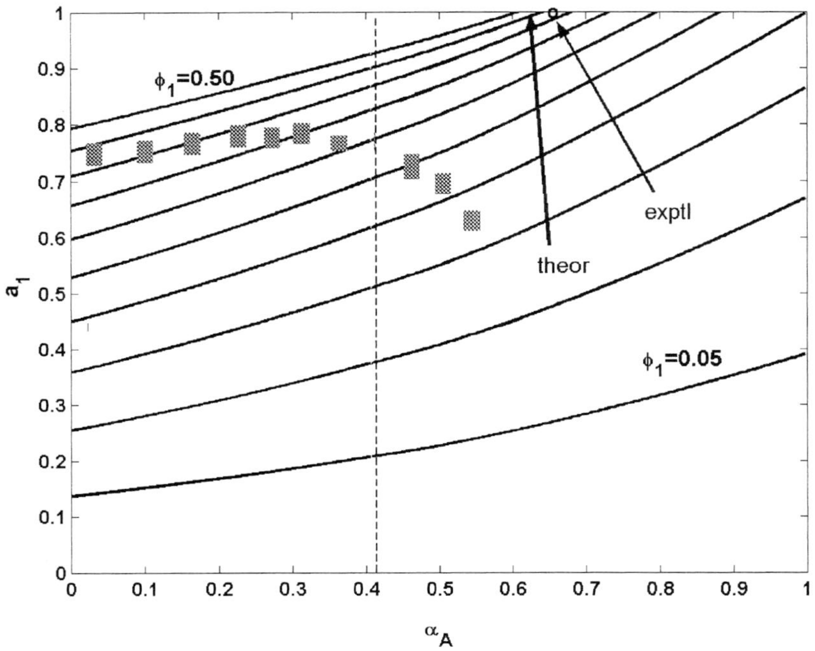

Fig. 4. Calculated dependence of solvent activity a_1 as a function of conversion for the polyurethane system, described in the text, at constant solvent (methyl amyl ketone) concentrations. The curves correspond to volume fraction of the solvent ranging from $\phi_1 = 0.50$ to 0.05 by steps 0.05. The bars correspond to the activity for experimental conversions and solvent contents in evaporating films shown in Fig. 5. Arrows indicate experimental and theoretical onsets of phase separation for systems containing 40 wt.-% solvent ($\phi_1 = 0.475$).

In the range of conversion up to 30-45%, the conversion of NCO groups measured in transmission (whole sample) is practically equal to that measured in the surface layer by ATR FTIR which means that the diffusion in the film is fast and the gradient in the film is not the evaporation rate determining factor. Determining the conversion and solvent content as a function of time, one can calculate from Eq. (17) in conjunction with Eqs. (9) – (16) the corresponding solvent activity as a function of conversion. The result of calculation is displayed in Fig. 4 by bars. It is seen that the solvent activity slightly increases with conversion which is consistent with the shape of evaporation curves in Fig. 5. Normally, the evaporation rate (slope of the evaporation curve) is proportional to the vapor pressure of the solvent over the film surface and would decrease with decreasing concentration of solvent in the polymer-solvent system. However, due to structure build-up and changes in thermodynamic interactions, the evaporation rate increases with decreasing concentration of the solvent in the film.

Onset of phase separation, detected experimentally by appearance of a liquid phase separated from the gel was observed in closed systems. Appearance of the liquid phase in a system containing 40 wt.-% solvent ($\phi_1 = 0.475$) was recorded repeatedly to occur at 65±3% conversion of NCO which is close to the value calculated (62%).

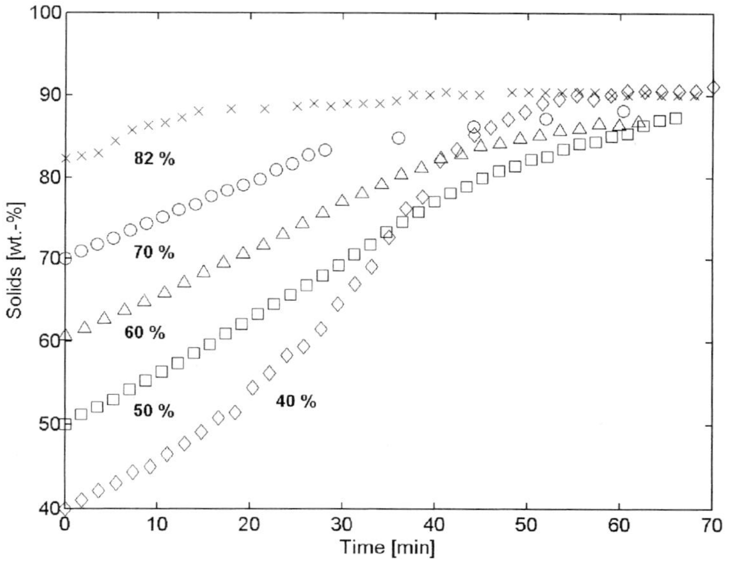

Fig. 5. Evaporation of methyl amyl ketone from films as a function of time for different initial concentration of solids in wt.-% indicated at curves; wet thickness 200μm, 200 p.p.m. catalyst (dibutyltin dilaurate), 25°C.

Conclusions

Branching and network build-up and change in polymer solvent interactions play an important role in solvent-based coating film formation by simultaneous crosslinking and solvent evaporation. Chemical changes – consumption of functional groups and formation of bonds and increase in structure connectivity – affect the solvent activity (increase in the activity at constant solvent concentration) and assist steady evaporation. However, the danger of phase separation during film formation should be avoided. The combination of branching theory with simple thermodynamics of mixing offers a reasonable explanation of experimental observations. Theoretically, the quasiternary system solvent-sol-gel is very challenging because of coupling of branching with thermodynamics of polymer-solvent mixtures and existing singularities at the gel point.

Acknowledgement

The authors thank Grant Agency of the Czech Republic (grant No. 203/99/D062) and DuPont Automotive, Inc. for support of this work.

[1] M. Dušková-Smrčková, K. Dušek, *J. Mater. Sci.*, **2002**, *37*, 4717.
[2] K. Dušek, M. Dušková-Smrčková, *Progr. Polym. Sci.* **2000**, *25*, 1215.
[3] K. Dušek , *J. Polym. Sci., Pt. C,* **1965**, *16*, 1289.
[4] K. Dušek, in "*Polymer Networks. Structure and Mechanical Properties*" A. J. Chompff, S. Newman, Eds., Plenum Press, New York 1972, p. 245
[5] K. Dušek, J.-P. Pascault, in: *"Wiley Polymer Networks Group Review. Vol. 1. Chemical and Physical Networks. Formation and Control of Properties"*, K. te Nijenhuis, W. Mijs, Eds. J. Wiley & Sons 1998, p. 277
[6] R.H.H. Williams, B.A. Rozenberg, J.-P. Pascault, Adv. Polym. Sci. **1996**, *128*, 96.
[7] J. Huybrechts and K. Dušek, Surface Coat. Int. **1998**(3), 117.
[8] K. Dušek, M. Dušková-Smrčková, Polym. Bull. **2000**, *45*, 83.
[9] P.J. Flory, " Principles of Polymer Chemistry", Cornell Univ. Press, Ithaca 1951.
[10] P.J. Flory, Polymer **1979**, *20*, 1317.
[11] B. Erman, P.J. Flory, *Macromolecules* **1982,** *15,* 806.
[12] R. Koningsveld, W.H. Stockmayer, E. Nies, "Polymer Phase Diagrams. A Textbook", Oxford University Press 2001.
[13] H. Tompa, "Polymer Solutions" , Butterworth, London 1956.
[14] K. Dušek, W. Prins, *Adv. Polym. Sci.* **1969**, *6*, 1.
[15] K. Binder, H.L. Frisch, J. Chem. Phys. **1984**, *81*, 2126.
[16] K. Dušek, *Adv. Polym. Sci.* **1986**, *78*, 1.
[17] K. Dušek, in „*Telechelic Polymers: Synthesis and Application*", E.J. Goethals, CRC Press, Boca Raton 1989, p. 89.

Macromol. Symp. **2003**, *198*, 271-281

Ternary Phase Separation in Polyurethane Elastomers with Immiscible Soft Segments

Werner Mormann,[*1] *Chenxi Zhang*[2]

[1] Universität Siegen, FB 8, Laboratorium für Makromolekulare Chemie, Adolf-Reichwein-Str. 2, 57068 Siegen, Germany
[2] Bayer Polyurethanes Asia Pte Ltd, Singapore Technical Center, 4 Tuas Avenue 2, 639444 Singapore

Summary: Polyurethane elastomers with two immiscible hard-segment polyols have been synthesized from homogeneous prepolymer mixtures, to obtain incompatibility on a molecular scale from reaction driven phase separation. Miscibility of a series of different soft-segment polyols and of their prepolymers with 4,4'-diisocyanatodiphenyl methane (MDI) has been investigated by cloud point analysis. UCST-behavior was observed for all binary combinations of polyols and of prepolymers; prepolymer formation lowered UCST with respect to polyols with the option to adjust it by further addition of MDI. Polyurethane elastomers have been synthesized and characterized with respect to their thermal, mechanical and viscoelastic (DMTA) behavior. Broad temperature ranges of loss factor above 0.1 were observed from polyurethanes with two immiscible soft-segment polyols. Rebound resilience, however, was still significant at room temperature, because of the low transition temperatures of the polyols used. Introduction of a third partially miscible component, a hard-segment from tetraethyleneglycol and MDI gave materials with a loss factor above 0.1 over a temperature range from –50 to 120 °C and higher. The studies proved that the concept of intimate mixing of immiscible soft-segments in polyurethanes works for the design of viscoelastic materials with good damping properties.

Keywords: damping, PUR-elastomers, ternary phase separation, viscoelasticity

Introduction

The use of energy absorbing and vibration damping materials has been continuously growing in the last decades. Applications of these materials include insoles for athletic shoes,[1] seismic isolation pads to protect buildings from earthquake damage, shock-absorbing encapsulates for under water sonar devices, vibration insulating systems for high-speed railway trucks[2] and the damping of vibrations and noise from the moving parts of machinery.[3]

Damping performance is related to the structure and to the molecular weight of materials. One mechanism of damping is based on material that absorbs mechanical or acoustic energy and transforms it into thermal energy, which is dissipated into the matrix. This mechanism of energy absorption can be described by the relation $\tan \delta = E'' / E'$, where E'' is the loss

 DOI: 10.1002/masy.200350823

modulus, E' the storage modulus and tan δ damping factor, dissipation factor or loss factor. At the glass transition condition, which involves both temperature and frequency effects, the cooperative molecular motion of chain segments transforms a maximum of mechanical energy into heat.

The damping temperature is defined as wide as the glass transition temperature of the polymer. Common damping materials are homopolymers or copolymers, which exhibit efficient damping in a temperature range of $T_g \pm 15$ °C.[4] The acoustic spectrum spans from 20 Hz to 20,000 Hz three decades of frequencies. For many polymers, according to the WLF-equation, the time-temperature equivalence is 6 °C to 7 °C per decade of frequency or about 20 °C for the acoustic range.[5] Thus polymers often suffice to damp the full acoustic range at one temperature, they fail, however, when temperature changes of more than 20 °C occur. Therefore, it is necessary to widen the temperature range of the glass transition.

One solution to this problem lies in selecting multicomponent polymers with a controlled degree of partial miscibility. Incompatible polymer alloys[6,7] display separated T_g's and therefore exhibit several damping ranges corresponding to the glass transition temperatures of the components. Homogeneous polymer alloys[8] show only one T_g, which may be slightly broader than that of the individual components. Semi-miscible polymer blends where mixing between the polymers is extensive, have a very broad damping range that covers a wider frequency and temperature range.

Semi-miscible polymer blends can be made by several methods among them formation of interpenetrating networks (IPNs). IPNs are intimate mixtures of two (or more) network polymers in which at least one of the networks is formed in the presence of another[9]. Ideally, bonding between the networks is absent, chains of one network are threaded through those of the second network; they are catenated. Advantages of IPNs over other multicomponent polymer blends are the forced miscibility and increased surface area. IPN's are one of the newer materials in energy absorption and vibration damping.[9-14] An approach, which has the advantages of IPNs, is the manufacture of a network structure from two different network polymers, which use the same branching units. These polymers should be miscible at the temperature of manufacture and undergo reaction driven phase separation to give a multiphase polymer blend with physical or chemical cross-links. This is schematically shown in Figure 1.

The present contribution reports on the synthesis and characterization of polyurethane elastomers made from two different soft-segment polyols that are inherently immiscible.

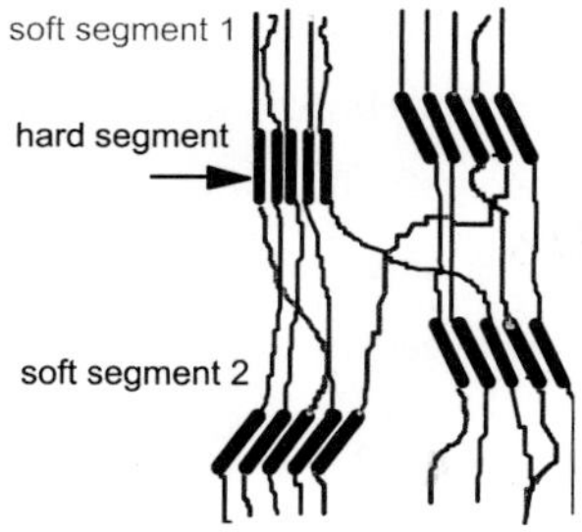

Fig. 1. Polyurethane with immiscible soft-segment polyols (schematic).

Results and Discussion

The strategy used in the present study is based on the synthesis of essentially linear multiblock polyurethane elastomers consisting of hard- and soft-segment blocks, which are immiscible at service temperature. The hard-segments are oligourethane units, e.g. from 4,4'-diisocyanatodiphenyl methane (MDI), and a "chain extender", e.g. 1,4-butanediol, the soft-segment blocks are α,ω-dihydroxy polyethers or aliphatic polyesters (cf. Scheme 1). The hard-segment domains are crystalline and function as multivalent branch points.

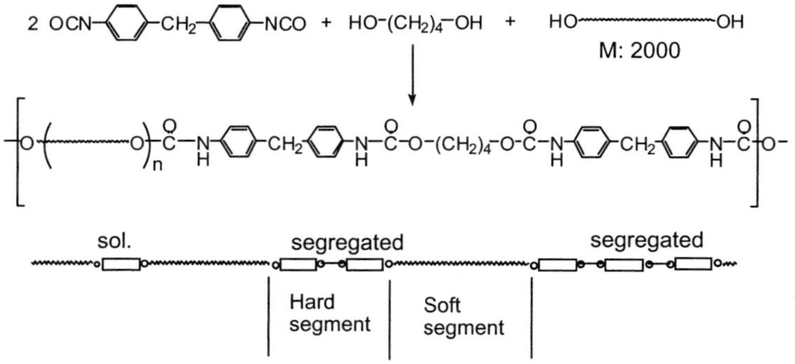

Scheme 1

Intimate mixing of incompatible soft-segment polyols, a prerequisite of good damping properties, should be possible with isocyanate terminated prepolymers of these polyols for the synthesis of polyurethane elastomers. Polyols, their structure and functionality are included in Table 1. As no literature data on miscibility of these materials was available, miscibility was studied measuring the cloud point upon cooling of homogeneous mixtures.

Miscibility of the polymethyl acrylate was observed with the polybutadienediol (PBDD), with polyoxypropylenediol (PPG), and with polyethylene glycol (PEG) above 185 °C. PPG is also miscible with polyoxytetramethylenediol (PTMO), while PPG and PEG are immiscible. All of the combinations studied over a wider range of composition showed typical upper critical solution behavior (UCST), which can be seen from cloud point curves in Figure 2.

Table 1. Miscibility of polyols (50 / 50 wt. %).

Polyol (functionality)	**PEG**	**PPG**	**PMA** H-4903	**PMA** H-4904	**PBDD** 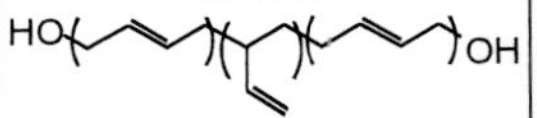
PTMO (2) 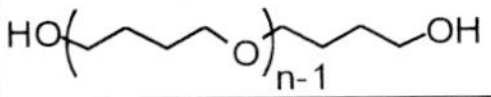	>215 + <215 -	+	+	>190 + <190 -	>140 + <140 -
PEG (2) 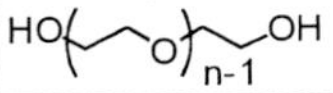		-	>185 + <185 -	-	-
PPG (2) 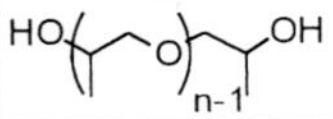			+	+	+
PMA H-4903, H-4904 (2.5) 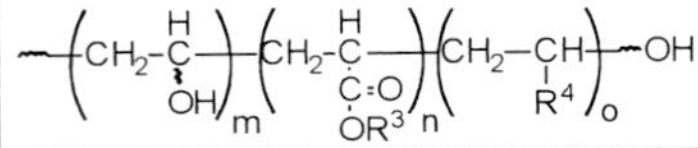				+	-

+ means miscibility at 25 °C, - means immiscibility, numbers stand for temperatures

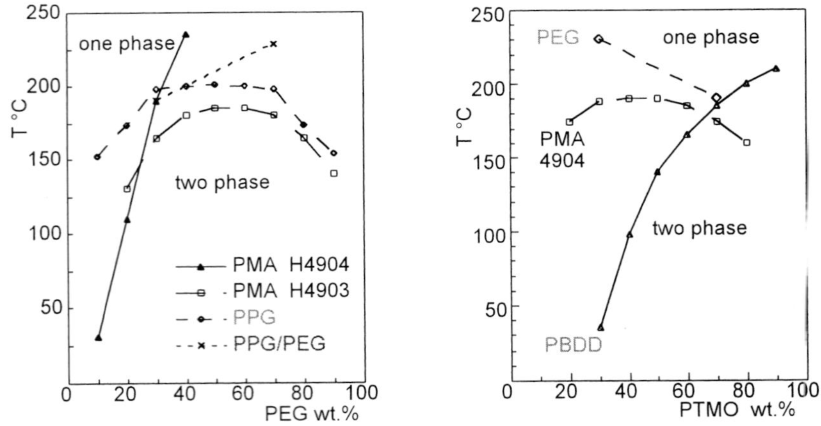

Fig. 2. Cloud point curves of mixtures of polyols with PEG (left) and PTMO (right).

Prepolymer formation, i.e. reaction of soft segment polyols with excess diisocyanate, adds identical MDI-monourethane units to the polyols, which should be a means to improve miscibility. Miscibility of prepolymers was studied as described for the polyols; the results are included in Table 2. Prepolymer formation has a huge effect on compatibilization in some combinations and almost no effect in others. PEG/PPG, which are miscible as polyols above 185 °C, become miscible as prepolymers above 95 °C, a temperature range that will be reached during elastomer formation. The same is true for PMA H-4904/PTMO (190 °C and 131 °C). On the other hand miscibility of PTMO/PEG is practically unchanged, even if a 1 : 8 ratio of polyol/isocyanate is used (215 °C vs. 197 °C).

Table 2. Miscibility of prepolymers (50 / 50 wt.%).

Polyol	**PEG**	**PPG**	**PMA H-4903**	**PMA H-4904**	**PBDD**
PTMO	1/8* >197 +	+	+	1/6* >131 +	+
PEG		>97 +	+	1/12* >182 +	-
PPG			+	+	+
PMA H-4903				+	1/6* >70 +
PMA H-4904	*: Molar ratio of polyol/MDI; (otherwise 1/4)				-

The effects of prepolymer formation and of further dilution with MDI are illustrated in Figure 3. Again UCST-behavior is observed and the strong decreasing effect of prepolymer formation on the UCST. Additional MDI, the amount of which is limited, however, by the desired hard-segment content of the polyurethane, is a method to adjust miscibility within a wide range. As the isocyanate is consumed during PUR-formation, reaction driven phase separation of the polyols is expected to occur.

Polyurethanes were synthesized to have a hard-segment content of 40 and 50 weight percent, respectively. PUR-elastomers with a single polyol were included for comparison. Prepolymers with the appropriate amount of MDI were mixed in a 1 : 1 ratio, heated to maximum temperature of 100 °C, the amount of chain extender was added sufficient to have a 2 percent excess of isocyanate groups (index 102). After homogenization, the mixture was transferred to a mold and cured at a temperature were initially homogeneous conditions were ensured. All molded sheets of 80 x 300 x 2 mm were annealed for 6 h at 120 °C to ensure comparable thermal history. To slightly enhance compatibility, apart from mixing prepolymers that had been separately synthesized, prepolymers were also made from the mixture of two polyols, which should give block-copolymers of the different polyols linked by a MDI unit.

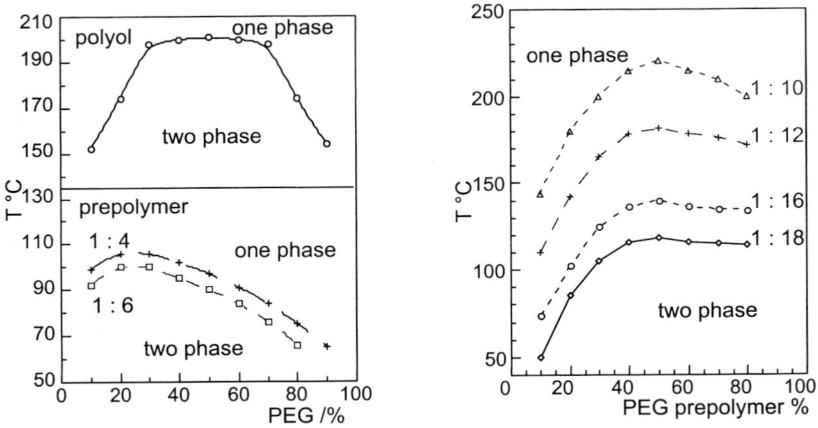

Fig. 3. Cloud point curves of mixtures of prepolymers from PPG / PEG (left), from PMA H4904 / PEG (right).

The PUR-elastomers were characterized by differential scanning calorimetry (DSC), dynamic mechanical thermal analysis (DMTA), tensile testing, and rebound resilience. In Table 3 properties of polyurethanes from a single polyol and of some binary combinations of these have been collected. In terms of elongation and tensile stress normal behavior is observed, only the rather small elongation at break of No. **103-2** does not fit. The result of mixed soft segments is a decrease in rebound resilience, which is to be expected. Somewhat surprising is the decrease of permanent set (permanent deformation) of the tensile specimen which is smaller for mixed soft segments than for the individual elastomers.

Glass transition temperatures from DSC are higher than those of the pure polyols, which indicates some phase mixing with the hard-segment blocks. DSC is not very sensitive to detect glass transition temperatures, in particular, there is a rather big difference between Tg from DSC and from DMTA, which was run at a frequency of 1 Hz. DMTA-values are by 20 K higher in some cases and almost equal in others. Tan δ values increase at the temperature of Tg from DSC, the maximum, which was taken as Tg, is at the temperatures given in column 8 of Table 3.

Table 3. Influence of soft segment composition on properties of polyurethane elastomers.

No.	Polyol	σ MPa	ε %	Rebound resil.[a)] %	Perm. set[b)] %	Tg °C DSC	Tg °C DMTA	Tm °C	Tg °C polyol
90-1	PTMO	43.0	510	53	120	-65	-61	229	-65
98-4	PEBA	25.5	475	39	115	-37	-17	201	-56
111-1	PPG	27.1	500	39	90	-47	-33	226	-66
110-2	PPG-PEBA	24.4	200	34	10	-48	-16	223	-66 -56
103-4	PEG-PPG	24.3	600	27	75	-46	-23	181	-46 -66
104-4	PEG-PTMO	38.8	525	46	105	-69	-58	226	-46 -65

All polyurethane elastomers have 50 % hard-segment content
a) rebound resilience according to DIN 53512
b) residual of 300 % elongation after 48 h

Table 4. Influence of the synthetic procedure on thermal properties of polyurethane elastomers.

No.	Polyols Method	HS %	σ MPa	ε %	Rebound resil. %	Perm. set %	Tg °C DSC	Tg °C DMTA	tan δ	tan δ > 0.1 (range)
103 -1	s	39	24.7	675	40	60	-46.7	-26.3	0.51	-52 – 1 (53)
-3	d **PEG**	39	25.7	700	34	50	-49.4	-25.4	0.46	-54 – 7 (61)
-2	**PPG** s	50	23.3	575	36	85	-47.4	-25.4	0.29	-53 – 5 (58)
-4	d	50	24.3	600	28	75	-46.4	-23.6	0.34	-50 –33 (83)
104 -1	s	39	36.7	550	57	50	-68.0	-55.5	0.27	-85 – 4 (89)
-3	d **PEG**	39	37.0	550	53	35	-69.0	-52.9	0.32	-81 – 1 (82)
-2	**PTMO** s	50	37.8	525	45	125	-58.8	-67.1 -26.6	0.12	-76 – -13 (63)
-4	d	50	38.8	525	47	105	-69.1	-58.9	0.19	-83 – - 1 (82)

s: prepolymer from mixture of two polyols
d: mixture of two prepolymers
PEG and PPG with weight ratio 1 : 4
PEG and PTMO with weight ratio 1 : 4

As mentioned above the method of synthesis, i.e. either from a mixture of two prepolymers or from a prepolymer made from a mixture of two polyols should have an influence on phase mixing or compatibilization of polyols in the final elastomer. Mixtures of PEG with PPG and of PEG with PTMO both with 75 wt. % PEG were used, because the prepolymer mixtures had suitable UCST (cf. Fig. 2). The mechanical and damping properties of this series of experiments together with the composition of the polyurethane elastomers are summarized in Table 4.

There is no influence of the synthesis on tensile strength or elongation at break in both series. Rebound resilience seems to be slightly lower for double prepolymer PUR with the exception of **104-2** and **104-4**, where they are equal within experimental error. Permanent set values are lower for the double prepolymer elastomers. Glass transition temperatures from DSC again are on average 20 K lower than those from DMTA for PEG/PPG (series **103**), while they are equal in the series **104** with the exception of **104-2**, where two maxima of tan δ are detected, one in the order of the DSC-value, the other with 26.6 °C as low as in the other series. The maxima of tan δ are in the range from 0.3 to 0.5 for PEG/PPG polyurethanes and from 0.12 to 0.32 for PEG/PTMO elastomers. The temperature range in which tan δ is larger than 0.1 is considered to be the range of damping. It is broader for the PEG/POTM series (approximately 80 K) and of the order of 60 for the PEG/PPG elastomers. This behavior is further illustrated by tan δ vs. time curves, which are shown in Figure 4.

The concept of immiscible soft-segment polyols obviously can be used to synthesize polyurethanes with good damping properties, a drawback of the materials used is the temperature range in which damping occurs. To introduce a further component in the polyurethanes that can contribute to damping at slightly higher temperature, tetraethyleneglycol was used as chain extender besides 1,4-butanediol. Again PUR-elastomers were made as described above.

From the data in Table 5 can be seen that the maximum of the loss factor in general is below 0.5. The only exception is the elastomer **113-1**, which is based on a polyesterdiol and tetraethyleneglycol. This sample has a maximum value of 0.90 for tan δ and in addition the damping maximum at room temperature, which also explains the very low rebound resilience of only 9 % for this sample.

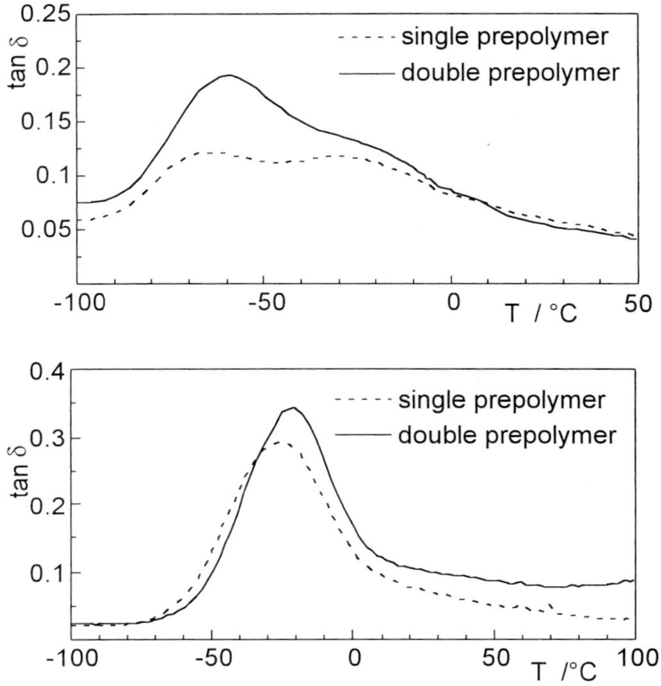

Fig. 4. Damping factor vs. temperature for elastomers 103-2 and -4 (left), for 104-2 and -4 (right).

Table 5. Influence of the hard-segment structure on viscoelastic properties of polyurethanes with immiscible soft-segments.

No.	Composition	tan δ max	tan δ > 0.1 °C	Range °C	Rebound resilience[a)] %
106-1	PTMO-TEG	0.18 0.34	-73 – 200	273	44
90-1	PTMO-BDO	0.17	-85 – -4	81	53
113-1	PEBA-TEG	0.90	-8 – 200	208	9
98-4	PEBA-BDO	0.35	-36 – 23	59	39
110-2	PPG-PEBA-BDO	-0.28	-42 – 15	56	34
114-1	PPG-PEBA-TEG	0.38 0.47	-42 – 200	242	18
112-2	PO / EO-TEG	0.29 0.34	-70 – 119	189	34

TEG: tetraethyleneglycol; BDO: 1,4-butanediol

a) rebound resilience according to DIN 53512

The temperature range of damping (tan δ > 0.1) is very large for elastomers with TEG as chain extender (190 K and more), which can also be seen from Figure 5, where storage modulus and loss factor of a polyurethane based on PPG and polyester with TEG as chain extender are shown as a function of temperature. For this sample the loss factor is bigger than 0.2 from -40 °C up to the melting of the hard-segments near 200 °C.

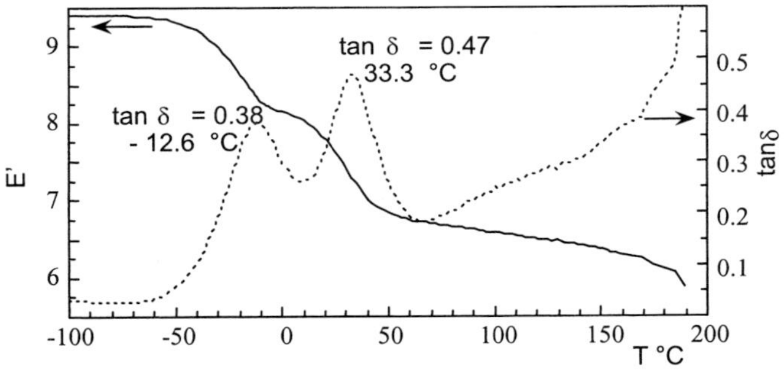

Fig. 5. DMTA of polyurethane 114-1 (frequency 1Hz).

Conclusions

The concept of immiscible soft-segment polyols in intimate contact has been applied to study polyurethane elastomers for mechanical and sound vibration damping. Synthesis from a homogeneous mixture of components proceeds via reaction driven phase separation to the expected morphology (cf. Figure 1). As glass transition temperatures of standard commercial soft-segment polyols are at low temperature, an additional phase to provide damping was introduced by means of a hard-segment consisting of oligourethanes from MDI and tetraethyleneglycol. This provided some phase mixing with increase of TG of the different soft-segment phases and a transition from the amorphous parts of the hard-segment. Efficient damping over a temperature range of more than 200 K was achieved by this method. The results prove that reaction driven phase separation has the same efficiency in terms of mixing of immiscible material on a nearly molecular level as IPNs. Work to extend this concept to polyurethane foams is under progress.

Acknowledgment

Financial support of this work from the Fonds der Chemischen Industrie gratefully acknowledged.

[1] B. Mareschal , H. Stilling, Cellular Polymers, **8(2)**, 144, 1989.
[2] G.M. Stack, J.M. Miller, E.Y. Chang, J. Appl. Polym. Sci. **42**, 911, 1991.
[3] V.P. McConnell, Plastics Design Forum, March/April, 1990.
[4] L.H. Sperling, Introduction to physical polymer Science, Chapt. 6, Wiley, New York, 1986.
[5] T. Murayma, Dynamic Mechanical Analysis of Polymer Material, Chapt. 1, Elsevier New York. 1978.
[6] D. Klempner, H.L. Frisch, K. C. Frisch, J. Polym. Sci. A-2, **8**, 921, 1970.
[7] K.C. Frisch, D. Klempner, H.L. Frisch, S. Migdal, H. Ghirdella, Polym. Eng. Sci., **15**, 339, 1975.
[8] E.E. Ungar, Noise and Vibration Control, I.I.Beranek Ed., Chapt. 14, McGraw Hill, New York, 1971.
[9] M.C.O. Chang, D.A. Thomas, L.H .Sperling, J. Polym. Sci. 13, Polym. Phys. **26**, 1627, 1988.
[10] M.C.O. Chang, D.A.Thomas, L.H. Sperling, J. Appl. Polym. Sci. **34**, 409, 1987.
[11] J.J. Fay, C.J. Murphy, D.A. Thomas, L.H. Sperling, Polym. Mater. Sci. Eng., **60**, 648, 1989.
[12] M.C.O. Chang, D.A. Thomas, L.H. Sperling, J. Polym. Mater. **6** 61, 1989.
[13] U. Sorathia, W. Yeager, T. Dapp, Sound and Vibration Damping with Polymers, R.D. Corsaro, L.H. Sperling Eds., Chapt. 21, ACS Sympos. Ser. **382**, Washington, DC 1990.
[14] D.A.Greehill, D.J.Hourston, J.A.Waters, Sound and Vibration Damping with Polymers, R.D. Corsaro, L.H. Sperling Eds., Chapt. 22, ACS Sympos. Ser. **382**, Washington, DC 1990.

Reactive Compatibilization of Epoxy/Polyorganosiloxane Blends

*S. G. Prolongo, J. C. Cabanelas, J. Baselga**

Instituto Tecnológico de Química y Materiales Álvaro Alonso Barba. Universidad Carlos III de Madrid. Avda. Universidad, 30, 28911 Leganés, Madrid, Spain
E-mail: jbaselga@ing.uc3m.es

Summary: A new thermoset material based on DGEBA with polyaminosiloxane curing agents is presented. The system shows reaction-induced compatibilization which prevents coalescence of polysiloxane and DGEBA rich domains, leading to gradient structured morphologies. The influence of curing temperature and/or chemical nature of the siloxane on the morphology and surface microhardness were examined. When siloxane is pre-reacted with epoxypropylphenylether (EPPE), a more homogeneous material is obtained. Microhardness profiles on the material are strongly influenced by the extension of the compositional gradients.

Keywords: blends, epoxy resin, poly(3-aminopropylmethylsiloxane), reactive compatibilization, thermosets

Introduction

Fracture toughness and impact resistance of highly crosslinked systems, as epoxy resins, are usually low limiting the use of these materials in applications requiring high fracture strength. The problem is usually solved introducing elastomeric (ATBN, CTBN) and/or thermoplastic (PSF, PEI, PPO, PMMA) materials, which form a segregated phase in the cured state, improving the mechanical properties of the blend.[1]

Polysiloxanes have very low glass transition temperature and good thermal stability, and may be used as epoxy toughening agents, especially for high temperature applications. Since they are immiscible with epoxy resins, different compatibilization approaches, including copolymer formation and reactive end groups insertion, have been proposed in literature.[2-5]

In this work, we follow a new and different approach. We have used a highly functionalised organosiloxane polymer, poly(3-aminopropylmethyl-siloxane) (PAMS), as a hardener for a

 DOI: 10.1002/masy.200350824

DGEBA base resin. The blend is completely immiscible at room temperature, but from our early studies,[6,7] it becomes partially miscible as polymerisation proceeds, due to the high functionality of the PAMS. A careful selection of the curing temperature allows tuning different final morphologies ranging from macroscopically homogeneous transparent samples to partially phase segregated opaque samples. This new system has some good properties like low water absorption, good adhesion and transparency but it is relatively brittle. In addition, curing reaction rate is fast and similar to that of DGEBA with other aliphatic amines and does not seem to be influenced by the low conversion gelation process.

The definition of homogeneity in thermosets is not completely closed. It is well known that it is impossible to reach complete conversion; if heterogeneity is defined in a low enough scale, cured resins would present intrinsic heterogeneities. If the monomers are immiscible, as in the present case, the increase of viscosity and crosslinking degree during cure should limit the diffusion controlled processes and severe concentration gradients should be observed. The objective of this work consists of the characterization of the microscopic homogeneity and to study the influence of the polysiloxane chemical composition on it. The polysiloxane functionality and the lateral chain composition is modified pre-reacting the amine groups with epoxypropylphenylether. Semi-quantitative SEM X-ray analysis will be used to analyse the presence of concentration gradients. The observed microstructures will be correlated with the microhardness of the different cured products.

Experimental

Materials. Diglycidyl ether of bisphenol-A based resin (DGEBA) was supplied by Gairesa (Spain) with an epoxy equivalent of 188 g (equiv. epoxy)$^{-1}$, as determined by acid titration. The ratio of secondary hydroxyl groups to epoxy group was 0.069 as determined by size exclusion chromatography (SEC) (Shimadzu LC9A). Poly(3-aminopropylmethylsiloxane) (PAMS) was synthesised from 3-aminopropylmethyldiethoxysilane monomer (from ABCR) by hydrolysis and condensation reaction.[6] The oligomer was characterized by SEC, vapour pressure osmometry (VPO, Knauer) and ^{29}Si NMR spectroscopy (Gemini). Number-average molecular weight measured by VPO was 1620 ± 90 g mol^{-1} and 1570 g mol^{-1} by ^{29}Si NMR, with an amine equivalent of 58.5 g (equiv. active hydrogen)$^{-1}$. Tg of PAMS was -65 ± 2°C (DSC, Perkin Elmer, 10 K min^{-1}, midpoint).

Lateral chain modification of PAMS was accomplished by reaction of its amine groups with 2,3-epoxypropylphenylether (EPPE) at different molar ratios. Reactions were carried out in a flask with continuous stirring at 333 K for 3 hours under N_2 atmosphere. Table 1 shows the characteristics of samples prepared with different PAMS/EPPE ratios. The progress of the lateral reaction was checked following the extinction of the primary amine band (4936 cm^{-1}) in the near IR (Perkin-Elmer GX2000, 4 cm^{-1} resolution). Tg of the samples was also analysed by DSC.

Table 1. Characteristics of the polysiloxane hardeners.

Sample	Mn G/mol	r_{EPPE} [a)]	T_g °C
PAMS	1600	0	-67
PAMS-EP10	1805	10.2	-54.6
PAMS-EP20	2010	20.0	-39.8
PAMS-EP40	2420	40.4	-18.8

[a)] Number of EPPE groups per 100 monomer units of PAMS.

Sample preparation and curing. Stoichiometric DGEBA and PAMS / PAMS-EPPE mixtures were prepared at room temperature and placed in a mould between two glass slides separated by a Teflon sheet, which were pre-treated with a demoulding agent. Curing was performed in an oven for 2 hours at different temperatures, as it is shown in Table 2. Tg of cured samples were obtained from DMTA measurements (Rheometric Scientific MkIII), using bars of 10 × 5 × 1 mm that were tested in bending mode at 1 Hz and 2 $Kmin^{-1}$ heating rate. Tg was determined as the temperature at the maximum in a tan δ scan. Curing was checked by FT-NIR (Perkin-Elmer GX2000) following the area of the epoxy band centered at 4530 cm^{-1}.

Determination of morphology. A Labophot microscope (Nikon) provided with photographic camera was used for transmission optical microscopy (TOM) and epifluorescence microscopy (EFM) imaging observations of the initial mixtures. Objectives of ×10 and ×40 with numeric apertures of 0.25 and 0.65, respectively, were used. For fluorescent labelling, 1,5-dimethylaminonaphtalen-5-sulphonylchloride (dansyl chloride, DNS; Molecular Probes) was added to PAMS (1:5000 molar ratio) and heated at 60 °C for 16 h. Reaction between DNS and PAMS was complete as checked by thin layer chromatography (TLC). Chromophore concentration was 10^{-4} M, as determined by UV/VIS spectrophotometry. DNS labelled PAMS was stored in a desiccator preserved from the light. EFM observations were taken with a barrier filter (λ_{max} < 420 nm).

Morphology of cured material was analysed measuring the X-ray signal of silicon atoms on Au/Pd-coated samples in an XL-30 SEM microscope (Phillips). X-ray semi-quantitative microanalysis was performed with an EDAX DX4i detector. Surface was analysed by mapping a 15×15 point matrix over a surface of 2 mm^2.

Microhardness. Vickers microhardness (square-based diamond pyramid) measurements were performed using a Micro Hardness tester DMT HVS-1000 (Time Technology), which allows displacement of the indenter along the x-length. Two different loads were employed: 2 N for indentations of about 0.2 mm^2 area, and 0.25 N for indentations of about 4.5 10^{-4} mm^2. Measurements with the low load were carried out every 40 μm, over an axis-length of 1 mm, totalling 25 consecutive measurements.

Table 2. Epoxy conversion and glass transition of cured samples.

Code	Hardener	Curing temp. (°C)	Epoxy conversion	Tg[a)] (°C)
C60	PAMS	60	0.86	98
C80	PAMS	80	0.88	-
C100	PAMS	100	0.93	155
C120	PAMS	120	0.96	-
C150	PAMS	150	0.97	163
C60P20	PAMS-EP20	60	0.81	90
C120P20	PAMS-EP20	120	0.86	109
C150P20	PAMS-EP20	150	0.92	118
C120P10	PAMS-EP10	120	0.96	125
C120P40	PAMS-EP40	120	0.74	73 [b)]

[a)] DMTA

[b)] This sample presents a shoulder at 110°C.

Results and Discussion

Curing. From an inspection of data presented on Table 2, epoxy conversion increases with curing temperature up to 150°C, at which the system reaches the maximum conversion. The high value of the Tg for DGEBA/PAMS samples reflects the high crosslinking density due to the high PAMS functionality; this point is especially notable if take into account the very flexible chain hardener used. When PAMS is pre-reacted with EPPE the initial mixtures with DGEBA remain also immiscible. The measured Tg of the pre-reacted systems is lower than the Tg of DGEBA/PAMS systems cured at the same conditions, as a consequence of the decrease in functionality of the modified PAMS. Samples with the same epoxy conversion show also lower Tg if they are pre-reacted with EPPE and the resulting decrease in crosslinking degree.. It is noticeable that the sample with the highest EPPE content, PAMS-EP40, showed a high temperature shoulder (at around 110°C) in the tan δ scan; it should be expected a highly heterogeneous microstructure for this sample.

Morphology development during curing. Room temperature mixtures have cloudy appearance in all the studied systems showing that the components are not miscible. A brief analysis of the initial morphology was done on DGEBA/PAMS mixtures immediately after mixing. TOM image just after mixing is presented in Figure 1a. Two phases are clearly observed. The dispersed phase forms quasi-spherical domains of about 2-5 μm surrounded by the continuous phase. Figure 1b shows the same region by epifluorescence microscopy (EFM). Bright areas correspond to the blue emission of DNS ($\lambda_{max} \approx 460$ nm). Since DNS was only attached to PAMS, these areas belong to PAMS-rich domains. PAMS constitutes therefore the continuous phase.

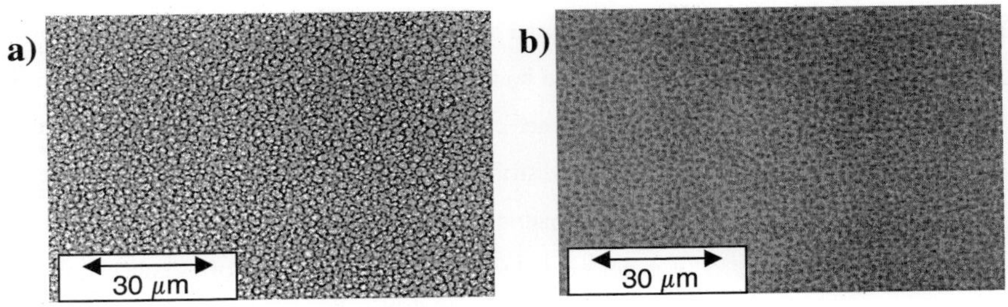

Fig. 1. a) TOM image of the DGEBA/PAMS mixture at room temperature. b) The same area observed by epifluorescence microscopy (EFM).

Sample is then heated for 2 minutes at 60°C and then cooled down again to room temperature. The resulting morphology is presented in Figure 2a. At this time, conversion is very low ($\alpha<0.05$) and main species are partially reacted PAMS with DGEBA and pure DGEBA; tertiary amine concentration is negligible at this conversion and so the crosslinking degree. A coalescence phenomena and phase inversion are observed. The dispersed phase size now is high but, as deduced from the EFM image (Figure 2b), it is composed mainly by PAMS. The polysiloxane is now the dispersed phase in an epoxy-rich (low fluorescence) continuous phase. A clear explanation for this behaviour is not available at the moment but it seems that the driving force may rely on the high functionality of PAMS. According to condensation reaction statistics [8,9], despite the low conversion reached, the number of non-reacted "free" PAMS molecules is now very low, becoming the minority component of the mixture.

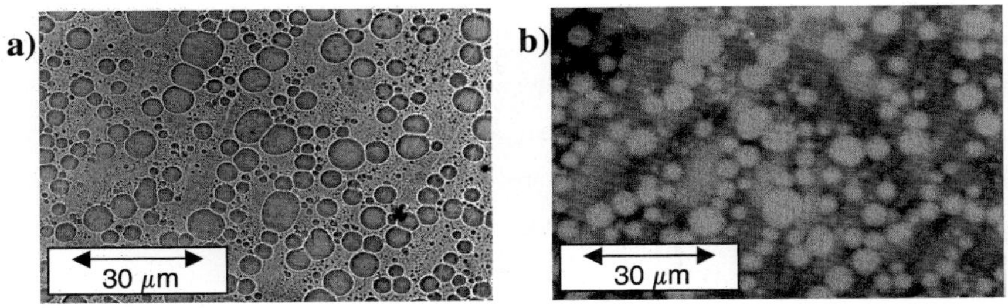

Fig. 2. a) TOM image of the DGEBA/PAMS mixture quenched at room temperature after 2 min. at 60°C. b) The same area observed by epifluorescence microscopy (EFM).

Observation of the morphology of cured samples by TOM or EFM is not easy because the refractive index of the components in the cured system are very similar to resolve the images by TOM, and because there are no clear phase boundaries to obtain good contrast by EFM. Since silicon atoms are only present in the hardener component, PAMS, we have performed a semi-quantitative analysis of the silicon surface distribution to study PAMS distribution in the cured specimens. Due to the uncertainty in the penetration depth of the electron beam and the lack of a Si standard in a similar matrix, it is not possible to give absolute values of Si concentration. For quantification purposes, the signal obtained in each of the 225 matrix points was normalized with the average Si signal measured in the total scanned area.

Figure 3 shows the concentration contour graphs for silicon distribution in the DGEBA/PAMS blend for three curing temperatures. A gradient structure is observed, especially for low conversion degrees, in which Si-rich domains of about 100-150 μm in size can be observed. With respect to the initial room temperature morphology, it can be observed that PAMS severely coalesces. Since the scanned area is large enough, it can be assumed that the surface average Si signal is proportional to the bulk Si concentration. Therefore, for sample cured at 80 °C (sample with highest compositional heterogeneity) the PAMS:DGEBA equivalent ratio in the domains of highest PAMS concentration is about 4:1, and about 1:2 for the regions were PAMS concentration is minimal. For samples cured at higher temperatures, compositional heterogeneity is lower but the same morphology remains (figure 3b). Modification of the lateral composition of PAMS with EPPE should have the benefit of increasing initial miscibility with DGEBA.

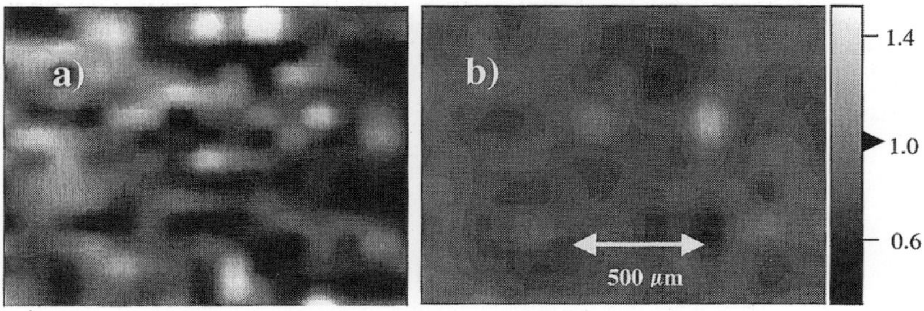

Fig. 3. Silicon concentration mapping of the surface of stoichiometric DGEBA/PAMS samples and cured at a) 80°C, b) 150°C (2 h).Grey scale corresponds to normalized Si signal.

In Figure 4 the same concentration map is presented for the DGEBA/PAMS-EP10 blend. Using PAMS pre-reacted with EPPE as hardener, concentration gradients seem to be smaller, even for the lower conversions, but surface morphology does not change appreciably. Coalescence of PAMS was not avoided.

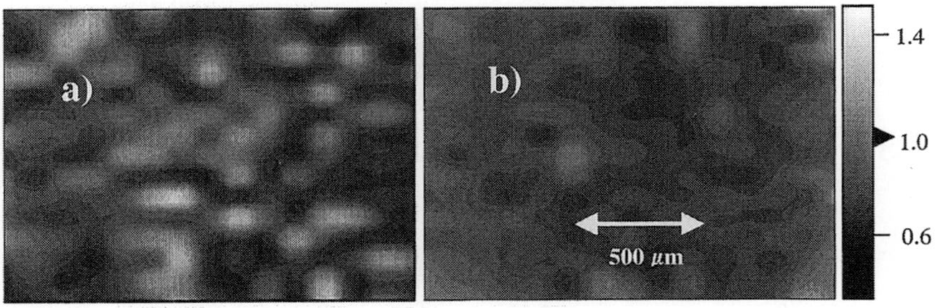

Fig. 4. Silicon concentration mapping of the surface for stoichiometric DGEBA/PAMS-EP10 samples and cured at a) 60°C, b) 150°C (2 h). Grey scale corresponds to normalized Si signal.

Data presented in Figures 3 and 4 needs to be quantified. We have divided the silicon concentration values into classes. The probability to find a region with the silicon concentration of the n class is defined by $p(i) = A(i) \Big/ \sum_{j=1}^{n} A(j)$ where A(i) is the area occupied by the range of silicon concentration from the nth class and the sum corresponds with the total area analyzed. In Figure 5 we present the probability function for Si distribution for the cured samples and in Table 3, the distribution function parameters are presented. It can be observed a Gaussian like distribution curve around the mean silicon concentration. The distribution sharpens as curing degree increases.

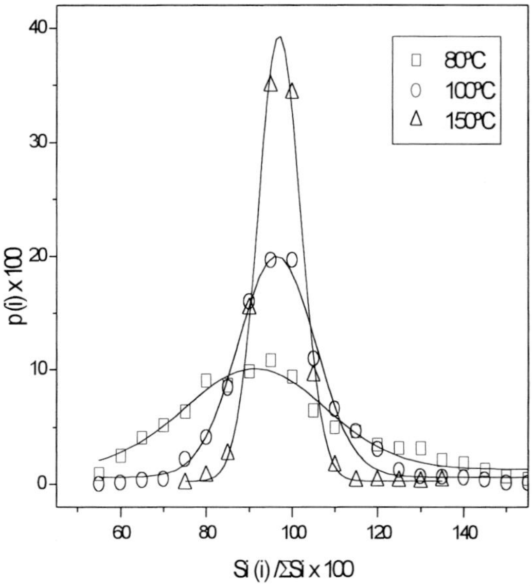

Fig. 5. Si distribution function for the DGEBA/PAMS system at different curing temperatures.

Several important conclusions can be obtained from Table 3 and Figure 5. It becomes evident that as curing degree increases samples become more homogeneous, but the analysis of the standard deviation reflects an interesting feature. Samples cured with PAMS containing 20% of EPPE pending units are more homogeneous, with sharp and symmetric distribution functions despite the lower conversions achieved. For the sample with 40% EPPE content, which presented a high temperature shoulder in the tan δ scan, a very wide standard deviation is observed suggesting a bimodal Si distribution. These results allow speculating about the morphology evolution. Coalescence and curing seem to be two simultaneous competing processes for fixing microstructure. At low temperatures, curing rate is low enough and coalescence of PAMS creates big domains. At higher temperatures, crosslinking rate limits the coalescing process resulting in a more homogenous microstructure.

Table 3. Statistical analysis of the cured samples.

Sample	Conversion	Standard Deviation[a)]	Asymmetry
C80	0.88	22.1	0.62
C100	0.93	12.7	0.93
C150	0.97	6.0	1.08
C60P20	0.81	13.8	0.90
C120P20	0.86	11.4	0.98
C150P20	0.92	7.4	1.04
C120P10	0.96	9.9	1.02
C120P40	0.74	43.9	0.22

a) Over a mean Si concentration normalized to 100.

Table 4. Vickers Hardness (MPa) of DGEBA/PAMS cured resins.

Sample	Curing T (°C)			
	60	80	100	120
DGEBA/PAMS	176±2	164±1	155±5	152±3
DGEBA/PAMS-EP10	194±4	185±3	173±2	161±3
DGEBA/PAMS-EP20	196±4	191±3	173±3	161±2
DGEBA/PAMS-EP40	124±5	144±4	146±3	149±3

Hardness measurements. In Table 4 the overall Vickers hardness is presented. The values presented are similar to literature data for epoxy novolacs and other brittle resins.[10,11] Hardness data for DGEBA/PAMS blends show a clear correlation with curing temperature being lower as curing (conversion) increases probably due to an increase in brittleness. It is interesting to note the behaviour for EPPE pre-reacted systems. For these systems, hardness is systematically higher than for pure PAMS systems but its variation with curing temperature shows the same trend.

Since hardness is a surface property hardness data may reveal the presence of depth heterogeneity which accumulates to the two-dimensional surface heterogeneity. Micro-hardness measurements were made along and axis and results are displayed in Figure 6. The values observed fluctuate around the mean hardness and the only explanation is the heterogeneous distribution of epoxy and hardener over the surface. Peak to peak distances range from 50 to 100 μm, comparable to the PAMS-rich domain size observed by SEM X-ray analysis. The fluctuations are in the order of ±20% for the lowest cured sample, and become small as curing and homogeneity increases, confirming the importance of these heterogeneities on the mechanical behaviour of the material. Si content was analysed in each of the impressions and no clear correlation was found between Si content and micro-hardness.

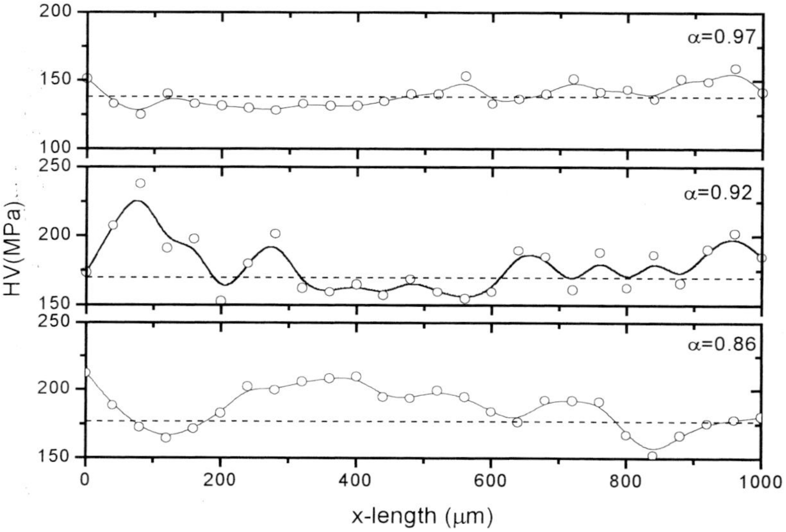

Fig. 6. Micro-hardness profiles in function of curing degree for DGEBA/PAMS samples. Mean hardness is also presented by an horizontal dot line.

Conclusions

The morphology and surface properties of a new epoxy/silicone system based in a polyfunctional aminosiloxane (PAMS) hardener have been studied. These components are immiscible at room temperature and they are initially phase separated. At curing temperature, PAMS coalesce forming large domains and as reaction proceeds a gradient structure is formed with PAMS and DGEBA rich "domains", but no phase boundaries are present. Quantification of these domains in cured resin was possible by using SEM X-ray microanalysis. Because of the competition between two simultaneous phenomena, curing and coalescence, cured resins show different heterogeneity as function of curing degrees, being more homogeneous at the highest curing temperatures employed. Homogeneity can be improved if lateral chemical composition of PAMS is varied pre-reacting with a monoepoxide monomer. PAMS becomes more compatible with DGEBA, but final Tg falls off because of the drop in crosslinking density. Microhardness measurements revealed the presence of a gradient structure and the length of the domain sizes coincides with SEM observations.

[1] D.R. Paul, C.B. Bucknall, *"Polymer Blends"*, Wiley Interscience,New York 2000, chap. 13, 22 and 26.
[2] T. Kasemura, S. Takahashi, K. Nishihara, C. Komatu, *Polymer* **1993**, *34*, 3416.
[3] S. Hou, Y. Chung, C. Chan, P. Kuo, *Polymer* **2000**, *41*, 3263.
[4] J. L. Hedrick, T. P. Russell, B. Haidar, A.C.M. Yang, *Macromolecules* **1989**, *22*, 4470.
[5] M. Ochi, *Polymer* **1999**, *40*, 1305.
[6] J.C. Cabanelas, B. Serrano, J. Gonzalez-Benito, J. Bravo, J. Baselga, *Macromol. Rapid. Comun* **2001**, *22* 694.
[7] J.C. Cabanelas, S.G. Prolongo, B. Serrano, J. Bravo, J. Baselga, *J. Mat. Process. Tech.* **2002**, to be published.
[8] W.H. Stockmayer, *J. Polym. Sci* **1952**, *9*, 69.
[9] W.H. Stockmayer, *J. Polym. Sci* **1953**, *11*, 424.
[10] M. Krumova, C. Klingshirn, F. Haupert, K. Friedrich, *Compos. Sci. Techol.* **2001**, *61*, 557.
[11] D.A. Tilbrook, R.L. Clarke, N.E. Howle, M. Braden, *Biomaterials* **2001**, *21*, 1743.

Thermodynamic Analysis of the Reaction-Induced Phase Separation in Model Systems Based on Blends of a Rubber in Diepoxide / Monoamine Precursors

*Ileana A. Zucchi, María J. Galante, Julio Borrajo, Roberto J.J. Williams**

Institute of Materials Science and Technology (INTEMA), University of Mar del Plata-National Research Council (CONICET). J. B. Justo 4302, 7600 Mar del Plata, Argentina

Summary: The reaction-induced phase separation in blends based on a carboxyl-terminated poly(butadiene-*co*-acrylonitrile) rubber (CTBN), dissolved in diglycidyl ether of bisphenol A (DGEBA) – benzylamine (BA) monomers, was studied. The polymerization kinetics was followed by size exclusion chromatography, for both the neat DGEBA-BA system and for blends containing 10 wt % CTBN. No effect of CTBN addition on the polymerization rate was observed within experimental error. The kinetics could be fitted with a model based on the presence of non-catalytic and OH-catalyzed reactions and assuming equal reactivity of primary and secondary amine hydrogens. Cloud-point conversions were determined at 60, 70 and 80 °C. The Flory-Huggins model provided a reasonable fitting of experimental data using an interaction parameter depending exclusively on temperature, and taking polydispersities of both linear polymers into account. Linear epoxy/amine systems may be used to test the reliability of thermodynamic theories in more complex situations (e.g., modifiers with a broad distribution of molar masses or mixtures of several modifiers).

Keywords: cloud point conversion, kinetics, phase separation

Introduction

Morphologies generated by reaction-induced phase separation in initially homogeneous blends of a modifier in thermoset precursors, are the resultant of both thermodynamic and kinetic factors.[1,2] A correct qualitative description of the phase separation process was obtained with the use of the Flory-Huggins thermodynamic model, assuming a blend of two monodisperse pseudocomponents, and using an interaction parameter depending exclusively on temperature.[3-5] Further refinements of the model were to consider the molar mass distribution of the thermosetting polymer as a function of conversion,[6-9] and the fact that the interaction parameter could vary with conversion due to the change in the chemical structure of the thermosetting polymer along reaction.[10,11] A quantitative fitting of experimental results required to take into account the polydispersities of both components and the fact that the

 DOI: 10.1002/masy.200350825

interaction parameter had to be described as a function of both conversion and temperature.[12-14] At conversions close to the gel point, the mathematical solution becomes increasingly difficult due to the large number of terms necessary to describe the distribution of the thermosetting polymer. After gelation it is necessary to take into account the distribution of finite species present in the sol, and to include the elastic contribution to the free energy.[15,16] Therefore, the modeling of formulations including modifiers that become phase separated at high conversions is relatively complex.

In order to avoid the problems introduced by gelation, the reaction-induced phase separation in the course of a linear polymerization may be conveniently analyzed. These model systems can be used to test the usual assumptions of thermodynamic theories employed to describe the reaction-induced phase separation process. A blend of a carboxyl-terminated poly(butadiene-*co*-acrylonitrile) (CTBN) rubber in a stoichiometric mixture of diglycidyl ether of bisphenol A (DGEBA) / benzylamine (BA) monomers, was selected for this study. The DGEBA/BA polymerization follows an ideal path when using stoichiometric proportions of both monomers or when employing a BA excess.[17] With a DGEBA excess, a polyetherification of epoxy groups was observed after completion of the epoxy/amine reaction.[17] The reliability of the Flory-Huggins model, with an interaction parameter depending exclusively on temperature, to fit cloud-point conversions in CTBN-stoichiometric DGEBA/BA blends, will be the focus of the present analysis.

Experimental

Materials

The chemical structures of the diepoxide, monoamine and rubber are shown in Figure 1 The epoxy monomer (Der 332, Dow), was based on diglycidylether of bisphenol A (DGEBA, mass per mole of epoxy groups = 174.3 g/mol, mass density = 1.17 g/cm^3). Its initial content of secondary hydroxyls was equal to 0.015 per mole of epoxy groups. Benzylamine (BA, Riedel-De Haën, mass density = 0.981 g/cm^3), was used in stoichiometric proportion with respect to DGEBA.

The selected rubber was a carboxyl-terminated poly(butadiene-*co*-acrylonitrile) copolymer (CTBN), containing 26 wt % acrylonitrile (Hycar 1300x13, BF Goodrich, mass density = 0.960 g/cm^3). Its number-average molar mass was M_n = 3200 g/mol and the polydispersity index was M_w/M_n = 2.13.[18]

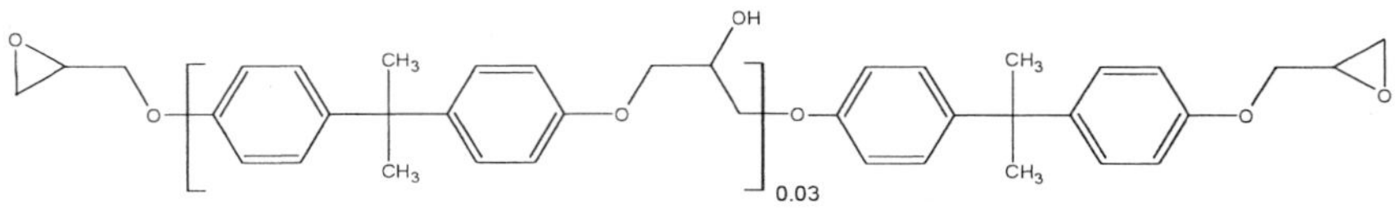

DGEBA

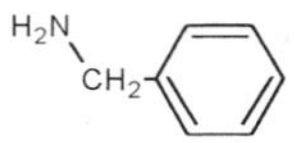

BA

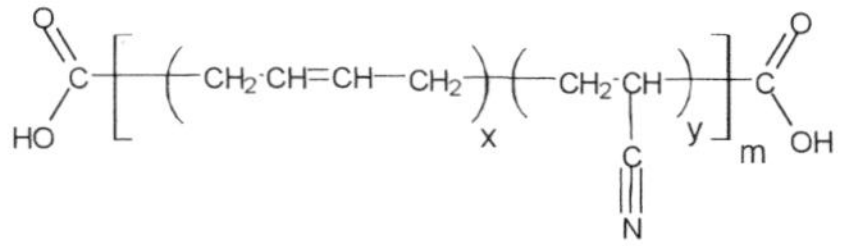

CTBN

Fig. 1. Chemical structures of the diepoxide (DGEBA), monoamine (BA) and rubber (CTBN).

Blends containing 5, 10 and 15 wt % CTBN in DGEBA/BA, were prepared by first dissolving the corresponding CTBN amount in DGEBA at about 80 °C, cooling to room temperature and adding the stoichiometric amount of BA. Blends were homogeneous at room temperature.

Techniques

Transmission optical microscopy (TOM) was employed to determine cloud-point times during polymerization at constant temperature. A Leica DMLB microscope provided with a video camera (LEICA DC 100) and a hot stage (Linkam THMS 600), was used for these purposes. Samples were placed between two glasses using a 0.5 mm stainless-steel spacer. The small average size of the generated dispersed domains, made it difficult to precise the cloud-point time by focusing on one plane of the sample through the microscope. The value could, however, be obtained by visual observation of the sample placed in the hot stage (it became opalescent at the cloud-point time).

Differential scanning calorimetry (DSC, Perkin-Elmer Pyris 1), was used to determine the total reaction heat of the neat DGEBA/BA system and the blends with CTBN. Runs were carried out at 10 °C/min under a nitrogen atmosphere. Glass transition temperatures of materials cured to complete conversion were determined during a subsequent heating scan at 10 °C/min. They were defined at the onset value of the change in the specific heat.

The polymerization kinetics was determined following the height of the DGEBA peak in size exclusion chromatograms, as a function of reaction time. A set of glass tubes containing known masses of the reacting mixture, in the range of 8-25 mg, were placed in a thermostat at the desired temperature. Tubes were removed at pre-determined times, their contents dissolved in 5 ml tetrahydrofuran (THF), and 25 μL of the resulting solution injected in the size exclusion chromatograph (SEC, Waters 510 with HR 0.5, 1 and 3 ultrastyragel columns, UV detector at 254 nm, and a THF flow rate of 1 ml/min). The height of the peak corresponding to the DGEBA monomer was measured and expressed per unit mass of the initial mixture. Assuming that both epoxy groups have equal reactivity and there are no substitution effects, the fraction of unreacted DGEBA at a particular epoxy conversion, p, is given by the simultaneous probability that both epoxy groups remain unreacted:[19]

$$c/c_0 = h/h_0 = (1 - p)^2 \qquad (1)$$

where h/h_0 is the ratio of the actual height of the peak with respect to the initial one (both expressed per unit mass of sample). The conversion is then given by:

$$p = 1 - (h/h_0)^{1/2} \qquad (2)$$

CTBN-DGEBA/BA blends were also polymerized at constant temperatures in aluminum molds. The resulting plaques, of about 5 mm thickness, were fractured after immersion in liquid air. Fracture surfaces were coated with a fine gold layer and observed by scanning electron microscopy (SEM, Jeol JSM 35CF).

Results and Discussion

Polymerization Kinetics

The effect of the CTBN addition on the polymerization kinetics was first studied. Figure 2a shows DSC thermograms for the neat epoxy and for a blend containing 10 wt % CTBN. The reaction heat was 96.3 kJ/mol for the neat epoxy system and 100.7 kJ/mol for the CTBN-modified epoxy. These values are within the range of reported reaction heats for epoxy-amine formulations. The peak temperature is the same for both systems, meaning that no significant effect of CTBN addition on reaction rate may be expected (the dilution effect of a 10 wt % CTBN addition is not relevant).

Glass transition temperatures obtained in the course of second scans were 54 °C for the neat epoxy/amine system and 46 °C for the CTBN-modified epoxy (Figure 2b). The decrease in the glass transition temperature is ascribed to the CTBN fraction that remained dissolved in the epoxy/amine matrix at the end of the polymerization.

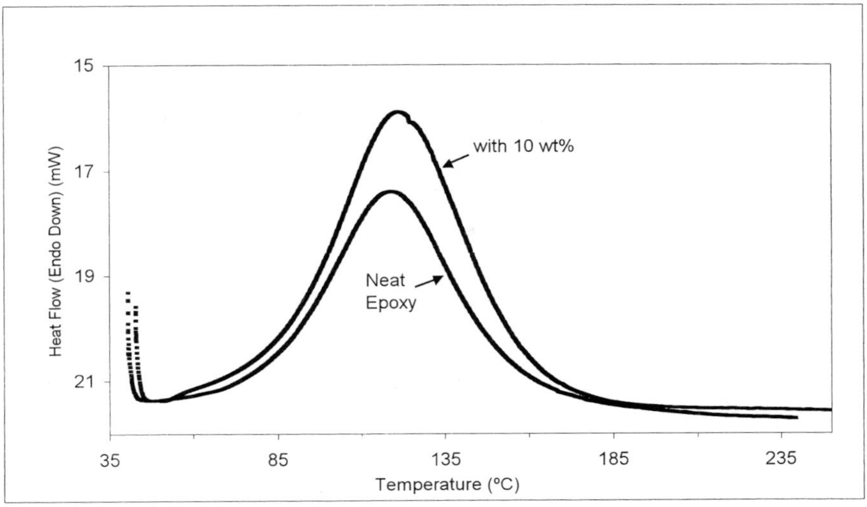

Fig. 2a. DSC scans for the neat epoxy and for the blend containing 10 wt % CTBN.

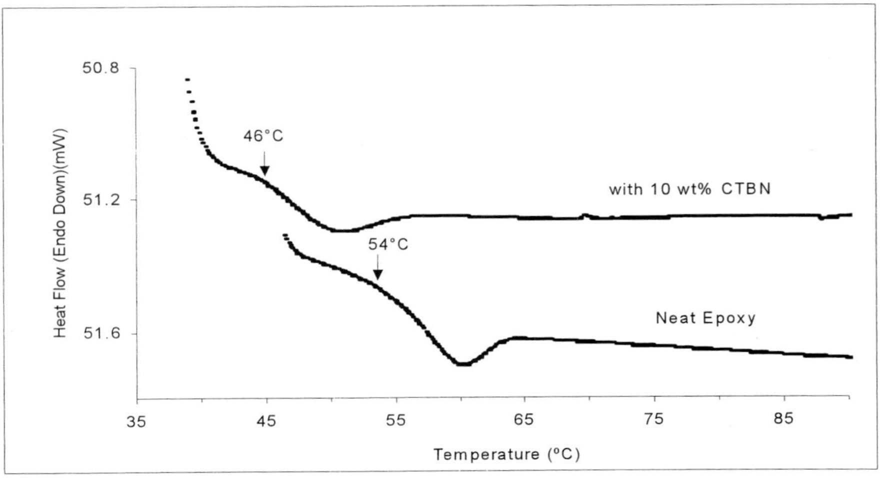

Fig. 2b. Glass transition temperatures of the neat epoxy reacted to complete conversion and of the epoxy-rich phase of the blend with 10 wt % CTBN, after polymerization.

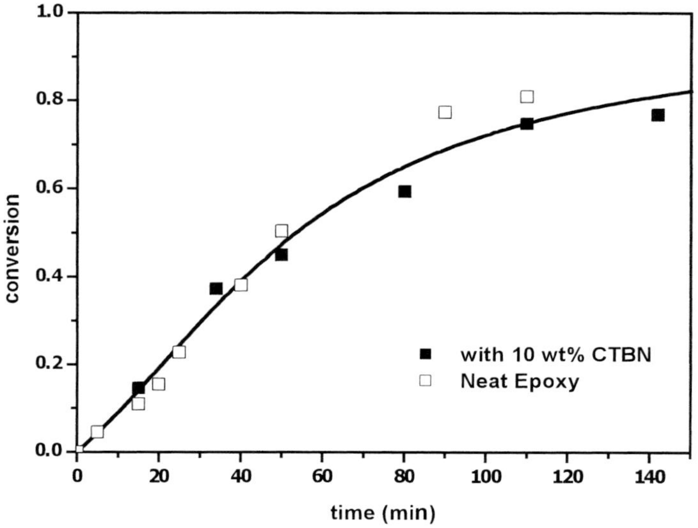

Fig. 3a. Conversion vs. time curves at 60 °C.

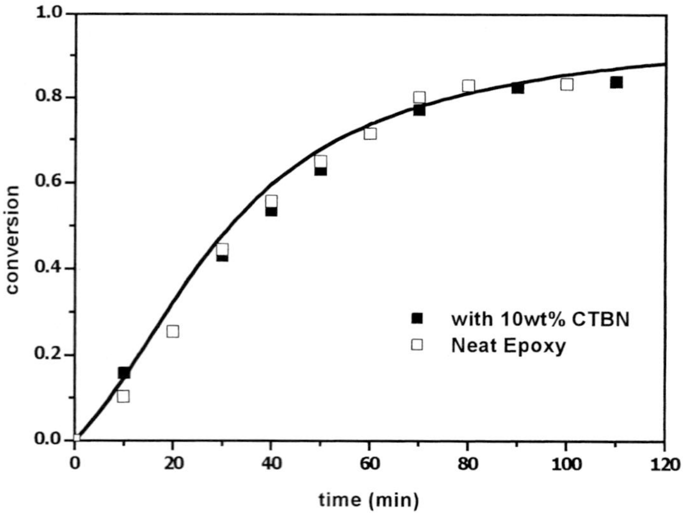

Fig. 3b. Conversion vs. time curves at 70 °C.

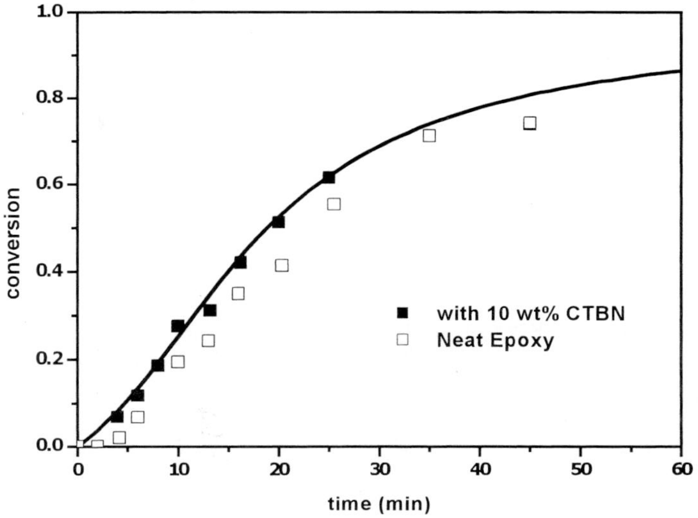

Fig. 3c. Conversion vs. time curves at 80 °C.

Conversion vs. time curves, arising from the determination of the residual DGEBA monomer by SEC, were obtained at 60, 70 and 80 °C, for both the neat epoxy/amine system and for a blend containing 10 wt % CTBN. Results are shown in Figures 3a-c. Within the experimental error of this determination there was no influence of the CTBN addition on the polymerization kinetics.

The following rate equation was used to fit the experimental results:

$$dx/dt = [k_A + k_B\,(p + 0.015)](1 - p)^2 \qquad (3)$$

This rate equation results from a mechanistic model assuming the presence of a non-catalytic path (rate k_A), and a path catalyzed by the OH groups initially present (0.015 per mole of epoxy groups), and those generated by reaction (rate k_B). Equal reactivities of primary and secondary amine hydrogens, was also assumed.

An excellent fitting of the proposed model with experimental results, shown by the curves plotted in Figures 3a-c, resulted from the following values of the specific rate constants:

$$k_A\,(min^{-1}) = 7.733\ 10^2\ exp(-\ 3848.4/T) \qquad (4)$$

$$k_B\,(min^{-1}) = 3.465\ 10^8\ exp(-\ 7607.1/T) \qquad (5)$$

The activation energy of the catalytic reaction (E_B = 63.2 kJ/mol), lies within the range of values reported in the literature for this type of mechanistic model.[19-23]

Morphologies Generated

The initial CTBN-DGEBA/BA blends were homogeneous but phase separated in the course of polymerization. Figure 4 shows SEM micrographs of blends with 5, 10 and 15 wt % CTBN, polymerized at 70 °C to a high conversion ($p > 0.9$).

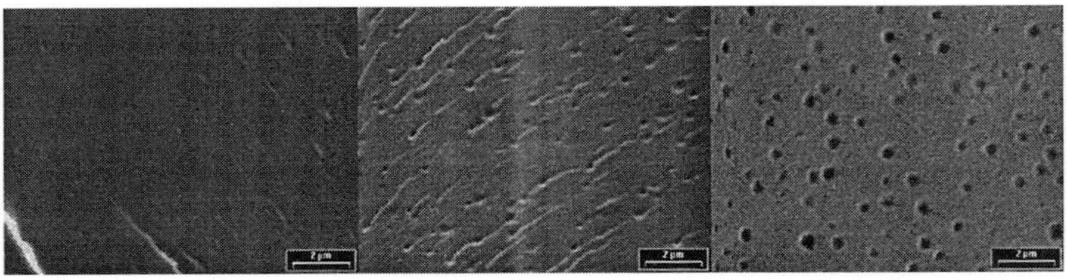

Fig. 4. SEM micrographs of blends containing, from left to right, 5, 10, and 15 wt % CTBN.

The blend modified with 5 wt % CTBN exhibits a small concentration of particles with sizes in the range of 100 – 200 nm. These blends were slightly opalescent and it was therefore not possible, to obtain cloud-point times with enough accuracy. Both the concentration and the average size of dispersed domains increased with the CTBN concentration, as expected.

Cloud-Point Conversions

Cloud-point times could be obtained for blends containing 10 and 15 wt % CTBN. These values were converted to cloud-point conversions using the kinetic model (it was assumed that the kinetics was the same for both concentrations). Table 1 shows the corresponding results.

Table 1. Cloud-point conversions for CTBN-DGEBA/BA blends.

wt % CTBN	Cloud-point conversion		
	60 °C	70 °C	80 °C
10	0.56	0.61	0.65
15	0.54	0.60	0.64

The selected CTBN rubber was very soluble in the epoxy/amine formulation and became phase separated at relatively high conversions. The cloud-point conversion increased with temperature, consistently with an upper-critical-solution-temperature (UCST) behavior. Cloud-point conversions decreased slightly by increasing the initial CTBN amount from 10 to 15 wt %.

Thermodynamic Analysis

The possibility of using the Flory-Huggins (FH) model to fit these experimental data will be now assessed. As the range of cloud-point conversions is relatively narrow, the interaction parameter, χ, will be considered as only dependent on temperature:

$$\chi = a + b/T \qquad (6)$$

The search will be focused on the existence of such a function and on the ability of the FH model to predict a slight dependence of the cloud-point conversion with composition in the range of 10 – 15 wt % CTBN.

In order to state the FH equation it is necessary to define the volume of the unit cell. This will be taken as the average molar volume of the CTBN "monomer". Taking into account the molar masses of butadiene (54 g/mol) and acrylonitrile (53 g/mol), and their mass fractions in the copolymer, the molar mass of the "monomer" is 53.74 g/mol, and the reference volume is $V_r = 53.74/0.96 = 55.98\ \text{cm}^3/\text{mol}$.

The molar-mass distribution of CTBN is described by the Schulz-Zimm (SZ) equation:[24]

$$w_x = [g^{h+1}/\Gamma(h+1)]\, x^h \exp(-gx) \qquad (7)$$

where w_x is the mass fraction of the x-mer, $h = [x_w/x_n - 1]^{-1} = 0.885$, $g = h/x_n = 0.01486$, and $\Gamma(h+1) = 0.9567$. The SZ equation for CTBN is finally written as:

$$w_x = 3.746\ 10^{-4}\, x^{0.885} \exp(-0.01486\, x) \qquad (8)$$

The volume fraction of a generic x-mer is given by:

$$\phi_x = w_x\, \phi_{20} \qquad (9)$$

where ϕ_{20} is the initial volume fraction of CTBN in the blend.

The molar-mass distribution of the epoxy/amine linear polymer may be calculated from the Stockmayer equation applied for a stoichiometric $A_2 + B_2$ polymerization, up to a conversion p of functional groups.[24] The concentration of a generic species, $E_{m,n}$, containing m BA units and n DGEBA units, is given by:

$$E_{m,n} = 2\,(BA)_0\,(1-p)^2 p^{m+n-1} \quad (m = n) \qquad (10)$$

$$E_{m,n} = (BA)_0\,(1-p)^2 p^{m+n-1} \quad (m = n+1 \text{ or } n = m+1) \qquad (11)$$

where $(BA)_0 = (DGEBA)_0$, is the initial concentration of one of the monomers. The corresponding volume fractions of $E_{m,n}$ species in the reaction mixture, is obtained from:

$$\phi_{m,n} = E_{m,n}\,(m\,V_{BA} + n\,V_{DGEBA})\,\phi_{10} / [(BA)_0\,V_{BA} + (DGEBA)_0\,V_{DGEBA}] \qquad (12)$$

where $\phi_{10} = 1 - \phi_{20}$, is the initial volume fraction of the epoxy/amine polymer in the blend, and $V_{BA} = 109.2$ cm^3/mol, $V_{DGEBA} = 297.9$ cm^3/mol, are the molar volumes of both monomers.

A pseudocomponent with a generic degree of polymerization $y = m + n$, may be defined. Its concentration and volume fraction are given by:

$$E_y = 2\,(BA)_0\,(1-p)^2 p^{y-1} \qquad (13)$$

$$\phi_y = y(1-p)^2 p^{y-1}\,\phi_{10} \qquad (14)$$

The free energy of mixing per mol of unit cells is given by the FH equation:

$$\Delta G/RT = \Sigma\,(\phi_x / x)\,\ln \phi_x + (1/Z)\,\Sigma\,(\phi_y / y)\,\ln \phi_y + \chi\,\phi_1\phi_2 \qquad (15)$$

where $Z = (1/2)(V_{BA} + V_{DGEBA}) / V_r$.

Cloud-point curves were calculated from equation (15) following the procedure developed by Kamide et al.[25] A detail of these calculations for the particular case of a monodisperse modifier is reported elsewhere.[6] Continuous distributions were discretized using 40 species for CTBN and 30 species for the DGEBA/BA linear polymer. It was verified that average values (number, mass and zeta), were the same for the discrete and continuous distributions. The numerical solution was obtained using Mathcad 8.

Figure 5 shows the predicted cloud-point conversions plotted as a function of the initial rubber volume fraction, for the three reaction temperatures. The FH model provides a reasonable fitting of the experimental cloud-point conversions, with an interaction parameter depending exclusively on temperature:

$$\chi = -0.155 + 75.273/T(K) \qquad (16)$$

The small decrease of the cloud-point conversion with composition in the range 10 – 15 wt % CTBN, is indeed predicted by the model.

Every couple of points located on both branches of a cloud-point curve, at a particular conversion, correspond to a particular blend with an associated cloud-point temperature vs. composition curve. Therefore, there is a different critical composition for any couple of points located at the same conversion. However, the minimum of each one of the curves represented in Figure 5, corresponds to a critical composition that pertains to the curve. These critical volume fractions are 0.241 at 60 °C, 0.251 at 70 °C and 0.263 at 80 °C. As the initial CTBN volume fractions (0.115 and 0.171), are located in the off-critical region at the left of the cloud-point, the generated morphologies consisted of a dispersion of rubber particles in an epoxy/amine matrix, as expected.[2]

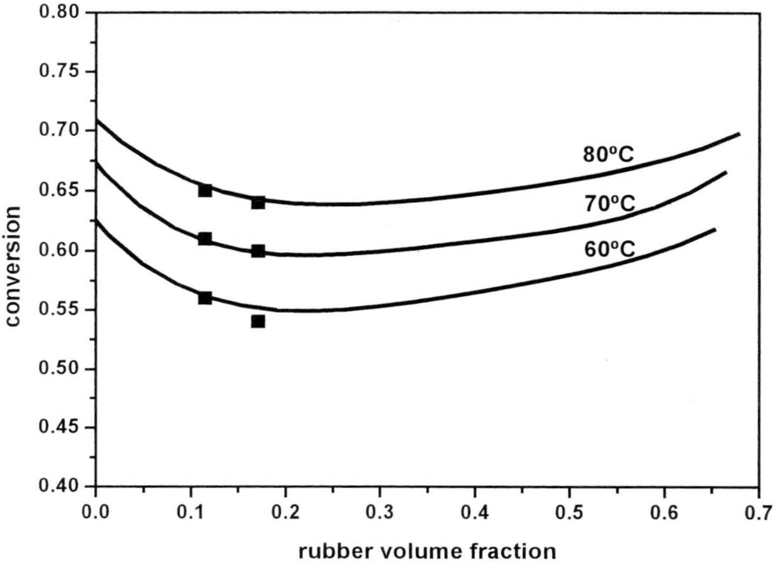

Fig. 5. Cloud-point conversion vs. the initial rubber volume fraction, for three reaction temperatures.

The location of the critical point can be calculated from expressions derived by Koningsveld and Kleintjens:[26]

$$1/(1 - \phi_{2c}) = 1 + (Z^{1/2} y_w x_z^{1/2})/(x_w y_z^{1/2}) \qquad (17)$$

$$\chi_c = 0.5[1/x_z^{1/2} + 1/(Z y_z)^{1/2}][x_z^{1/2}/x_w + y_z^{1/2}/(Z^{1/2} y_w)] \qquad (18)$$

For the initial formulation ($p = 0$), $y_w = y_z = 1$. The Schulz-Zimm distribution leads to the following average values for CTBN: $x_w = 126.83$ and $x_z = 194.12$. As $Z = 3.636$, equations (17) and (18) lead to the following critical values: $\phi_{2c} = 0.173$ and $\chi_c = 0.189$. From equation (16), $T_c = 218.8$ K. This means that the model predicts that no phase separation will occur in the initial blend above 218.8 K. This agrees with the fact that no phase separation was observed by keeping the initial formulations during prolonged times at 253 K.

Conclusions

The use of a model system consisting of a blend of a rubber in diepoxide /monoamine monomers (leading to a linear epoxy/amine polymer), enabled to test the reliability of the FH model to fit experimental data without the mathematical complexities introduced by gelation. This is particularly useful for modifiers (e.g. a rubber) that are very soluble in the mixture of monomers and, consequently, phase separate at high conversions as was the case of the blend selected for this study. Model systems may be used to test thermodynamic models in more complex situations, such as modifiers exhibiting a very broad distribution of molar-masses, or formulations including two different modifiers.

Acknowledgements

The financial support of CONICET, ANPCyT and University of Mar del Plata, is gratefully acknowledged.

[1] R. J. J. Williams, B. A. Rozenberg, J. P. Pascault, *Adv. Polym. Sci.* **1997**, *128*, 95.
[2] J. P. Pascault, R. J. J. Williams, in: "*Polymer Blends, Vol. 1: Formulation*", D. R. Paul, C. B. Bucknall, Eds., J. Wiley & Sons, New York 2000, p.379.
[3] R. J. J. Williams, J. Borrajo, H. E. Adabbo, A. J. Rojas, in: "*Rubber-Modified Thermoset Resins*", C. K. Riew, J. K. Gillham, Eds., *Adv. Chem. Ser. 208*, American Chemical Society, Washington DC 1984, p.195.
[4] A. Vazquez., A. J. Rojas, H. E. Adabbo, J. Borrajo, R. J. J. Williams, *Polymer* **1987**, *28*, 1156.
[5] S. M. Moschiar, C. C. Riccardi, R. J. J. Williams, D. Verchère, H. Sautereau, J. P. Pascault, *J. Appl. Polym. Sci.* **1991**, *42*, 717.
[6] C. C. Riccardi, J. Borrajo, R. J. J. Williams, *Polymer* **1994**, *35*, 5541.
[7] N. Clarke, T. C. B. McLeish, S. D. Jenkins, *Macromolecules* **1995**, *28*, 4650.
[8] S. Elliniadis, J. S. Higgins, N. Clarke, T. C. B. McLeish, R. A. Choudhery, S. D. Jenkins, *Polymer* **1997**, *38*, 4855.
[9] C. J. G. Plummer, R. Mezzenga, L. Boogh, J. A. E. Manson, *Polym. Eng. Sci.* **2001**, *41*, 43.
[10] R. Mezzenga, L. Boogh, J. A. E. Manson, *J. Polym. Sci., B: Polym. Phys.* **2000**, *38*, 1883.
[11] R. Mezzenga, L. Boogh, J. A. E. Manson, *J. Polym. Sci., B: Polym. Phys.* **2000**, *38*, 1893.
[12] J. Borrajo, C. C. Riccardi, R. J. J. Williams, Z. Q. Cao, J. P. Pascault, *Polymer* **1995**, *36*, 3541.
[13] C. C. Riccardi, J. Borrajo, R. J. J. Williams, E. Girard-Reydet, H. Sautereau, J. P. Pascault, *J. Polym. Sci., B: Polym. Phys.* **1996**, *34*, 349.
[14] P. A. Oyanguren, C. C. Riccardi, R. J. J. Williams, I. Mondragon, *J. Polym. Sci., B: Polym. Phys.* **1998**, *36*, 1349.
[15] J. Borrajo, C. C. Riccardi, R. J. J. Williams, H. Masood Siddiqi, M. Dumon, J. P. Pascault, *Polymer* **1997**, *39*, 845.
[16] C. C. Riccardi, J. Borrajo, R. J. J. Williams, H. Masood Siddiqi, M. Dumon, J. P. Pascault, *Macromolecules* **1998**, *31*, 1124.
[17] J. Klee, H. H. Hörhold, W. Tänzer, in: "*Crosslinked Epoxies*", B. Sedlácek, J. Kahovec, Eds., Walter de Gruyter, Berlin 1987, p. 47.
[18] D. Verchère, H. Sautereau, J. P. Pascault, S. M. Moschiar, C. C. Riccardi, R. J. J. Williams, *Polymer* **1989**, *30*, 107.
[19] D. Verchère, H. Sautereau, J. P. Pascault, C. C. Riccardi, S. M. Moschiar, R. J. J. Williams, *Macromolecules* **1990**, *23*, 725.
[20] C. C. Riccardi, H. E. Adabbo, R. J. J. Williams, *J. Appl. Polym. Sci.* **1984**, *29*, 2481.
[21] G. Wisanrakkit, J. K. Gillham, *J. Appl. Polym. Sci.* **1990**, *41*, 2885.
[22] N. Galego, A. Vazquez, R. J. J. Williams, *Polymer* **1994**, *35*, 857.

[23] C. W. Wise, W. D. Cook, A. A. Goodwin, *Polymer* **1997**, *38*, 3251.
[24] L. H. Peebles, "*Molecular Weight Distributions in Polymers*", Wiley-Interscience, New York 1971.
[25] K. Kamide, S. Matsuda, H. Shirataki, *Eur. Polym. J.* **1990**, *26*, 379.
[26] R. Koningsveld, L. A. Kleintjens, *J. Polym. Sci., Polym. Symp.* **1977**, *61*, 221.

A New Class of Epoxy Thermosets

Emmanuel Girard-Reydet,[1] *Jean-Pierre Pascault,*[1] *Anthony Bonnet,*[2] *François Court,**[2] *Ludwik Leibler*[3]

[1] Laboratoire des Matériaux Macromoléculaires – UMR CNRS 5627, Bât. J.Verne, INSA, 20 av. Albert Einstein, 69621 Villeurbanne Cedex, France
[2] CERDATO, ATOFINA, 27470 Serquigny, France
[3] Laboratoire Matière Molle et Chimie, UMR ESPCI-CNRS-ATOFINA, 10 rue Vauquelin, 75231 PARIS CEDEX 05, France

Summary: The reinforcing strategies of epoxy thermosets rely on the control of the phase separation between the additive and the growing thermoset. With standard additives, such as reactive liquid rubbers, the length scale of the resulting domains is the micrometer. Here, we present a route that enable a control of the morphology down to the nanometer scale. This strategy is based upon the self-assembly process of blends of epoxy and SBM triblock copolymers, namely Poly(Styrene-b-1,4 Butadiene-b-Methyl methacrylate). It relies on the respective affinities between the epoxy precursors and each of the three blocks. Liquid epoxy has a strong affinity for PMMA, whilst it is not miscible with polystyrene nor polybutadiene at standard processing temperatures. Thus, within the reactive system, microphase separation leads to a regular network of S-B domains. This nanostructure is governed by thermodynamics. The size and geometry of the dispersed domains are controlled by the concentration and the ratio between blocks lengths. The domain size is of the order of magnitude of the chain length, ranging typically from 10 to 30 nanometers. What controls the blend's morphology throughout the curing process of the thermoset was one topic on which we focused our interest. Nanostructured thermosets have been obtained. These supramolecular architectures yield significant toughness improvements while preserving the transparency of the material. The reinforcing mechanisms are not yet fully understood : it is intriguing to induce significant toughening with elastomer domains smaller than 30 nanometers in diameter. Besides being efficient epoxy tougheners, SBM can broaden the scope of applications of thermosets due to specific rheological behaviors. Thanks to the self assembly process taking place in the blend of the SBM block copolymers with the epoxy thermosets precursors, the reactive solvent can be turned into a reactive gel or solid (before curing). This physical gelation is induced by the microphase separation and is thus thermoreversible. At relatively moderate loadings of block copolymers the reactive blend behaves like a thermoplastic material, with adjustable modulus and tackiness. These results evidence that SBM block copolymers open a broad area for designing new class of thermoset materials.

Keywords: nanostructures, rheology, thermosets

 DOI: 10.1002/masy.200350826

Introduction

Epoxy thermosets are used in many demanding applications in which stiff, high glass transition temperature, T_g, low creep and high heat resistant materials are required. On the other hand, compared to most materials, epoxies are very brittle, and toughening of such thermosets has been a major objective of polymer science for many years. Several types of impact modifiers have been developed for this purpose. Among them, reactive liquid rubbers, like CTBN (Carboxyl Terminated Butadiene acryloNitrile) are the most widespread and, to a lesser extent, high T_g thermoplastics and core-shell particles. Nevertheless, significant progress is still highly desirable in this field. Users are seeking additives that would toughen epoxy without adversely reducing their T_g or stiffness. Furthermore, in some cases, the tougheners commercially available are just not efficient. As far as processing is concerned, thermosets have peculiarities that have to be taken into account when designing a new reinforcing strategy. In most thermoplastics toughening is obtained by melt blending together the polymer and the impact modifier, the toughening of thermosets can only be achieved by the *in situ* strategy. The impact modifier is added to the epoxy precursors (epoxy oligomer and hardener) prior to curing, hence, process wise, the solubility of the additive in the precursors is required.

Thanks to their composition (butadiene + acrylonitrile) and low molar mass, CTBN are initially soluble in epoxy oligomers. When the material is being cured, the miscibility of CTBN with the growing network decreases which leads to phase separation.[1] In the fully cured epoxy thermoset, micron size rubber domains can be observed. The adhesion between the domains and the matrix is ensured by the carboxyl functionality. Several authors[1] have shown that a careful control of the morphology of these systems was necessary to obtain a good balance of properties and an efficient toughening. The preformed core-shell particles used in epoxy range typically from 200 to 400 nm.[2] In both cases, the geometry of the dispersed rubbery phase is spherical or droplet like. Our motivation was to elaborate a new class of thermoset materials with completely unique morphologies, both, in terms of geometry and in terms of length scale. As far as the latter is concerned, a breakthrough would be to gain one order of magnitude, to control the formation of dispersed domains down to 10-20 nm., i.e. the order of magnitude of the chain length. The strategy is based upon the self-assembly process of blends of epoxy systems and linear A-B-C triblock copolymers, namely Poly(Styrene-b-Butadiene-b-Methylmethacrylate), SBM. Such a strategy, first proposed by Leibler et al. [3,4] was shown to be successful in thermoplastics. It relies on the selective affinity of one of the block for the host matrix, PMMA and the epoxy precursors, in the present case. With this approach trully nanostructured thermosets have been obtained.[5-10] We will show that the final morphologies are governed *ab initio* by the SBM molecular characteristics. This illustrates both the versatility and the robustness of this route. Another aspect of robustness is the capacity to elaborate nanostructured materials with industrial samples. The toughness of SBM modified epoxy has been investigated and compared with standard macrophase separated systems. Another interesting aspect of SBM /epoxy is the blend behaviour before curing, its rheology will be the focus of this paper last section.

Experimental

Techniques used in this paper, materials and samples preparation are given in reference.[10]

SBM Samples description. SBM are linear triblock copolymers. They are produced via anionic polymerization. Since its discovery in the sixties, living anionic polymerization has been used as the ideal synthetic route towards well defined block copolymers. For many years, anionic polymerization was limited to three monomers: styrene, butadiene and isoprene. In the eighties, it became possible to synthesize block copolymers with acrylates and methacrylates such as MMA.[11-14] Thanks to ATOFINA patented chemistry and process technology, the synthesis of SBM triblock copolymers can be performed at relatively elevated temperatures and in a non polar solvent, yielding a 1,4-microstructure of the polybutadiene block higher than 85% and a PMMA block with over 75% syndiotactic triads. The incompatibility between the three chemical species constituting SBM and the connectivity between the blocks lead to microphase separation as depicted in Figure 1. In the bulk, the resulting well ordered morphology can be evidenced by transmission electron microscopy as shown in Figure 2a. In this example, the SBM sample has a lamellar morphology. Depending on its composition, i.e. the percentage of styrene, butadiene and methylmethacrylate in the chain, a broad spectrum of morphologies and mechanical behaviors can be covered by SBM triblock copolymers. This variety of morphologies has been investigated in depth by R. Stadler and V. Abetz, who have established a SBM ternary phase diagram.[15-16] Figure 2, represent the morphologies of two SBM samples with different compositions, 30/30/40 and 20/25/55 respectively (expressed in terms of mas fraction of S , B and M).

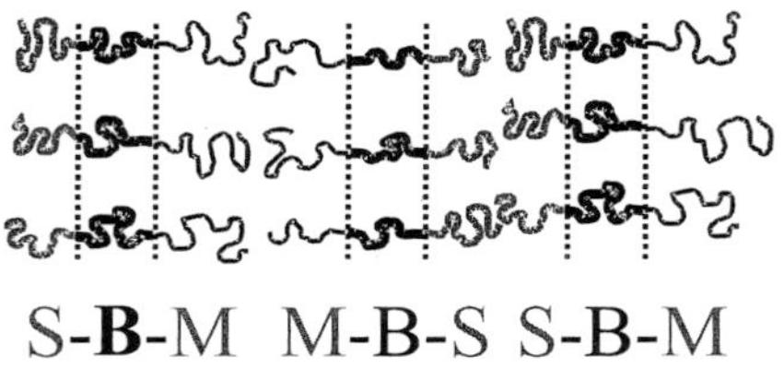

Fig. 1. Schematic representation of the chain organization of SBM triblock copolymers. Incompatibility between blocks leads to a self-organization process into a microphase separated morphology.

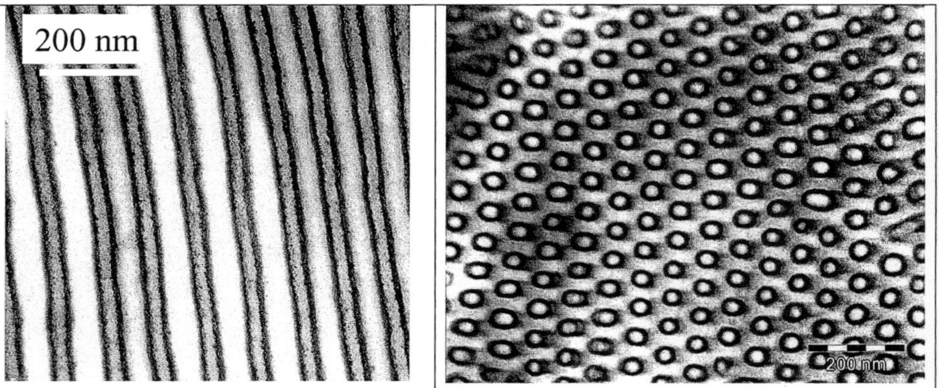

a- SBM-30/30/40 : alternating lamellar morphology	b- SBM-20/25/55 : PMMA matrix with hexagonally packed PS-PB core-shell cylinders.

Fig. 2. Transmission electron micrograph of pure SBM films solvent casted from chloroform. The polybutadiene blocks are selectively stained with osmium tetraoxide and appear black.

Synthesis of a Nanostructured Epoxy Thermoset

The molecular characteristics of the pure SBM sample used in this section are the following : the number average molar mass of the PS block is 27000 g/mol, the number average molar mass of the PB block is 11000 g/mol and the number average molar mass of the PMMA block is 84000 g/mol, i.e. the SBM composition is 22/9/69 expressed in weight fraction of S, B and M, respectively. This SBM sample is noted $S_{22}^{27}B_9M_{69}$. The epoxy precursors were diglycidyl ether of bisphenol A, DGEBA (LY556 from VANTICO) and 4,4'-Methylenebis-(3-chloro-2,6-diethylaniline), MCDEA as hardener, commercialized by LONZA.

Both, SBM and epoxy precursors (DGEBA + MCDEA) are soluble in chloroform. This solvent was used as a casting solvent to obtain a 1mm thick film of a 50/50 blend of SBM in the epoxy precursors. After full curing of the sample, for several hours under pressure at 190°C, a transparent film is obtained. This film was immersed for several days in toluene, the swelling observed remained smaller than 3% . Toluene is a good solvent of SBM, the non-swelling of the sample is a signature that this 50/50 blend consists of a continuous thermoset phase. This result in itself is worth noticing as in most thermoplastic/thermoset blends phase inversion occurs when the weight fraction of the thermoplastic exceeds 20%.[1] The transmission electron microscopy of the cured sample is presented in Figure 3a. The PB domains are all spherical in shape, their diameter is close to 10 nm., with a very narrow size distribution, and they are extremely well distributed within the matrix. One can observe that PB domains seem to gather by groups of three or four around a central brighter sphere. This central sphere hardly presents any contrast with the matrix. As schematically represented in Figure 3b, this central sphere is proposed to consist of the PS blocks of the SBM chains while the PB blocks form spheres decorating the PS sphere, the PMMA block being trapped within the epoxy matrix. This description is consistent with the macroscopic non-swelling behavior of the sample. Our conclusion is supported by dynamic mechanical analysis[10] which reveals the presence of a mixed phase PMMA+cured epoxy and two pure phases: PB and PS. Furthermore, the fact that the PB domains fail to cover continuously the surface of the PS sphere is consistent with the composition of the SBM sample used (22/9/69). We would like to emphasize the fact, that a similar morphology has already been reported in the literature.[15-16] Stadler et al. have observed such a morphology in a pure SBM sample with a highly asymmetric composition, 15/5/80, they described it as the *Spheres on Spheres* morphology. Considering that the epoxy swells selectively the PMMA phase, a mass balance would lead to a composition of 11/5/84 for the blend investigated here. This composition falls into the stability range of the *Spheres on Spheres* morphology of pure SBM, which is consistent with the morphological observation. As an additional test, the epoxy reactive system was modified with 10% of SB diblocks. The SB used are the precursors of the SBM 22/9/69 collected during the SBM synthesis. After curing, the film was turbid and transmission electron microscopy revealed micron size, macrophase separated, domains of SB diblocks within the epoxy matrix.[10] The adhesion being so poor between the nodules and the matrix, most of the domains were torn off during the ultra-microtoming procedure.

Thanks to the affinity between the PMMA block of SBM triblocks and the epoxy system used in this study a nanostructured thermoset was obtained. The complex *Spheres on Spheres* morphology is the template of a morphology of a pure SBM, nevertheless it presents the characteristics of a thermoset material. Due to the length scale of the structure the resulting material is transparent.

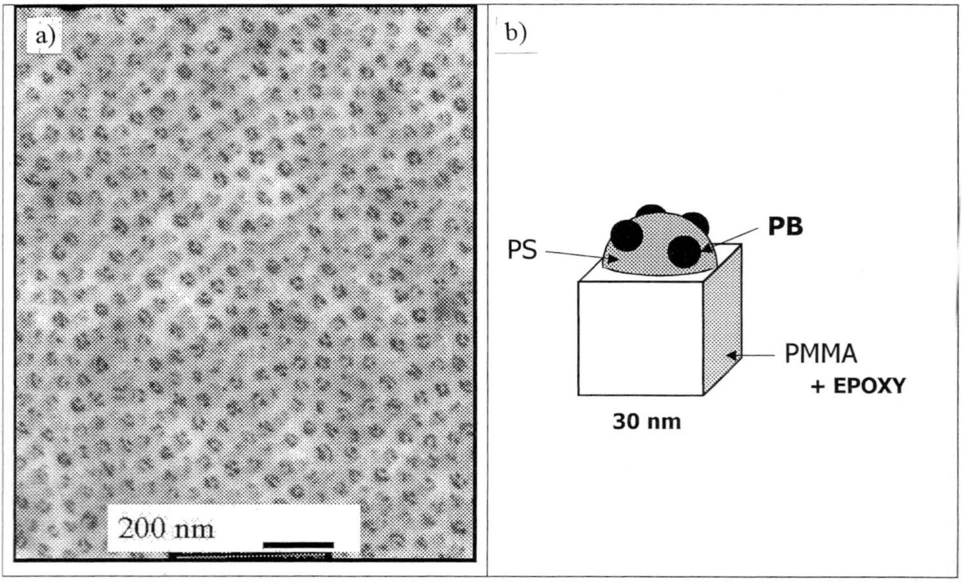

Fig. 3. a) Transmission electron micrograph of a blend Epoxy/SBM-50/50 after curing. Purified SBM composition is 22/9/69 in mass fraction of S, B and M respectively ($S_{22}^{27}B_9M_{69}$).The polybutadiene blocks are selectively stained with osmium tetraoxide and appear black. b) Schematic representation of the spheres on spheres morphology.

A Robust Route towards Nanostructured Materials

As mentioned, the synthesis technology used in this study enables the anionic polymerization of the methacrylic block at relatively elevated temperatures. This is a key to the industrial production of SBM. Though, in the given conditions, some of the living SB precursors of the triblocks are „killed“ during the polymerization of MMA. In other terms, the product obtained is not a pure SBM triblock but a blend of SBM triblocks and their SB diblock precursors. In this section two different SBM+SB mix were used. One of them is the blend of the SBM triblock and the SB diblock investigated in the previous section. The second system is also composed of an SBM triblock copolymer with a large PB block and a similar PMMA block. The numbers between brackets are the number average molar mass of the corresponding blocks:

SBM1 : $S_{22}^{27}B_9M_{69}$ (27000-11000-84000) + 21 % SB diblock (27000-11000)

SBM2 : $S_{12}^{14}B_{18}M_{70}$ (14000-27000-85000) + 10 % SB diblock (14000-27000)

Both products were added to the epoxy precursors to obtain epoxy/SBM-70/30 blends. TEM observations were conducted on the cured transparent samples, the results are presented in Figures 4a and 4b for SBM1 and SBM2, respectively. In both cases, a regular morphology, with dispersed domains smaller than 100 nanometers, is obtained. These results are to be compared with the experiment conducted on pure SB diblocks in the previous section. In the presence of SBM chains, the SB diblocks do not macrophase separate. Thus, nanostructured thermosets can be obtained from blends of SB and SBM, in other terms from unpurified industrially produced triblocks. A result which illustrates the robustness of this route toward nanostructured epoxy.

Let us now focus on the morphologies generated with SBM1 and SBM2. The dispersed domains in Figure 4a and 4b are nearly identical in size, though they have significantly different geometries. In the latter system, the B blocks, which appear dark, form a continuous corona on a central PS sphere (Figure 4b). In the former case, the outer limit of the PS sphere does not appear clearly as the contrast between the epoxy matrix and PS is very low. The presence of the central sphere is revealed by the respective positions of the darker and smaller B spheres that decorate its surface. This difference between the morphologies of the epoxy thermosets obtained from SBM1 and SBM2 is a direct consequence of the molecular characteristics of the SBM used and of the relative interaction parameters between the various chemical species. SBM2 has a PB block longer (larger molar mass) than its PS block, thus a negative curvature (convex) is expected for the S/B interface. The formation of a continuous B shell on the S sphere is energetically favored. The PB block of SBM1 is shorter than its PS block, the natural curvature of the S/B is thus positive (concave). A core-shell structure would impose a curvature that is not favored. The chains prefer to self-assemble in a *Spheres on Spheres* type of morphology that respects the natural concave curvature.

The TEM pictures provide another interesting piece of information concerning the effect of the SB diblocks. Compared to the morphology obtained with the pure SBM1 triblock (Figure 3a) Figure 4a reveals that there are more PB spheres decorating the central PS sphere, that this sphere has "swollen" from 30 nm up to 100 nm. Nevertheless, the curvatures of the different interfaces, S/B and B/matrix, are globally maintained. This swelling is the consequence of the incorporation of the SB diblocks within the structure. This is also evidenced in Figure 4b: in the clear sphere formed by the PS blocks, a darker sphere can be observed like an inner core, this sphere is constituted of the B blocks of the residual SB diblocks. A schematic representation of the chains organization is presented in Figure 4c. It illustrates how the SBM triblocks act as stabilizers for the SB diblocks in the epoxy system preventing a macrophase separation.

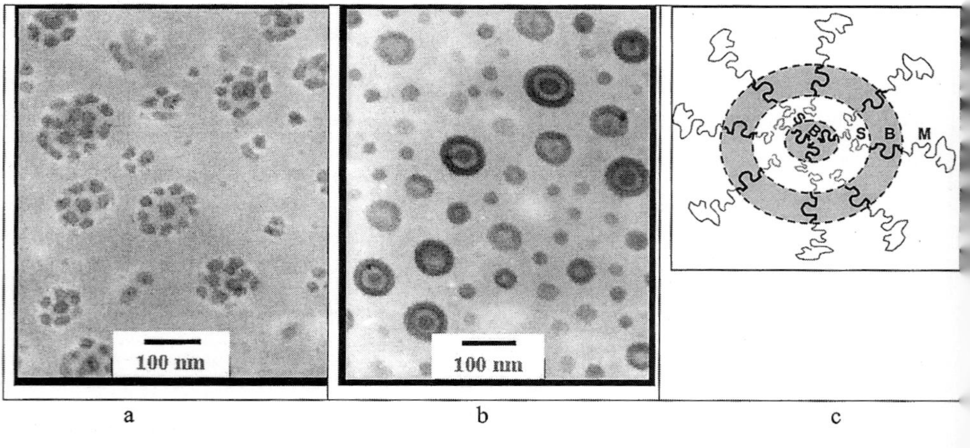

Fig. 4. a) Transmission electron micrograph of a blend epoxy/SBM1-70/30 after curing. b) Transmission electron micrograph of a blend epoxy/SBM2-70/30 after curing. c) Schematic representation of chains organization within the dispersed domains.

Such refined geometries of the nano-scaled domains are the consequence of the microphase separation of both, the PS and the PB blocks. It proves the versatility offered by ABC triblock over AB diblocks in controlling blends morphologies. It is worth recalling that another weakness of the diblock architecture is its abitility to incorporate defects. Nanostuctured thermosets can be obtained from pure BM diblock copolymers, but not from „industrial“ BM. The BM diblocks fail to stabilize their residual homopolybutadiene.[10] Macrophase separation is observed and it adversely affects the mechanical properties.

Toughened Nanostructured Epoxy Thermosets

The properties of various nanostructured thermosets, including those described in the previous section, have been evaluated and compared to the reference unmodified epoxy system and to epoxy toughened with standard additives, such as liquid rubbers (Carboxy Terminated Butadiene-co-acryloNitrile rubbers). The results obtained on fully cured samples are reported in Table 1.

Table 1. Properties of modified epoxies.

Sample DGEBA/MCDEA +...	K_{Ic}^* ($MPa.m^{1/2}$)	T_g (°C)	Transparency	Morphology
Neat	0.65 ± 0.02	187	yes	x
30% SBM1	1.53 ± 0.12	171	yes	Figure 4.a
30% SBM2	1.27 ± 0.11	177	yes	Figure 4.b
15% CTBN8 (18% AN)	1.15 ± 0.10	161	no	x
15% CTBN13 (27% AN)	1.23 ± 0.11	149	no	x
10% SBM2	1.05 ± 0.10	184	yes	Figure 5

K_{Ic}^* : The critical stress intensity factor, K_{Ic} was obtained from three-point bending test performed on single-edge notched specimens (SEN). The procedure proposed by Williams and Cawood [17] was strictly followed with a crosshead speed of 10 $mm.min^{-1}$. K_{Ic} was calculated as the mean value of 10 tests at least.
Materials: DGEBA (LY556) with hardener MCDEA; CTBN8 and CTBN13 from BFGoodrich ; SBM1 and SBM2 from ATOFINA as described in the previous section.

All the samples listed in Table 1 were immersed in toluene for a week. Toluene is a good solvent for SBM and CTBN. None of the samples underwent a swelling larger than 3%, which assesses that these materials consist of an epoxy matrix and a discrete, non-continuous, thermoplastic phase. With liquid rubbers phase inversion occurs at loadings close to 25%, whereas with the SBM used in this study no phase inversion was observed below 50% of SBM. This is a consequence of the material structure and thus a specificity of SBM triblocks compared to the other additives used to modify epoxy thermosets. The results in Table 1 show that both, SBM and

CTBN are efficient tougheners of the epoxy thermoset. The best increase of fracture toughness (K_{Ic}) was observed on the first two samples containing 30% SBM. At this stage, we have no explanation for the fact that the "raspberry" morphology (Figure 4a) seem to induce a better toughening than the "core-shell" or "onion" type morphology (Figure 4b). Substantial toughness improvements while minimizing the effect on other important properties is a key issue. Epoxy thermosets are used for their high heat resistance and low creep behavior, thus reducing the T_g of the epoxy matrix is detrimental to the material overall performance. It is observed that SBM induces a limited reduction of T_g while the addition of CTBN reduces T_g significantly. From Table 1, it can be assessed that the compromise of properties is clearly in favor of SBM over CTBN. Furthermore, it was observed that the fully cured thermoset keeps its optical clarity, which can be of interest in some applications. At this point, we would like to focus on the last sample tested. At 10% SBM the material remains as transparent as the non-toughened epoxy. The transmission electron micrograph (Figure 5) is consistent with the macroscopic observation.

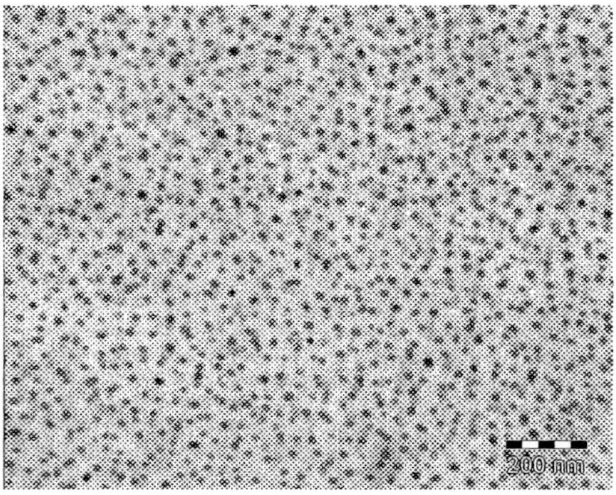

Fig. 5. Transmission electron microsgraph of a blend Epoxy/SBM1-90/10 after curing. The B block has been selectively stained with Osmium tetraoxide and appears dark.

The SBM domains are extremely evenly distributed and regular in size. As the diameter of the butadiene particles is 15 nm light cannot be scattered, hence the toughened material remains transparent. But, the most striking result is the toughening itself. It is indeed a major finding of this work that an epoxy thermoset can be toughened with such small particles. TEM investigations were carried on a fractured specimen. In the area just beneath the crack propagation, cavitation of the nano-spheres was observed. Several authors have reported that elastomer particles cannot cavitate if their radius is smaller than a 100 to 200 nanometers.[18-20] These results are further detailed and discussed in a separate paper.[21]

Processing of Nanostructured Thermosets

In this section, we focus on the rheological behavior of SBM/epoxy formulations before curing. Thanks to their nano-scaled ordered structures, the epoxy thermosets investigated here are tough and transparent, this was the motivation of this work. However, ordered structures have a well-known drawback, they are difficult to process. Several studies on the rheological behavior of pure AB and ABA block copolymers have shown that microphase separation hinders flow. Below T_{ODT} (Order-Disorder Transition Temperature), i.e. when the material is in an ordered state, it remains highly elastic ; above T_{ODT} the material is disordered, it flows easily.[22] Block copolymers are known to have the ability to form micelles or ordered structures in selective solvent. Thus, SBM/epoxy blends are expected to present complex rheological behaviors, comparable to those of ordered block copolymers. For the sake of clarity, only the results obtained on non-reactive systems are presented here.

The blends of DGEBA liquid epoxy and SBM (without diamine hardener) are prepared by mechanical stirring in a glass reactor at a temperature of 160°C. After a couple of minutes, a complete incorporation of the triblock copolymer into the liquid DGEBA is obtained. At this temperature, the solution is fluid and transparent. Upon cooling the solution remains perfectly transparent. The influence of the SBM concentration on the viscoelastic properties of the solution was investigated using a plate/plate rheometer. The temperature range investigated goes from -20°C, the T_g of the liquid epoxy used, up to 200°C. The results obtained at three different SBM concentrations, 5%, 10% and 15%, are reported in Figure 6 and compared to the viscoelastic behavior of the pure liquid epoxy. The latter presents the simple rheological response of a low molar mass material. Above its T_g (-20°C) it flows easily, like a Newtonian fluid, the loss modulus, G'', is larger than the storage modulus, G'. The solution containing 5% of SBM shows a similar rheological behavior, it is just slightly more viscous than the pure epoxy solution. Complementary frequency sweeps measurements show that it also presents a Newtonian behavior at room temperature, G' scales with ω^2 and G'' with ω. A major difference is observed when the SBM concentration is raised to 10%. The blend behaves like a gel, at room temperature and up to 135°C, its storage modulus is larger than its loss modulus. On complementary flow measurements (not presented) it is evidenced that its complex viscosity is highly shear rate dependent. On this temperature range, the 10% SBM blend is extremely viscous and does not present a Newtonian behavior.

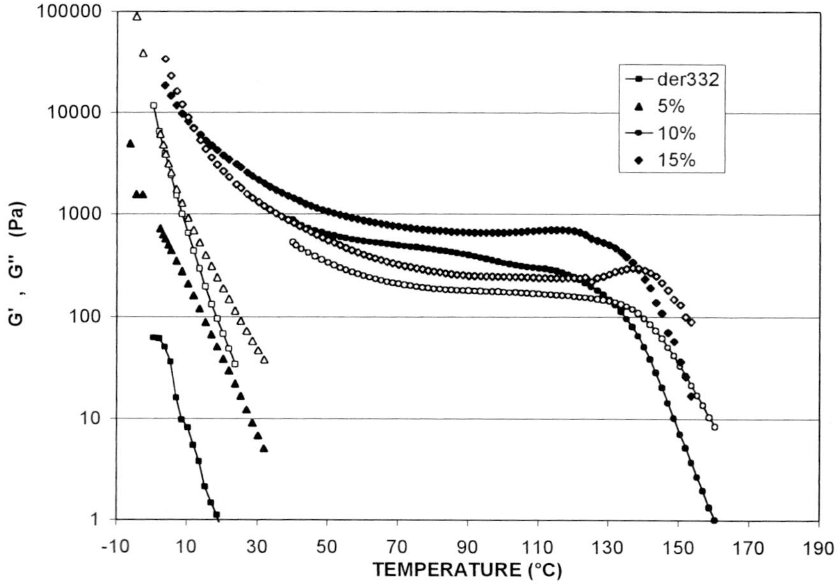

Fig. 6. Rheological behaviour of liquid epoxy modified with SBM. Elastic and shear moduli, G' and G'' respectively (G': filled symbols; G'': open symbols); squares: pure liquid DGEBA; triangles: DGEBA + 5% SBM; circles: DGEBA + 10% SBM; DGEBA + 15% SBM.

This gel-like behavior is thermoreversible. When the temperature is raised above 135°C, the material starts to flow, both G' and G'' drop sharply, G'' becomes larger than G' (Figure 6). In these conditions as evidenced by the flow measurement done at 150°C, the solution recovers a Newtonian behavior. Above 135°C, the scaling laws, G' α ω^2 and G'' α ω are obtained. The solution containing 15% SBM presents a similar viscoelastic signature : a sharp thermoreversible sol-gel transition. The plateau level of the storage modulus is slightly raised and the crossover temperature is shifted 10°C higher.

Based on these experimental results and others obtained on various SBM/DGEBA blends, we propose that the physical sol-gel transition observed here is the consequence of a microphase separation transition, or Order-Disorder Transition. When the temperature is low DGEBA acts as a selective solvent for the PMMA block, thus a regular network of SB nanodomains is created within the material. When the temperature is raised, the segregation power decreases, the whole SBM chain becomes soluble in DGEBA, the structure is lost, the viscosity drops. For entropic reasons, it is expected that below a given concentration of SBM in DGEBA the system will remain disordered at any temperature. In the system investigated here, this critical concentration is located between 5% and 10% of SBM in DGEBA. The SBM loading is not the only parameter affecting the crossover temperature. It can be set at will by adjusting the molecular characteristics of the SBM chain. This specificity of SBM/DGEBA solutions can thus be turned into a real processing advantage and put to use in various applications.

Conclusions

Thanks to its PMMA block, SBM can be easily incorporated in epoxy precursors. SBM chains can be either dispersed at the molecular level, or form microphase separated structures. In the latter case, the liquid epoxy acts as a selective solvent of the SBM triblock copolymer, S and B blocks are not soluble and self assemble in regular domains. The transition from one state to the other depends on the segregation power, i.e. the temperature or the SBM concentration or the components molar masses. This transition was evidenced by its rheological signature, a thermoreversible sol-gel transition. The gel like behaviour is the consequence of the ordered structure present in the blend. The reactive liquid can thus be turned into a reactive gel or even a reactive solid at higher SBM loadings, in a switchable and controlable manner. We believe that important new applications for thermosets are expected from this unique behavior. It opens the possibility of handling and processing reactive systems like thermoplastic materials. Depending on the affinity between PMMA and the growing epoxy network,[10] the macrophase separation of the additive can be avoided and the initial morphology can be preserved througout the curing process. The choice of the SBM molecular characteristics and the blend composition enables one to prepare nanostructured thermosets. A template of the morphology of a pure SBM has been obtained. Besides being versatile tools for finely tuning the thermoset morphology, we have shown that SBM is a robust route towards nanostructured epoxies. SBM act as a stabilizer, or surfactant, for the residual SB diblocks originating from the industrial synthesis. Controlling the morphology of epoxy thermosets was shown to have another advantage besides preserving transparency, significant toughnening has been obtained with SBM. The toughening mechanism deserves a carefull study as it is to our knowledge the first time that toughening of thermosets is reported with rubbery domains smaller than 20 nanometers. This is a major specificity of SBM over other styrenic block copolymers such as SB, SBS and SEBS that cannot be dissolved in DGEBA and are, thus, not used as modifiers for epoxy thermosets.

Acknowledgments

The authors are especially endebted to Sophie Ritzenthaler, Régis Cipriani and Christophe Navarro. The financial supports of Atofina is gratefully acknowledged.

[1] R.J.J.Williams, B.A.Rozenberg, J.P.Pascault, *Adv. Polym. Sci.* **1997**, *128*, 95.
[2] H.J.Sue, E.I.Garcia-Meitin, D.M.Pickelman, and P.C.Yang, in Rubber Toughened Plastics I, C.K.Riew ED, Advances in Chemistry Series 233, American Chemical Society, Washington DC, **1993**, p. 405
[3]Patent,WO9929772, F.Court, L.Leibler, A.Mourran, C.Navarro, V.Royackkers
[4] L.Leibler, Conference, Materials Research Society, Spring Meeting **2000**, San Francisco.
[5] M.A.Hillmyer, P.M.Lipic, D.A.Hadjuk, K.Almdal, F.S.Bates, J. Am. Chem. Soc. **1997**; 119: 2749-2750.
[6] P.M.Lipic, F.S.Bates, M.A.Hillmyer, J. Am. Chem. Soc. **1998**; 120: 8963-8970.
[7] R.B.Grubbs, J.M.Dean, M.E.Broz, F.S.Bates, Macromolecules **2000**; 33: 9522-9534.
[8] J.Mijovic, M.Shen, J.W.Sy, I.Mondragon, Macromolecules **2000**; 33: 5235-5244.
[9] H.Kossonen, J.Ruokolainen, P.Nyholm, O.Ikkala, Macromolecules **2001;** 34: 3046-3049.
[10] a) S.Ritzenthaler, F.Court, L.David, E.Girard-Reydet, L.Leibler, and J.P.Pascault, Macromolecules **2002**; 35: 6245-6254; b) S.Ritzenthaler, F.Court, E.Girard-Reydet, L.Leibler, and J.P.Pascault, Macromolecules **2002** to appear.
[11] G.Riess, M.Schlienger, S.Marti, J. Macromol. Sc., Phys. **1980**, B17(2), 355-374.
[12] R.Fayt, R.Jérome, P.Teyssié, J. Polym. Sc. : Polym. Letters, vol 19, **1981**, p 79-84 ; J. Polym. Sc. : Polym. Phys., vol 19, **1981**, p 1269-1272 ; J. Polym. Sc. : Polym. Letters, vol 24, **1986**, p 25-28 ; Makromolecular Chemie, 187, **1986**, p 837-852 ; J. Polym. Sc.: Polymer Physics, vol 27, **1989**, p 775-793.
[13] T.Ouhadi, R.Fayt, R.Jérome, P.Teyssié, J. Polym. Sc. : Polym. Phys., vol 24, **1986**, 973-981.
[14] R.Stadler, C.Auschra, J.Beckmann, U.Krappe, I.Voigt-martin, L.Leibler, Macromolecules, 28, **1995,** 3080
[15] C.Aushra, R.Stadler, Macromolecules, 26, **1993**, 6364-6377.
[16] V.Abetz, T.Goldacker, Macromolecular Rapid Communication, 21, **2000**, 16-34.
[17] J.G.Williams, M.J.Cawood, Polym. Test., **1990**; 9; 15.
[18] Y.Huang, D.L.Hunston, A.K.Kinloch, and C.K.Riew in Rubber Toughened Plastics I, C.K.Riew ED, Advances in Chemistry Series 233, American Chemical Society, Washington DC, **1993**, p.1-38.
[19] J. P. Pascault, R. J. J. Williams, in: "*Polymer Blends, Vol. 1: Formulation*", D. R. Paul, C. B. Bucknall, Eds., J. Wiley & Sons, New York **2000**, p.379-416.
[20] A.F.Yee, J.Du, M.D.Thouless, in: "*Polymer Blends, Vol. 2: Performance*", D. R. Paul, C. B. Bucknall, Eds., J. Wiley & Sons, New York **2000**, p.225-264.
[21] A.Bonnet et al, to be submitted.
[22]F.S.Bates, J.H.Rosendale, G.H.Fredrickson, J.Chem.Phys. 92, 1990, 6255

Macromol. Symp. **2003**, *198*, 323-333

Blends of an Epoxy Thermosetting Polymer with Aromatic Liquid-Crystalline Polyesters

Wojciech Jasinski,[1] *Véronique Carlier,*[2] *Roger Legras,*[2] *Werner Mormann*[1,]*

[1] Universität Siegen, Fachbereich 8, Laboratorium für Makromolekulare Chemie, Adolf-Reichwein-Str. 2, D-57068 Siegen, Germany
[2] Université Catholique de Louvain, Unité de Chimie et de Physique des Hauts Polymères, place Croix du Sud 1, B-1348 Louvain-la-Neuve, Belgium

Summary: A series of aromatic liquid-crystalline polyesters with different composition have been synthesized to adjust transition temperatures and molecular weight. Miscibility of polyesters with bisphenol-A-diglycidyl ether (DEGBA), 4,4'-methylene-bis(3-chloro-2,6-dimethylaniline) (MCDEA) and the influence on transition temperatures has been studied. Miscibility of binary and ternary mixtures was found over the whole range of composition depending on the temperature. Thermoset formation by curing of LC-polyester / DEGBA / MCDEA mixtures containing different amounts of polyester resulted in reaction-induced phase separation with polyester content from 30 to 50 wt.%. Cloud point techniques, scanning electron microscopy (SEM) and dynamic mechanical thermal analysis (DMTA) have been applied.

Keywords: blends, epoxy resins, liquid-crystalline polyesters (LCP), reaction-induced phase separation

Introduction

Epoxy resins are important thermosetting polymers used in numerous applications because of their outstanding properties like chemical resistance, low creep, and low shrinkage on cure. These properties require a high level of crosslinking, which makes the materials brittle and results in poor resistance to crack propagation.[1] Various attempts have been made to overcome these inherent drawbacks by modification with thermoplastics or reactive rubbers. Most of these were synthesized from an initially homogenous mixture of modifier and thermoset precursors; phase separation occurs as the crosslinking reaction proceeds.[2] Kim at al. used reactive and non-reactive poly(ether sulfone) and tried to control morphology of the blends.[1] Swier et al. using the same thermoplastic polymer have investigated the reaction-induced phase separation with

DOI: 10.1002/masy.200350827

modulated different scanning calorimetery.[3] Butadiene-acrylonitrile rubber as modifier increases the fracture toughness at the expense of modulus.[4] Other polymers like polystyrene, poly(methyl methacrylate), polycarbonate, polyetherimide or poly(ε-caprolactone) have also been used as modifiers for epoxy resins.[4-7] Liquid crystalline polymers, to our knowledge, have not been described in combination with thermosetting polymers; only low molar mass liquid crystalline compounds have been used as polymer dispersed liquid crystals which however, serve other purposes.[8,9]

The influence of liquid-crystalline polyesters in cured epoxies has not been investigated. The reason can be their high transition temperatures and poor solubility, which makes blending rather difficult. The goal of our studies of blends from tractable LC polyesters and epoxy resins is to find out whether they can have advantages over non-LC polyester based modifiers.

In the present communication we report on the synthesis of aromatic LC polyesters with properties suitable for use as modifiers in thermosetting epoxy resins and on first results of heterogeneous blends obtained by reaction driven phase separation.

Results and Discussion

Synthesis of Polyesters

Suitable LC-polyesters were synthesized from dicarboxylic acid chlorides and trimethylsilylated bisphenols as described previously in the literature.[10]

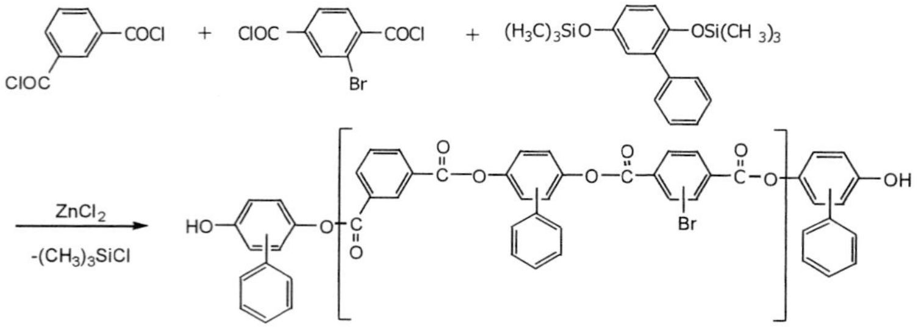

Scheme 1

2-bromoterephthaloyl chloride together with small amounts of isophthaloyl chloride (0-10mol%) and 1,4-bis(trimethylsiloxy)-2-phenylbenzene (silylated phenylhydrquinone) were used as monomers. Molecular weight was adjusted with a slight excess of silylated phenol, which also ensured that only hydroxy terminated polyesters were obtained (Scheme 1). Isophthaloyl chloride was incorporated to interrupt the rigid chain by introduction of kinks, thus decreasing transition temperatures and improving solubility.

Polycondensation reactions were carried out either in *o*-dichlorobezene or in bulk with zinc chloride as a catalyst.[11,12] Polyesters were diluted or dissolved in *o*-dichlorobezene and precipitated in methanol. LC-properties were studied by polarized light microscopy and thermal analysis. Decrease of molecular weight alone was not sufficient to obtain polyesters with clearing temperatures below 300 °C nor did it improve solubility (Figure 1, Table 1). Modification of the polyesters by incorporation of isophthaloyl chloride helped to decrease transition temperatures (Figure 2).

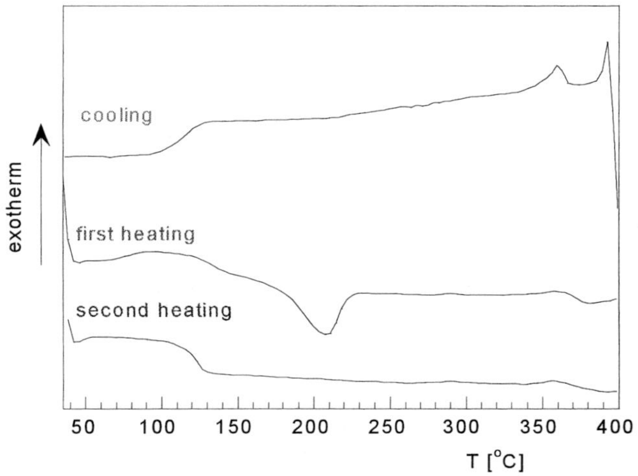

Fig. 1. DSC curves of the linear LC polyester 1 (heating rate 20 K/min).

The solubility of polyesters in common solvents was improved as well. The polyester with 2.5mol% of isophthaloyl chloride was chosen for modification of epoxy resins because of the low clearing temperature and a broad liquid-crystalline range. The results of polycondensation reactions are summarized in Table 1.

Table 1. Composition and properties of LC polyesters.

No.	Polyester	Mole ratio	GPC[a)] Mn	Mw	T_g [°C]	T_m [°C]	T_{iso} [°C]
1	Linear	10:11	37190	88150	121	207	386
2	Linear	9:10	28870	51580	121	192	344
3	Kinked 2.5%	9:10	1361	11220	118	130	250
4	Kinked 5%	9:10	2968	16220	112	135*	230*
5	Kinked 10%	9:10	6803	44010	112	123	210*

* temperatures obtained from polarized light microscopy

a) GPC eluent: DMAc (No. 1, 2); THF (No. 3-5)

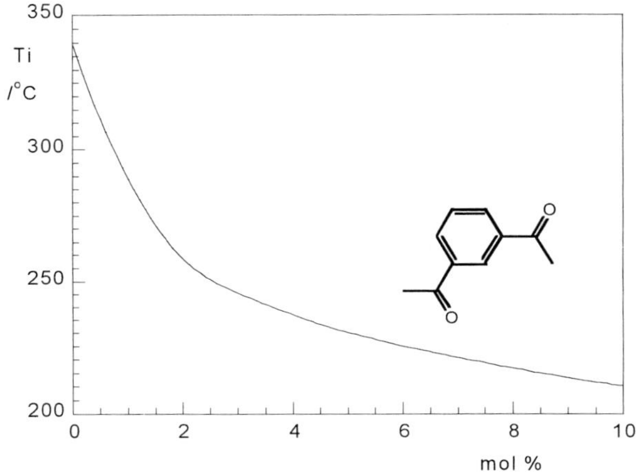

Fig. 2. Influence of isophthaloyl moieties on the clearing temperature of polyesters.

Polarized light microscopy in some cases gave the nematic schlieren texture with a biphasic region typical of step growth polymers. As annealing did not improve textures, the LC phase was identified from mixtures with a monomeric nematic substance (1,4-bis-(4-methoxybenzoyl-oxy)benzene).[13]

Epoxy/LC-Polyester Blends

Miscibility of the LC polyester with DEGBA and with the diamine MCDEA was studied with binary systems. MCDEA and the polyester after homogenization at 150 °C were miscible in any ratio. Mixtures of the polyester with DEGBA prepared in a similar manner were miscible at higher temperature in the composition range from 50 to 100 wt.% polyester, but formed two phases upon cooling. Figure 3 shows the temperature of phase separation or isotropization and the glass transition temperature versus composition of the polyester/DEGBA system. Data were obtained from DSC and optical microscopy. In the microscope two phases are visible, a macroscopic sample is transparent, which is probably due to refractive index matching.

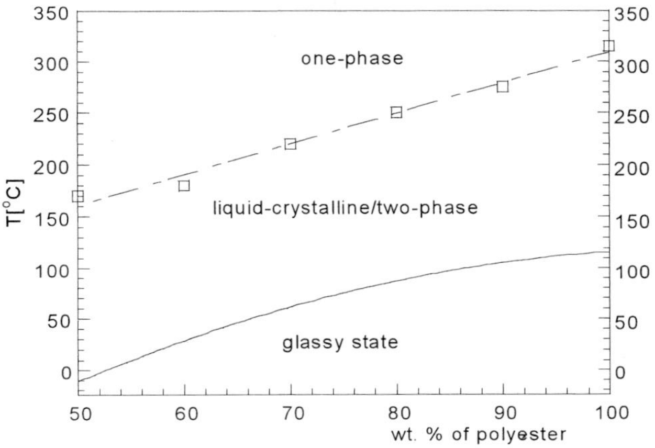

Fig. 3. Phase separation and glass transition temperature of DEGBA/polyester (2) mixtures.

Blends were prepared in a two-stage process: first, a homogeneous mixture of polyester and epoxy resin was made at 160 °C and the diamine was added to this mixture (Figure 4).

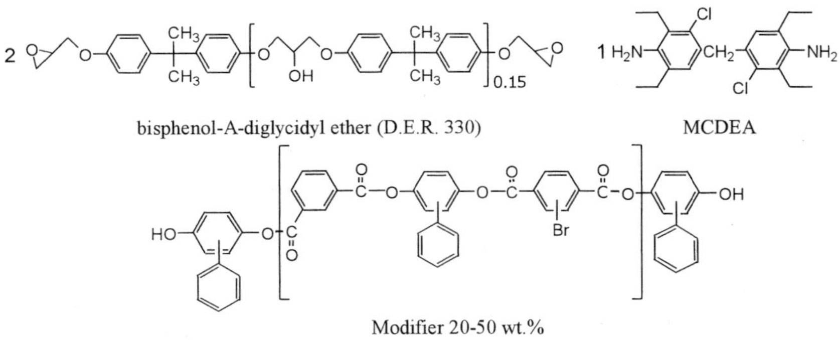

Fig. 4. Composition of polymer blends.

Side Reaction between Epoxide and Polyester

Polymerisation of epoxy groups or reaction with hydroxy terminal groups in the polyester chain can occur during preparation of samples. This was checked with DSC and GPC for a mixture of 60 wt.% polyester and 40 wt.% epoxide, which were mixed as powders in a mortar, heated to 140 °C for 5min, transferred into aluminium pans and isothermally heated to 160, 200, 220, and 250 °C for 45 minutes in the DSC cell. The results from isothermal DSC (Table 2) and GPC (Figure 5) confirm that no significant conversion of epoxy groups occurs up to 200 °C.[14]

Table 2. Enthalpies of reaction between polyester and DEGBA from DSC.

Isothermal process [°C]	$\Delta H^{a)}$ [kJ/mol of epoxy group]
160	0.54
200	2.13
220	41.3
250	$75.2^{b)}$

a)ΔH± 5kJ/mol [14], b) insoluble in THF

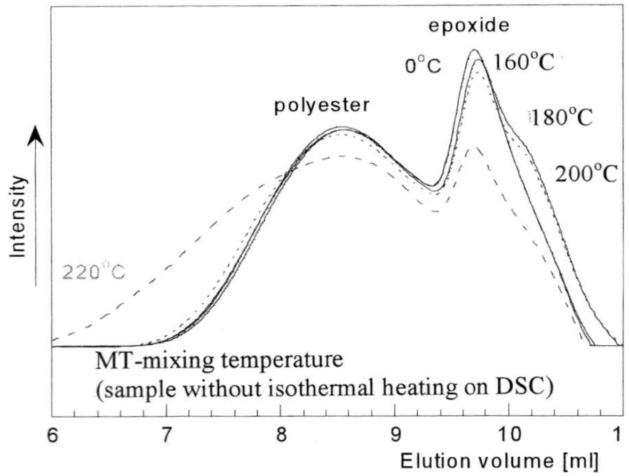

Fig. 5. GPC-elution volume curves of polyester/epoxide mixtures after heating.

Cloud Point Measurements

Phase separation in an epoxy resin/LC polyester mixture was detected by cloud point measurements, i.e. change of intensity of light transmitted and scattered through the sample. This method detects phase separation when dispersed particles have a diameter of at least 0.1 μm.[15]

Table 3. Cloud point for different amount of polyester and curing temperatures.

Polyester [wt.%]	Curing temp. [°C]	t_{CP} [min]	$Conv._{CP}$ [%]
20	160	-	-
30		50	89
40		35	69
50	160	30	59
	180	20	69
	200	-	-
	220	-	-

Samples with 20, 30, 40 and 50 wt.% of polyester reacted at 160 °C. The dependence of the phase separation on the curing temperature was investigated with a mixture with 50% of polyester at 160, 180, 200 and 220 °C.

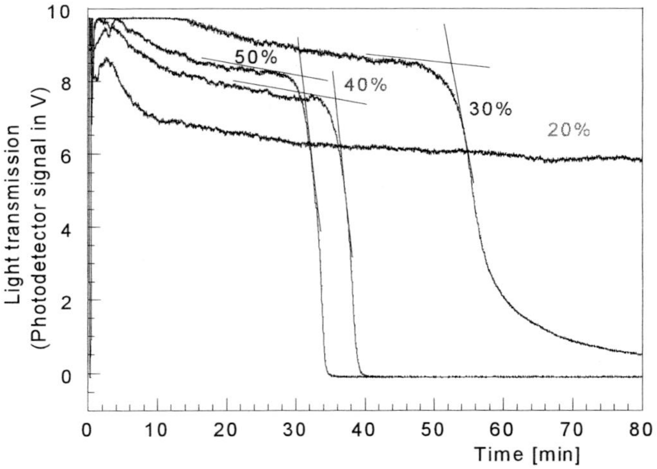

Fig. 6. Cloud point for different amount of LC-polyester in epoxy blends (curing at 160 °C).

The results obtained in these experiments show that phase separation occurred for sample with 30, 40 and 50% of LC-polyester when cured at 160 °C, the sample with 20% of polyester remained homogeneous (Figure 6). Phase separation occurs after shorter times with increasing amount of polyester. The sample with 50 wt.% polyester was also cured at 180, 200, and 220 °C. Phase separation was observed 10 minutes earlier for the sample cured at 180 °C, conversion was 10% higher than at 160 °C which is not in agreement with results from other groups that phase separation starts at given extent of reaction.[15] Further studies will be necessary to clarify this. For samples cured at 200 and 220 °C, however, no phase separation was observed (Table 3). At 220 °C this can be due to reaction of terminal groups with epoxy moieties as described above. Another explanation could be the domain size of the dispersed phase which is to small to be detected by the cloud point method.

Morphology

The existence of a phase-separated structure of the cured epoxy LC-polyester blends was proven by dynamic mechanical thermal analysis and scanning electron microscopy. Samples for DMTA were prepared as described above and cured in a mould at elevated temperatures. Curing conditions and DMTA-results are given in Table 4.

Table 4. Curing regime and viscoelastic analysis.

Curing regime[a]	% polyester	T_g [°C]		tan (δ)	
		polyester	epoxy	$h^{b)}_{\alpha\ polyester}$	$h_{\alpha\ epoxy}$
1	50	105	125	0.25	1.25
2		105	135	0.2	1.1
1	40	105	143	0.12	1.07
2		105	148	0.12	0.96
1	30	105	148	0.7	1.05
2		105	155	0.7	0.95
1	0	-	125	-	1.25
2		-	160	-	0.95
-	100	95	-	2.3	-

a) curing regime: **1** – 2h at 150 °C, next 2h at 190 °C; curing regime **2** – plus 30min. at 240 °C
b) peak-height of α-transitions

Dynamic viscoelastic properties can be used to yield information on relaxations and thermal transitions of individual phases in multiphase polymers, hence also on the microstructure of thermosetting resins.[16] The glass transition temperature of the neat polyester is 95 °C, that of the neat epoxy resin 160 °C.

Figure 7 shows the temperature dependence of the loss factor of modified epoxy networks between 40 and 180 °C at a frequency of 1Hz. A strong maximum between 125 and 155 °C and a weak maximum around 105 °C are visible in each curve. Peak temperatures and peak height of the maxima are included in Table 4. The low α–relaxation peak corresponds to the glass transition temperature of polyester, it is 10 °C above T_g of the pure polyester. The peak at higher temperature is the glass transition temperature of the epoxy resin. Tg of the epoxy network decreases with the increasing polymer content. Hence we deal with a polyester rich and an epoxy rich phase and the amount of polyester in the epoxy phase increases with increasing polyester content in the mixture.

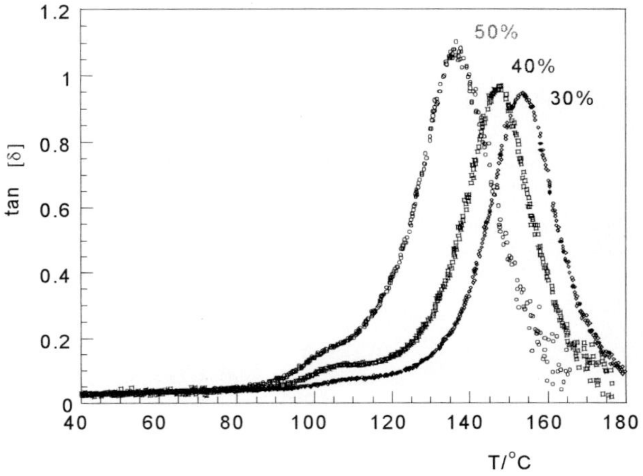

Fig. 7. DMTA of samples with different concentration of polyester (curing regime 2).

For SEM investigation cured samples were fractured in liquid nitrogen, the fracture surface was coated with gold in a sputter coater and observed in the microscope (SEM – Philips XL20, tungsten filament, 20kV acceleration voltage).

The SEM pictures (Figure 8) show nodules and holes. The sample with 20% polyester looks quite homogeneous with only few spots that could indicate the presence of a second phase (Figure 8). With increasing amounts of modifier phase separation can easily be distinguished. The finest structure is observed with 50 wt.% of polyester.

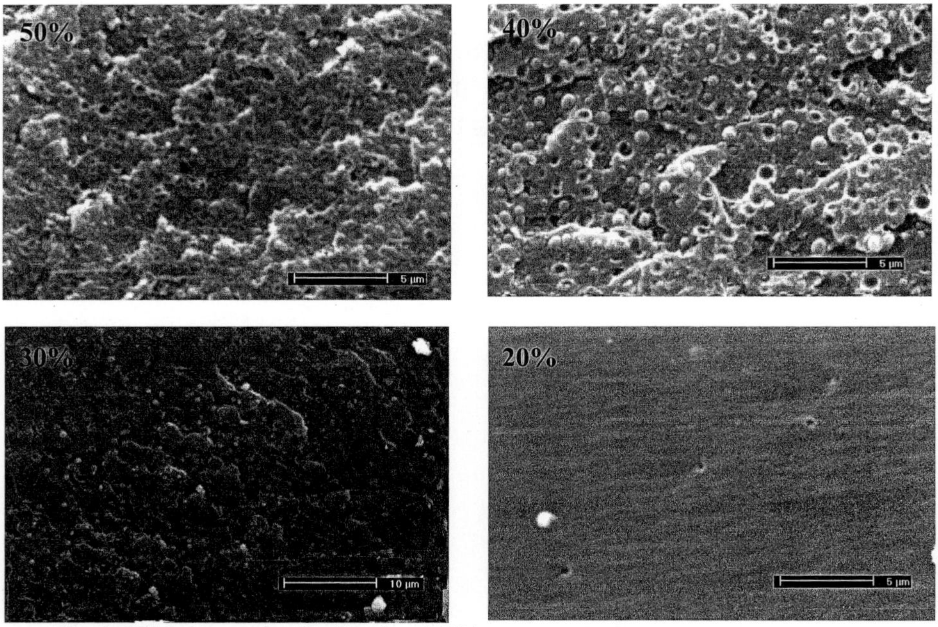

Fig. 8. SEM pictures for 50, 40, 30 and 20%.

Conclusions

Liquid-crystalline precursors for the modification of epoxy resins have been synthesized by step polycondensation. The products have lower transition temperatures and higher solubility than commercially available LC-polyesters. The polyesters were miscible with epoxy resin and diamine used in experiments. The initially homogenous reaction mixture of epoxy-amine and LC-polyester transformed to a two-phase system by reaction driven phase separation. DMTA-measurements indicated that the epoxy rich phase contained significant amounts of polyester and the polyester rich phase significant amounts of epoxy resin. Scanning electron microscopy proved the existence of a two phase system above a polyester content of more than 20 wt.%. The smallest particles of the dispersed phase were observed with 50 % polyester.

Acknowledgments

Financial support from the European Commission in the framework of the Research Training Network (contract no. HPRN-CT-2000-00146) is gratefully acknowledged.

[1] B. S. Kim, T. Chiba, I. Takashi, *Polymer*, **1995**, *36 (1)*, 43-47
[2] R. J.J. Williams, B. A. Rozenberg, J. P. Pascault, *Adv. Polym. Sci.*, **1997**, *128*, 95-156
[3] S. Swier, B. Van Mele, *Termochimica Acta*, **1999**, *330*, 175-187
[4] D. Hao, X. Tang, X. Wang, Y. Luo, *J. Appl. Polym. Sci.*, **2001**, *82 (4)*, 833-838
[5] M. J. Galante, J. Borrajo, J. J. Williams, E. Girard-Reydet, J. P. Pascault, *Macromolecules*, **2001**, *34*, 2686-2694
[6] A. Bonnet, J. P. Pascault, H. Sautereau, *Macromolecules*, **1999**, *32*, 8524-8530
[7] J. L. Chen, F, Ch. Chang, *Macromolecules*, **1999**, *32*, 5348-5356
[8] J. W. Doane, J. L. West, *US Patent 4994204*, **1991**
[9] C. C. Riccardi, J. Borrajo, J. J. Williams, M. Dumon, J. P. Pascault, *Macromolecules*, **1998**, *31*, 1124-1132
[10] H. R. Kricheldorf, *Silicon in Polym. Synth.*, **1996**, 323-324
[11] W. Heitz, N. Nießner, *Makromol. Chem.*, **1990**, *191*, 225-235
[12] H. R. Kricheldorf, G. Schwarz, *Makromol. Chem.*, **1983**, *184*, 475-496
[13] H. Han, P. K. Bhowmik, *Prog. Polym. Sci.*, **1997**, *22*, 1431-1502
[14] J. Goins, G. P. Simon, W. D. Cook, *J. Appl. Polym. Sci.*, **1999**, *72* 1479-1488
[15] C. Hedruel, J. Galy, J. Dupluy, M. Delmotte, C. More, *J. Appl. Polym. Sci.*, **2001**, *82 (5)*, 1118-1128
[16] I. Mondragon, P. M. Remiro, M. D. Martin, A. Valea, M. Franco, V. Bellenguer, *Polym. Int.*, **1998**, *47*, 152-158

Macromol. Symp. **2003**, *198*, 335-344

Blends of a New Thermoplastic in a Thermoset Epoxy Matrix

*François Constantin,**[1] *Françoise Fenouillot,*[1] *Jean-Luc Guillaume,*[2] *Raymond Koenig,*[2] *Jean-Pierre Pascault*[1]

[1] Laboratoire des Matériaux Macromoléculaires, IMP UMR 5627 CNRS, Institut National des Sciences Appliquées de Lyon, 69621 Villeurbanne Cedex, France

[2] Epoxy Business, DOW Chemicals Research Center, Rheinmünster, Germany

Summary: Properties of thermoplastic modified epoxy network have been studied. The particularity of this work is the use of new thermoplastic epoxies whose structure is close to the final matrix. Blends of thermoset epoxy (Diglycidyl Ether of Bisphenol A/4-4' methylenebis [3-chloro 2,6-diethylaniline]) with a thermoplastic content from 5 to 40%w have been realised. Initial miscibility in the thermoset precursors shows an UCST behaviour with a maximal value near 130°C for a thermoplastic content of 10%. Due to the presence of tertiary amine and pendant hydroxyl groups on the thermoplastic backbone, epoxy amine reactions are faster than for the neat system but the thermoplastic seems not to have reacted with the thermoset network. The final blends are transparent but toughening increase is rather low.

Keywords: blends, kinetic, miscibility, thermoplastic epoxy, toughness

Introduction

The improving of the toughness of a thermoset epoxy network has been intensively studied with many thermoplastic modifiers such as high Tg polymers (poly(ether sulfone) PES or poly(ether imide) PEI).[1] The main issue is to avoid a dramatic decrease of the glass transition temperature of the reinforced thermoset epoxy. Generally the thermoplastic is dissolved in the initial reactive epoxy system. As the reaction goes on, epoxy molar mass increases and phase separation appears[2] at a relatively low conversion leading to a thermoplastic rich phase in an epoxy rich matrix. The size of the minor phase and the adhesion between phases determine the final properties such as mechanical ones.

A fully miscible system, i.e. with no phase separation but a miscible additive in the final network is an unusual behaviour. The homogeneity of polycarbonate based epoxy blends[3]

 DOI: 10.1002/masy.200350828

was attributed to extensive trans-reactions between carbonate group of polycarbonate and the hydroxyl groups of the epoxy.

For our study, the thermoplastic was a poly(hydroxy-amino)-ether, from a new family of polymers.[4] These thermoplastic polymers, named BLOX* (* Trademark of the Dow chemical company) thermoplastic resins, are synthesized by reaction between epoxy monomer(s) and ethanolamine. In appropriate conditions, a linear chain is obtained.[5-6] These amorphous materials exhibit exceptional gas barrier protection properties, excellent adhesion on a variety of substrates, good mechanical toughness and stiffness.

BLOX present a similar chemical structure to Phenoxy. Phenoxy are epoxy homopolymers that have already been studied as a modifier of thermoset epoxies. Phenoxy are miscible over the entire composition range with Diglycidyl Ether of Bisphenol A (DGEBA) monomer. Miscibility and phase behaviour, morphology and then properties, are dependent on the choice of the curing agent.[7] The apparition of phase separation is controlled by both thermodynamic and kinetic factors.[7-8] The influence of the hydroxyl pendant groups during curing have been discussed.[9]

Our purpose was to investigate the initial miscibility of these new BLOX thermoplastics in an epoxy monomer with a close chemical structure and to appreciate its behaviour during the reaction leading to the epoxy network. Miscibility, kinetic and mechanical properties are presented and discussed.

Experimental

The epoxy monomer used in this study was Diglycidyl Ether of Bisphenol A (D.E.R. 330* liquid epoxy resin) with a polydispersity of $\bar{n}$= 0,15. The hardener, a bifunctionnal aromatic amine, was 4-4' methylenebis [3-chloro 2,6-diethylaniline] (LONZA). They were used without further purification.

Table 1. Characteristics of the BLOX.

Name	Resorcinol Content (%w)	Molar Mass (g/mol)	M_w/M_n	Glass Transition Temperature (°C)
BLOX00	0	~ 40,000	~ 3	~ 75
BLOX40	40	~ 60,000	~ 4	~ 55

Two amorphous BLOX thermoplastics have been used. The first one is a copolymer (BLOX00) obtained by reaction of DGEBA with ethanolamine. The second is a copolymer

(BLOX40) synthesised from the same monoamine but with two epoxy monomers, DGEBA and Diglycidyl Ether of Resorcinol. The characteristics and chemical structure of these products are described in Table 1 and Figure 1.

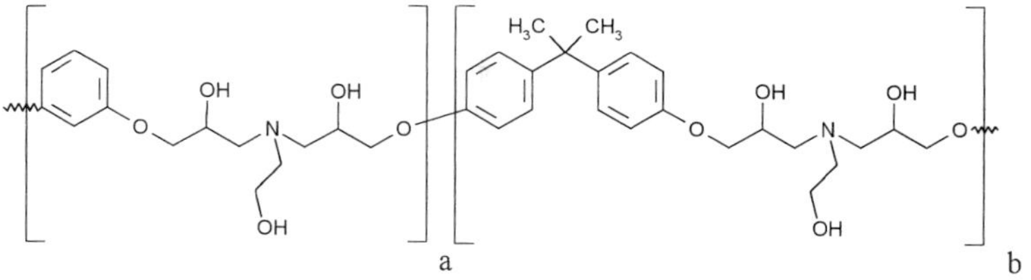

Fig. 1. Chemical Structure of BLOX, a and b corresponding respectively to resorcinol and bisphenol A unit.

During BLOX synthesis, the ethanolamine is used in a small excess to control chain ends (no epoxy groups) and molar masses. This means that a secondary amino group is on both BLOX chain ends: BLOX-N*H*-C_2H_4OH.

BLOX thermoplastic was ground into fine powder to increase contact surface with DGEBA. Initial epoxy monomer/BLOX blends have been prepared in a twin-screw extruder (CLEXTRAL BC21). 30/70 and 50/50 blends (DGEBA/BLOX weight content) have been prepared with a classical thermoplastic/liquid monomer screw profile as shown in Figure 2. Extrudates were quenched in water and dried two days at room temperature.

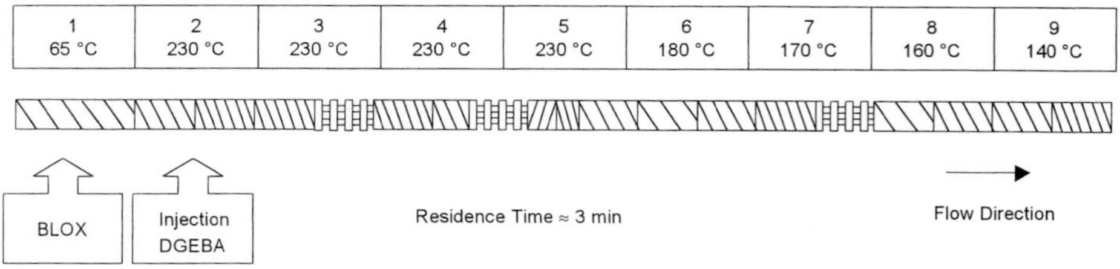

Fig. 2. Temperature and screw profile for BLOX/DGEBA blends preparation.

For the determination of the cloud point curve, samples were prepared at 135°C during 5 min from extrudates.

For the kinetic study, DGEBA/BLOX blends were heated during 4 min at 135°C, the hardener was then stoechiometrically added and mixed 1 min more.

For the mechanical tests, 100g samples were blended at 135°C for 15 min in a glass reactor and then poured into a mould. The concentration of BLOX was varied between 5% to 40%w/w.
Cloud-point setup was home-made and used visible incident light and a photo transistor.[10]
Steric exclusion chromatography in tetrahydrofuran (THF) was carried out to follow the kinetic of curing. Every 15 min, samples were removed from the regulated oven and quenched in ice. THF was then added to reach a concentration around 3 g/mol. The solution is introduced in the apparatus (2 columns WATERS, pores diameter ~500nm and ~100nm). Evolution of the DGEBA main peak (peak n=0) may be expressed by: $X = 1-(A_t / A_0)$ with A_0 and A_t respectively the peak area for time=0 and t, X the conversion in molecules. As we suppose that the epoxy groups exhibit the same reactivity,[10] conversion can be calculated by the following expression: $x = 1-(A_t / A_0)^{1/2}$ with x the conversion in functions. Kinetic of the neat DGEBA-MCDEA system has been previously published.[11]
Rheological measurements have been performed in order to estimate the gel time. The sample was introduced between two parallel plates of 25 mm diameter, heated at 135°C and a constant strain is applied. A frequency sweep was used between 1 and 100Hz. The moduli G' and G'' and their ratio (tan δ) were registered as a function of time.
The critical stress intensity factor, K_{1C} was obtained from three-point bending tests performed on single-edge notched specimens (SEN). The procedure proposed by Williams and Cawood[12] was strictly followed with a crosshead speed of 10 mm·min^{-1}. K_{1C} were calculated as the mean values of 7 tests.

Results

Initial Miscibility

The cloud point curve (CPC) of the initial thermoplastic/DGEBA system is shown in Figure 3. The vitrification curve (FOX equation [13]) has been plotted on the same graph. The curves obtained are typical of Upper Critical Solution Temperature (UCST) behaviour. For a temperature higher than the cloud point curve (CPC), the system is transparent for a fixed BLOX content. Characteristic points of such CPC are the threshold point, the maximum of the CPC and the intersection of the CPC curve with the vitrification curve. The threshold is around (5%, 110°C) for BLOX00 and higher for BLOX40 (10%, 135°C). Despite the similarity of their chemical structure and the eventual development of hydrogen bonds, the affinity between DGEBA monomer and BLOX thermoplastic is poor. Blends with a BLOX content higher than 45% are miscible and transparent over the whole temperature range.

The miscibility window in DGEBA for the copolymer BLOX00 is wider than that for the copolymer BLOX40. As there is no noticeable difference between the two BLOX molar masses, the difference in miscibility essentially results from the difference in the chemical structures. The presence of the resorcinol groups appears to decrease the miscibility of the thermoplastic.

It has to be noticed that when a sample containing 10%w of BLOX in DGEBA is heated at 135°C during 1 hour and then cooled at room temperature in the cloud point apparatus, the temperature obtained for the cloudy/transparent transition is lower than for sample without any pre-heating. As observed by SEC measurements, the DGEBA slowly homopolymerizes in presence of BLOX and miscibility is therefore increased. Apparently no DGEBA grafting on BLOX was detected. This surprising result will be further investigated and discussed in another paper.[14]

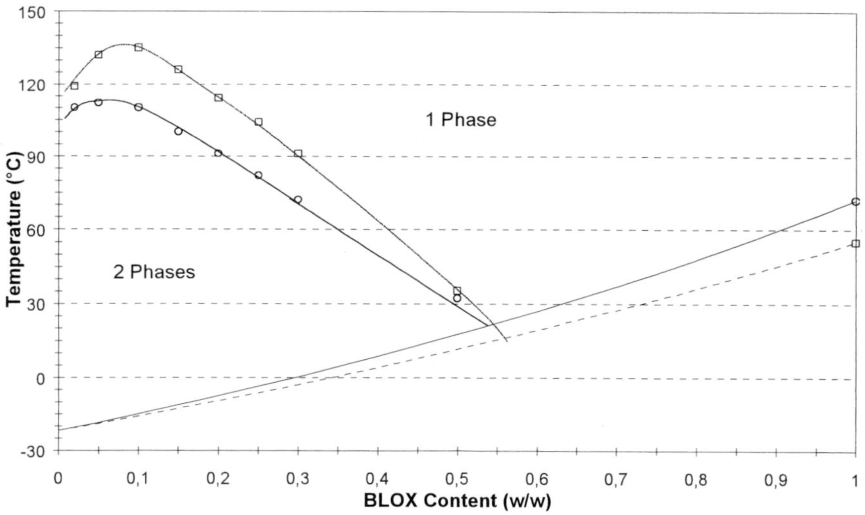

Fig. 3. Initial cloud point and vitrification curves for DGEBA/BLOX00 (○)(-) and DGEBA/BLOX40 (□)(---).

From DSC measurements, a single glass transition temperature near 30°C for a 50w/50w blend has been evidenced. These results are in agreement with the visual observations of the blends at the end of the extruder: 30/70 blends were cloudy whereas 50/50 were transparent at room temperature.

We have to notice that during blending (3min at 190°C in the extruder), the amine secondary end-groups of BLOX have probably reacted with the DGEBA in excess leading to epoxy end-groups without chain extension.

Kinetic

The epoxy/amine reactions have been carried out at 135°C for the neat system and for three blends containing 2, 10 and 20 BLOX00 (%w/w) respectively. The evolutions of epoxy conversion with time are presented in Figure 4.

Whatever the initial composition, the kinetic is faster in presence of BLOX compared to the neat system even if the epoxy and amino groups are diluted by the presence of the thermoplastic. The higher is the BLOX content, the faster is the kinetic. This behaviour can be explained by the presence of hydroxyl groups on the BLOX backbone and also by tertiary amines which catalyse the amine/epoxy reactions. Molecules containing tertiary amine are often used as accelerator for epoxy/amine reaction.

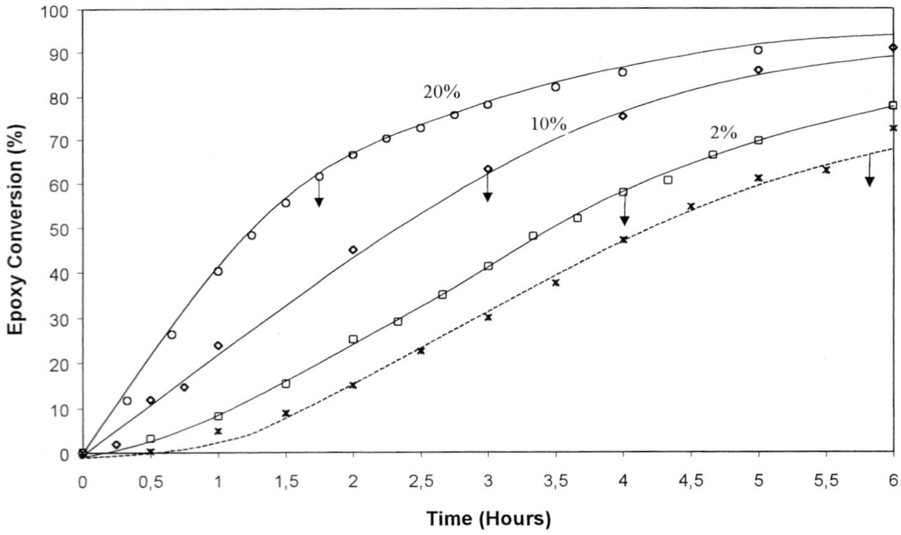

Fig. 4. Conversion versus time for different DGEBA/MCDEA/BLOX00 blends cured at 135°C: Neat (×), 2%TP (□), 10% TP (◇) and 20% (○). Arrows (↓) indicate gelation.

Gel times were then determined by two different measurements. Firstly the apparition of insolubles in THF and secondly the iso-frequency point on dynamical rheological measurement. The following Table 2 summarizes all the results.

Table 2. Gel time for DGEBA/MCDEA/BLOX00.

Thermoplastic content (%w)	Gel time		Conversion at the gel
	SEC	Rheology	
0 %	5 h ¾	6 h	~ 0.6
2 %	4 h	3 h 50	~ 0.6
10 %	3 h	3 h	~ 0.57
20 %	1 h ¾	1 h 40	~ 0.6

The initial composition of the blend is an epoxy-terminated BLOX in a stoechiometric epoxy monomer/hardener blend. During the epoxy/amine polyaddition, BLOX may react with its few epoxy terminal functions and/or the numerous hydroxyl groups of its backbone. Generally, the kinetic constant of epoxy/amine is greater than the one of epoxy/hydroxyl groups in the previous stage of curing when amino hydrogens are is still abundant.

Gelation appears for the same epoxy conversion (x=0.6) as the neat system. This result indicates that the growing thermoset is formed by polyaddition between a diepoxyde and tetrafunctional diamine. BLOX was supposed not to react with its pendant OH groups in our curing conditions; any reaction between hydroxyl groups of BLOX and DGEBA would have led to a decrease of the gel conversion. Similar result has been often obtained in thermosetting epoxy blends with phenoxy, a thermoplastic containing numerous hydroxyl pendant groups.[9] For a quite low curing temperature, no etherification appears in the previous stage of the reaction.

BLOX probably participate to the reaction with its epoxy chain ends. It may react as a diepoxyde with a higher molar mass than DER330. In this case, the functionality is not changed and the gel conversion is not affected as in our experiments.

An interesting point is that, in those conditions (curing reaction at 135°C and MCDEA as a hardener), no phase separation appears despite an initial UCST behaviour. It also means that the affinity and miscibility of BLOX is higher when the epoxy conversion increases. Two phenomena may explain this result.

i) The state of miscibility of any mixture may be described by the Flory-Huggins equation:[15]

$$\Delta G_m = RTV_r\,(\rho_1.\varphi_1.\ln\varphi_1/M_1 + \rho_2.\varphi_2.\ln\varphi_2/M_2) + RT\chi_{12}.\varphi_1.\varphi_2$$

with R: gas constant; T: temperature; V_r: reference molar volume; ρ_i: specific mass; φ_i: volume fraction; M_i: molar mass; χ_{12}: interaction parameter.

For thermoplastic/growing thermoset blends, the combinatorial first term decreases due to the increasing molar mass of thermoset precursors. The χ_{12} interaction parameter is usually rather constant and finally the Gibbs free energy of mixing increases leading to a lower miscibility of the thermoplastic in the growing thermoset and for a given conversion, a phase separation occurs. In our case the result is quite different. An explanation could be that the enthalpy may decrease as the conversion increases. Indeed, when the epoxyde react with the amino hydrogens, hydroxyl groups are formed. Numerous hydrogen bonds may appear between -OH groups of the growing thermoset and BLOX leading to specific interactions, to a possible decrease of χ_{12} up to negative value and an increasing solubility of the thermoplastic in the growing thermoset.

ii) Another explanation would be that a few amount of epoxy grafted BLOX or growing thermoset grafted BLOX may be generated due to the high concentration of BLOX hydroxyl pendant groups. Such grafted thermoplastic may increase the miscibility of BLOX in the final network.

Fracture Toughness

Samples with BLOX00 were cured at 135°C during 11h and post-cured 3h at 190°C (Cycle 1) to ensure a total conversion. As the maximum of the UCST curve for BLOX40/DGEBA is close to 140°C for 10%w, samples with BLOX40 were heated 11h at 150°C and also post-cured 3h at 190°C (Cycle 2) to avoid an initial immiscibility of the reactive blend.

The critical stress intensity factor is plotted as a function of the BLOX00 content in Figure 5. Only the values for BLOX00 are presented because results do not depend on the nature of BLOX. The K_{IC} can be seen to moderately increase with the thermoplastic concentration. This poor toughness increase may be explained by the miscibility of final blends. As no separation phase appears, no useful heterogeneous morphology in term of mechanical properties is produced during the cure. However, an interesting point is that final samples were transparent whatever the BLOX content and BLOX nature.

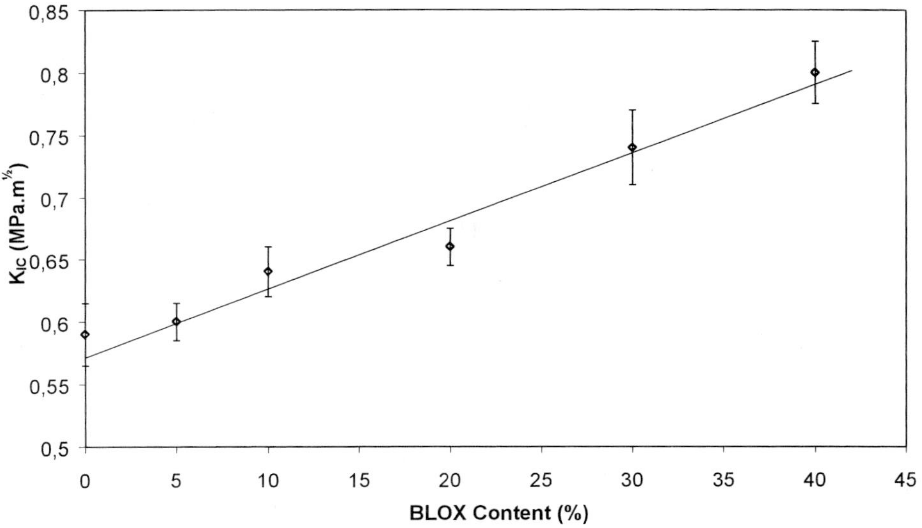

Fig. 5. Fracture toughness vs BLOX content for DGEBA/MCDEA/BLOX00 blends (○).

Conclusions

New thermoset/thermoplastic epoxy blends have been prepared by using a new thermoplastic poly(hydroxy-amino ether). The initial miscibility of the thermoplastic in DGEBA has been studied and an UCST behaviour has been highlighted with a high temperature threshold around 100°C. These results are definitely different from those obtained with Phenoxy/DGEBA system that is miscible over the entire composition range. But heating BLOX/DGEBA blends increases the miscibility. This will be discussed in another publication.[14].

After adding the MCDEA hardener, the thermoplastic was observed to catalyse the epoxy/amine reaction due to the presence of hydroxyl and tertiary amine groups. The higher is the thermoplastic content, the faster is the kinetic. No phase separation is observed during the reaction at 135°C and transparent blends are generated. Whatever the BLOX content, gelation appears for the same gel conversion as the neat system indicating that the pendant hydroxyl groups of the thermoplastic doesn't significantly react with the growing thermoset. Increasing of hydrogen bonds between blend components, few reactions between amine end-groups of BLOX and the growing network and/or the possible formation of epoxy grafted BLOX may explain the final miscibility of this system.

The toughness has been shown to moderately increase with thermoplastic content.

Acknowledgement

This work was financially supported by DOW Deutschland GmbH and Co. OHG. The authors also whish to thank DOW for providing all polymers and monomers used in this study. Authors would like to thank the Prof. R.J.J Williams for his interesting comments.

[1] E. Girard-Reydet, V. Vicard, J.P. Pascault, H. Sautereau, *J. Appl. Polym. Sci.*, **1997**, 65, 2433
[2] J.P. Pascault, R.J.J. Williams, In: *"Polymer Blends Volume1: Formulation"*, Edited by D.R. Paul, C.B. Bucknall, **2000**, 379
[3] C.C. Su, E.M. Woo, *Macromolecules*, **1995**, 28, 6779
[4] J.E. White, D.J. Brennan, H.C. Silvis, M.N. Mang, In: *"Specialty Monomers and polymers; Synthesis, properties and applications"*, Edited by K.O. Havelka, C.L. McCormick, ACS Symposium Series, **2000**, 755, 132
[5] H.C. Silvis, J.E. White, *Polymer News*, **1998**, 23, 6
[6] H.C. Silvis, J.E. White, US Patent 5,275,853 (**1994**)
[7] Q. Guo, *Polymer*, **1995**, 35, 25, 4753
[8] E.M. Woo, H.K. Hseih, *Polymer*, **1998**, 39, 1, 7
[9] H.K. Hseih, C.C. Su, E.M. Woo, *Polymer*, **1998**, 39, 11, 2175-2183
[10] D. Verchere , H. Sautereau, J.P.Pascault, C.C. Riccardi, S.M. Moschiar, R.J.J. Williams, *Macromol.*, **1990**, 23, 725
[11] E. Girard-Reydet, C.C. Riccardi, H. Sautereau, J.P. Pascault, *Macromolecules*, **1995**, 28, 7599
[12] J.G. Williams, M.J. Cawood, *Polym. Test.*, **1990**, 9, 15
[13] T.G. Fox, *Bull. Am. Phys. Soc.*, **1956**, 1, 123
[14] F. Constantin et al, to be submitted
[15] J.P. Flory, *J. Chem. Phys.*, **1942**, 10, 51; M.L. Huggins, *Ann. N.Y. Acad. Sci.*, **1942**, 43, 1

Macromol. Symp. **2003**, *198*, 345-353

An Investigation of sPS/Epoxy Blends by Means of X-Ray Scattering Techniques

N. E. Zafeiropoulos,[1] *J. Schut,*[1,2] *A. Pohlers,*[1] *M. Stamm,*[1] *J.-F. Gerard*[2]

[1]Institut für Polymerforschung Dresden e.V., Hohe Strasse 6, Dresden 01069, Germany
[2]National Institute of Applied Sciences (INSA), Laboratory of Macromolecular Materials (IPM/LMM), UMR CNRS 5627, Avenue Jean Capelle, 69621, Villeurbanne Cedex, France

Summary: SPS is a fast crystallisable polymer that possesses a number of interesting properties due to its semicrystalline nature. However, one major problem is the necessity of applying high temperature for processing sPS (above 290°C), as its melting point is approximately 270°C, and its degradation threshold is at 310-320°C A possible solution to the processing problem may be blending of the sPS with another polymer to lower the viscosity, and subsequently, the processing temperature. SPS, however, exhibits polymorphism like many other semicrystalline polymers. Due to its polymorphism it is necessary to investigate the crystallisation of sPS in sPS blends. Such an investigation is conducted in the present study for the system sPS/epoxy by means of x-ray scattering techniques. The results from WAXS revealed that the degree of crystallinity increases by decreasing the sPS content in the sPS/DGEBA blends. Investigations by means of SAXS revealed a trend of increasing the long period by decreasing the sPS content in the blends.

Keywords: blends, crystallisation, SAXS, syndiotactic polystyrene, WAXS

Introduction

Syndiotactic polystyrene (sPS) is a relatively new engineering thermoplastic that has been introduced by Ishihara et al. in 1986.[1] In contrast to the other two polystyrenes, the commonly used atactic one and the isotactic one, sPS crystallises very fast, at rates comparable to the ones for polyolefines.[2,3] Furthermore, sPS like many other semicrystalline polymers exhibits polymorphism and four distinct crystal forms have been reported until today. The first two crystal modifications are the α and β crystal forms and are the most common ones. In both these forms, sPS chains have an all-trans planar zigzag conformation.[4] The α crystal phase is the

 DOI: 10.1002/masy.200350829

kinetically stable crystal form for sPS, i.e. it forms upon fast cooling of sPS from the melt, and possesses a hexagonal structure.[4] On the other hand, the β crystal phase is the thermodynamically stable crystal form of sPS, i.e it forms mostly under slow cooling and isothermal crystallisation conditions, and possesses an orthorhombic structure[4]. The other two crystal forms of sPS are the δ and γ phases, which are usually formed upon crystallisation of sPS from solution and are also termed as 'clathrate'[5,6]. Both the δ and γ phases posses a helical symmetry and crystallise in the monoclinic system.[5,6]

As may be expected the sPS has properties that are very different from the commonly used atactic PS, due to its high crystallinity. However, even though many of the sPS properties make it very attractive for use in certain applications (e.g. high chemical resistance, high stiffness, high heat deflection point), it is not totally free of problems. In particular, sPS due to its semicrystalline structure has a high melting point (≈275°C), which creates many problems in terms of processing as high temperatures need to be applied to lower the viscosity of the melt. The situation becomes more complicated by the fact that sPS being a styrenic polymer degrades very fast at temperatures above 300°C. One method to improve the processability of sPS is to blend it with another polymer. Very recently, our team has studied the system sPS/epoxy resin and a notable depression of the melting point of about 40-50°C was observed.[7] The present study focuses on a preliminary investigation of the sPS crystallinity in the presence of epoxy resin, conducted by means of wide angle x-ray scattering (WAXS) and small angle x-ray scattering (SAXS).

Experimental

Syndiotactic polystyrene (Questra QA 101™), supplied by Dow Chemical Co., of minimum 99% syndiotacicity (^{1}H and ^{13}C-NMR), with a number average polar mass of 94,100 g/mol, and weight average molar mass of 192,000 g/mol was used. The epoxy resin used was a diglycidylether of bisphenol-A (DER 330™), purchased by Dow Chemical Co., with n=0.15, and molecular weight of 383.1 g/mol. The aromatic diamine (MCDEA) used as a hardener was 4,4-methylenebis(3-chloro-2,6-diethylaniline) (Lonzacure M-CDEA™, 98%) was kindly supplied by Lonza Co., Germany.

The low content sPS blends (below 30wt% sPS) were prepared in test tubes in the following manner. sPS was added to the epoxy resin (DGEBA) at 290°C and stirring was employed for 10 min to ensure full dissolution. The DGEBA had been previously degassed in a vacuum oven. Subsequently, the sPS/DGEBA unreacted blend was mixed with MCDEA at 220°C and the ternary blend was left to cure at the same temperature. Five different concentration (sPS) blends were prepared (5, 8, 15, 20, 26 wt% sPS). The higher content sPS blend (64wt%) due to its high viscosity was prepared in a Clextral co-rotating twin screw extruder at temperatures between 240 and 290°C. sPS plates were prepared through injection moulding at 290°C and the moulded specimens were air-cooled to room temperature. Afterwards, thin slices (approximately 1mm) of the cured blends were cut and were used for the scattering experiments.

WAXS (wide angle x-ray scattering) experiments were performed on a four circle diffractometer (SIEMENS Kalsruhe, now BRUCKER AXS Kalsruhe) operated at 40 mV and 30 mA, with CuKα radiation (λ=1.542 Å). The primary x-ray beam was monochromatised by the use of a graphite crystal. The scattering range measured was: $2\theta = 1.5 \ldots 40.0°$ in transmission by area gas detection system HiStar / GADDS. The beam at the detector pin hole had a diameter of 0.5 mm (distance sample-to-detector 12 cm), and data were obtained at measuring times of Δt = 1800 s (accumulation). The SAXS experiments were performed on a three pinhole camera with CuKα radiation ((λ=1.542 Å) generated from a 18kW Rigaku rotating anode, operating at 40 mV and 100mA. The primary x-ray beam was monochromatised by the use of a multilayer focusing mirror. The sizes of the first and second pinholes were 0.9 and 0.1 mm, respectively, and the size of the guard pinhole before the sample was 0.6 mm. The scattered x-rays were collected by a two-dimensional position-sensitive gas detector HiStar/GADDS (20x20cm^2) with approximately 1-mm resolution. The sample to detector distance was 1.76m. The data were corrected for background and detector response and then azimuthally averaged to one dimensional form of intensity (I) vs. the scattering wavevector ($q=(4\pi/\lambda)\sin(\theta)$). All scattering measurements were performed at room temperature. All calculated degrees of crystallinity are normalised to the sPS content in the blends, and do not refer to the whole specimens.

Results and Discussion

In Figure 1 the WAXS patterns for pure sPS and four sPS/DGEBA blends are shown.

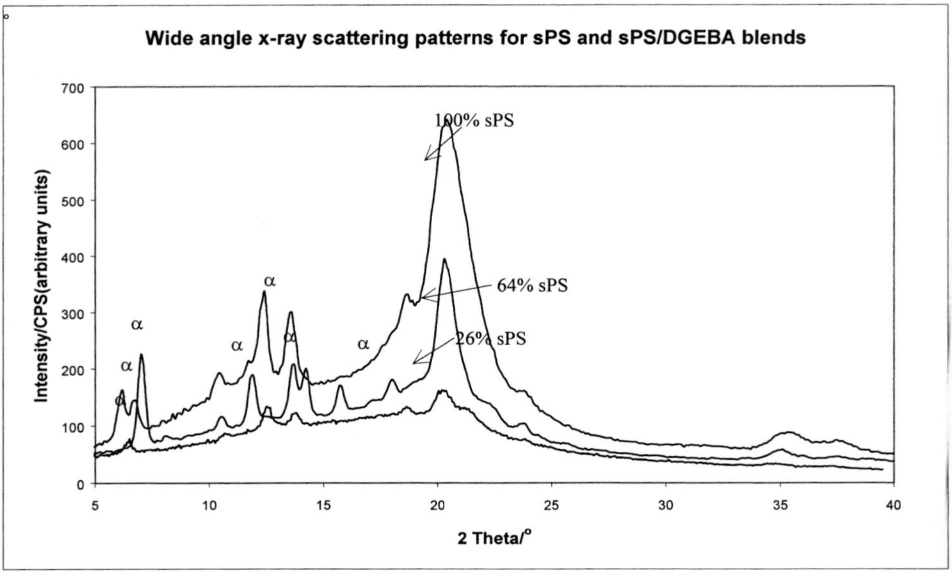

Figure 1. WAXS patterns for sPS and sPS/DGEBA blends.

From the WAXS pattern one can see that the low sPS blends (15, 20, 26wt% sPS) are mostly dominated by the β crystal form of sPS, which is the thermodynamically stable crystal form. Only for the 26wt% sPS blend one peak at 6.5° 2θ appears. On the other hand, the high sPS content blend (64wt% sPS) and the pure sPS exhibit a mixture of α and β crystal forms. It is recalled that the α crystal form is the kinetically stable crystal form of sPS. The difference of the crystal forms between the low sPS content and the high sPS content blends, as well as the pure sPS is probably due to the processing route, rather than the presence of the DGEBA resin. In fact, for the low sPS content blends, crystallisation took place at 220°C, while curing of DGEBA was carried out. Hence, the thermodynamically stable β crystal form was allowed to develop. On the other hand, sPS crystallised under air-cooling for the cases of the injection moulded 100% sPS and the 64wt% sPS/DGEBA blend. Under air cooling the kinetically stable α crystal form is more favoured to develop. The absence of the δ and γ crystal forms, which have been reported in the literature to form in gels of sPS with good solvents[8] indicates that the blend is fully phase separated after curing and little, if any, DGEBA may be found in the sPS rich areas of the blend. It should also be noted that for the 8wt% sPS blend (not shown in Figure 2) only two very weak peaks were located, and for the 5wt% sPS blend, no crystalline material was detected. For the 8wt% sPS blend it was possible to subtract the amorphous scattering (and also the incoherent scattering) from the crystalline scattering curves, but such an operation was not successful for the 5wt% sPS blend. It should similarly be noted that the width of the crystalline peaks does not significantly change fore different concentrations. It, thus, can be concluded that a nearly perfect phase separation occurs and that the DGEBA compound is mostly expelled from the crystalline sPS phase.

It is interesting to investigate the effect of the presence of DGEBA on the overall degree of crystallinity of sPS. In Figure 2 the overall degree of sPS crystallinty (x_c) is shown.

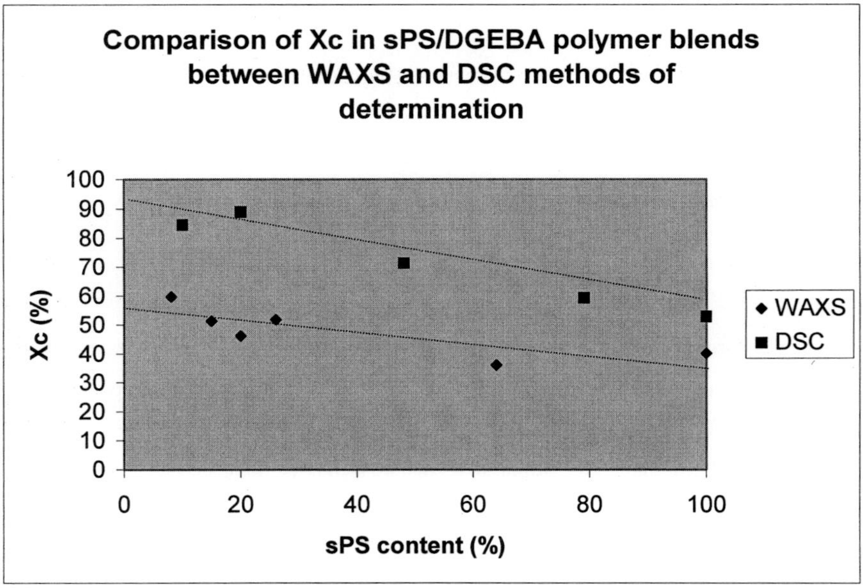

Figure 2. Comparison of x_c in sPS/DGEBA polymer blends between WAXS and DSC methods of determination.

The surprising result is that the degree of sPS crystallinity shows a trend of increasing with increasing the DGEBA content. This trend has also been found after examination of sPS/DGEBA blends by means of DSC as is also shown in Figure 2. At this stage it is rather difficult to explain, but one possible explanation may be that during phase separation the sPS chains are more extended at smaller concentrations, and the conditions are more favourable for a higher degree of crystallisation to develop. This is also supported by the increase in the long period, as is explained below, with decreasing concentrations. Another more trivial explanation might be just the different undercooling at different concentration, since the crystallisation of sPS start at 270°C and for the blends at 220°C only. Crystallinity and the long period are known to be dependent on crystallisation temperature. More research is currently ongoing in an attempt to shed more light on this interesting behaviour. It should also be noted at this point that the calculations concerning the degree of crystallinity were performed for the sPS part of the blend (i.e. the calculated x_c values are for the sPS only and not for the whole blend). The overall degree of crystallinity of the whole blend, is obviously reduced as the content of DGEBA increases.

The SAXS experiments have also revealed an interesting behaviour for the low content sPS blends. In Figure 3 the $\log(I*q^2)$ vs. q is shown for the low sPS content blends. Note that the SAXS intensity profiles were corrected by the subtraction of the background intensity arising from the thermal density fluctuation (I_b) by the use of the Porod-Ruland model:[9]

$$I(q) = K\frac{e^{-\sigma^2 q^2}}{q^4} + I_b \qquad (1)$$

were K is the Porod constant and σ is related to the thickness of crystalline/amorphous interphases. The values of K, σ and I_b may be calculated by curve fitting the intensity at high q range.

As seen in Figure 4 for the 5 and 8wt% sPS content blends no long period was detected. However, for the other three blends, a broad peak is seen at q values between 0.3 and 0.5 nm^{-1}. These peaks are quite broad and denote the long period of the semicrystalline sPS. The long period can be calculated from Bragg's law as $L=2\pi/q^*$, were q^* is the peak position. In Table 1 the calculated values for the log period is given.

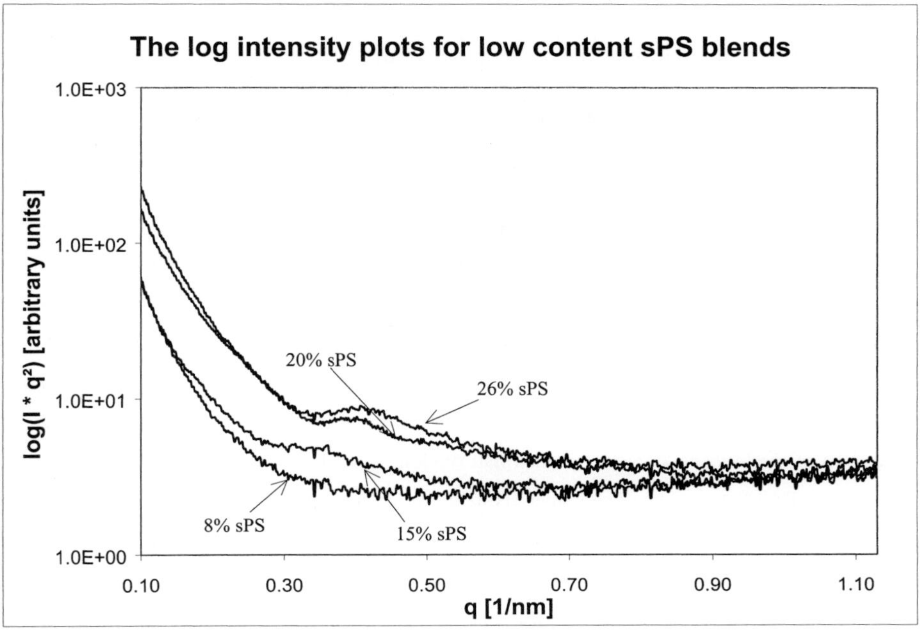

Figure 3. The logarithmic intensity plot for the low sPS content blends.

Table 1. Calculated values for the long period.

Grade of Material	Long period (L) (nm)
15 wt% sPS blend	17.6 ±1.1
20 wt% sPS blend	16.2 ±0.9
26 wt% sPS blend	14.9 ±0.8
100% sPS	15.2 ±0.9

It is interesting to note that that the long period appears to increase by decreasing the sPS content. This increase again may be due to an extended chain conformation of sPS at low concentrations as mentioned before. SAXS data, however, also indicate strong phase separation between sPS and DGEBA giving rise to crystallisation of sPS. SAXS experiments are also ongoing for the high sPS content blends. It should also be noted that for all the tested blends, extensive scattering at small q was found, which may be attributed to the phase separated sPS domains.

Conclusions

The investigation of sPS/DGEBA blends by means of WAXS revealed that the low sPS content blends are mostly dominated by the β crystal form of sPS, which is the most thermodynamically stable form. On the other hand, the high sPS content blend and the pure sPS were found to have both the β and α crystal forms. However, the non-existence of the α crystal form in the low sPS content blends is probably due to the processing route followed to prepare the samples, rather than to the presence of DGEBA. The occurrence of the largely unperturbed crystalline phase is, however, a strong indication that nearly perfect phase separation between sPS and DGEBA takes place. The calculation of the degree of crystallinity for the sPS domains, revealed that sPS becomes more crystalline with increasing of the DGEBA content. This result has also been found by the calculation of the degree of crystallinity for the blends by means of DSC. The examination of the low sPS content revealed that the long period is shifting to higher values by increasing the DGEBA content. The higher crystallinity and increase of the long period at small concentrations of sPS might be explained by two effects: the decreased undercooling due to the decrease of the crystallisation temperature and a possible change in chain conformation at different concentrations. The reasons for this shift of the long period to higher values is also at the moment unknown. More research is currently underway in an attempt to shed more light to the mechanisms of sPS crystallisation involved during phase separation. Synchrotron x-ray scattering experiments will be performed in the future to study the crystallisation of sPS during phase separation in real time.

Acknowledgements

The research undertaken in the present study has been funded by the Research Training Networks (RTN) programme under the name “PolyNetSet”, contract number HPRN-CT-2000-00146, under the auspices of the Directorate-General for research of the European Commission. We would also like to acknowledge Dow Company for kindly supplying the sPS used in the present study.

[1] N. Ishihara, T. Seimiya, M. Kuramoto, M. Uoi, Macromolecules, **1986**, 19, 2465
[2] S. Cimmino, E. DiPace, E. Martuscelli, C. Silvestre, Polymer, **1991**, 32, 1080
[3] N. Tomotsu, N. Ishihara, T. Newman, M. Malanga, J. Mol. Catal., **1998**, 128, 167
[4] Y. Chatani, Y. Shimane, T. Ijitsu, T. Yukinari, Polymer, **1993**, 34, 1625

[5] S. Moyses, P. Sonntag, S. J. Spells, O. Laveix, Polymer, **1998**, 39, 3537
[6] M. Kobayashi, T. Nakoaki, N. Ishihara, Macromolecules, **1990**, 23, 78
[7] J. Schut, M. Stamm, M. Dummon, J.-F. Gerard, Macr. Symp., submitted
[8] T. van Hooy-Corstjens, Polymer-Solvent Compounds, PhD Thesis, Technical University of Eindhoven, 2002
[9] W. J. Ruland, J. Appl. Cryst., **1971**, 4, 70

Macromol. Symp. **2003**, *198*, 355-362

DGEBA Monomer as a Solvent for Syndiotactic Polystyrene

Jaap Schut,[1,2] *Manfred Stamm,*[2] *Michel Dumon,*[1] *Jean-François Gérard**[1]

[1] National Institute of Applied Sciences, Laboratory of Macromolecular Materials (IMP/LMM), UMR CNRS 5627; 17, Avenue Jean Capelle, 69621 Villeurbanne Cedex, France
E-mail: jfgerard@insa-lyon.fr
[2] Institute of Polymer Research Dresden, Physical Chemistry and Physics of Polymers; Hohe Strasse 6, 01069 Dresden, Germany

Summary: Syndiotactic polystyrene (sPS) has to be processed at high temperatures (i.e. >290°C due to its melting point of 270°C), which approaches its degradation temperature. We aim to facilitate the processing of sPS by lowering its melt temperature and viscosity with a curable epoxy/amine model system as reactive solvent, which will result in a thermoplastic-thermoset polymer blend. As a first step we therefore investigated the melting behaviour of sPS in epoxy monomer, established its phase diagram, and investigated the crystalline form of sPS in these mixtures. DGEBA epoxy monomer is found to be a solvent for syndiotactic polystyrene at temperatures above 220°C. The DGEBA-sPS phase diagram was established by means of DSC, on the basis of crystallization and melting peaks. The form of the curve in the phase diagram indicates that DGEBA is a poor solvent for sPS. In WAXS studies of blends only the β crystalline form was detected, not the δ form, thus no sPS-DGEBA polymer-solvent compounds (clathrates) were detected. However, DGEBA can still serve as a monomer for improved processing as it depresses the crystallization temperature by 20 to 60 K upon addition of 20 to 90 wt% DGEBA respectively, while a 16 to 45 K melting peak depression can be observed by adding 20 to 90 wt % DGEBA.

Keywords: blends, crystallization, phase diagram, reactive solvent, syndiotactic polystyrene

Introduction

Syndiotactic polystyrene (sPS) is a relatively new and promising semi-crystalline engineering thermoplastic with a melting point around 270°C, which was first synthesized in 1985 by Ishihara *et al.* with the use of a titanium metallocene catalyst.[1] In addition to the advantages of atactic polystyrene (attractive price, low specific gravity, low dielectric constant, and good

 DOI: 10.1002/masy.200350830

processability), sPS has very good solvent, chemical and steam resistance, high modulus of elasticity, high heat resistance and good dimensional stability, all due to its semi-crystalline nature. Syndisotactic polystyrene crystallizes about two orders of magnitude faster than iPS (which has not been commercialised due to its slow rate of crystallization), and displays a crystallinity between 15 to 60%. sPS is therefore expected to fill the price-performance gap that currently exists between performance and commodity plastics, and has been commercialised by Dow Chemical Co. (Questra™) and Idemitsu (Xarec™).

However, due to its high melting point sPS has to be processed at temperatures exceeding 290°C, which approaches its degradation temperature, and is thus a disadvantage.

Our goal in the present study is twofold. Firstly to facilitate the processing of sPS by lowering its melt temperature and viscosity with a curable epoxy/amine model system as reactive solvent, which will result in a thermoplastic-thermoset polymer blend. Secondly we aim to generate new blend morphologies, using the crystallization of sPS as an adjustable parameter during the crosslinking of the epoxy/amine monomers. The present study is concerned with the characterisation of the solvent quality of epoxy monomer for syndiotactic polystyrene. We have therefore investigated the melting behaviour of sPS in epoxy monomer, established its phase diagram, and studied the crystalline form of sPS in these mixtures.

Syndiotactic polystyrene has two stable chain conformations, the all-trans, planar zigzag (or T_4) conformation, and the helical (or T_2G_2) conformation, which give rise to 4 crystalline modifications: α, β, δ, and γ.[2] Fast cooling from the melt leads to the α form crystalline (kinetically stable), while slow cooling leads to the β crystalline form (thermodynamically stable), with sPS having the planar zig-zag conformation in both cases. The δ form is a polymer-solvent complex (clathrate) where sPS is crystallised in its helical conformation and solvent molecules are incorporated in the crystalline unit cell. When the δ form is carefully heated to remove the solvent molecules, the δ_e or γ forms result, which are solvent free forms of sPS in the helical conformation.[3],[4],[5] Figure 1 shows the schematic phase diagrams of sPS in a good and poor solvent,[6] with the superposition of the aPS-good solvent phase diagram for comparison.

Figure 1a shows that upon *slowly cooling* a weakly concentrated, hot solution of sPS in a good solvent (such as chloroform, benzene, toluene, or chlorobenzene) the δ form is generated, which is a stable thermoreversible gel. At higher concentrations ór in the case of a poor solvent (such as decalin, cyclohexane, or cyclohexanol, Figure 1b at the whole concentration range, this gel can only be formed by *rapid quenching* of the hot solution, which now yields a *metastable* thermoreversible gel. However, when these solutions are *slowly cooled*, sPS directly crystallises in the β crystalline form. Heating a stable gel causes it to melt directly to a homogeneous solution, while heating a metastable gel gives first rise to a transition from δ to β -observed as a horizontal line in the phase diagram- followed by a melting of the β crystals to a homogeneous solution.

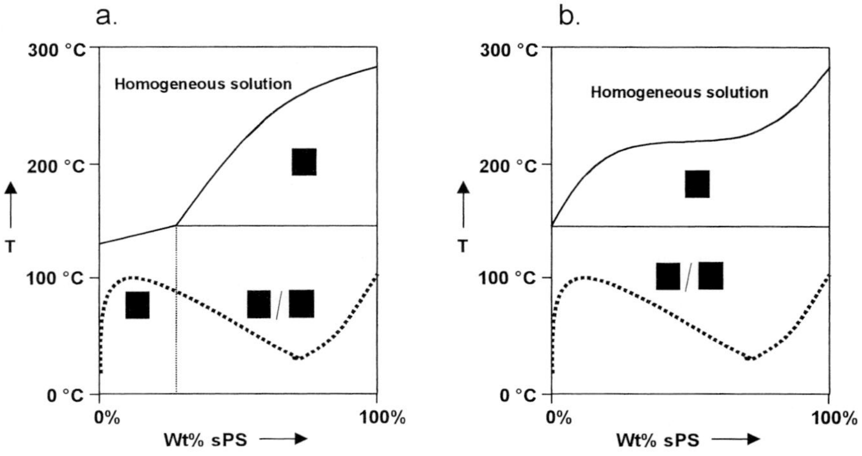

Fig. 1. Schematic temperature-composition diagram of syndiotactic polystyrene and solvent systems. a) Good solvent b) Poor solvent. The dotted curves represent the atactic PS-good solvent phase diagram for comparison.

Compared to pure sPS a melting point depression is observed in sPS-solvent mixtures upon the addition of both a good and a poor solvent. The normal Upper Critical Solution Temperature phase of aPS in a good solvent diagram is "hidden" by the crystallization of the syndiotactic polymer.

Experimental

Syndiotactic polystyrene (Questra QA 101™, supplied by Dow Chemical Co.), denoted sPS, of 99+% syndiotacticity (^{1}H and ^{13}C-NMR), with a number average molar mass of 94,100 g.mol^{-1}, and weight average molar mass of 192,000 g.mol^{-1} was used. The epoxy resin system used was a diglycidylether of bisphenol-A (Dow DER 330™, n=0.15, M=383.1 $g.mol^{-1}$, purchased from Dow chemical Co.), denoted DGEBA.

sPS-DGEBA blends at concentrations below 30 wt% sPS were prepared in a large test tube equipped with a mechanical stirrer. Powdered sPS was added to vacuum degassed (liquid) DGEBA and heated to 290°C in a Wood's metal bath to ensure dissolution within 10 min. Due to their high viscosity, blends containing 60 wt% sPS or above were prepared in a Haake internal mixer or a Clextral co-rotating twin screw extruder at 240-290°C.

DSC measurements were performed on a water-cooled Perkin-Elmer DSC-7 under nitrogen atmosphere at a temperature scan rate of 10 $K.min^{-1}$. All samples were held for 5 min at 300°C to erase thermal history prior to the crystallization and subsequent melting scans.

Results

Syndiotactic polystyrene was found to be soluble in DGEBA epoxy monomer above 220°C. To establish the temperature-composition diagram of the non-reactive sPS-DGEBA system, DSC was

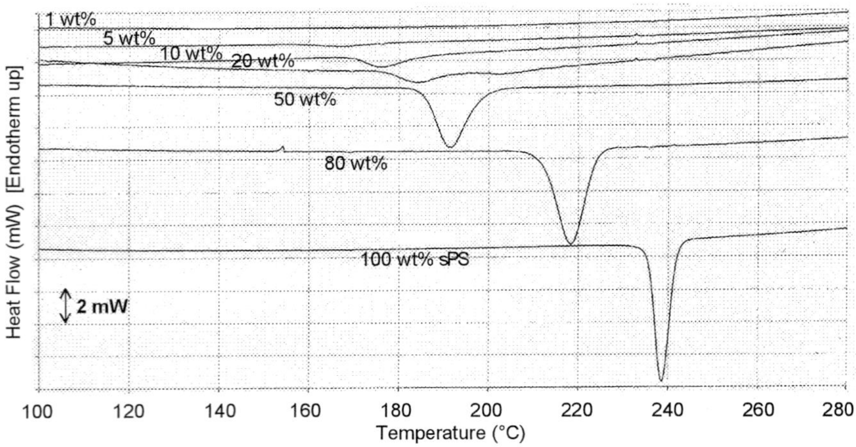

Fig. 2. DSC cooling scans (10 $K.min^{-1}$) of different sPS-DGEBA mixtures. Samples were held for 5 min at 300°C before cooling.

employed to investigate crystallization and melting behaviour. Figure 2 shows the crystallization scans of several of these blends. The crystallization temperature is depressed by 20 to 60 K upon addition of 20 to 90 wt% DGEBA, respectively.

The melting point depression can be observed from the DSC heating scans in Figure 3. Pure sPS exhibits multiple melting peaks, attributed to melting and re-crystallization of different lamella crystals.[7],[8],[9] While a 16 to 45 K melting peak depression can be observed by adding 20 to 90 wt % DGEBA.

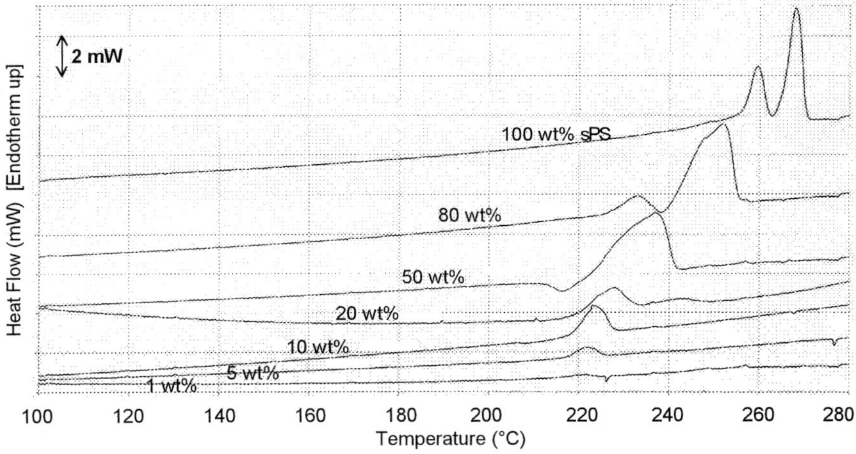

Fig. 3. DSC heating scans (10 K.min^{-1}) of different sPS-DGEBA mixtures. Samples were previously crystallised by cooling from 300°C at 10 K.min^{-1}.

The onset of crystallization and melting peak temperatures were used to construct the phase diagram shown in Figure 4.

Scrutinising the onset of crystallization one may see that 220°C is the minimum temperature at which low concentration blends (<40 wt% sPS) can be safely processed in homogeneous conditions, i.e. without premature crystallization. On the other hand the minimum temperature needed for the reactive extrusion of high concentration blends (>60 wt% sPS) is 230-250°C.

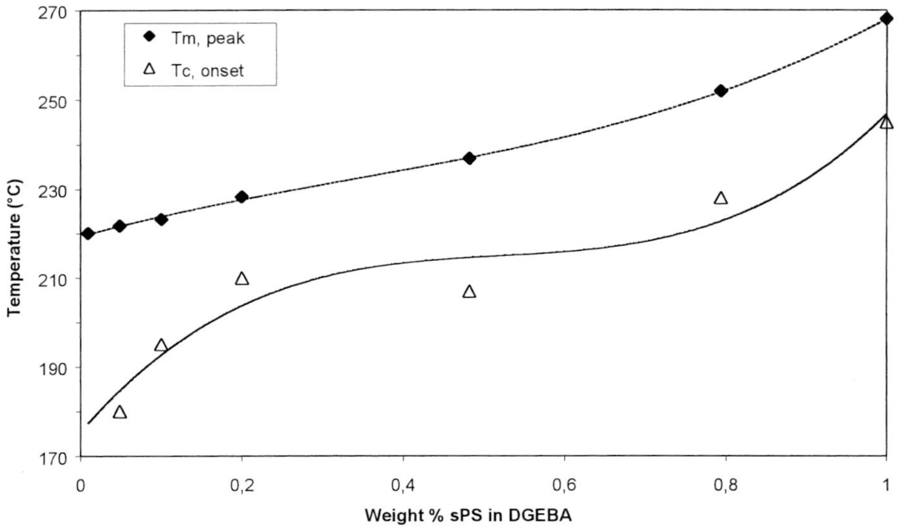

Fig. 4. Temperature-composition diagram of the syndiotactic polystyrene-DGEBA system.

These high temperatures needed to obtain homogenous mixtures of sPS in DGEBA, cause some concern as to the degradation or thermal homopolymeriization of the epoxy monomer. TGA and SEC/GPC however showed that there was no significant degradation or polymeriization even after 6h at 220°C.[10]

The shape of the crystallization curve resembles that of a poor solvent system as shown in Figure 1b. When homogeneous solutions of sPS in DGEBA were quenched into liquid nitrogen no gel formation could be observed, but only crystallization of the sPS into spherulites.

Figure 5 shows the WAXS patterns of pure and slowly crystallised sPS compared to its blends with DGEBA. The pure sPS pattern is characteristic for the β crystalline form, with peaks at 6.3, 7.0, 10.6, 12.5, 13.7, 18.7, 20.4, 24.0, 35.2, and 37.7° (2θ scale).[11]

Although the patterns have a low signal to noise ratio due to the low sPS concentration, the major peaks at 6.3, 12.5, and 20.4° are seen not to shift in the 20, 10, and 3% mixtures, nor do any additional peaks arise when compared to the β form of pure sPS, although the amount crystallinity is low due to low sPS concentrations.

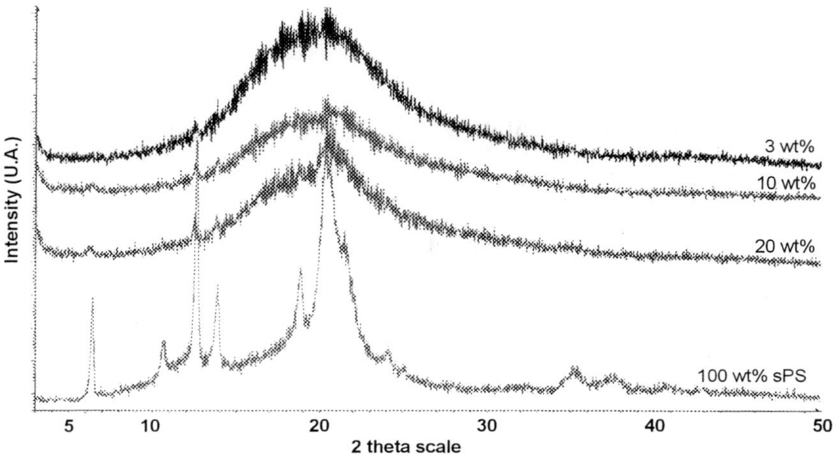

Fig. 5. Wide angle X-ray diffraction patterns of pure (β) sPS and its mixtures with DGEBA monomer.

Conclusions

DGEBA epoxy monomer is found to be a solvent for syndiotactic polystyrene at temperatures above 220°C. Although these are relatively high temperatures for using epoxy resin, no significant degradation or thermal homopolymeriization of DGEBA could be detected by TGA and SEC/GPC. The DGEBA-sPS phase diagram was established by means of DSC. The form of the crystallization curve in the phase diagram indicates that DGEBA is a poor solvent for sPS. This is also supported by the fact that no other crystalline modifications other than β could be detected by WAXS at various concentrations of DGEBA. No sPS-DGEBA polymer-solvent compounds (clathrates) can therefore be detected, and relatively pure sPS and DGEBA phases are formed during phase separation. However, DGEBA can still serve as a monomer for improved processing as it depresses the crystallization temperature by 20 to 60 K upon addition of 20 to 90 wt% DGEBA respectively, while a 16 to 45 K melting peak depression can be observed by adding 20 to 90 wt % DGEBA.

Additional work is in progress on the effect of the presence of DGEBA monomer on the crystallization kinetics of sPS, as well as on ternary blends cured with a diamine hardener processed on laboratory scale and by reactive extrusion. Future work will include the optimisation of the reactive extrusion of sPS with DGEBA/MCDEA, studying the reaction induced phase separation and crystallization by WAXS/SAXS, and relating the mechanical properties of the cured blends to their compositions and thermal histories.

Acknowledgments

This research was funded by the Research Training Networks (RTN) program under the name « PolyNetSet »; contract number HPRN-CT-2000-00146, under auspices of the Directorate-General for Research of the European Commission.
Our thanks go to Mr. Ruben Vera (Claude Bernard University, Villeurbanne) for performing the WAXS measurements, and Dr. Nick Zafeiropoulos for the useful discussions.

[1] N. Ishihara, T. Seimiya, M. Kuramoto, M. Uoi, *Macromolecules*, **1986**, *19*, 2464-2465.
[2] G. Guerra, V.M. Vitagliano, C. De Rosa, V. Petraccone, P. Corradini, *Macromolecules*, **1990**, *23*, 1539-1544.
[3] S. Moyses, S. Spells, *Polymer*, **1999**, 40(*12*), 3269-3274.
[4] G. Guerra, C. Manfredi, P. Musto, S. Tavone, *Macromolecules*, **1998**, *31*, 1329-1334.
[5] E. Reverchon, G. Guerra, V. Venditto, *J. Appl. Pol. Sc.*, **1999**, *74*, 2077-2082.
[6] S. Rastogi, J.G.P. Goossens, P.J. Lemstra, *Macromolecules*, **1998**, *31*, 2983-2998.
[7] Y .S. Sun, E.M. Woo, *Macromolecules*, **1999**, 32, 7836-7844.
[8] Y .S. Sun, E.M. Woo, *J. Pol. Sci. part B: Pol. Phys.*, **2000**, 38, 3210-3221.
[9] Y .S. Sun, E.M. Woo, *Pol. Mat. Eng.*, **2001**, 84, 551.
[10] Unpublished results.
[11] S. Duff, S. Iwamoto, T. Iwamoto, F. Fujibayashi, C. Birkinshaw, **2001**, 42(3), 991-1000.

Macromol. Symp. **2003**, *198*, 363-375

In Situ Monitoring of Reaction-Induced Phase Separation with Modulated Temperature DSC

*Steven Swier, Bruno Van Mele**

Vrije Universiteit Brussel (VUB), Department of Physical Chemistry and Polymer Science, Pleinlaan 2, B-1050 Brussels, Belgium

Summary: A linearly polymerizing and network forming epoxy-amine system, DGEBA-aniline and DGEBA-MDA, respectively, will be modified with 20 wt% and 50 wt% of a high-T_g thermoplastic poly(ether sulphone) (T_g=223°C), respectively, both showing LCST-type demixing behavior. Reaction-induced phase separation (RIPS) in these modified systems is studied using Modulated Temperature DSC (MTDSC) as an in situ tool. Phase separation in the linear system can be probed by vitrification of the PES-rich phase, occurring at a higher conversion than the actual cloud point from light scattering measurements. The negative slope of the cloud point curve in a temperature-conversion-transformation diagram unambiguously shows the LCST-type demixing behavior of this system, while the relation between the composition/glass transition of the PES-rich phase and the cure temperature is responsible for the positive slope of its vitrification line. Phase separation in the network forming system appears as reactivity increases at the cloud point due to the concentration of reactive groups. Different mixture compositions alter the ratio between the rate of phase separation and the rate of reaction, greatly affecting the morphology. Information about this in situ developed structure can be obtained from the heat capacity evolutions in non-isothermal post-cures.

Keywords: curing of polymers, epoxy-amine, modulated temperature differential scanning calorimetry, poly(ether sulphones), reaction-induced phase separation

Introduction

Reaction-induced phase separation (RIPS) is the result of a decrease in the entropic contribution to the free energy of mixing during polymerization.[1] The increase in conversion leads to a 'chemical quench' from the miscible to the immiscible region of the phase diagram by either a decrease of the lower critical solution temperature (LCST: case of poly(ether sulphone), PES) or an increase of the upper critical solution temperature (UCST: case of poly(ether imide), PEI).[2-7] Depending on the composition of the quasi-binary blend, the metastable or unstable region will be entered during this process, in principle resulting in the nucleation and growth or spinodal

 DOI: 10.1002/masy.200350831

decomposition mechanism, respectively. However, especially in the case of these high-T_g modifiers, thermodynamics is merely the driving force for phase separation, while the kinetics of interdiffusion between the coexisting phases will become determining at some point.[8,9] Crucial in this respect is the ratio between the rate of phase separation and that of chemical reaction. For a certain modifier chemistry and composition, the cure temperature (T_{cure}) will be the most important parameter influencing both rates at the same time.[10] The interdependence of these rates reveals the complexity in controlling the morphology in these systems. An increase in T_{cure} in a system showing LCST type demixing behavior, for example, will, together with the resulting lower conversion at phase separation (x_{ps}), increase the interdiffusion rates at x_{ps}. However, the corresponding higher cure rate also results in a faster molecular weight increase of the epoxy-amine beyond this point, reducing the diffusion rate of the components in the reactive mixture. Additional points of interest are the gel point of the epoxy-amine and possible vitrification of either of the coexisting phases. The former corresponds to the onset of network formation of the epoxy-amine rich phase, arresting phase separation.[11] The latter occurs in case T_{cure} is below the full cure glass transition of either the epoxy-amine rich phase, T_{gfull} or the modifier-rich phase, T_{gMOD}. Vitrification of either or both these phases can fix the morphology at a certain conversion, thus freezing in a thermodynamically unstable polymer system. To clarify these aspects, Gillham proposed cure diagrams to optimize processing and material properties of epoxy resins.[12] By including the onset of phase separation, the succession of chemical and physical events during isothermal RIPS can be predicted.[13,14] These time-temperature-transformation (TTT) and time-conversion-transformation (TxT) diagrams have to be used with caution, since, in principle, the two phases originating at the cloud point have their own cure diagrams. Thus, cure diagrams can be used to pinpoint the crucial events during RIPS, but adequate control over the morphology requires a measuring technique capable of studying the evolution in chemorheology in situ. Modulated Temperature DSC (MTDSC) has the potential to measure both the reaction advancement and morphological evolution in one experiment with excellent temperature control.[15,16] This was illustrated for PES-modified epoxy-amine systems, for which RIPS can be detected *indirectly* as a stepwise decrease of the heat capacity due to vitrification of the segregating PES-rich phase. The applicability of MTDSC for in situ monitoring of cure advancement (non-reversing heat flow signal) and chemorheological effects (heat capacity signal) will be extended in this work, in combination with optical microscopy as a complementary technique.

Experimental

Two epoxy-amine systems were studied: a bifunctional epoxy, diglycidyl ether of bisphenol A (DGEBA, Epon825 from Shell), with an epoxy equivalent weight of 180 $g.eq^{-1}$, in combination with appropriate amounts of a bifunctional amine (aniline from Fluka), or a tetrafunctional amine hardener (methylene dianiline, MDA from Janssen Chimica), with an amine equivalent weight of 46.5 $g.eq^{-1}$ and 49.5 $g.eq^{-1}$, respectively. Mixtures of these pure systems were modified, respectiviely, with 20 wt% and 50 wt% of low molecular weight, M_w=20,000 $g.mol^{-1}$, poly(ether sulphone) (PES from Aldrich).

Cure experiments were performed on a TA Instruments 2920 DSC with MDSC™ option and a Refrigerated Cooling System (RCS). Helium was used as a purge gas (25 $ml.min^{-1}$). Cure was performed in hermetic aluminum pans (TA Instruments), with sample weights between 5 and 10 mg. Modulation conditions were: amplitude of 1°C and a 60s period. Cloud points were detected by measuring the light transmitted through thin samples (held between glass slides) in a Mettler Toledo FP82HT hot stage equipped with a photodetector. A Spectratech optical microscope was used with a magnification of 4.

Results and Discussion

RIPS as Measured by Vitrification of the PES-Rich Phase

RIPS in the epoxy-amine systems DGEBA+aniline and DGEBA+MDA, modified with the high-T_g engineering thermoplastic PES, can be probed as vitrification of the PES-rich phase.[15,16] The former system modified with 20wt%PES is shown at three cure temperatures in Figure 1. The non-reversing heat flow and the initial increase in the heat capacity signal contain complementary information on the epoxy-amine reaction.[15] The decrease in the heat capacity signal accompanied by a relaxation peak in the heat flow phase marks the time at which the T_g of the PES-rich phase (T_g pure PES =223°C) rises above the cure temperature and the mobility of this phase is frozen in. In case the cure temperature is below the full cure glass transition of the epoxy-amine (T_{gfull}=94°C), vitrification of the segregating phase rich in this component can also take place. This clearly occurs in Figure 1 for the reaction at 80°C, while only partial vitrification is seen at 90°C. Upon curing an unmodified epoxy-amine in a temperature range corresponding to the glass transition region of the fully cured system this effect was also seen.[15]

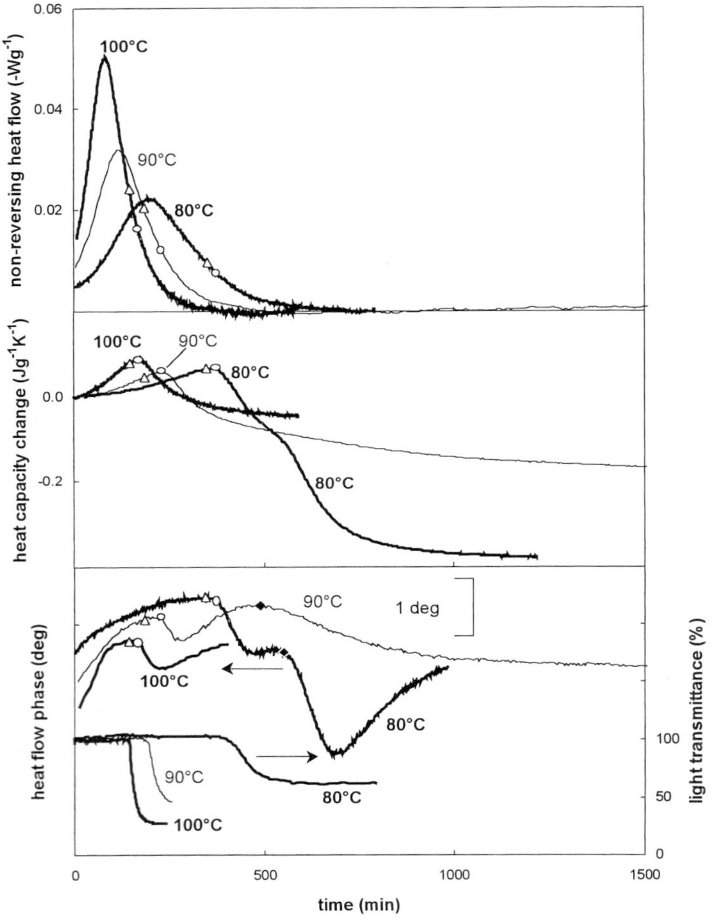

Fig. 1. Non-reversing heat flow, heat capacity change and heat flow phase signal from MTDSC and % of light transmittance from optical microscopy (OM) for the reactive blend DGEBA+aniline(r=1)/20wt%PES cured at 100°C, 90°C and 80°C; the cloud point from OM (Δ), onset of heat flow phase relaxation corresponding to vitrification of PES-rich phase (○) and epoxy-amine-rich phase (♦) are also shown (see text).

The fact that RIPS is not measured directly with MTDSC in this system can be inferred from light transmittance experiments, also performed at 80°C, 90°C and 100°C in Figure 1: the onset of phase separation (cloud point) obtained in this way (Δ) occurs prior to the onset of vitrification of the PES-rich phase calculated from the heat flow phase (○). The latter signal is also used to pinpoint the vitrification of the epoxy-amine rich phase (♦).

Time-Conversion-Transformation (TxT) Diagram for DGEBA+Aniline/20wt%PES

Partial integration of the non-reversing heat flow signal gives the epoxy conversion corresponding to the three major events discussed in Figure 1. A wide range of cure temperatures (60°C-155°C) was studied in this way to construct the TxT diagram of the DGEBA+aniline system modified with 20wt% PES (Figure 2). Cure temperatures higher than 140°C were measured without modulation to decrease the equilibration time for these fast cures. Therefore only the cloud point conversion could be calculated. The negative slope of this curve (Δ) holds proof of the LCST-type demixing behavior of this system: for a higher cure temperature the onset of phase separation will be encountered at lower conversions.[1,10,17]

The vitrification line of the epoxy-amine-rich phase has also been included in Figure 2 (♦) and reflects the expected T_g-x behavior.[12] Further phase separation accompanied by morphology changes will be inhibited beyond this point. This line is also shown for the unmodified epoxy-amine system (×) and matches that of the modified one, meaning that the epoxy-amine-rich phase contains little or no PES. Otherwise, this high-T_g component would increase the glass transition of this phase and decrease the conversion at which vitrification sets in.

In contrast to the cloud point curve, a positive slope is also found for the vitrification line of the PES-rich phase (○). This can be understood by realizing that the T_g of this phase is determined by its *composition* and by the T_g of its constituents. At higher cure temperatures, vitrification occurs for a phase richer in PES. This means that the LCST will have decreased more to achieve purer coexisting phases in comparison to lower cure temperatures. In case a pure PES-phase would be formed at phase separation, vitrification of this phase would occur immediately, resulting in a coinciding cloud point curve and vitrification line.

While this information could also be obtained from the combination of dynamic rheometry or dynamic mechanical analysis, providing the onset of vitrification, and DSC, providing the conversion,[18] the advantage of using one technique that can measure both is evident. By using these complementary techniques, the TxT diagram of an epoxy-amine system modified with PEI has been obtained, where an UCST-type demixing behavior is observed.[13] In this case both the cloud point curve and the vitrification curve were found to exhibit a positive slope, characteristic for this specific demixing behavior.

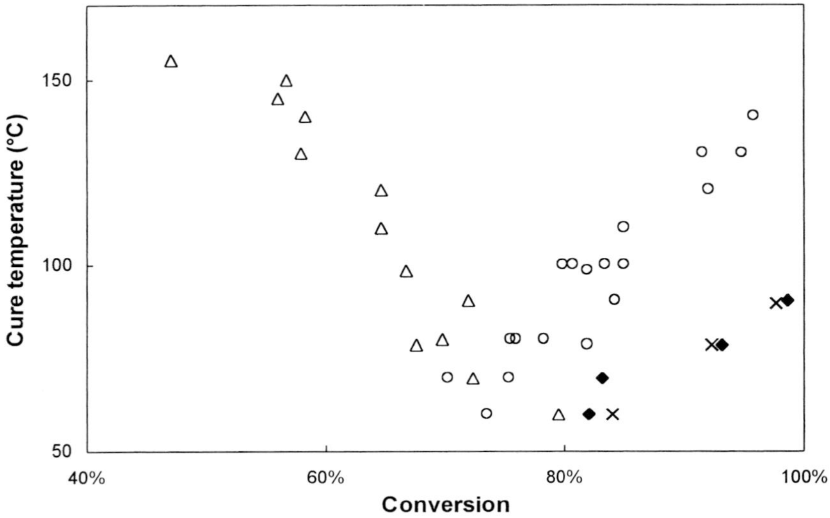

Fig. 2. TxT diagram for a stoichiometric DGEBA+aniline/20wt%PES mixture with: cloud points from optical microscopy (Δ), onset of heat flow phase relaxation corresponding to vitrification of PES-rich phase (○) and epoxy-amine-rich phase (♦); the cloud point at 60°C was obtained by using UV/Vis spectrometry in transmittance at 450nm; the onset in the heat flow phase relaxation corresponding to vitrification of the pure DGEBA+aniline(r=1) is given for reference (×).

The combination of an almost pure epoxy-amine rich-phase and a PES-rich phase changing in composition, reveals the asymmetry of the phase diagram: one branch will be close to the pure epoxy-amine, while the other branch will gradually evolve to purer PES as conversion rises. This has also been simulated within the framework of the Flory-Huggins theory and is related to the difference in molecular weight between PES and the epoxy resin.[7] Phase diagrams constructed for mixtures of DGEBA and PES with different molecular weights also confirm this idea.[2,20] Moreover, non-isothermal post-cure experiments after isothermal cure above T_{gfull} have already shown that the epoxy-amine-rich phase does not attain a glass transition higher than 95°C [16]. Curing at higher temperatures just results in a higher fraction of this phase. While the proposed TxT diagram is essential to control the curing of blends with appropriate pre-curing and post-curing schedules, more information is present about the time-dependency of phase separation and morphology development in the MTDSC experiments itself. This can be seen, for example, in the slow heat capacity and heat flow phase evolutions with time in Figure 1.

Network Forming DGEBA+MDA/50wt%PES: Effect of Reactive Mixture Composition

Reactivity Increase as a Probe for Phase Separation

When epoxy-amine network systems are cured with higher thermoplastic contents, sudden reactivity increases were found at phase separation.[21,22] This can be seen in the DGEBA+MDA (T_{gfull}=175°C for r=1) system, modified with 50wt% PES in Figure 3. In case a stoichiometric mixture of the epoxy-amine is introduced ①, RIPS results in a shoulder in the non-reversing heat flow signal after 20 and 7 min for T_{cure} equaling 130°C and 150°C, respectively. Note that high cure temperatures had to be used here to induce phase separation at low enough conversions. Otherwise vitrification of the relatively high-T_g homogeneous reactive blend occurs prior to the cloud point, and no RIPS can be seen (not shown). The disadvantage of using these high cure temperatures is, however, that the first collected data points are unreliable (dashed line in Figure 3). Instead of using different cure temperatures, the rate of phase separation and reaction can be changed more independently by using different mixture compositions for the reactive epoxy-amine component. This only holds in case this change in epoxy-amine ratio does not affect the position and shape of the phase diagram too much, resulting in a similar driving force at the onset of phase separation, while the reaction rate will increase for higher amounts of the amine. Note, however, that adding an excess of the amine reactive component could increase the miscibility with PES and therefore delay the onset of RIPS. Three mixture compositions of DGEBA+MDA are shown in Figure 3, all cured at 150°C. A decrease in reactivity is seen in the order: excess amine (r=1.4, ②), stoichiometric mixture (r=1, ①) and excess epoxy (r=0.7, ③). The shoulder in the non-reversing heat flow, indicating a reactivity increase accompanying RIPS, is absent in the first mixture.

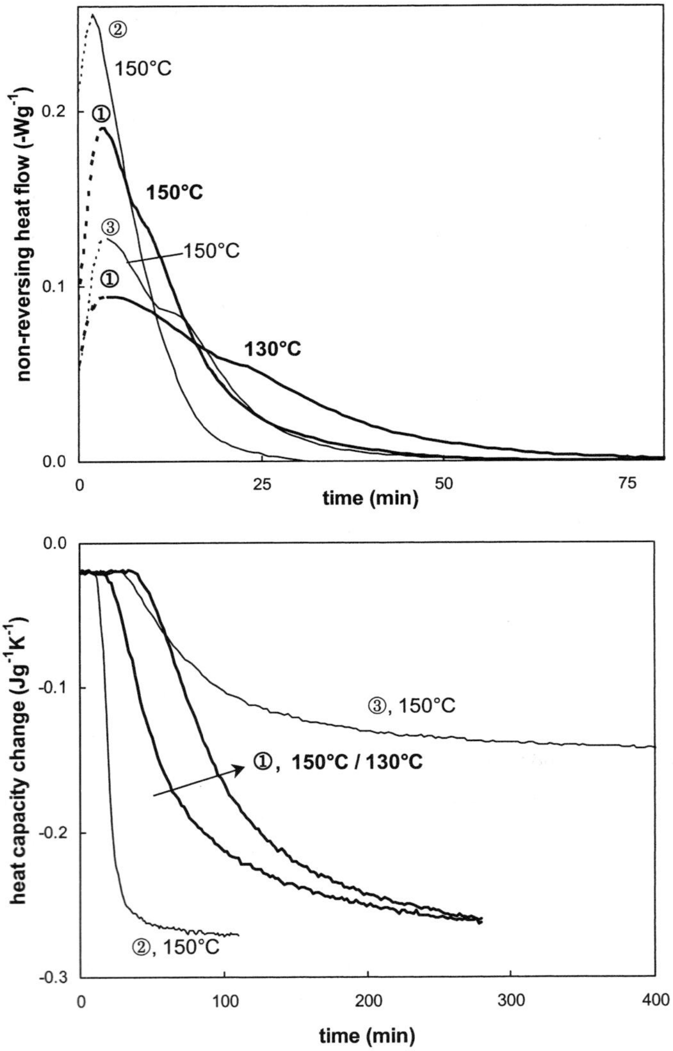

Fig. 3. Non-reversing heat flow and change in heat capacity (note the longer time scale) for the reactive blend DGEBA+MDA/50wt%PES: stoichiometric mixtures of the epoxy-amine are cured at 150°C and 130°C (r=1: ①); an excess amine (r=1.4: ②) and excess epoxy (r=0.7: ③) mixture is cured at 150°C; dashed line corresponds to the time the sample temperature needs to reach the cure temperature within 0.5°C.

Chemorheological Information Obtained in quasi-Isothermal Conditions

The change in heat capacity, also shown in Figure 3, does not exhibit the double step-wise decrease as obtained in the linearly polymerizing system in any of the studied compositions. The fact that the glass transitions of both phases are close together could be responsible for this insensitivity in the stoichiometric system, which was already commented for the same epoxy-amine modified with 20wt% PES.[16] The glass transition of PES (223°C) can be compared to the full cure glass transition of both epoxy-amine systems (T_{gfull}): 175°C for the network-forming system in comparison to 95°C for the linearly polymerizing one. On the other hand, the lower glass transitions of both off-stoichiometric systems means that no vitrification of a *pure* epoxy-amine phase can occur: T_{gfull} for the excess epoxy and amine mixture is 91°C and 140°C, respectively (compare with T_{cure}=150°C). The total decrease in heat capacity can shed some light on these issues. This value is related to the amount of material that has vitrified during the isothermal cure step at, for example, 150°C. For the stoichiometric system with 50wt%PES, this corresponds to vitrifying almost all the material in the mixture. In case of the excess epoxy system, a close correspondence is found with vitrification of the PES phase only. When the excess amine is used, however, the decrease in heat capacity is more than what would be expected for the PES phase alone. This therefore indicates vitrification of a homogeneous, partially reacted mixture of epoxy-amine and PES having a higher glass transition in comparison to the unmodified system.

Derivative of the Heat Capacity Signal in Relation to the in situ Formed Morphology

To clarify the isothermally obtained information, Figure 4 depicts the derivative of the heat capacity signal as a function of temperature for different preceding cure schedules. This method was already proposed to study interphase behavior in interpenetrating polymer networks.[23,24] After the isothermal cure of the stoichiometric system at 150°C this signal has a maximum at 172°C with a shoulder about 10°C lower (a, ①), indicating two phases with a similar glass transition and thus composition. To initiate phase separation at a lower conversion, thus increasing diffusion rates, cure can be performed non-isothermally at a high heating rate (c, ①). It is clear that this method results in a better segregation of the coexisting phases, pointing out the diffusion rate as the rate-determining step. Another issue which is frequently addressed in literature is that the network fixes the morphology, therefore freezing in a thermodynamically unstable system.[1,2] This can be easily shown here by adding a post-cure step at 230°C to the system which was isothermally cured at 150°C and measuring the resulting thermal properties (b, ①). While a better separation of both T_g's is achieved after this post-cure step, the system does not attain the segregation of the non-isothermal cure. Thus looking at the derivative of the heat capacity signal can be useful in the design of cure schedules to achieve a desired end material in view of interphase and morphology. As indicated, curing at lower temperatures increases the interphase, which corresponds to a decrease in particle size[25] For the excess amine system, the non-isothermal post-cure experiment in Figure 4 after the isothermal cure at 150°C only shows one glass transition at 172°C (a, ②), corresponding to the absence of phase separation as envisioned before. The higher reactivity seen in Figure 3 can be held responsible for a reduction in the time available for the development of phase separation. Another possible explanation would be that diffusion rates are lower in comparison to the stoichiometric system, which does show phase separation. This can however be negated by realizing that the highest glass transition is found for the stoichiometric DGEBA+MDA system. As indicated before, a better miscibility of the excess amine mixture could also postpone RIPS in this system. By applying the non-isothermal cure, phase separation can be initiated at a higher temperature (c, ②) and thus lower conversion, resulting in higher diffusion rates. For the excess epoxy system, a much lower glass transition in combination with the lower reactivity will respectively result in higher diffusion rates and more time for morphology evolution at 150°C. The resulting marked segregation is translated in two distinct glass transitions already seen after the isothermal cure at 150°C (a,③: at 110°C and 160°C). Note that while a T_{gEPAM} higher than T_{gfull} is found, due to residual PES present, no vitrification of this phase can occur at 150°C, as concluded from Figure 3. Other cure schedules (non-isothermal at 10°Cmin^{-1} and post-cure at 230°C after cure at 150°C) are also shown here, again pointing out the importance of designing cure schedules with the required morphology in mind. The fact that both schedules result in similar end properties for the excess epoxy means that further interdiffusion to attain the thermodynamically stable end state is not restricted to the same extent as in the other mixtures. The lower crosslink density of this system can be responsible for this effect.[26]

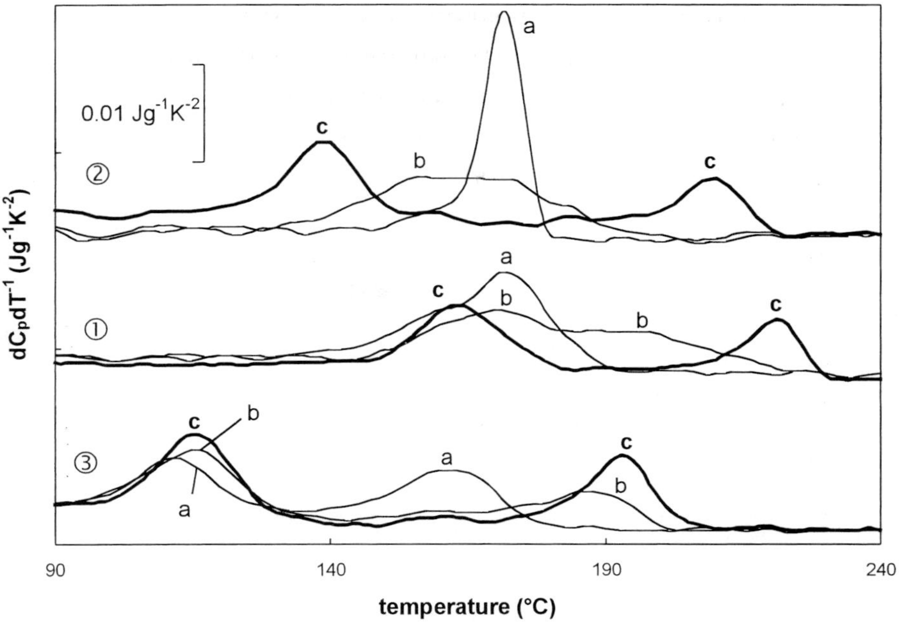

Fig. 4. Derivative of heat capacity to temperature (dCpdT^{-1}) for non-isothermal post-cure measurements at 2.5°Cmin^{-1} following: isothermal cure at 150°C (see Figure 3) (a); isothermal cure at 150°C followed by post-cure at 230°C for 30 min (b) and non-isothermal cure at 10°Cmin^{-1} to 230°C (c); studied systems: stoichiometric (r=1: ①), excess amine (r=1.4: ②) and excess epoxy (r=0.7: ③) DGEBA+MDA/50wt%PES; T_{gfull} equals 175°C, 137°C and 95°C for r=1, r=1.4 and r=0.7, respectively.

Conclusions

The TxT diagram of RIPS in the linearly polymerizing DGEBA+aniline modified with 20wt% PES can be constructed by using MTDSC in conjunction with optical microscopy. This diagram can be used to design appropriate cure schedules. Vitrification times of both the high-T_g PES-rich phase and the low-T_g epoxy-rich phase can be obtained in quasi-isothermal conditions from step-wise decreases in the heat capacity signal, while the corresponding conversions are calculated from the simultaneous non-reversing heat flow information. The cloud point exhibits a negative slope, confirming the LCST-type demixing behavior, while a positive slope is found for the vitrification lines of the PES-rich and epoxy-rich phases, corresponding to their T_g-x behavior. Vitrification lines for the low-T_g phase and the unmodified epoxy-amine coincide, indicative for an asymmetric phase diagram with one branch close to the pure epoxy-amine component.

Both the time-dependency of phase separation and the morphology advancement influence the evolution in the heat capacity signal. This was further illustrated in the network forming system DGEBA+MDA/50wt% PES where the amount of material that freezes in during RIPS can be estimated. When the amine is used in excess (T_{gfull}=140°C), more material is found to freeze in at 150°C than the fraction expected in case only the PES-rich phase would vitrify. The high reactivity of this system, however, does not allow enough time for phase separation to occur. A postponed RIPS can also be held responsible for this effect. Thus, vitrification of a higher T_g homogeneous reactive blend takes place. Lower reactivities are found for the stoichiometric and excess epoxy mixtures (also cured at 150°C), where the non-reversing heat flow does show a reactivity increase in isothermal conditions, indicating a concentration effect of reactive groups at RIPS. The derivative of the heat capacity signal in non-isothermal post-cure experiments confirms the statements made in this paragraph and can also be used to study the effect of cure temperature on the final morphology.

The proposed methodology in this work is further extended for low-T_g triblock copolymer modifiers, where higher interdiffusion rates result in an additional excess contribution in the heat capacity signal.[27] In this way MTDSC can be used as a direct probe for phase separation, instead of measuring an indirect effect like vitrification of the modifier-rich phase or noticing a reactivity increase at the cloud point.

[1] Williams RJJ, Rozenberg BA, Pascault J-P. Advances in Polymer Science 1997;95:128.
[2] Yamanaka K, Inoue T. Polymer 1989;30:622.
[3] Riccardi CC, Borrajo J, Williams RJJ, Girard-Reydet E, Sautereau H, Pascault J-P. Journal of Polymer Science: Part B 1996;34:349.
[4] Jansen BJP, Meijer HEH, Lemstra PJ. Polymer 1999;40:2917.
[5] Yamanaka K, Takagi Y, Inoue T. Polymer 1989;60:1839.
[6] de Graaf LA, Hempenius MA, Möller M. Polymer Preprints 1995;36:787.
[7] Clarke N, McLeish TCB, Jenkins SD. Macromolecules 1995;28:4650.
[8] Kiefer J, Hilborn JG, Hedrick JL. Polymer 1996;37:5715.
[9] Rajagopalan G, Gillespie JW, McKnight SH. Polymer 2000;41:7723.
[10] Tran-Cong Q, Shibayama M. In:"Structure and Properties of Multiphase Polymeric Materials", Ed. Araki T, Marcel Ackker, 1998.
[11] Kim BS, Chiba T, Inoue T. Polymer 1993;34:2809.
[12] Gillham JK, Enns JB. Trends in Polymer Science 1994;2:406.
[13] Girard-Reydet E, Vicard V, Pascault J-P, Sautereau H. Journal of Applied Polymer Science 1997;65:2433.

[14] Boogh L, Pettersson B, Månson J-A. Polymer 1999;40:2249.
[15] Swier S, Van Assche G, Van Hemelrijck A, Rahier H, Verdonck E, Van Mele B. Journal of Thermal Analysis 1998;54:585.
[16] Swier S, Van Mele B. Thermochimica Acta 1999;330:175.
[17] Inoue T. Progress in Polymer Science 1995;20:119.
[18] Ishii Y, Ryan AJ. Macromolecules 2000;33:167.
[19] Bucknall CB, Gomez CM, Quintard I. Polymer 1994;35:353.
[20] Bucknall CB, Gomez CM, Quintard I. Polymer 1994;35:353.
[21] Bonnet A, Pascault J-P, Sautereau H, Taha M., Camberlin Y. Macromolecules 1999;32:8517.
[22] Ritzenthaler S, Girard-Reydet E, Pascault JP. Polymer 2000;41:6375.
[23] Hourston DJ, Song M, Pollock HM, Hammiche A. Journal of Thermal Analysis 1997;49:209.
[24] Hourston DJ, Song M, Schafer F-U, Pollock HM, Hammiche A. Polymer 1999;40:4769.
[25] Martinez I, Martin MD, Eceiza A, Oyanguren P, Mondragon I. Polymer 2000;41:1027.
[26] Oleinik EF. Advances in Polymer Science 1986;80:49.
[27] Swier S, Van Mele B. submitted in Polymer.

Dielectric In Situ Sensor Monitoring of Phase Separation and Changes in the State of Each Phase

D. Kranbuehl,[1] J. Rogozinski,[1] J. Warner,[1] P. Best,[1] J.P. Pascault,[2] H. Sautereau,[2] G. Boiteux,[3] G. Seytre[3]*

[1]Departments of Chemistry and Applied Science, College of William and Mary, Williamsburg, VA, USA
[2]Laboratoire des Matériaux Macromoléculaires
[3] Laboratoire des Matériaux Polymères et Biomatériaux, UMR-CNRS 5627 Villeurbanne Cedex, France

Summary: Frequency dependent dielectric measurements have been used to monitor and characterize the phase separation process and changes in state of each phase. The measurements are made in situ using a micro planar sensor. They can be made both in the laboratory as well as in an industrial production or use environment. Two examples are presented. The first is monitoring the onset of phase separation, the buildup in Tg and change in composition of each phase during "reactive processing" of a high performance thermoplastic (TP) PPI, thermoset precursors (TS) DGEBA-MCDEA intially homogeneous blend. The second example involves monitoring the stability, onset of phase separation, as a function of temperature on a mineral oil, stearyl alcohol, water, sufactant emulsion used in the cosmetic industry.

Keywords: blend, cure monitoring, dielectric, PEI; reactive processing, thermoset-thermosplastic

Introduction

Thermoset (TS)/Thermoplastic(TP) blends are materials resulting from the mixing of the TP polymer with the TS precursor (generally a diepoxy-diamine system).[1-8] Usually the mixture is initially homogeneous. With time and a temperature below the Tg of the TP, there is an increase in the molar mass of the TS precursor as the reaction advances. A liquid-liquid phase separation occurs at a given conversion. The phase separation process generally continues until the TP rich phase becomes rich enough to form a glass. The overall process is stopped when the TS is fully reacted or both phases have achieved the glassy state.

These TS/TP blends have two main applications 1) toughness improvement of TS networks with high performance ductile thermoplastics and 2) new processing routes which reduce the initial viscosity of otherwise intractable high temperature TP polymers such as polyphenylene ether and

 DOI: 10.1002/masy.200350832

polyetherimides. The first application occurs at low concentrations (<20 wt %) of TP. The second application involves TP concentrations generally greater than 30 wt %. In this case phase inversion occurs resulting in a TP rich continuous (β) phase with occluded TS rich (α) phase particles.

Experimental

The TS precursor consisted of an epoxy prepolymer and diamine curing agent. The epoxy prepolymer used was a diglycidyl ether of bisphenol, a liquid at room temperature and with a low degree of polymerization, DGEBA $\bar{n} = 0.15$. (Ref. LY556 from Ciba Geigy). The curing agent was an aromatic diamine with a low reactivity, 4,4' methylene-bis (3-chloro-2,6-diethylaniline), MCDEA supplied by Lonza. The diamine was used at the stoichiometric ratio, epoxy to amine hydrogen groups equal to 1. The TP was Polyetherimide, PEI Ultem 1000 supplied by General Electric. The structures are shown in Figure 1. Blends with concentration higher than 30wt% of TP (33, 48 and 64 wt % of PEI) were prepared using a co-rotating twin screw extruder in a one stage process.

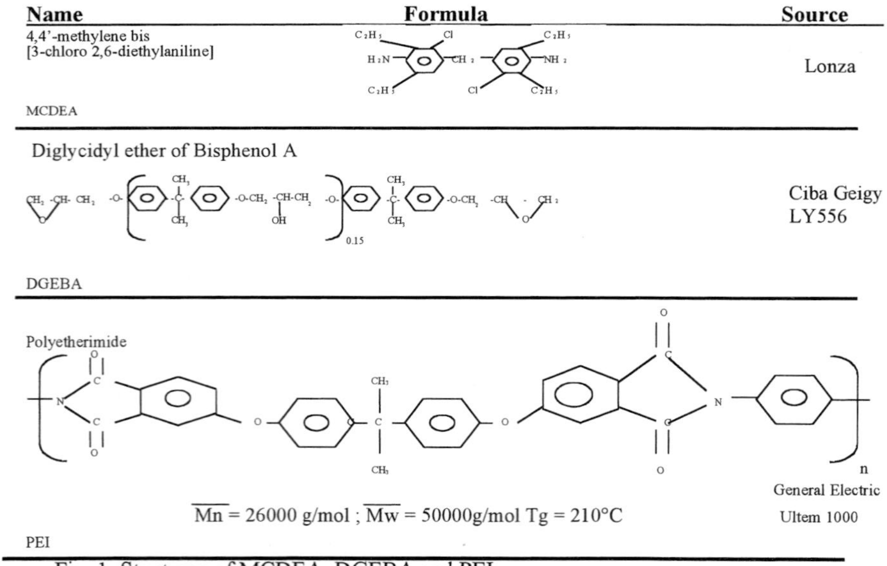

Fig. 1. Stuctures of MCDEA, DGEBA and PEI.

The emulsion, used in the cosmetics industry and supplied by ICI Wolrdwide Industries, was a non reactive mixture of a mineral oil, stearyl alcohol and two commercial surfactants Brij 721 and Brij 72 in water.

Background

Frequency dependent measurements of the dielectric impedance of a material as characterized by its equivalent capacitance, *C*, and conductance, *G*, are used to calculate the complex permittivity, $\varepsilon^*=\varepsilon'-i\varepsilon''$, where $\omega=2\pi f$, f is the measurement frequency and C_o is the equivalent air replacement capacitance of the sensor.[9,10]

$$\varepsilon'(\omega) = \frac{C(\omega)\,\mathrm{material}}{C_o} \qquad \varepsilon''(\omega) = \frac{G(\omega)\,\mathrm{material}}{\omega\, C_o} \tag{1}$$

This calculation is possible when using the planar sensor with interdigitated gold electrodes.[9,10] Thereby it's geometry is invariant over all measurement conditions. Both the real and the imaginary parts of ε^* can have dipolar (d) and ionic (i) charge components.

$$\varepsilon' = \varepsilon'_d + \varepsilon'_i \qquad \varepsilon'' = \varepsilon''_d + \varepsilon''_i \tag{2}$$

Plots of the product of frequency (ω) multiplied by the imaginary component of the complex permittivity $\varepsilon''(\omega)$ make it relatively easy to visually determine when the low frequency magnitude of ε'' dominated by the mobility of ions and when at higher frequencies the rotational mobility of bound charge dominates ε''. Generally, the magnitude of the low frequency overlapping values of $\omega\varepsilon''(\omega)$can be used to measure the change with time of the ionic mobility through the parameter σ where

$$\sigma\ (\mathrm{ohm^{-1}cm^{-1}}) = \varepsilon_o\omega\varepsilon''_i\ (\omega) \tag{3}$$

$$\varepsilon_o = 8.854 \times 10^{-14}\ \mathrm{C^2\ J^{-1}\ cm^{-1}}$$

The changing value of the ionic mobility is a molecular probe which can be correlated with viscosity in simultaneous laboratory experiments. In turn this calibration of σ with viscosity can be used to monitor changes in the viscosity by measuring $\sigma(t)$ during cure.

The dipolar component of the loss at higher frequencies can the be determined by subtracting the ionic component. The peaks in ε" dipolar (which are usually close to the peaks in ε") can be used to determine the time or point in the cure process when the "mean" dipolar relaxation time has attained a specific value $\tau = 1/\omega$. The dipolar mobility, as measured by the mean relaxation time τ, is monitoring the α–relaxation process associated with vitrification and it can be used as a molecular probe of the buildup in T_g. For example, the time of occurrence of a given dipolar α- relaxation time as measured by a peak in a particular high frequency value of $\varepsilon''(\omega)$ can be quantitatively related to the attainment of a specific value of the glass transition temperature.
Interfacial polarization processes which occur in heterogeneous dielectrics will be present at the onset of phase separation, that is beginning at the onset of the transition from a homogeneous to a heterogeneous two phase system.[11-13] These effects arise at the interface of two media having differing permittivities and conductivities. These interfacial effects can be quite strong sometimes creating increases in ε' by factors of 10 and a charge polarization relaxation time. The effect is dependent on the magnitude of the conductivity difference (σ_2-σ_1) between the occluded phase σ_2, and the continuous phase, σ_1.
The most well known theoretical model of this effect is the Maxwell-Wagner-Sillars model.[11-13] Using the Maxwell-Wagner-Sillars model, the complex dielectric permittivity of the mixture $\varepsilon^*(\omega)$, of orientated occluded ellipsoids with complex dielectric constant, $\varepsilon^*_2(\omega)$, at a volume fraction v_2 dispersed in a continuous matrix (v_1) with a complex dielectric constant, $e^*_1(\omega)$, can be calculated from the following equation:

$$\varepsilon^* = \varepsilon^*_1 \frac{\varepsilon_1^*(1-v_2)(1-A)+\varepsilon_2^*[v_2+A(1-v_2)]}{\varepsilon'_1+A(1-v_2)(\varepsilon_2^*-\varepsilon_1^*)} \qquad \mathbf{(4)}$$

where A($0 \leq A \leq 1$), is the depolarization factor of the ellipsoidal filler particles. It depends on the shape of the particles (length of the long a to b short axis ratio for spheroids) and of the orientation of the field relative to the particle.

Discussion

Figure 2a shows for 33% PEI blend the changing value of ε′ during the 135°C cure over the frequency range of .005, .05, .125, .5, 5, 25, 50, and 250 KHz. The lowest frequency is the top line. The value of ε′ shows a sharp rise at 128 minutes indicating the onset of phase separation.

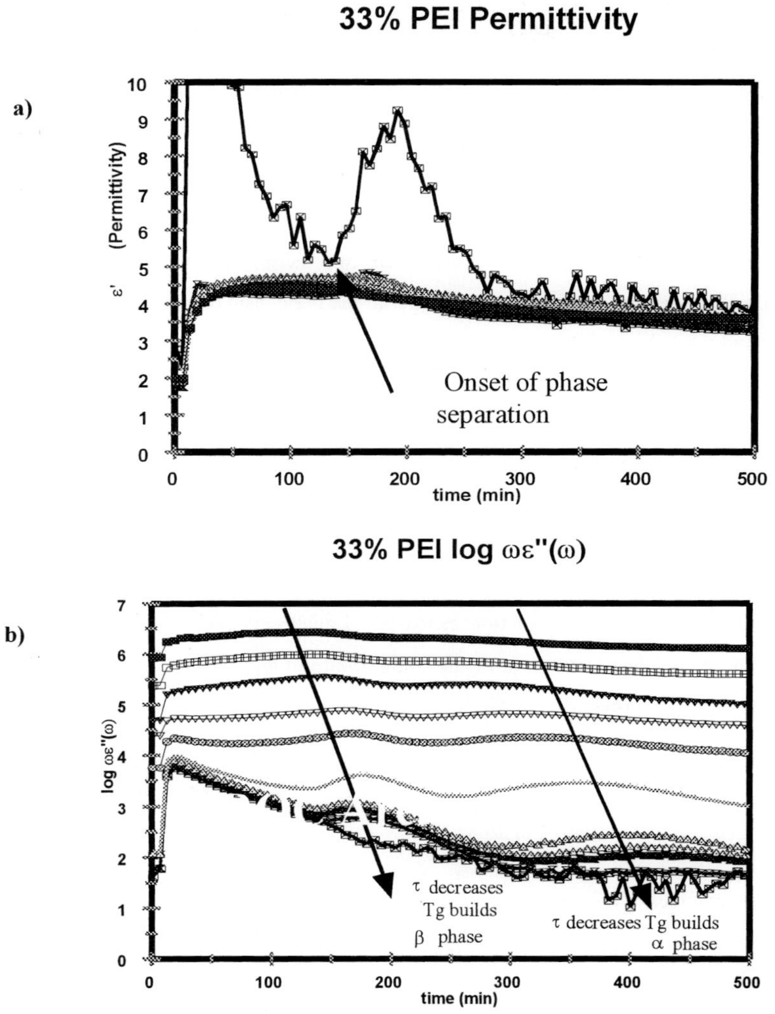

Fig. 2. a) Dielectric Sensor output during cure at 135°C; b) Dielectric sensor output during cure at 135°C, highest frequency is top line.

Figure 2b displays for the 33% PEI blend the loss component ε'' of the permittivity scaled by frequency with time during isothermal cure at 135°C. Two sets of peaks in ε'' occur. The first set occurs both before and after phase separation. They occur at the higher frequencies before phase separation and at the lower frequencies after phase separation. A second set of peaks starting at the higher frequencies occur only after phase separation. These peaks can be used to determine the achievement of a relaxation time τ where $\tau = (2\pi f)^{-1}$ at the time of a peak in $\varepsilon''(\omega)$ at that frequency, f. The imaginary loss, component ε'' shows overlapping values at the lower frequencies. These values of ε'' can be used to determine a frequency independent conductivity using equation 3. In the following discussion these two dielectric quantities (τ, σ) and the viscosity (η) will be measured and analyzed along with the previously reported values of extent of reaction x and the values of T_g to characterize the phase separation behavior of the reactive TP/TS blend. The relations can be used to build a fundamental basis from which the transformations of the reactive TP/TS blend can be monitored in situ during polymerization-fabrication using dielectric measurements.

Before phase separation. The TP/TS blend is a homogeneous mixture in which one observes, at different levels of TS precursor dilution, the changing conductivity of the epoxy-amine i-mers as the viscosity and the extent of reaction increases. Equation 3 was used to calculate the conductivity from the dielectric data. In Figure 3, the values of $\log(\sigma(t))$ are plotted versus the log of viscosity, reported previously.[6,7] This correlation in figure 3 permits the sensor monitored value of $\sigma(t)$ before phase separation to be used as a means to monitor the changing values of viscosity prior to phase separation. Examples of the relationship between the changing $\sigma(t)$ versus x(t) and viscosity are reported in Table 1.

Table 1. Slopes, conductivity, viscosity and extent of reaction.

% PEI	Slopes		
	log (σ) vs x	log (η) vs x	log (σ) vs log (η)
0	-1.6	3.2	-0.7
33	-4.9	4.8	-1.3
48	-6.9	3.9	-1.8

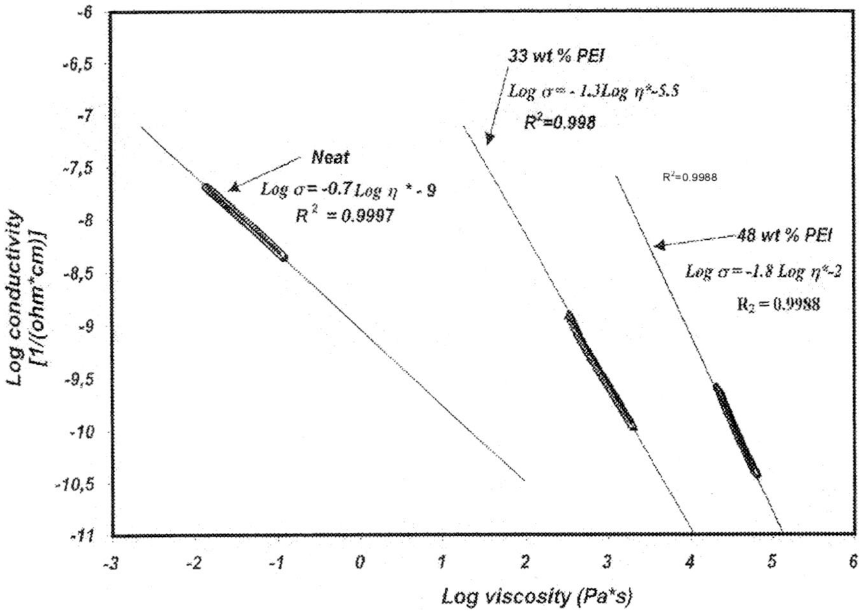

Fig. 3. Log(conductivity) versus log(viscosity).

The onset of phase separation is characterized by the creation of an interface between forming phases. From the point of view of dielectric behavior, the initial onset of the phase separation process has the potential to generate a large change in the permittivity ε' since interfacial surfaces with charge separated by molecular distances represent a large increase in capacitance i.e., $C_o \approx$ area/distance. Furthermore, once the phase separation process has generated an occluded phase of conductive epoxy rich i-mers, the % phase, surrounded by a PEI rich nonconducting continuous phase, there is the potential as previously described for a Maxwell-Wagner-Sillars relaxation effects. Figure 2a (33 wt% PEI) clearly shows a large sudden increase in the value of ε' at 128 minutes. Another sample showed this rise at 105 minutes. The cloud point was observed at 118 minutes. For the 48% PEI blend the sharp rise in ε' occurred at 211 and 202 minutes in two separate runs. For the 48% PEI system the cloud point was observed at

235 minutes. Rheological measurements showed the phase separation transition at 113 minutes for 33% PEI and 216 minutes for 48% PEI [6,7]. Clearly as summarized in Table 2 the onset of the dielectric interfacial polarization, as seen by a sharp rise in ε', is a good instrumental means to monitor in situ with a capacitance sensor the onset of the phase separation process.

Table 2. Comparison of detected times (min)of phase separation for PEI / TS blends.

	33 wt. %	48 wt. %
Dielectric	115	205
cloud point	118	235
DSC (Ref 8)	118	235
Rheology	113	216

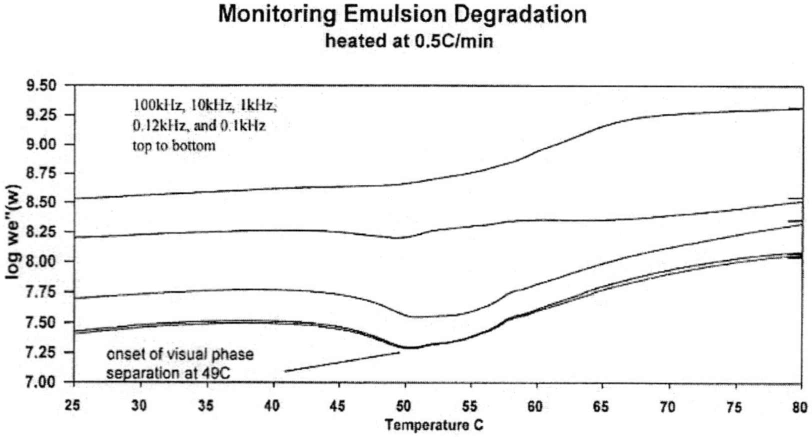

Fig. 4. Monitoring Emulsion Degradation.

Determination of the stability of emulsions is an important property for a wide variety of industries, particularly for cosmetics. Often the temperature or aging time at which the emulsion phase separates is determined visually. Dielectric sensing is an automated instrumental method which can also detect the maximum stable temperature zone for a particular surfactant system. Figure 4 displays the changing value of $\log[\omega * \varepsilon''(\omega)]$ versus temperature at .1, .12, 1, 10, and 100 kHz. A distinct drop in the otherwise expected increase in ε″ occurs at 45°C. In the 45°C to 50°C region the emulsion is phase separating, initially into a continuous phase with a lower dielectric loss, but which experiences a larger temperature dependence during the 0.5°C / minute temperature ramp. The phase separation process was detected visually in the 50 to 53 degree temperature range. Dielectric sensing detects changes on molecular level and when the A/d ratio of the interface is large such as at the earliest stages of phase separation. For these reasons changes in ε″ are detected before phase separation is observed visually.

Relaxation peaks. They are observed in Figure 2b both before, during and after the onset of phase separation. The focus of this discussion is an analysis of these dipolar peaks and their basis as a consequence of the phase separation MWS process, the buildup in T_g of the PEI rich continuous β-phase and the buildup in T_g of the epoxy rich occluded α-phase. From the beginning, before, during and after the onset of phase separation, the epoxy groups in the homogeneous, and in both the α and β phases continue to react. With increasing extent of reaction more epoxy i-mers become insoluble in the PEI and diffuse in to the α, occluded epoxy rich phase. As the TS i-mers diffuse out of the β-phase, the β-phase becomes richer in PEI. As a result the T_g of the continuous β–phase increases. Since T_g of the neat PEI is well above the 135°C cure temperature, a point in time is reached where the increase in weight percent of PEI in the β-phase causes the T_g of the β-phase to approach 135°C. At this time, the β-phase enters into the glassy state and the epoxy-amine oligomer diffusion process into the α-phase is quenched. Throughout this time and continuing after achievement of the glassy state in the β-phase, the TS i-mers in the occluded α–phase continues to react. As in the β-phase, but due to reaction advancement, the Tg of the α-phase increases until its T_g also reaches 135°C. Again once this buildup in T_g in the α-phase has occurred, the α-phase reaction is quenched by formation of a glass. Post cure at a higher temperature can of course increase T_g of both phases up to completion of the epoxy reaction in both the α and β-phases.

The relationship between the time of occurrence of peaks in $\varepsilon''(\omega)$ and the corresponding relaxation time τ, and the extent of reaction x(t) in the TP/TS blend before phase separation and after phase separation in the TS rich occluded phase is shown in Figure 5.

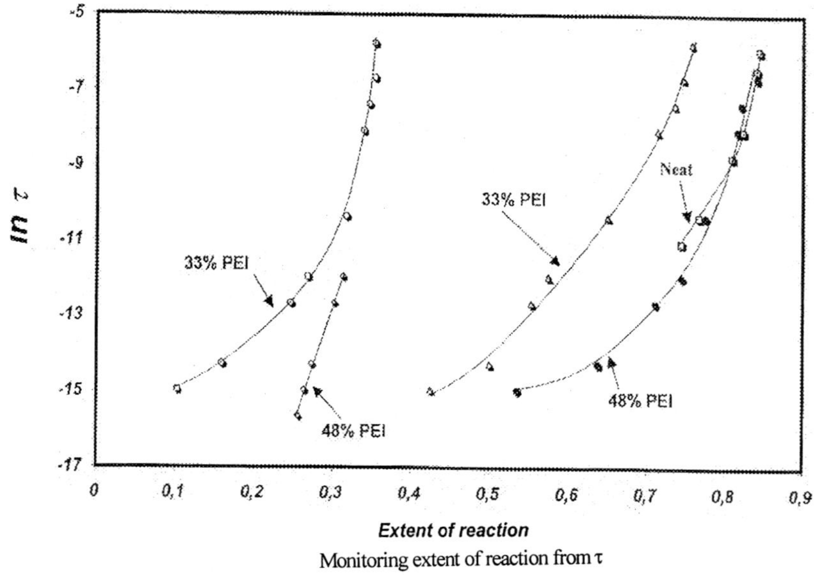

Fig. 5. Monitoring correlation of extent of reaction with τ for PEI / TS blends at 135°C.

Conclusions

Frequency dependent dielectric measurements are an excellent technique for monitoring the complex chemical, composition and morphological changes occurring during cure of high temperature TP/TS blends. The T_g alpha-relaxation peaks can be used to monitor the buildup in T_g and thereby reaction advancement in the homogeneous phase before phase separation and in the occluded phase after phase separation.

The onset of phase separation is readily detected by a sharp MWS charge polarization rise in ε'. The dielectric detected time is in close agreement with mechanical and cloud point techniques. The conductivity in these blends does not follow the generally expected inverse viscosity power dependence and decreases much more rapidly then expected based on the concentration of thermoplastic and the change in viscosity. The results support a hydrogen bond exchange mechanism which is strongly affected by the concentration-proximity of the proton exchange groups.

Acknowledgement

Support from a joint CNRS and the NSF Foundation INT-0003760 grant made this work possible.

[1] R.J.J. Williams, B.A. Rozenberg, J.P. Pascault, *Adv. Polm. Sci.* 1996, *128*, 95.
[2] E. Girard-Reydet, H. Sautereau, J.P. Pascault, P. Keates, P. Navard, G. Thollet, G. Vigier, *Polymer* 1998, 39, 2269-2280.
[3] R.A. Pearson, in: "Rubber toughened Plastics I," C.K. Riew, A.J. Kinloch, Eds., *Adv. Chem. Sev. ACS* 1993, *233*, 407-425.
[4] R.W. Vanderbosch, H.E.E. Meijer, P.J. Lemstra, *Polymer*, 1995, *36*, 2903.
[5] H. Elkind, F. Mauer, P. Stechman, *Polymer*, 1997, *38(5)*, 1047.
[6] A. Bonnet, J.P. Pascault, H. Sautereau, M. Taha, Y. Camberlin, *Macromolecules*, 1999, *32(25)*, 8517.
[7] A. Bonnet, J.P. Pascault, H. Sautereau, Y. Camberlin, *Macromolecules*, 1999, *32(25)*, 8524.
[8] A. Bonnet, J.P. Pascault, H. Sautereau, J. Rogozinski, D. Kranbuehl, *Macromolecules*, 2000, *33(10)*, 3833.
[9] D. Kranbuehl, "*Dielectric Spectroscopy of Polym. Mats.* ", J. Runt, Ed., *Am. Chem. Soc.*, *1997*, 303.
[10] C. Mathieu, G. Boiteux, G. Seytre, R. Villain, P. Dublineau, *J. Non-Cryst. Solids*, 1994, *172-174*, 1012-1016.
[11] R.W.J. Sillars, *Inst. Electr. Eng.*, 1937, *80*, 378.
[12] L.H.K. Van Beek, "*Dielectric Behaviour of Heterogeneous Systems.* ", Progress in Dielectrics, 1967, *7*, 69-114.
[13] B. Lestriez, A. Maazouz, J.F. Gerard, H. Sautereau, G. Boiteux, G. Seytre, D. Kranbuehl, *Polymer*, 1998, *39(26)*, 6733-42

In Situ Real-Time Monitoring of a Diglycidyl Ether of Bisphenol A Epoxy Resin Modified with Poly(etherimide)/Polysulphone Blends by Simultaneous Dielectric/Near-Infrared Spectroscopy

*G. Kortaberria, P. Arruti, I. Mondragon**

"Materials + Technologies" Group, Escuela de Ingenieria Tecnica Industrial, Universidad del Pais Vasco/Euskal Herriko Unibertsitatea, Avda. Felipe IV, 1B. 20011, Donostia-San Sebastian, Spain

Summary: Simultaneous dielectric and near infrared measurements were performed in "real –time" to follow polymerisation reactions on blends of a diglycidyl ether of bisphenol A (DGEBA) epoxy resin with 4,4'-diaminodiphenylmethane (DDM) hardener and a mixture of polysulphone (PSU) and polyetherimide (PEI) as modifier. All the blends had a 10 wt% of PSU/PEI mixture. The effect of the PEI/PSU ratio in the mixture was studied. Monitoring of the α-relaxation (related to vitrification) was performed by dielectric measurements, while epoxy conversion was followed by near infrared spectroscopy. The effect of the PEI/PSU ratio on this behaviour was studied, as well as that of the curing temperature. Obtained results were compared with that of the blends with neat PSU and PEI as modifiers.

Keywords: curing, dielectric spectroscopy, epoxy, near infrared

Introduction

In situ real-time monitoring for chemical and physical changes during processing of reactive polymer-forming materials is of crucial importance to scientist and engineers.

Dielectrometry has been developed by many authors as a non-destructive and convenient method for monitoring of thermosets curing. Many dielectric studies of epoxy-amine systems have been reported.[1-8] Several groups[9-12] have tried to relate the changes in dielectric properties to chemical and physical phenomena that occur during reaction.

Among a host of spectroscopic, dielectric and acoustic monitoring techniques, Fourier transform spectroscopy represents a particularly attractive choice owing to its possibility to offer molecular level information contained in the infrared portion of the electromagnetic spectrum. And while the use of conventional (off-line) FTIR spectroscopy to study reactive polymeric systems has been well documented in the literature[13,14] the use of *remote* FTIR for

 DOI: 10.1002/masy.200350833

in situ monitoring of process has a much more recent origin. Only a handful of investigations of reactive systems have been conducted in the near infrared range, which extends from about 14.000 to 4000 cm^{-1} and which contains weaker overtones of the fundamental absorptions.[15-23]

So the combined use of near infrared and dielectric spectroscopy becomes of particular interest in monitoring of the reactive processes. Both techniques working together enable to generate complementary information about chemical changes (from near infrared) and physical changes (from dielectric spectroscopy) associated with various phenomena that occur during processing of thermosets including vitrification, gelation and various relaxations.

Epoxy resins are often modified with thermoplastics or elastomers in order to improve their fracture toughness. Usually, a single modifier is employed but the possibility of exists for the simultaneous using of two or even more modifiers.

The main objective of this preliminary work is to study the effect of the modification of the DGEBA/DDM resin with PEI/PSU mixture at 10 wt% with different PEI/PSU ratios on the curing reaction, by using dielectric and infrared spectroscopy working simultaneously. Blends were cured at 100°C and 120°C in order to study the effect of the curing temperature.

Experimental

The bifunctional epoxy matrix consisted of stoichiometric amounts of diglycidyl ether of bisphenol A (DGEBA) epoxy resin (DER 332, n=0.03, kindly supplied by Dow Chemical) and 4,4'-diaminodiphenylmethane DDM, HT 972), kindly provided by Ciba. The modifiers were polysulphone of bisphenol A (PSU Udel P-1700), kindly supplied by Amoco Chemical, and polyetherimide (Ultem 1000), provided by GE Plastics.

DGEBA-(PEI/PSU) mixtures containing different PEI/PSU ratios (25/75, 50/50, 75/25) were prepared by dissolution in methylene chloride. The solvent was evaporated in an oil bath at 90°C. Last solvent traces were removed in a vacuum oven at 90°C overnight.

Dielectric measurements were performed on a Solartron 1260 impedance gain-phase analyzer operable in the range from 0,1 Hz to 1 MHz, interfaced with a computer to the reaction cell used.

Fourier transform infrared spectroscopy was performed using a Nicolet Nexus-IR System 750 Spectrometer with spectral range coverage from 15.800 to 50 cm^{-1}. Near infrared data were obtained using a calcium fluoride beamsplitter, a white light source and PBSe or MCT-A detector. For simultaneous measurements, a large core (600/630 μm) low OH fused silica type 3M's multimode optical fibre was used.

Results and Discussion

Dielectric Spectroscopy

The advancement of the curing reaction of DGEBA/DDM resin modified with 10 wt% PEI/PSU mixture is followed. The effect of the PEI/PSU ratio is studied as well as that of the curing temperature. Obtained data are compared with those for blends of DGEBA/DDM with 10 wt% PSU and 10 wt% PEI. The main α-relaxation (related with the vitrification phenomena) is followed in time and frequency domains in order to study the effect of the curing in it.

Vitrification occurs when the curing resin changes from the gel to the glassy state. When the cure proceeds at constant temperature, as in this case, the time at which the increasing Tg of the system reaches the cure temperature defines this physical event. When this change occurs, the permittivity decreases until it reaches an assymptotic value. At the same time, the loss-factor curve presents a peak depending on the frequency, and after this peak it also reaches an assymptotic value. This peak, corresponding to the α-relaxation, occurs when $\varpi\tau$ (frequency-relaxation-time product) equals unity. This α-relaxation is the material relaxation process associated with glass transition, and is linked with cooperative segmental motions and with microscale density fluctuations.[24]

Figure 1 show permittivity and loss-factor vs cure time plots for the DGEBA/DDM resin modified with 10 wt% PEI/PSU (50/50) at 100°C. There can be clearly seen the fall in permittivity at the same time of the appearance of the peak, corresponding to the slowing down of the relaxations of the permanent electrical dipoles.[25]

Figures 2a, 2b and 3 show the effect of the modifier in this behaviour. Blends of DGEBA/DDM with 10 wt% PEI/PSU at several ratios (25/75, 50/50, 75/25) are compared with blends of DGEBA/DDM with 10 wt% PEI and 10 wt% PSU.

It can be seen that reaction advances faster in the case of the matrix modified with neat PEI, slowing down with the addition of PSU, until it is the slowest in the case of the matrix modified with neat PSU. This slowing down can be seen in the loss-peak and in the fall of permittivity. The reaction is retarded with the amount of PSU in the modifiers mixture.

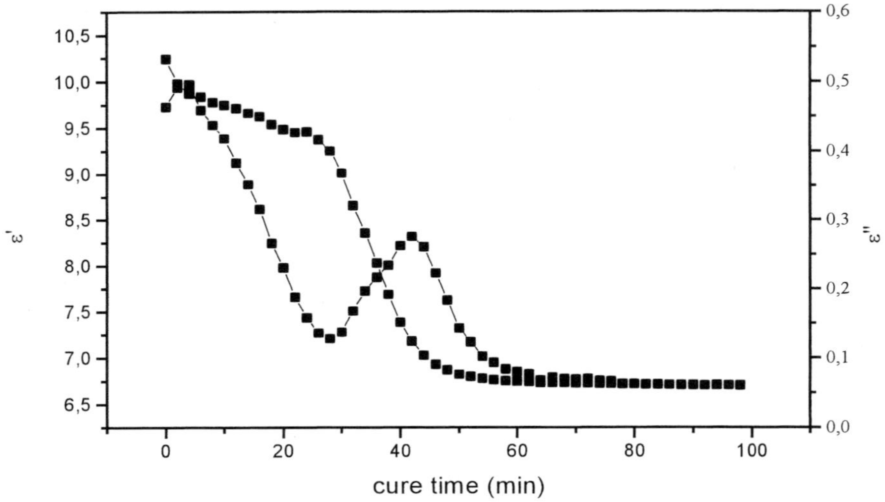

Fig. 1. dielectric loss and dielectric constant vs cure time plot for the DGEBA/DDM matrix modified with 10 wt% PEI/PSU (50/50).

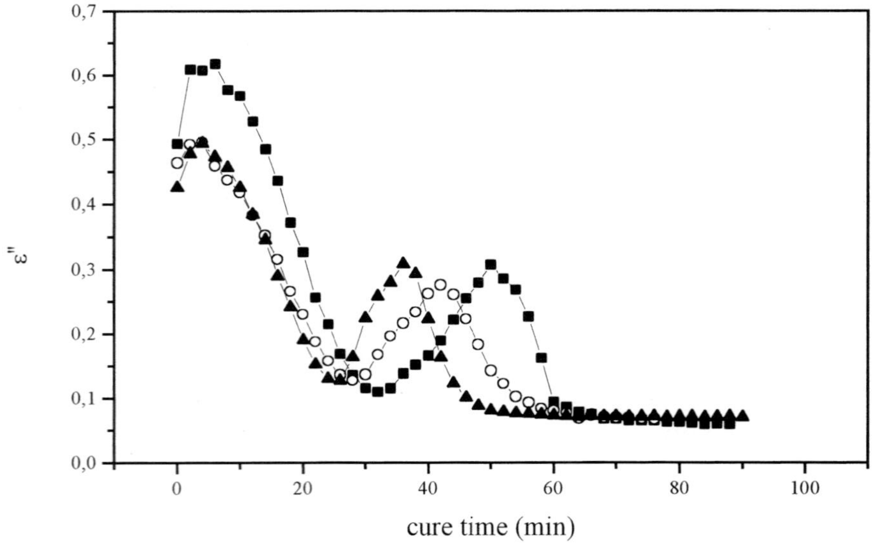

Fig. 2a. dielectric loss vs cure time plot for the DGEBA/DDM matrix modified with 10 wt% PEI (▲), 10 wt% PEI/PSU (50/50) (o), and 10 wt% PSU (■) respectively, at 100°C and 10 kHz.

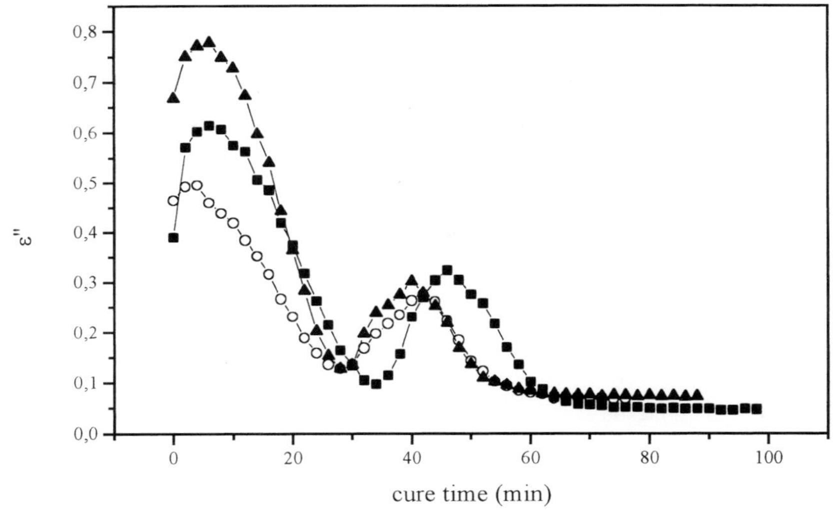

Fig. 2b. dielectric loss vs cure time plot for the DGEBA/DDM matrix modified with 10 wt% PEI/PSU (75/25) (▲), 10 wt% PEI/PSU (50/50) (o), and 10 wt% PEI/PSU (25/75) (■) respectively, at 100°C and 10 kHz.

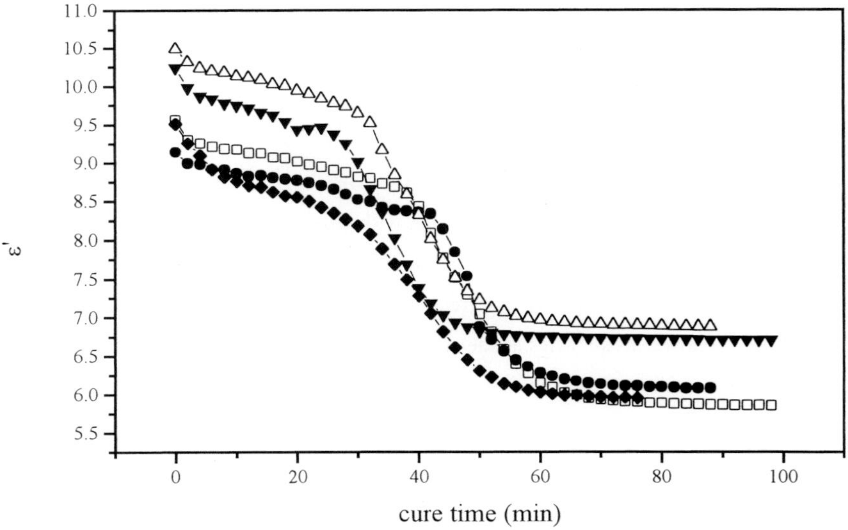

Fig. 3. dielectric constant vs cure time plots for DGEBA/DDM matrix modified with 10 wt% PEI (◆), 10 wt% PSU (●), 10 wt% PEI/PSU (75/25) (□), 10 wt% PEI/PSU (50/50) (▼), and 10 wt% PEI/PSU (25/75) (Δ) respectively, at 100°C and 10 kHz.

The effect of the cure temperature is shown in Figure 4, where loss factor and permittivity are measured for the DGEBA/DDM matrix modified with 10 wt% PEI/PSU (50/50) at 100° and 120°C. The reaction is going faster with higher curing temperature.

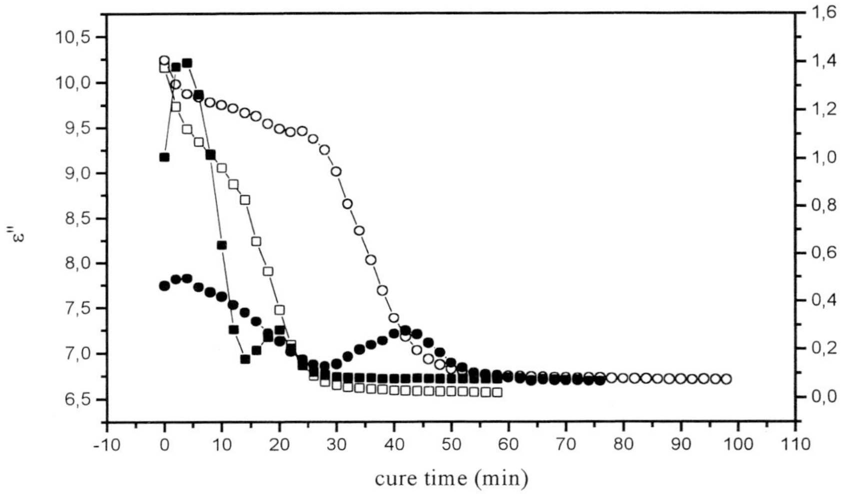

Fig. 4. dielectric loss and dielectric constant vs cure time plot. Dielectric loss at 100°C (●), dielectric loss at 120°C (■), dielectric constant at 100°C (o) and dielectric constant at 120°C (□).

The relaxational behaviour of the blends can be followed by studying the evolution of the imaginary impedance Z" in the frequency domain with cure time. Figure 5 shows the Z" vs frequency plot for the DGEBA/DDM matrix modified with 10 wt% PEI/PSU (50/50) at 100°C. The relaxation peak of the Z" vs frequency plot appears from the high frequency region and moves to lower frequencies as the cure advances.[26]

During the cure, as the Tg of the resin becomes higher the medium becomes more viscous and so the chain segments have more difficulty in moving, and consequently dipoles need longer times for their orientation and relaxation. At the same time, the height of the peak increases and the width decreases, the distribution of relaxation times becoming narrower.

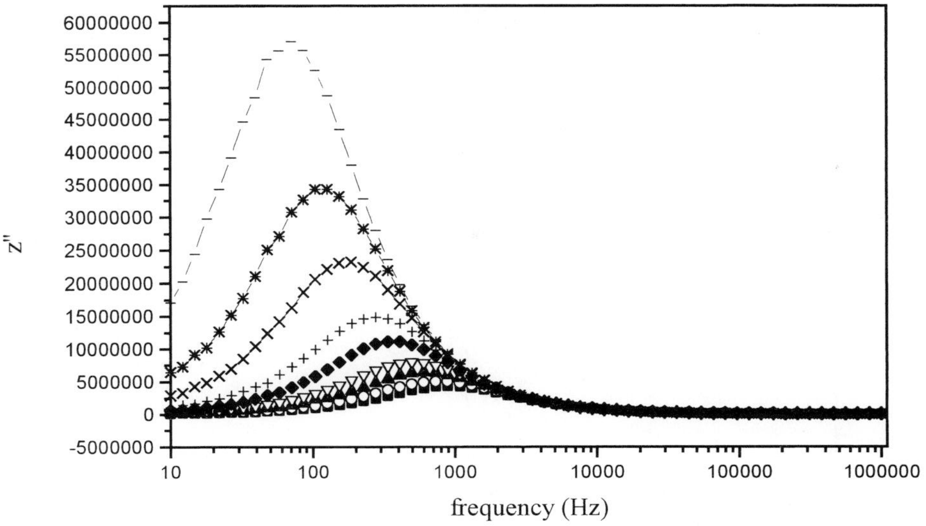

Fig. 5. imaginary impedance vs frequency plots for the DGEBA/DDM matrix modified with 10 wt% PEI/PSU (50/50) at several cure times: 18 min (■), 20 min (o), 22 min (▲), 24 min (▽), 26 min (◆), 28 min (+), 30 min (×), 32 min (*) and 34 min (-).

Near Infrared Spectroscopy

Figure 6 shows the evolution for the near infrared spectra with cure time, for the DGEBA/DDM matrix modified with 10 wt% PEI/PSU (50/50) at 100°. The trend displayed by major peaks of relevance in epoxy/amine systems[27-29] are evident: decrease in epoxy absorption (6070 cm^{-1}, epoxy group stretching vibrations, first overtone), a decrease in amine absorption (5056 cm^{-1}, $-NH_2$ combination band of stretching and bending vibrations, and 6670 cm^{-1}, $-NH_2$ combination of symmetric and asymetric stretching vibrations).

The peak at 6070 was used to monitor the disappearance of the epoxy group in this study. Various reference peaks were examined and the most reproducible results were obtained using the peak at 5980 cm^{-1}, due to aromatic –C-H stretching first overtone band.

Figure 7 shows the epoxy conversion vs cure time plot for all the blends studied, cured at 100°C. The trend of the reaction advancement is similar to that shown by dielectric spectroscopy, i.e. being the reaction of the matrix modified with 10 wt% PEI is the fastest and the modified with 10 wt% PSU is the slowest. In the case of matrices modified with PEI/PSU mixtures, reaction advances more slowly by increasing the content of PSU in the mixture.

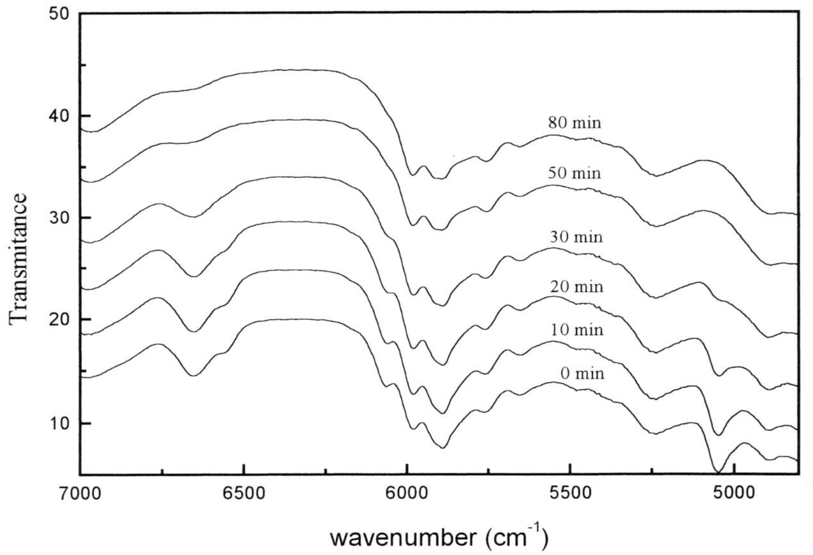

Fig. 6. near infrared spectra of the DGEBA/DDM matrix modified with 10 wt% PEI/PSU (50/50).

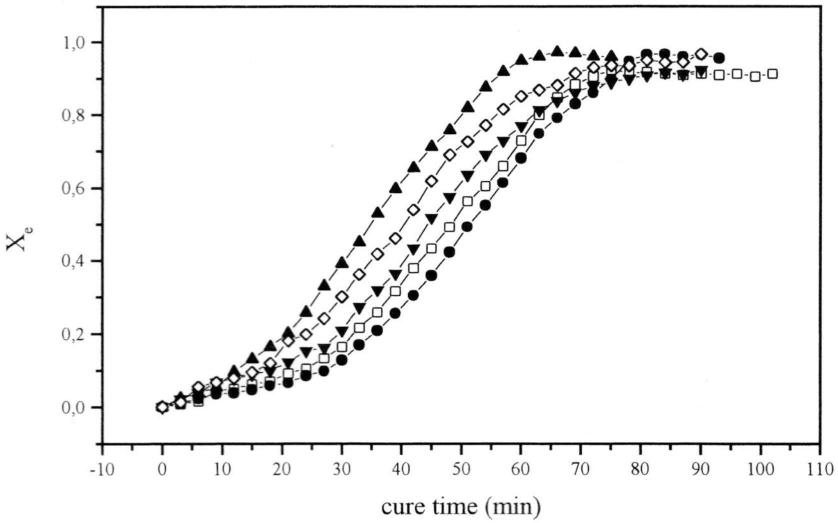

Fig. 7. epoxy conversion vs cure time plots for the DGEBA/DDM matrix modified with 10 wt% PEI (▲), 10 wt% PEI/PSU (75/25) (◇), 10 wt% PEI/PSU (50/50) (▼), 10 wt% PEI/PSU (25/75) (□), and 10 wt% PSU (●) respectively, cured at 100°C.

It is worth noting, however, that difference between the curing behaviour of all the systems studied is not very significant, so indicating that the reaction will not be very affected using PSU or PEI as modifier.

Conclusions

In situ "real-time" simultaneous dielectric and near infrared measurements have been performed to follow polymerisation of thermoplastic-modified epoxy mixtures. Mixtures of two thermoplastic with different ratios have been used as modifier.

Monitoring of cure advancement have been carried out in order to study the effect of the thermoplastic and its amount on it. Measurements have been made at 100°C and 120°C, in order to show the effect of the cure temperature.

Dielectric and near infrared measurements are in quite well agreement indicating that the PSU thermoplastic slows down the cure reaction more than PEI. The bigger the amount of PSU in the mixture is, the slower the cure reaction advances, as it could be seen for the mixtures with PEI/PSU ratios of 25/75, 50/50, and 75/25, and for the mixtures with 10 wt% PEI and 10 wt% PSU. It is worth noting, however, that the differences between all the systems are not very significant, thus indicating that the use of PEI or PSU as modifier does not affect too much the reaction kinetics. The effect of the curing temperature is more significant, advancing the reaction much faster at higher curing temperatures. It can be concluded that the cure temperature is more determining than the kind of thermoplastic in our case.

The relaxational behaviour with cure time have been also shown in the frequency domain, in order to study the effect of the cure on the main relaxation of the resin. This α-relaxation goes towards low frequencies with the cure advancement, because the medium is more viscous and dipoles need longer times to orient and relax.

1. J. O. Simpson and S. A. Bidstrup, J. Polym. Sci.: Part B: Polym. Phys., 33, 55 (1995)
2. G. Levita, A. Livi, P. A. Rolla, and C. Culicchi, J. Polym. Sci.: Part B: Polym. Phys., 34, 2731 (1996)
3. J. Fournier, G. Williams, C. Duch, and G. A. Aldridge, Macromolecules, 29, 7097 (1996)
4. J. P. Elondou, Eur. Polym. J., 35, 1473 (1999)
5. J. P. Elondou, Eur. Polym. J., 35, 1481 (1999)
6. J. P. Elondou, O. Ayina, H. N. Nga, and J. F. Gerard, J. Polym. Sci.: Part B: Polym. Phys., 36, 2911 (1998)
7. S. Pichaud, X. Deuteurtre, and A. Fit, Polym. Int., 48, 1205 (1999)
8. T. Koike, J. Appl. Polym. Sci., 50, 1943 (1993)
9. G. Boiteux, P. Dublineau, M. Feve, C. Mathieu, G. Seytre, and J. Ulanski, Polym. Bull., 30, 441 (1993)
10. M. B. M. Mangion and J. P. Johari, J. Polym. Sci.: Part B: Polym. Phys.., 29, 1127 (1991)
11. M. G. Parthun and J. P. Johari, J. Polym. Sci.: Part B: Polym. Phys., 30, 655 (1992)
12. T. Koike, J. Appl. Polym. Sci., 44, 679 (1992)
13. S. Andjelic, J. Mijovic, and F. Bellucci, J. Polym. Sci.: Part B: Polym. Phys., 36, 641 (1998)
14. C. Vinh-Tung, G. Boiteux, G. Seytre, G. Lachenal, and B. Chabert, Polym. Comp., 17, 761 (1996)
15. M. Franco, I. Mondragon, and C. B. Bucknall, J. Appl. Polym. Sci., 72, 427 (1999)

16. M. Franco, M. A. Corcuera, J. Gavalda, A. Valea, and I. Mondragon, J. Polym. Sci.: Part B: Polym. Phys. 35, 233 (1997)
17. D. R. Day, T. J. Lewis, H. L. Lee, and S. D. Senturia, J. Adhesion, 18, 73 (1985)
18. L. K. H. Van Beek, in Progress in Dielectric, 7, 69 (1967)
19. D. A. Wasylyshyn and G. P. Johari, J. Polym. Sci.: Part B: Polym. Phys., 36, 2703 (1998)
20. G. M. Maistros, H. Block, C. B. Bucknall, and I. K. Partridge, Polymer, 33, 4470 (1992)
21. C. G. Delides, D. Hayward, R. A. Pethrick, and A. S. Vatalis, Eur. Polym. J., 28, 505 (1992)
22. G. Korkakas, C. M. Gomez, and C. B. Bucknall, Plast. Rubber Comp. Process. Appl., 19, 285 (1993)
23. I. Partridge and G. Maistros, High Perform. Polym., 8, 1 (1996)
24. B. D. Fitz and J. Mijovic, Macromolecules, 32, 4134 (1999)
25. S. Andjelic and J. Mijovic, Macromolecules, 31, 8463 (1998)
26. G. M. Maistros and C. B. Bucknall, Polym. Eng. And Sci., 36, 1517 (1994)
27. J. Mijovic, S. Andjelic, and J. M. Kenny, Polym. Adv. Tech., 7, 1 (1995)
28. G. A. George, P. Cole-Clark, N. St John, and G. Friend, J. Appl. Polym. Sci., 42, 643 (1991)
29. N. St John and G. A. George, Polymer, 33 (13), 2679 (1991)

Phase Separation in a Poly(ether sulfone) Modified Epoxy-Amine System Studied by Temperature Modulated Differential Scanning Calorimetry and Dielectric Relaxation Spectroscopy

S. Montserrat, F. Roman, M. Basany, P. Colomer*

Departament de Màquines i Motors Tèrmics, E.T.S. d'Enginyeria Industrial de Terrassa, Universitat Politècnica de Catalunya, 08222-Terrassa, Spain

Summary: The phase separation induced by the curing reaction of an epoxy based on diglycidylether of bisphenol A (DGEBA) with methylene dianiline (MDA) modified with poly(ether sulfone) (PES) at a concentration of 20 wt% was studied by temperature modulated differential scanning calorimetry (TMDSC) and dielectric relaxation spectroscopy (DRS). The effect of phase separation on the curing kinetics and vitrification phenomena is analysed. The dependence of the log of the measuring frequency on the degree of conversion allows the correlation between the dipolar relaxation of each phase and the vitrification observed by TMDSC to be established.

Keywords: dielectric properties, kinetics, phase separation, temperature modulated DSC, vitrification

Introduction

The toughness of thermosetting networks can be improved by modifying the epoxy formulation with the inclusion of a thermoplastic.[1,2] One procedure for obtaining this modification involves dissolving the thermoplastic in the epoxy-hardener system and further curing. As the reaction proceeds, the molecular weight of the resin increases and a phase separation takes place at a given conversion. In the case of thermoplastics with a high glass transition temperature (T_g), at relatively low concentration, the phase separation generates a thermoplastic-rich phase dispersed in the epoxy-rich phase. Each phase undergoes the corresponding vitrification process when T_g reaches the curing temperature. The morphology of the multiphase system, which determines the final thermal and mechanical properties of the modified thermoset, depends on the curing conditions.

 DOI: 10.1002/masy.200350834

Poly(ether sulfone) (PES) is a high-T_g thermoplastic which is used as a modifier for thermosets. The phase separation induced by the curing of an epoxy-diamine system modified by PES has been monitored by optical light transmission[3] and light scattering.[4] The vitrification processes of this system have been studied using calorimetric techniques,[5-7] dynamic mechanical analysis[8-10] and dielectric relaxation spectroscopy (DRS).[6,11]

The objective of this paper is to analyse the effect of PES on the reaction kinetics and vitrification phenomena by temperature modulated differential scanning calorimetry (TMDSC). The relaxation process occurring in each phase is also studied using DRS and it is correlated with the vitrification observed by TMDSC.

Experimental

The unmodified system was an epoxy based on DGEBA (Araldite F from Ciba Speciality Chemicals), with an epoxy equivalent weight of 193.7 g.eq^{-1}, which was cured with methylene dianiline (MDA) (from Aldrich) in a stoichiometric ratio. The DGEBA-MDA system was modified with 20 wt% of commercial poly(ether sulfone) (PES) (Ultralon E2010 from BASF). In order to obtain a homogeneous PES modified epoxy-amine system, the PES was dried in a vacuum chamber at 90°C for 24h before being dissolved in methylene chloride at about 10 wt%. The epoxy was dissolved in the PES solution at room temperature with continuous stirring, and the solution was then heated to 80°C in an oil bath in order to evaporate as much solvent as possible. The residual solvent was removed after 14h under vacuum at 80°C. The appropriate amount of MDA was added to the clear viscous mixture of epoxy and PES and heated to 100°C for 5 min in order to melt and disperse the solid MDA. Finally, the samples were enclosed in aluminium pans or spread on ceramic cells for calorimetric or dielectric analysis respectively.

The temperature modulated DSC measurements were performed with a METTLER TOLEDO 821e equipped with an intracooler, and STAR© software was used for alternating DSC (ADSC) evaluation. The temperature and heat flow calibrations were performed using indium and zinc standards. The modulation conditions used in the quasi-isothermal curing reaction were an amplitude of 0.5 K and a period of 1 min. The sample weight was approximately 8-10 mg. The calorimetric scans were performed with a nitrogen gas flow of 50 mL min^{-1}.

A dynamic electric analyser DEA 2970 from TA Instruments was used to measure both the dielectric permittivity (ε') and the dielectric loss factor (ε'') of the isothermally cured epoxy samples at different frequencies in real time. Dielectric measurements were performed using a single-surface ceramic cell. The dielectric permittivity and the dielectric loss factor were measured at nine frequencies within the interval 10 Hz to 100 kHz. The time taken to scan at the different frequencies was 1.2 min. All dielectric measurements were performed under a nitrogen atmosphere with a gas flow of 500 mL min^{-1}.

Temperature Modulated DSC Results

The effect of the PES on the isothermal curing of the DGEBA-MDA system was analysed from the three TMDSC signals: the total heat flow <φ>, the modulus of the complex heat capacity $|C_p{}^*|$ and the phase angle δ. The fundamental aspects of ADSC signals, which appear during the cure, have been described elsewhere.[12] Figure 1a shows the total heat flow of the unmodified system and the PES modified system during the curing at 80°C. The <φ> curve of the modified system is shifted to longer times, which *a priori* indicates a slowdown of the global kinetics. The degree of conversion, α, for both systems at 80°C is shown in Figure 1b. Total conversion is not achieved due to the vitrification of both systems. The limiting conversion in the unmodified system is higher than in the modified system.

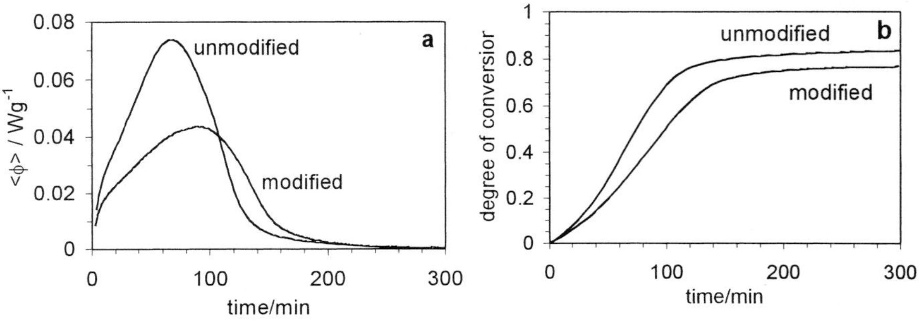

Fig. 1. Total heat flow (a) and degree of conversion (b) of the unmodified (DGEBA-MDA) and the modified (DGEBA-MDA/20%PES) systems during isothermal curing at 80°C. The modulation conditions were: amplitude 0.5K, and period 60s.

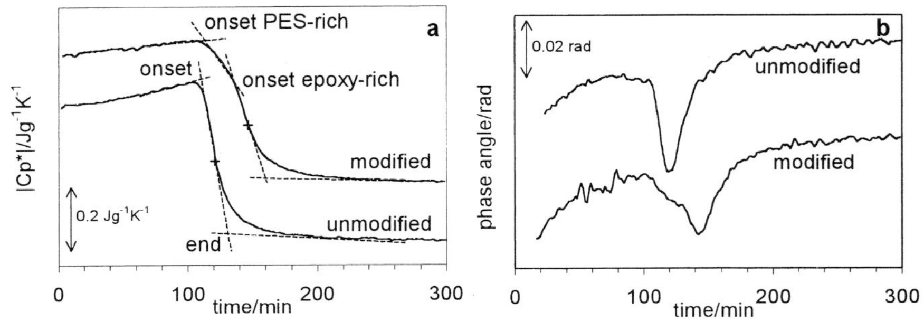

Fig. 2. Modulus of C_p* (a) and phase angle (b) for the unmodified (DGEBA-MDA) and the modified (DGEBA-MDA/20%PES) systems at a curing temperature of 80°C. Modulation conditions: 0.5 K amplitude, and 60s period. The onset time for vitrification is indicated in $|C_p^*|$ for each phase. The crosses show the vitrification midpoint for the epoxy matrix.

TMDSC allows the heat evolved during the cure to be measured in conjunction with the heat capacity of the system. The evolution of the modulus of the complex heat capacity $|C_p^*|$ for the unmodified DGEBA-MDA system (Fig. 2a) follows the general trend of other epoxy-amine systems.[5,12] Initially $|C_p^*|$ increases due to the increase in the number of configurational states or the vibrational modes during the growth of the linear chains. A drastic decrease in $|C_p^*|$ follows which may be attributed to the vitrification of the system.[5,12] During this process, the system goes from a rubber-like state corresponding to the highest $|C_p^*|$ value to a glassy state at the lowest value of $|C_p^*|$. As shown in Figure 2a, a time corresponding to the onset, midpoint and end of the vitrification may be defined. At the same time the phase angle for the unmodified system (Fig. 2b) shows a relaxation peak in the region where the drop in $|C_p^*|$ appears.

The evolution of $|C_p^*|$ for the modified system shows a two-step decrease. The first decrease may be attributed to the vitrification of a PES-rich phase, which results from the phase separation induced by the progress of the curing reaction. This step is followed by an abrupt decrease, whose slope is similar to that observed in the unmodified system. The second decrease in $|C_p^*|$ corresponds to the vitrification of the epoxy matrix. The relaxation process associated with each vitrification is also shown in the phase angle by a shoulder and a peak

(Fig. 2b). The vitrification of the epoxy-rich phase is shifted to longer times with the inclusion of PES. This shift has also been observed at the other curing temperatures (60, 100 and 120°C). These results are in good agreement with those of Swier and Van Mele.[5]

As in other thermosetting systems, the curing kinetics is controlled chemically by the reactivity of the functional groups, but as the reaction progresses, the viscosity of the medium increases and the systems transforms from a liquid (or gel) into a glass. When the vitrification sets in, the kinetics becomes diffusion-controlled. The chemical kinetics for this epoxy-amine system fits the autocatalytic model proposed by Kamal[13] well:

$$\left(\frac{d\alpha}{dt}\right)_{chem} = (k_1 + k_2\alpha^m)(1-\alpha)^n \quad (1)$$

In this equation k_1 and k_2 are rate constants which are dependent on the temperature assuming an Arrhenius form, $k(T)=A\exp(-E_a/RT)$, where A is the pre-exponential factor, R is the gas constant and E_a is the apparent activation energy. The exponents m and n indicate the empirical reaction order. The values for k_1 and k_2 from fitting the Kamal equation at different temperatures, together with the corresponding apparent activation energies, are shown in Table 1. These results confirm that the inclusion of PES slows down the chemical kinetics due to a dilution effect.[5]

Table 1. Rate constants and apparent activation energies for the curing reaction of the unmodified and modified system.

T(°C)	Unmodified system		Modified system	
	$k_1 \cdot 10^5$ (s^{-1})	$k_2 \cdot 10^4$ (s^{-1})	$k_1 \cdot 10^5$ (s^{-1})	$k_2 \cdot 10^4$ (s^{-1})
60	2.1	4.2	1.4	1.1
80	5.4	12.5	2.9	4.0
100	127.5	21.4	125.1	17.0
120	416.4	101.9	380.3	55.1
E_a ($kJmol^{-1}$)	$E_a(k_1)$=52.9	$E_a(k_2)$=54.8	$E_a(k_1)$=61.4	$E_a(k_2)$=71.3

The diffusion-controlled regime may be analysed by the chemical reaction rate given by equation (1) corrected by a diffusion factor. In the present epoxy-amine system this diffusion factor coincides with a mobility factor MF((T,α), which may be calculated from the values of the specific heat at vitrification:[14]

$$\mathrm{MF}(\mathrm{T},\alpha)=\frac{\left|C_{p}^{*}\right|(t,T)-\left|C_{p,g}^{*}\right|(T)}{\left|C_{p,l}^{*}\right|(t,T)-\left|C_{p,g}^{*}\right|(T)} \qquad (2)$$

where $|C_p^*|(t,T)$ is the actual value of $|C_p^*|$, and $|C_{p,g}^*|(T)$ and $|C_{p,l}^*|(t,T)$ are the values of $|C_p^*|$ in the glassy and the liquid state respectively. This coincidence allows the calculation of a simulated reaction rate from the chemical reaction rate corrected by the mobility factor. The simulated reaction rate is shown in Figure 3a for the unmodified epoxy system. As in other epoxy-amine systems,[12,14] the simulated reaction rate agrees quite well with the experimental reaction rate. A similar procedure has been applied to the modified system using an overall mobility factor calculated by $|C_p^*|$. The overall simulated reaction rate for the modified system is shown in Figure 3b.

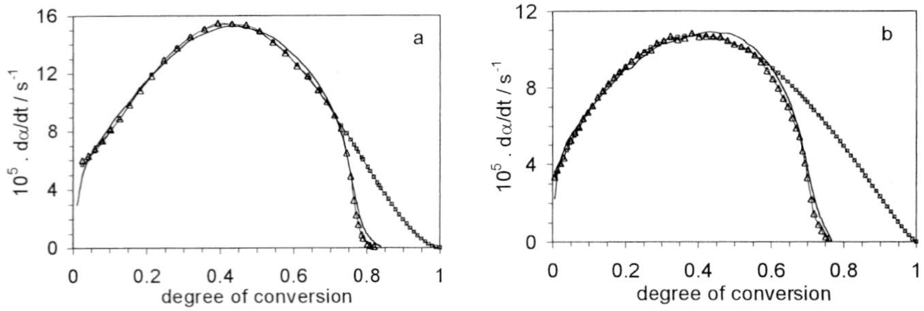

Fig. 3. Reaction rate (dα/dt) against the degree of conversion for the unmodified (a) and the modified (b) system at 80°C. The solid line shows the experimental reaction rate. The reaction rate obtained by Kamal model (□) and the simulated rate (Δ) are also plotted.

The deviation of the experimental reaction rate respect to the chemical reaction rate in the modified system corresponds to the onset of the vitrification of the PES-rich phase, the conversion values for which are shown in Figure 4 at different temperatures. As expected, the degree of conversion at this point increases with temperature. It is observed that the conversion between the onset of vitrification of the PES-rich phase and the onset of vitrification of the epoxy matrix tends to increase with temperature. One may assume that this change of conversion takes place in the epoxy matrix. It is observed that the conversion during vitrification of the epoxy matrix, which is given by the variation between the onset of vitrification of this phase and the limiting value of the conversion, tends to decrease with temperature. In the unmodified system, it is also observed that the restricted change in conversion during the diffusion-controlled step decreases with temperature. This effect is due to the increase of the crosslinking density of the network.

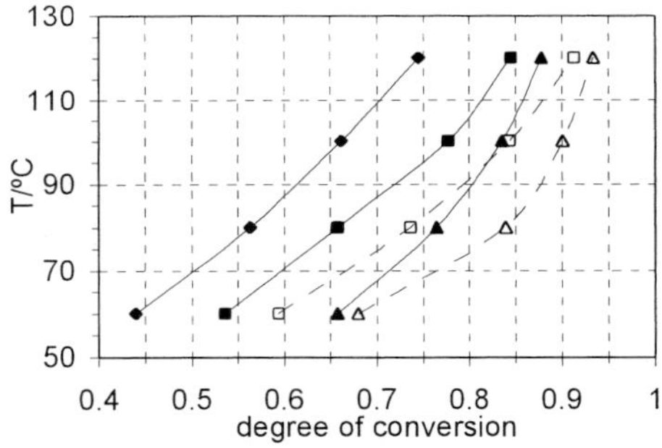

Fig. 4. Degree of conversion corresponding to the vitrification in both modified and unmodified systems: onset vitrification of the PES-rich phase (♦) and the epoxy matrix (■) in the modified system, onset vitrification in the unmodified system (□). The limiting conversion in the unmodified (Δ) and the modified (▲) system is also shown.

Dipolar Relaxation and its Correlation with Vitrification

At a fixed frequency ω, the complex permittivity is calculated by $\varepsilon^* = \varepsilon' - i\,\varepsilon''$, where the real component ε' is the permittivity, which is proportional to the capacitance, and the imaginary component ε'' is the loss factor, which is proportional to the conductance. The angular frequency ω is $2\pi f$, where f is the measuring frequency. Both components ε' and ε'', which are shown in Figures 5a and 5b, have dipolar and ionic, or charge, contributions. At low frequencies, the permittivity ε' (Fig. 5a) shows very large values and decreases drastically due to an electrode polarisation effect, caused by the ionic impurities present in the reactants. At frequencies higher than 1 kHz the curves for ε' show a decrease from the relaxed permittivity ε_o to the unrelaxed permittivity ε_∞. In the modified system, the variation of ε' shows a two-step decrease. However, an increase in this property is observed at low frequencies before the drop to the unrelaxed value. According to other authors[15] this peak may be attributed to an interfacial polarisation process, which occurs at the onset of the phase separation.

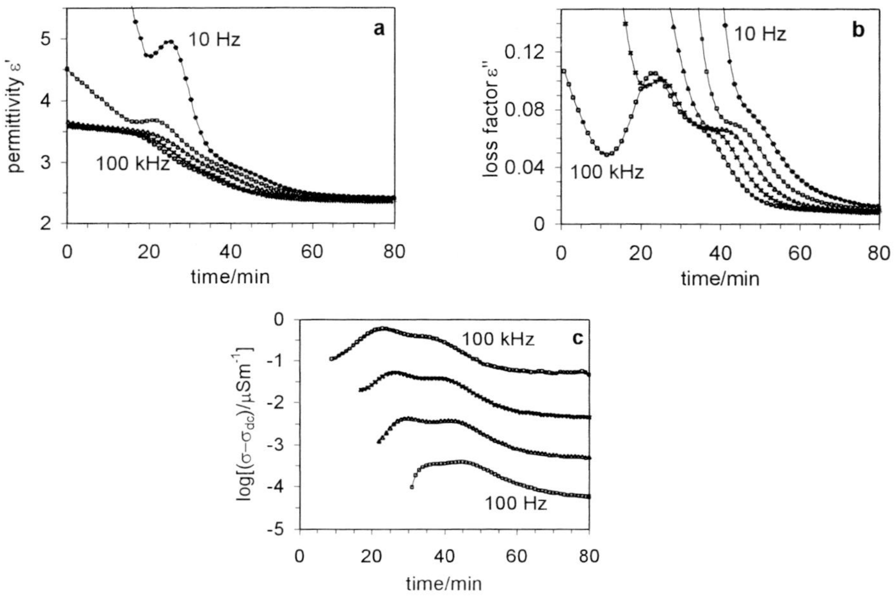

Figure 5. Permittivity ε' (a), loss factor ε'' (b) and conductivity after the subtraction of dc-conductivity (c) of the DGEBA-MDA/20%PES system at a curing temperature of 100°C, for the indicated frequencies: 10 Hz(◇), 100 Hz(□), 1 kHz(Δ), 10 kHz(x) and 100 kHz(o).

The loss factor in the neat epoxy-amine system shows only one dipolar peak, whereas the modified system shows two peaks (Fig. 5b). The first peak is well defined at high frequencies and corresponds to the relaxation of the PES-rich phase. The latter is less clearly defined, it appears as a shoulder, and corresponds to the relaxation of the epoxy-rich phase. This is supported by the lower mobility of the PES chains, which is due to the higher T_g of PES

(226°C) compared with the T_g of the neat epoxy (163°C). Other arguments are given below taking into account the correlation of the DRS and TMDSC data.

In order to study the dipolar relaxation during the curing reaction, the loss factor has been corrected from the ionic contribution $\varepsilon''_i(\omega)$. This ionic contribution is obtained from the dc-conductivity $\sigma_{dc}(T,t_c)$, which is independent of the frequency and decreases with increasing curing time t_c : $\sigma_{dc}(T,t_c) = e_o\omega\ \varepsilon_t''$, where e_o is the permittivity of free space (8.85 pF m^{-1}). The value of $\sigma_{dc}(T,t_c)$ was calculated by the method proposed by Mangion and Johari,[16] and others,[11] using an exponential equation. The dc-conductivity contribution is subtracted from the conductivity data, σ-σ_{dc}, which values are shown in Figure 5c. This quantity allows to calculate the loss factor of the dipolar relaxation, which is not shown in this work, by the relation $\varepsilon''_d(\omega)=(\sigma-\sigma_{dc})/e_o\omega$.

The time for the peak value of $\varepsilon''_d(\omega)$ at a given frequency, which is the same as in the peak of conductivity after the subtraction of dc-conductivity (σ-σ_{dc}), characterises the dipolar relaxation at a given frequency. As in other epoxy-amine systems, the logarithm of the measuring frequency has been observed to have a linear dependence on the time of the dipolar peak $\varepsilon''_d(\omega)$, t_m:[17,18]

$$\log f_m = \log f_m(0) - \frac{k_d\, t_m}{\ln 10} \qquad (3)$$

where f_m is the frequency of the maximum dipolar relaxation, and $f_m(0)$ and k_d are temperature dependent constants, whose values are shown in Table 2 for the neat epoxy and the modified epoxy-amine systems. Two sets of values are shown for the modified system corresponding to the relaxation of the PES-rich phase (first dipolar peak) and the epoxy-rich phase (second peak). Extrapolating this equation to a frequency of 16.7 mHz (the value used in TMDSC experiments) gives a value of t_m situated within the interval where the corresponding vitrification occurs. This correlation is better understood by plotting the log of frequency against the degree of conversion, as in Fig. 6.

Table 2. Dependence of the frequency on the time of maximum loss factor: temperature dependent constants $f_m(0)$ and k_d according to equation (3).

	Unmodified system		Modified system 1st peak		Modified system 2nd peak	
T/°C	$\log(f_m(0)/Hz)$	k_d $10^3/s^{-1}$	$\log(f_m(0)/Hz)$	k_d $10^3/s^{-1}$	$\log(f_m(0)/Hz)$	k_d $10^3/s^{-1}$
60	18.18	1.38	9.09	1.08	-	-
80	20.26	3.32	13.62	2.43	17.79	2.63
100	16.90	6.18	11.48	4.74	14.28	4.62

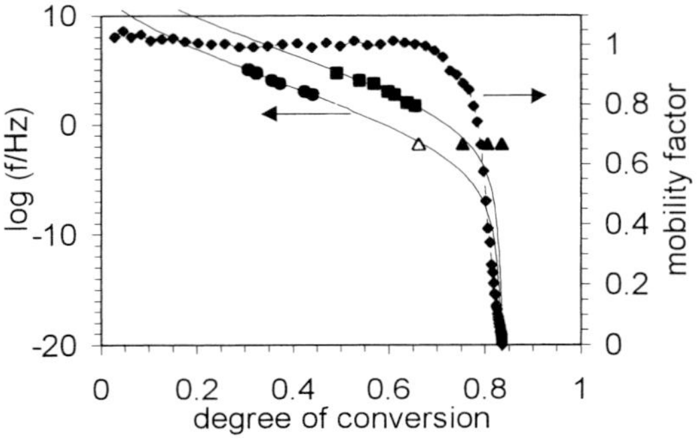

Figure 6. Plots of the logarithm of frequency against the degree of conversion for the first (•) and second (▪) dipolar relaxation peaks during the curing of the modified system at 100°C. The values at log(f/Hz)= -1.78 correspond to the TMDSC measurements: the triangle (Δ) corresponds to the onset of PES-rich phase vitrification and the three filled triangles indicate (from left to right) the onset, midpoint and limiting conversion values for the vitrification of the epoxy matrix. The solid lines correspond to the continuous variation of log(f) on α. The mobility factor under the same curing conditions is also shown (♦).

In the range of dielectric frequencies, $\log(f_m)$ decreases steadily with the degree of conversion, but in the region of the diffusion-controlled regime indicated by the decrease in the mobility factor, the change of log(f) with the conversion becomes steeper. In the modified system log(f) for the first dipolar peak correlates very well with the onset of vitrification of the PES-rich phase, which, using TMDSC, is observed at a frequency of 16.7 mHz. Similarly, log(f) for the second dipolar peak correlates with the vitrification of the epoxy-rich phase, which takes place at a slightly higher conversion. In the case of the unmodified system (not shown in this work) the results agree with those obtained in other neat epoxies.[17,18]

Acknowledgements

Financial support has been provided by the DGI (MAT2000-1002-C02-01). The authors are grateful to CIBA Speciality Chemicals and NORDISA for supplying the epoxy and the PES respectively. One of us (F.R.) acknowledges a grant from the MEC (SB99-002221691) to collaborate in this project.

[1] R. J. J. Williams, B. A. Rozenberg, J-P. Pascault, *Adv. Polym. Sci.* **1997**, *128*, 95.
[2] S. J. Shaw in:"*Chemistry and Technology of Epoxy Resins*", B. Ellis, Ed., Chapman and Hall, Glasgow 1996, Chap 4.
[3] C. B. Bucknall, C. M. Gómez, *Polymer*, **1994**, *35*, 353.
[4] I. Alig, communication O13 in 7th European Symposium on Polymer Blends, Villeurbanne, France, 2002.
[5] S. Swier, B. Van Mele, *Thermochim. Acta*, **1999**, *330*, 175.
[6] I. Alig, W. Jenninger, J. E. K. Schawe, *Thermochim. Acta*, **1999**, *330*, 167.
[7] W. Jenninger, J. E. K. Schawe, I. Alig, *Polymer*, **2000**, *41*, 1577.
[8] I. Mondragon, I. Quintard, C. B. Bucknall, *Plast. Rubb. Comp. Process. Appl.*,**1995**, *24*, 2038.
[9] I. Martinez, M. D. Martin, A. Eceiza, P. Oyanguren, I. Mondragon, *Polymer*, **2000**, *41*, 1027.
[10] K. Mimura, H. Ito, H. Fujioka, *Polymer*, **2000**, *41*, 4451.
[11] I. Alig, W. Jenninger, *J. Polym. Sci,: Polym. Phys.*, **1998**, *36*, 2461.
[12] S. Montserrat, J. G. Martín, *J. Appl. Polym Sci.*, **2002**, *85*, 1263.
[13] M. R. Kamal, *Polym. Eng. Sci.* **1974**, *14*, 231.
[14] G. Van Asche, A. Van Hemelrijck, H. Rahier, B. Van Mele, *Thermochim. Acta*, **1995**, *268*, 121.
[15] A. Bonnet, P-P. Pascault, H. Sautereau, J. Rogozinski, D. Kransbuehl, *Macromolecules*, **2000**, *33*, 3833.
[16] M. B. Mangion, G. P. Johari, *J. Polym. Sci,: Polym. Phys.*, **1991**, *29*, 1117.
[17] J. Fournier, G. Williams, C. Duch, G. A. Aldridge, *Macromolecules*, **1996**, *29*, 7097.
[18] S. Montserrat, F. Roman, P. Colomer, *Polymer* **2003**, *44*,101.

Processability and Chemical Resistance of the Polymer Blend of Thermoplastic Polyurethane and Polydimethylsiloxane

Siriporn Damrongsakkul,[1,*] *Ratirat Sinweeruthai,*[1] *Julia S. Higgins*[2]

[1]Polymer Engineering Laboratory, Department of Chemical Engineering, Faculty of Engineering, Chulalongkorn University, Bangkok 10330, Thailand
[2]Imperial College, University of London, London SW7 2BY, UK

Summary: This work is aimed to develop the melt blend of thermoplastic polyurethane (TPU) and polydimethylsiloxane (PDMS) and to study the effect of the chemical resistance on the tensile properties and morphology of the blends. The master batch blends at 2% of PDMS with 98% of TPU were firstly prepared by an internal mixer and then the blends of TPU/PDMS were prepared by melt mixing using a twin screw extruder. The maximum PDMS content that can be mixed with TPU was found to be no higher than 1%. Higher PDMS content leaves an unmelted TPU fraction in the blends due to the short residence time in the twin screw extruder. The resultant blends show an increase in the elongation at break up to 30% and in Young's modulus up to 40% at the optimum PDMS concentration of around 0.6%-0.8%, beyond which these properties diminish. The ultimate tensile strength and the energy to break are decreased by about 20% and 10%, respectively. The Scanning Electron Micrographs of the blends show dispersed phases of PDMS in a TPU matrix. The domain size of the PDMS phase becomes smaller when increasing PDMS content from 0.2% to 0.8%. The morphology of the fractured surface of TPU/PDMS blends shows less fibrous characteristics with increasing PDMS content in the blends.

For the study of the effects of chemical resistance on the tensile properties and morphology of TPU/PDMS blends, two chemical reagents, sulfuric acid (H_2SO_4, 3% v/v) and sodium hydroxide (NaOH, 10% w/v) are selected. The results on the relationship of chemical resistance to tensile properties and morphology of the blends show that NaOH solution has a stronger effect on the tensile properties and morphology of virgin TPU and the blends than H_2SO_4 solution. The ultimate tensile strength and the energy to break of virgin TPU after base immersion was found to be strongly decreased, which could be caused by the base hydrolysis of the polyester soft segment of polyurethane. The effect of PDMS content in the blends on the base resistance and tensile properties is similar to results before immersion, i.e. the effective PDMS content in the blends that can generally improve tensile properties of the blends after immersion in NaOH does not exceed 0.8%. The results are in agreement with the weight loss of TPU/PDMS blends after base immersion and the morphology of the fractured surface of TPU/PDMS blends after base immersion that exhibit very small amounts of a fibrous character. There are also some small particles detached at the surface. This could be the result of an occurrence of a corrosive reaction between the sample surface and NaOH solution.

Keywords: morphology, polydimethylsiloxane, polymer blends, tensile properties, thermoplastic polyurethane

 DOI: 10.1002/masy.200350835

Introduction

Polymer blends have emerged as an important group of polymeric materials since the late 1970s and have experienced substantial growth since the 1980s. Interests in polymer blending are reflected in the numbers of patent and research publications by both industry and academia.[1-9] Blending of polymeric materials has been shown to be a useful and cost effective route, in comparison with the synthesis of new polymers, for enhancing material properties and developing materials with desired performance.

During the past decades, thermoplastic polyurethane (TPU) has received considerable attention from both the scientific and industrial communities.[10-11] Applications of TPU include automotive exterior body panels, medical implants such as the artificial heart, membranes, ski boots and flexible tubing. TPU features the best physical properties of all elastomers exhibiting high modulus and abrasion resistance, with excellent resistance to aging (ozone), atmospheric factors, typical solvents and oils. Moreover, TPU is capable of bearing greater loads than other rubbers. But the properties of TPU can be changed by the environment in that a physical or chemical process may occur. There are three environmental effects considered particularly important for polyurethane elastomers: the effect of heat, swelling by immersion in certain fluids and hydrolytic resistance. In certain fluids such as alcohols, acids, bases, ketones and esters, polyurethanes tend to show swelling, degradation and changes in some mechanical properties.

From the discussion above, it is interesting to study polymer blends of TPU with other polymers. Polydimethysiloxane (PDMS) was chosen to blend with TPU because PDMS is a polymer with a unique combination of properties resulting from the presence of an inorganic siloxane backbone with organic methyl groups attached to the silicon. This chemical configuration produces polymers which have good chemical stability, in particular good oxidative resistance. Thus, this research is aimed to study the tensile properties and the morphology of TPU/PDMS blends. The effects of chemical reagents on the tensile properties and the morphology of TPU/PDMS blends are investigated.

Materials and Experiments

Materials: Thermoplastic polyurethane used in this study is S385A series with the trade name of Skythane, obtained from SK Chemicals Co, Ltd. S385A is a polyester-based thermoplastic polyurethane of which the hard segment is 4,4'-diphenylmethane diisocyanate (MDI) extended with 1,4-butanediol (BO). The soft segment is polyester polyol and 1,4-butanediol.

Polydimethylsiloxane used in this study was obtained from Dow Corning Co, Ltd; it is transparent fluid with a viscosity of 60,000 centistoke.

Sample preparation: The master batch of TPU/PDMS at 2% by weight of PDMS was prepared using the internal mixer of Haake "Rheomix3000p" at 190C for 10 min with a rotor speed of 30 rpm. After that the master TPU/PDMS batch was compressed and cut into pieces. The master batch was used to blend with the virgin TPU, using the counter-rotating twin screw extruder of Haake, to obtain the blends with the final PDMS concentration of 0.2%-1% in 0.2% increments. The extruded blends were cooled and finally cut into pellet form. The virgin TPU sample was treated by the same procedure for use as a reference sample.

Measurements: Tensile properties were measured following ISO 527-1 of the British Standards Institution (BSI) with "Lloyd LR 5K". Chemical resistance of all samples was tested following ASTM D543-95. Chemical reagents used were distilled water, sulfuric acid (H_2SO_4, 3% v/v) and sodium hydroxide (NaOH, 10% w/v). Samples were immersed in chosen chemical reagents at 25C for 168 hr, after which they were dried and tested for tensile strength. Scanning electron microscopy (SEM), using a JEOL JSM-6400 instrument, was used to study sample morphology.

Results and Discussion

Since PDMS is a liquid, the internal mixer is first employed for preparation of the master batch before final blending of the master batch with virgin TPU in the twin screw extruder. Using this procedure, the maximum concentration of PDMS in the TPU/PDMS blend is found to be 1% of PDMS; more than 1% excess PDMS resulted in unmelted TPU due to the short residence time in the twin screw extruder.

Comparison of the ultimate tensile strength, the elongation at break, Young's modulus and the energy to break, of the blended TPU at all compositions before and after immersion in various chemical reagents are shown in Figures 1-4.

Comparing to the virgin TPU, the prepared blends, before immersion, show an increase in the elongation at break up to 30% and in Young's modulus up to 40% at the optimum PDMS concentration of around 0.6-0.8%, beyond which these properties diminish. The ultimate tensile strength and the energy to break are decreased by about 20% and 10%, respectively. The optimum PDMS concentration on a similar system was also reported by Hill *et al*[5] but at higher values than our results. This could arise from differences in the structure and characteristics of the polymer components. The dependence of tensile properties on PDMS

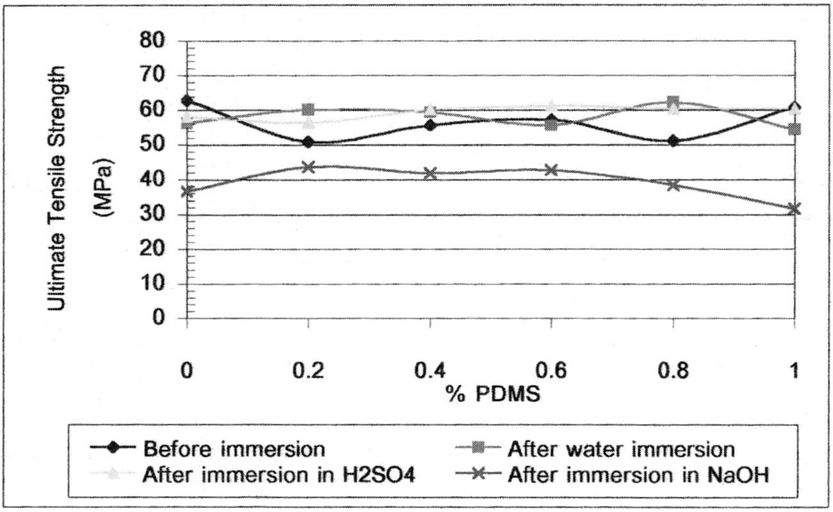

Fig. 1. Comparison on the ultimate tensile strength of the virgin TPU and blended TPU before and after immersion in various chemical reagents

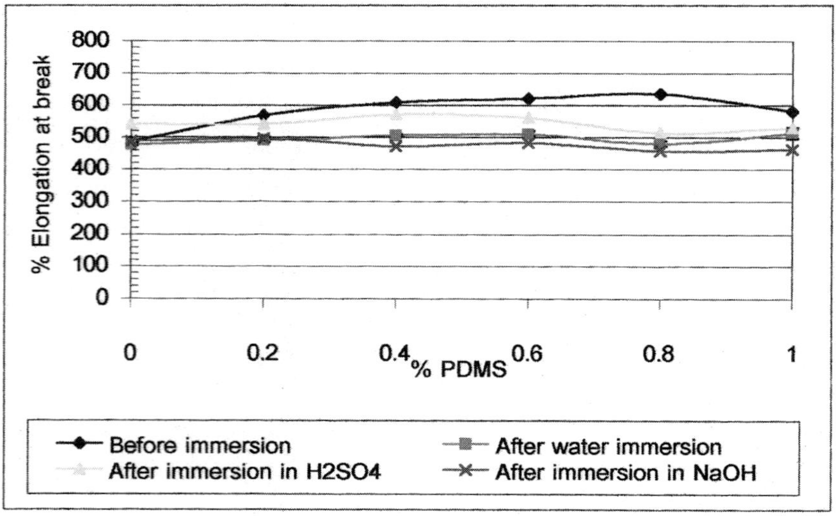

Fig. 2. Comparison on the elongation at break of the virgin TPU and blended TPU before and after immersion in various chemical reagents.

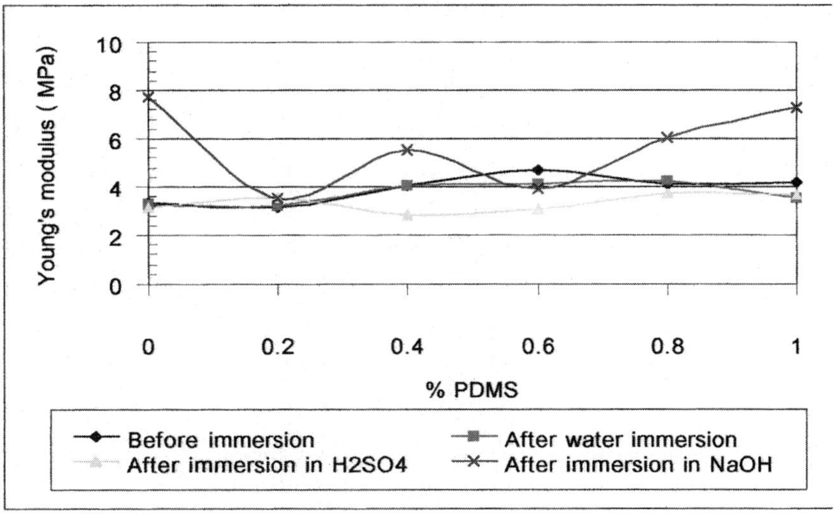

Fig. 3. Comparison on Young's modulus of the virgin TPU and blended TPU before and after immersion in various chemical reagents.

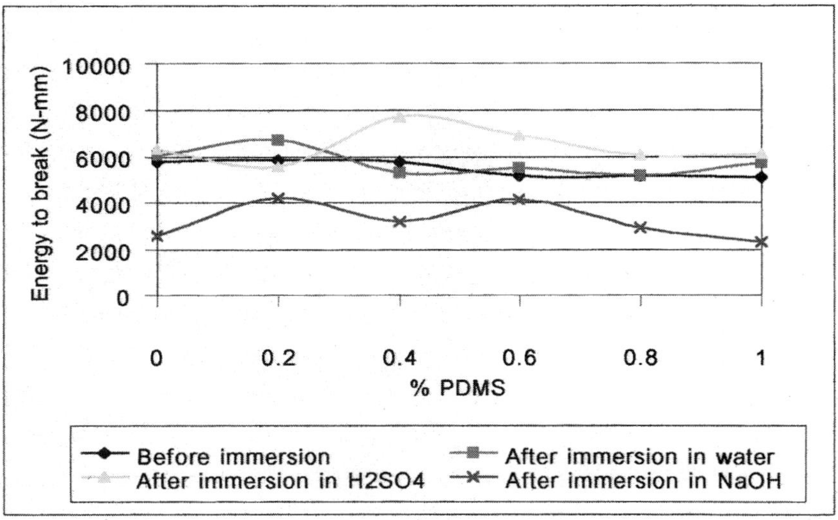

Fig. 4. Comparison on the energy to break of the virgin TPU and blended TPU before and after immersion in various chemical reagents.

concentration in the blends can be preliminary explained by the change of the morphology and the fractured mechanism of the blends, as observed by SEM (Figures 5-8).

Fractured surface of blended TPU at 0.2% and 0.4% PDMS concentrations in Figure 6 and Figure 7 is different from that of virgin TPU (Figure 5), with less fibrous characteristic in the blends and with, in addition, areas of craze structure. In contrast, the fibrous characteristic and craze formation in blended TPU at 0.8% PDMS concentration, shown in Figure 8, is not visible. In this case the fractured structure appears in a chunky crack form. It is evident that there are three different fracture characteristics. The first, of a fibrous nature, occurs with the virgin TPU. The second, craze formation with some fibrous characteristics, is observed in the blended TPU at 0.2% and 0.4% of PDMS. The third, the chunky formation without any of the fibrous characteristics observed in virgin TPU, occurs in the blended TPU with 0.8% of PDMS. A change in the phase morphology of the TPU blend at the optimum PDMS concentration corresponds to the results on tensile properties, where the reduced effect of PDMS content on the tensile properties of TPU blends occurred at around 0.6%-0.8% PDMS concentration.

The ultimate tensile strength, Young's modulus and the energy at break of the blends at all compositions after immersion in water and acid (3% H_2SO_4), comparing to those before immersion, are not significantly different. In the case of acid immersion, a slight effect of PDMS contents on the acid resistance to the tensile properties is noticed.

In the case of base immersion (10% NaOH), the results on tensile properties show that base immersion has a significant effect on tensile properties. The effect of PDMS concentrations on the base resistance to the tensile properties is marked, i.e. the effective PDMS contents in blended TPU does, more or less, not exceed 0.8%. At 1% PDMS concentration, the effect on each property diminishes, with the value of each property being close to that of pure TPU. This is similar to the results found in the case of TPU/PDMS blends before immersion.

Comparing immersion in different chemical reagents, NaOH (10% w/v) has the strongest influence on the tensile properties of virgin TPU and blended TPU. For virgin TPU after base immersion comparing to before base immersion the ultimate tensile strength and energy at break is strongly decreased by 42% and 55%, respectively. This could be caused by the base hydrolysis of the polyester soft segment of polyurethane[12]. The effect of PDMS concentration on the base resistance to tensile properties is similar to results before immersion, i.e. the effective PDMS contents in the blends that can generally improve tensile properties of the blends after immersion in NaOH does not excess 0.8%. The results are in agreement with the weight loss of TPU/PDMS blends after base immersion and the morphology of the fractured

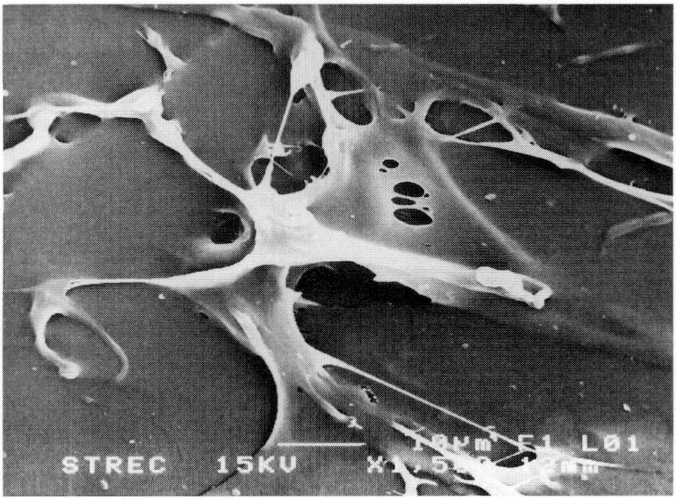

Fig. 5. SEM photograph of fractured surface of virgin TPU.

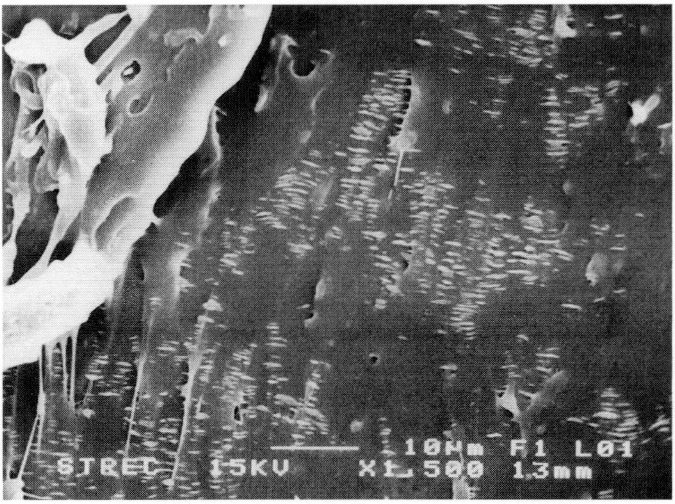

Fig. 6. SEM photograph of fractured surface of TPU/PDMS blends at 0.2% PDMS.

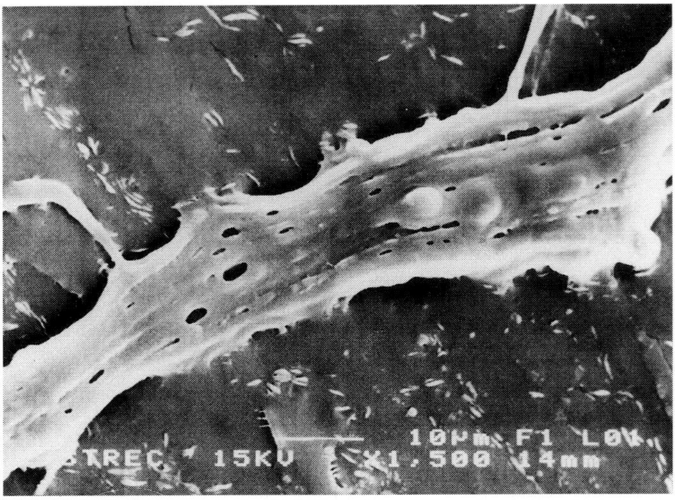

Fig. 7. SEM photograph of fractured surface of TPU/PDMS blends at 0.4% PDMS.

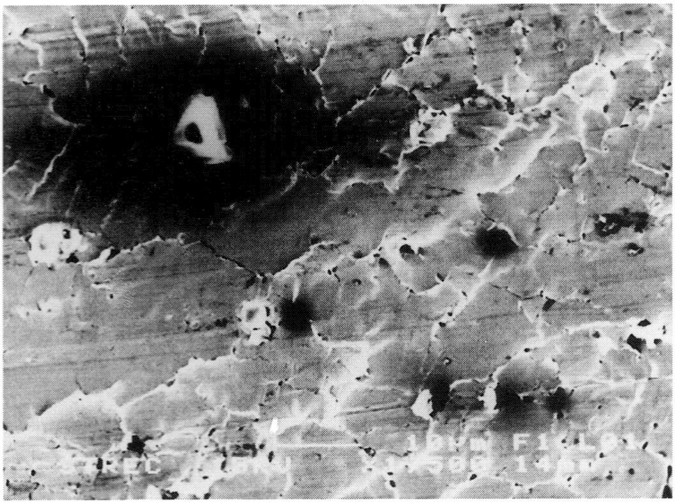

Fig. 8. SEM photograph of fractured surface of TPU/PDMS blends at 0.8% PDMS.

surface of TPU/PDMS blends after base immersion that exhibits very small amount of fibrous characteristic and there are some small particles detached at the surface (Figure 9). This could be the result of a corrosive reaction between the sample surface and NaOH solution or the degradation of the soft domain of polyurethane in base solution as mentioned previously. Further study will be performed to clarify this point.

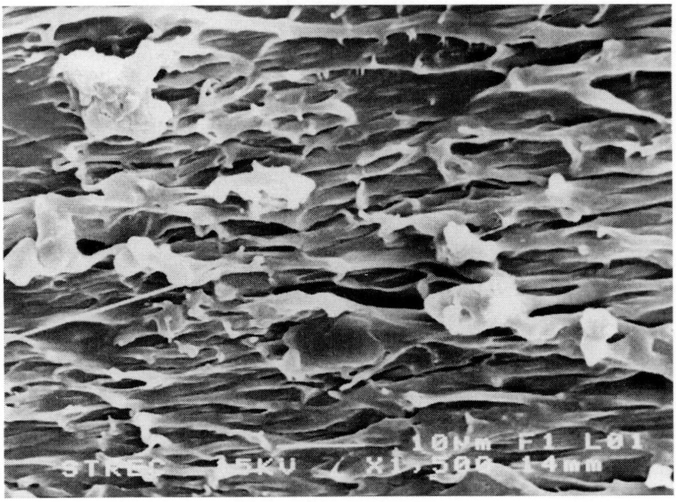

Fig. 9. SEM photograph of fractured surface of TPU/PDMS blends at 0.8% PDMS after base immersion.

Acknowledgements

S.D. gratefully acknowledges the financial support from the Thailand Research Fund (TRF).

[1] D.R. Paul, C.B. Bucknall, *"Polymer Blends volume 1 :Formulation"*, J. Wiley & Sons, New York 2000.
[2] L.A. Utracki, *"Polymer Alloys and Blends"*, Hanser Publishers, New York 1990.
[3] D.J. Walsh, J.S. Higgins, A. Maconnachie, *"Polymer Blends and Mixtures"*, Martinus Nijhoff, Dordrecht 1985.
[4] G. Oertel, *"Polyurethane Handbook"*, Hanser Publishers, Germany 1985.
[5] D.J.T. Hill, M.I. Killeen, J.H. O'Donnell, P.J. Pomery, D.S.T. John, A.K. Whittaker, *Journal of Applied Polymer Svience*, **1996**, *61*, 1757-1766.
[6] S.N. Jaisankar, G. Radhakrishnan, *Polymer Engineering and Science*, **2000**, *40*, 621-626.
[7] Y. Kim, W.J. Cho, C.S. Ha, *Journal of Applied Polymer Science*, **1999**, *71*, 451-422.
[8] P. Potschke, K. Wallheinke, H. Fritsche, H. Stutz, *Journal of Applied Polymer Science*, **1997**, *64*, 749-762.
[9] A. Stanciu, A. Airinei, D. Timpu, A. Ioanid, C. Ioan , V. Bulacovschi, *European Polymer Journal*, **1999,** *35*, 1959-1965.
[10] C. Hepburn, *"Polyurethane Elastomers"*, Applied Science Publication, England 1982.
[11] Z. Wirpaza, *"Polyurethanes (Chemistry, Technology and Applications)"*, Ellis Horwood, Great Britain 1993.
[12] M. Furukawa, T. Yokoyama, *Journal of Polymer Science: Polymer Letters Edition,* **1979**, 17, 175-180.

Bulk and Surface Properties of Blends with Semifluorinated Polymers and Block Copolymers

Doris Pospiech, Liane Häußler, Dieter Jehnichen, Wolfram Kollig, Kathrin Eckstein, Karina Grundke*

Institute of Polymer Research Dresden, Hohe Str. 6, 01069 Dresden, Germany

Summary: The goal of the investigation presented here is the development of extremely hydrophobic materials based on polysulfone that can be applied, for instance, as fouling-resistant membrane materials. The concept used is the addition of semifluorinated polymers to polysulfone in suitable blend compositions. The influence of molecular parameters like chain structure of the semifluorinated polymer (segmented block copolymers, random copolymers) and segment molecular weight on the state of phase separation in the bulk and its influence on the surface properties have been systematically examined. It could be shown that segmented block copolymers with semifluorinated polyester segments with intermediate segment molecular weight are more suitable in blends with polysulfone than random polysulfone copolymers having semifluorinated side chains with respect to form homogeneous thin films (coatings) with highly non-wetting properties.

Keywords: blends, fluorinated polymers, polysulfone, self-organization, wetting

Introduction

Compounds with ultrahydrophobic properties are of special interest for materials with drastically reduced wetting, for instance for the development of self-cleaning or bioinert surfaces for implants that suppress adhesion of proteins in contact with blood. Semifluorinated low molecular compounds as well as polymers have extremely low surface free energy[1-7] which is the prerequisite for ultrahydrophobicity. On the other hand, they are expensive and, therefore, often applied only as an additive, using the ability of fluorinated compounds to segregate to the surface.[8-9] The quality of the surface depends significantly on the state of miscibility between additive and matrix polymer. Here, we want to discuss the influence of semifluorinated polymers with different chain structure on the state of miscibility in binary blends with polysulfone (bulk properties) and the influence of the solid state structure on the surface properties of thin spin-coated films of these blends in order to achieve stable and homogeneous films with low surface free energy (i.e., high contact angles versus water). The influence of chemical composition

 DOI: 10.1002/masy.200350836

(concentration and distribution of side chains) and phase separation in the bulk on the phase separation in blends with polysulfone will be outlined for the pure semifluorinated polyester **1**, segmented block copolymers (BCP) having semifluorinated (SF) polyester segments and polysulfone blocks **2**, as well as for copolymers of polysulfone (PSU) having semifluorinated side chains attached to 4,4'-bis(oxyphenyl)pentanoic acid randomly distributed along the chain **3**.

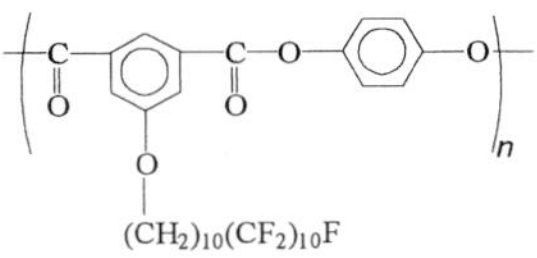

1: semifluorinated polyester (SF-PES)

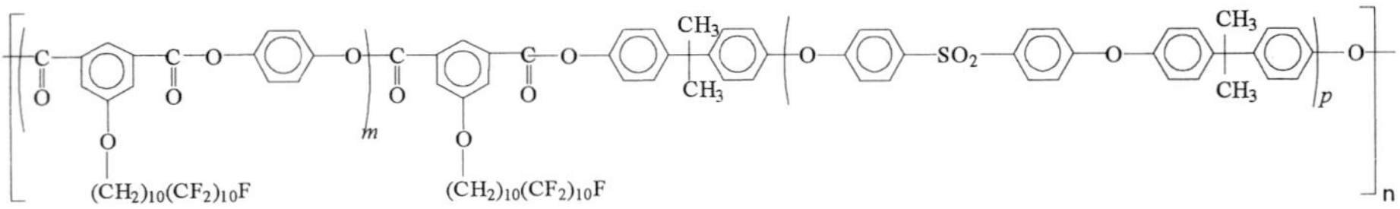

2: segmented semifluorinated polyester / PSU block copolymer (BCP)

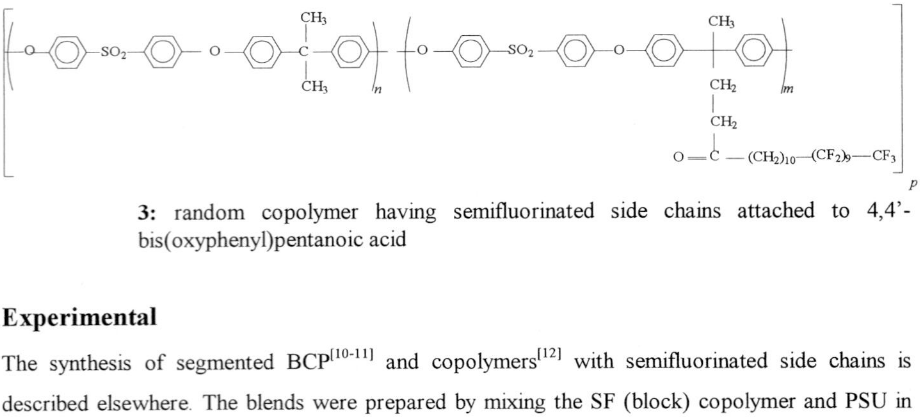

3: random copolymer having semifluorinated side chains attached to 4,4'-bis(oxyphenyl)pentanoic acid

Experimental

The synthesis of segmented BCP[10-11] and copolymers[12] with semifluorinated side chains is described elsewhere. The blends were prepared by mixing the SF (block) copolymer and PSU in the appropriate weight ratio in a trifluoroacetic acid / chloroform mixture (1/1 v/v, c_{pol}: 60 mg / 20 ml), precipitation into methanol, filtering off and drying in vacuum at 50 °C for 5 h. DSC measurements were carried out with a Perkin Elmer DSC 7 (heating and cooling rate: 20 K/min). The blend compositions in the text are given in wt/wt%.

Thin films for surface characterization were prepared by spin coating (Headway Inc., USA, 2000 rpm/60 s) solutions of the blends (c_{pol}: 1 wt%, solvents: pentafluorophenol/ chloroform for BCP/PSU blends and chloroform for random copolymer / PSU blends) on pre-cleaned silicon wafer. The films were dried at 50 °C for 4 h in vacuum and then annealed at 200 °C in vacuum for 4 h. Dynamic contact angles were measured using ADSA-P[13] and water as probe. AFM measurements of the same surfaces were carried out by means of a Nanoscope III (Digital Instruments, USA) in tapping mode for both height and phase imaging (tapping frequency 70 kHz, free amplitude 4 V). The area roughness was estimated from 10×10 μm^2 pictures by RMS. X-ray photoelectron spectroscopy was performed using an AXIS Ultra instrument (Kratos Analytical, UK) with AlK$\alpha_{1,2}$ radiation and pass energies of 160 eV or 20 eV, respectively.

Results and Discussion

Bulk Structure of the Semifluorinated Polyester and Block Copolymers

All polymers under investigation contain –oxydecylperfluorodecyl side chains ($-O(CH_2)_{10}-(CF_2)_9-CF_3$). The strong microphase separation in the pure semifluorinated polyester **1** is caused by the thermodynamic immiscibility between oxydecyl- and perfluorodecyl part of the side chains. The major reason for the extremely low surface free energy is the formation of a well-ordered surface layer of CF_3 groups which is supported by the highly ordered bulk structure.[14] This structure can be maintained in the segmented BCP **2** presumed a certain molecular weight of the semi-fluorinated blocks, i.e., the state of phase separation in the segmented block copolymers can be controlled by the molecular weights of both PSU and SF.[10,11] Three types of segmented BCP having different degrees of phase separation were examined with respect to their influence in blends with PSU (Figure 1).

Phase separated SF-PES/PSU block copolymers consist of domains containing the self-organized SF-side chains embedded in the amorphous PSU matrix (PSU segments).[10] This proposed structure model was confirmed by temperature-resolved SAXS/WAXS measurements.[10,11,20] A similar structure can be assumed based on WAXS/SAXS results for the random copolymers **3** independently on the concentration of SF-side chains.[12]

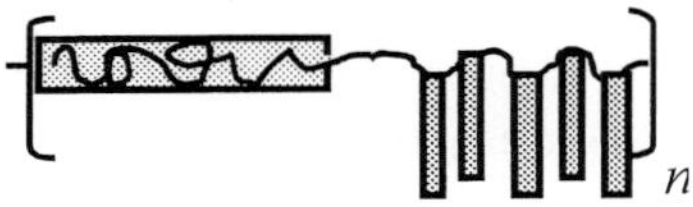

1) weakly phase separated BCP: SF-PES 3,600 g/mol / PSU 2,300 g/mol.
2) strongly phase separated BCP: SF-PES 9,000 g/mol / PSU 9,000 g/mol.
3) non-phase separated BCP: SF-PES 1,750 g/mol / PSU 9,000 g/mol.

Fig. 1. Schematic structure of the segmented polysulfone / semifluorinated polyester block copolymers (left: PSU block; right: SF-PES block).

Evaluation of Phase Separation in the Blends (Bulk Structure)

Blends of the semifluorinated polyester (poly(p-phenylene-oxydecylperfluorodecyl isophthalate) with PSU

Calculation of the Flory-Huggins interaction parameter χ between PSU and the semifluorinated polyester based on the solubility parameter concept [15,16] gave a χ of 13.6 referring to the strong demixing tendency between both polymers that should even be enforced by the self-organization of the SF-PES in domains. Accordingly, blends of **1** with polysulfone are completely immiscible, as illustrated in Figure 2 by the DSC curves (2nd heating) of solution precipitated blends.

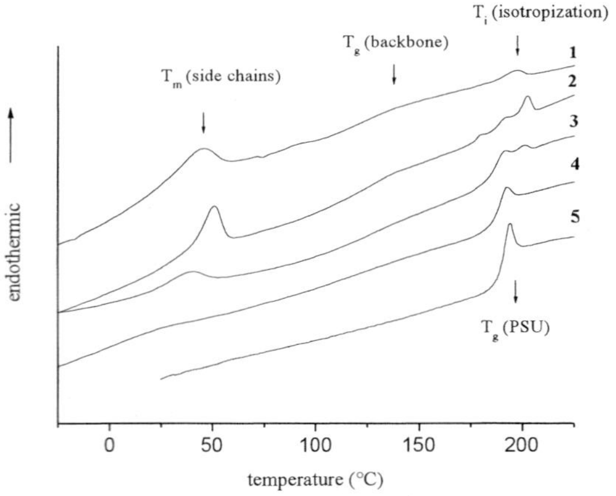

Fig. 2. DSC curves (2nd heating) of blends of semifluorinated polyesters and polysulfone with different composition (1: SF-PES; 2: SF-PES/PSU = 75/25; 3: SF-PES/PSU = 50/50; 4: SF-PES/PSU = 25/75; 5: PSU).

The SF-PES (curve 1) shows three distinct thermal transitions which were assigned recently by a combination of methods like DSC, temperature-resolved SAXS/WAXS, rheology and pVT measurements. [18,19] The first one is the melting of semifluorinated side chains that are then still trapped within the rigid lattice of backbones, the second one is the glass transition of the polyester backbone, causing simultaneously a smectic-smectic transition in the order, and the third one is the isotropization after that the polymer melt does not show signs of ordering, as detected by temperature-dependent SAXS. The blends of SF-PES and PSU show all transitions found in the SF-PES as well as the glass transition of PSU without change in temperature, referring to complete immiscibility.

Blends of segmented block copolymers with PSU

The influence of the segmented SF-PES / PSU block copolymers in binary blends with polysulfone was examined in dependence on the state of phase separation and the molecular weight of the block copolymer segments. It was expected that long PSU segments of block copolymers will mix with the PSU matrix while shorter ones form two phases (PSU matrix and PSU phase of the block copolymer). Semifluorinated polyester segments will, on the other hand, crystallize in their own domains if the BCP used in the blend is phase separated. DSC results obtained for blends with different composition are shown in Figures 3-5. The shift of phase behavior of the BCP in the blends is visible. Figure 3 shows the DSC curves of blends of PSU with the weakly phase separated BCP 1. The appearance of two glass transitions indicates the demixing of the amorpous phases of both PSU and SF-PES, whereas the crystallization (or self-organization in smectic layers) of semifluorinated side chains in the SF-block is suppressed due to the rather short block molecular weight. The blend of PSU with the strongly phase separated BCP 2 represents a partially miscible blend system. The DSC curves given in Fig. 4 show the typical melting transitions of side chain melting of SF-side chains at about 60 °C, which can additionally be detected by temperature-resolved SAXS. That means, extending the SF-block in comparison to BCP 1 induces an increased demixing not only in the block copolymer but also in the blend. Mixing PSU with a non-phase separated block copolymer (having only very short SF-blocks and long PSU blocks) results in a miscible blend, which can be detected by the appearance of only a single T_g.

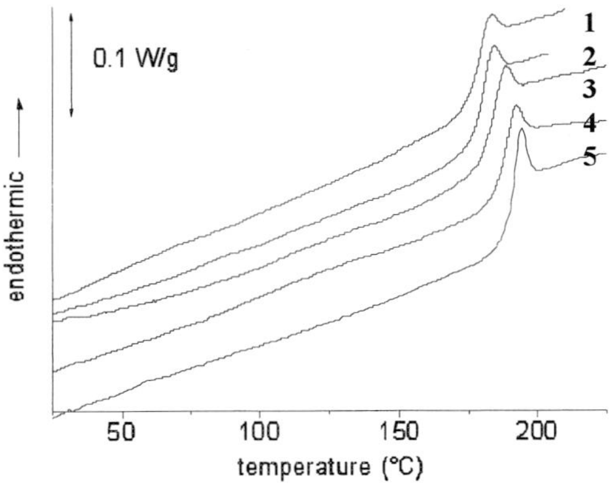

Fig. 3. DSC curves (2nd heating) of blends of a weakly phase separated block copolymer (SF-PES 3,600/PSU 2,300) and polysulfone with different composition (1: BCP; 2: BCP/PSU = 75/25; 3: BCP/PSU = 50/50; 4: BCP/PSU = 25/75; 5: PSU).

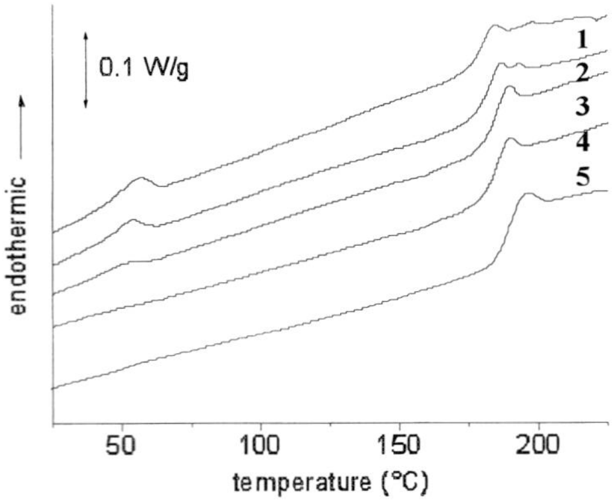

Fig. 4. DSC curves (2nd heating) of blends of a strongly phase separated block copolymer (SF-PES 9,000/PSU 9,000) and polysulfone with different composition (1: BCP; 2: BCP/PSU = 75/25; 3: BCP/PSU = 50/50; 4: BCP/PSU = 25/75; 5: PSU).

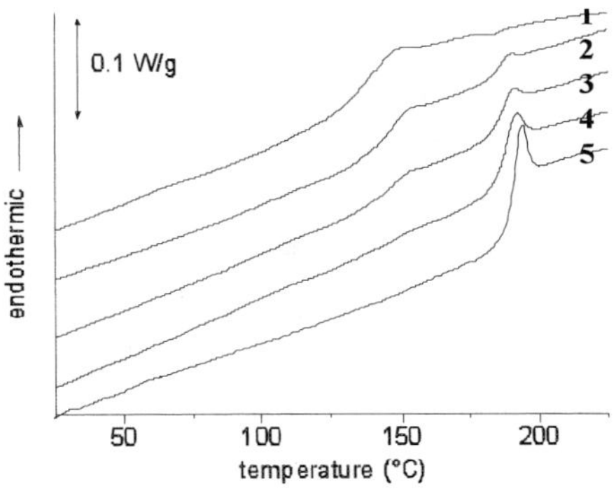

Fig. 5. DSC curves (2nd heating) of blends of a non-phase separated block copolymer (SF-PES 1,750/PSU 9,000) and polysulfone with different composition (1: BCP; 2: BCP/PSU = 75/25; 3: BCP/PSU = 50/50; 4: BCP/PSU = 25/75; 5: PSU).

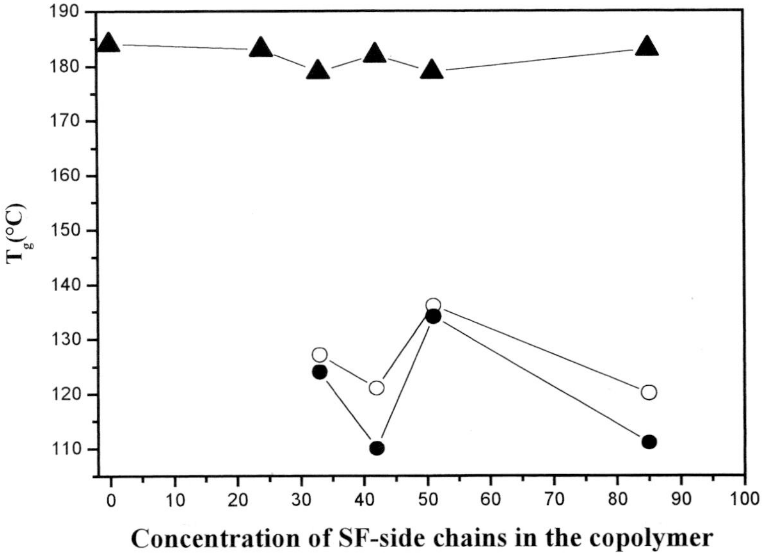

Fig. 6. T_g of blends (50/50 wt/wt) of polysulfone with copolymers of polysulfone with 4,4'-bis(oxyphenyl)pentanoic acid with -oxydecylperfluorodecyl side chains (○: T_g of copolymers; ● $T_{g,\,copolymer}$ in the blend; ▲: $T_{g,\,PSU\,matrix}$ in the blend).

Thus, it can be noted that the degree of phase separation in the blends can be controlled by the molecular weights of both, PSU and SF-PES segments in the BCP. The critical molecular weight of the PSU segments to achieve miscibility with the PSU matrix is in the region of 9,000 g/mol which is about four times the entanglement molecular weight of PSU and thus very high compared to other segmented PSU block copolymers investigated.[17] An explanation therefore is the self-organization tendency of the SF-PES enhancing the phase separation also in the blends enhancing the demixing in the blend.

Blends of semifluorinated random PSU copolymers with PSU

All blends of random PSU copolymers having different concentrations of SF-side chains are not miscible with PSU according to DSC results (Figure 6). Two glass transitions can be observed, the higher one caused by the PSU homopolymer and the second one by the random copolymer phase in the blend. Both T_g are only slightly shifted compared to the PSU matrix homopolymer and the pure random copolymers, indicating the strong immiscibility of the blend.

Surface Properties of the Blends

The surface properties of the blends were examined on thin films spin coated from a solution containing both blend partners on Si wafer. Thus, in most cases except the homopolymer blends smooth films with an averaged roughness between 5 and 30 nm were obtained. The roughness decreases after annealing above the glass transition of PSU (200 °C) to 2 – 16 nm. The roughness of these films does not influence the contact angles measured because it is well below the critical roughness in the order of several hundreds of nanometers up to 1 μm which was found in systematic contact angle investigations of polymer films.[21,22] The films were dried after spin coating at 50 °C and after measurement annealed to come closer to thermodynamic equilibrium. A real thermodynamic equilibrium in the samples will not be reached after procedure. The results obtained for the blends were interpreted using the results of wetting measurements of the block copolymer films reported recently. [10,11,20,22] For the segmented SF-PES/PSU block copolymers a dependence of the contact angle on the segment molecular weight of the semifluorinated segments was shown, i.e., the contact angle increases with raising molecular weight of the blocks.

The film thickness of the blend films were in the range of 150 to 250 nm, as determined by ellipsometry. We are well aware that in these thickness ranges influences of the film thickness, i.e., due to confinement effects, as described for diblock copolymers frequently (see, for instance, refs.[24,25] may play an important role. However, the thickness of all samples are comparable and the contact angles reported describe thus a certain, defined state after annealing. The situation may change if very thick films in the micrometer range are investigated.

Blends of the semifluorinated polyester (poly(p-phenylene-oxydecylperfluorodecyl isophthalate) homopolymer with PSU

Due to the high immiscibility of SF-PES and PSU the polymer films formed by spin coating on Si wafer are very heterogeneous. With only one wt% SF-PES, PSU films with high roughness having structural maxima of about 250 nm in height detected by AFM are obtained. The contact angle found here is 118° which is 13° higher than that of pure PSU and only 6° lower than that of the semifluorinated polyester. Higher concentrations of SF-PES in the blend cause highly rough surfaces with pretty low contact angles (see Figure 7) even after annealing which usually results in smoothing of spin coated films. Due to that reason films containing these semifluorinated aromatic polyesters **1** need indispensably compatibilization by block copolymers and cannot be used to get an exact information on surface free energies.

Blends of segmented block copolymers with PSU

Figure 7 summarizes the results of contact angle measurements (using water as probe) on surfaces of blends of block copolymers with PSU. It is obvious that the degree of phase separation in the blends also plays a major role for the wetting behavior of the films.

The wetting behavior of the completely immiscible PSU/SF-PES blend is strongly determined by the heterogeneous surface resulting from demixing during blend preparation, drying and annealing. The low contact angles are due to these heterogeneous surfaces. The miscible blend shows contact angles that almost follow an additivity rule between those of PSU and the respective block copolymer. Thus, the miscible blends behave like a copolymer. The good mixing of both polymers prevents obviously additional surface segregation of the SF-blocks of BCP 3. In contrast, higher contact angles than in the pure blend components are found on surfaces of partially miscible blends which can only be explained by segregation of SF-PES segments from their phase separated domains to the surface. A comparable effect was found earlier on films of the pure block copolymers having a phase separated surface morphology.[22]

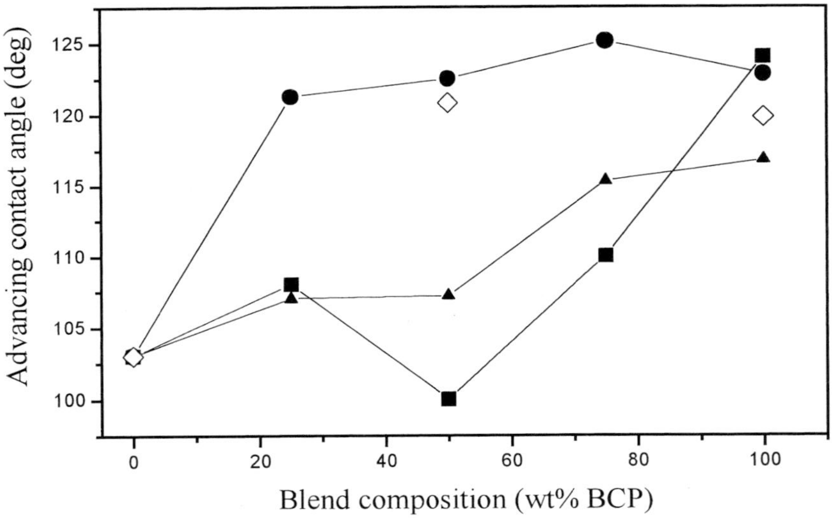

Fig. 7. Water contact angle on surfaces of thin films of blends consisting of PSU and segmented BCP (SF-PES/PSU), all samples annealed for 4 hrs at 200 °C in vacuum.

■: immiscible homopolymer blend of PSU with SF-PES
●: partially miscible blend of PSU with BCP 2
◇: immiscible blend of PSU with BCP 1
▲: miscible blend of PSU with BCP 3.

Surface segregation of the semifluorinated side chains was proven by XPS investigation. The results for blends of BCP 3 and PSU are given in Table 1. The surface segregation is stronger in the blends than in the pure BCP which might be due to the stronger fixation of SF chains in the ordered SF domains in the BCP than in the SF domains of the blend (lower degree of ordering). Surface segregation is fostered by consecutive *annealing* of the samples. Annealing (temperature and time) of the partially miscible blends causes the decrease of roughness of the films (see as an example the development of roughness in the blend (BCP 2 (SF-PES 9,000 / PSU 9,000) / PSU in dependence on the annealing temperature in Table 2), and a slight change of contact angle. In contrast, the contact angles of miscible BCP/PSU blends (for example BCP 3/PSU) are increased by annealing which seems to be induced by a dewetting process of the miscible film from the substrate, as observed by AFM. The dewetting process causes a higher roughness than in the original sample, as reported earlier in ref. [24] The film roughness can also be controlled by the composition of the blend as illustrated in Figure 8 for different binary and ternary blends of PSU, SF-PES and block copolymers. The addition of BCP always results in smoothing of the films.

Table 1. [F]/[C] ratio found by XPS at different take-off angles in blends of BCP 2 (SF-PES 9,000 / PSU 9,000) / PSU after annealing at 200 °C for 4 h.

[F]/[C] ratio	Blend composition (BCP 2 / PSU) 25/75	50/50	72/25	BCP 2
take-off angle 60°	0.582	0.531	0.554	0.480
take-off angle 45°	0.550	0.542	0.553	0.450
take-off angle 0°	0.564	0.556	0.590	0.500

It could be shown that addition of 5 wt% of the block copolymer to the homopolymer blend results in a significant decrease of roughness of the samples (#6 in Fig. 8).

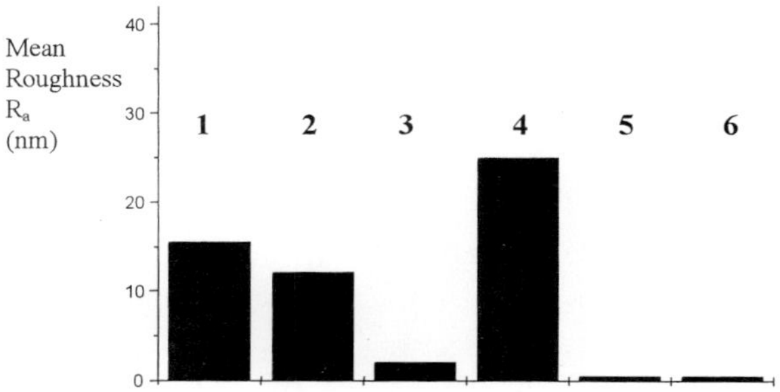

Fig. 8. Roughness R_a determined by AFM of binary and ternary blends of PSU, SF-PES and BCP with different composition.
(1: PSU / BCP (PSU 2300/SF-PES 1850), (25/75); 2: PSU / BCP (PSU 2300/SF-PES 1850), (50/50); 3: PSU / BCP (PSU 2300/SF-PES 1850), (75/25); 4: SF-PES / BCP (PSU 2300/SF-PES 1850), (95/5); 5: PSU / BCP (PSU 2300/SF-PES 1850 g/mol), (95/5); 6: PSU / SF-PES / BCP (PSU 2300/SF-PES 1850 g/mol), (75/20/5).

Table 2. Roughness of films of blends of BCP 2 (SF-PES 9,000 / PSU 9,000) and PSU in dependence on the annealing temperature (annealing time: 4 h).

R_a (nm)	Blend composition (BCP / PSU)		
	25/75	50/50	75/25
annealed at 90 °C	7.4	16.5	28.0
annealed at 150 °C	8.0	13.5	27.0
annealed at 200 °C	2.1	12.2	15.6

Blends of semifluorinated random PSU copolymers with PSU

In spite of the strong immiscibility of the blends smooth films could be prepared by spin coating the blend solutions. However, contact angles obtained on dried and annealed blend films showed a significant difference to the pure copolymers with semifluorinated side chains (Figure 9).

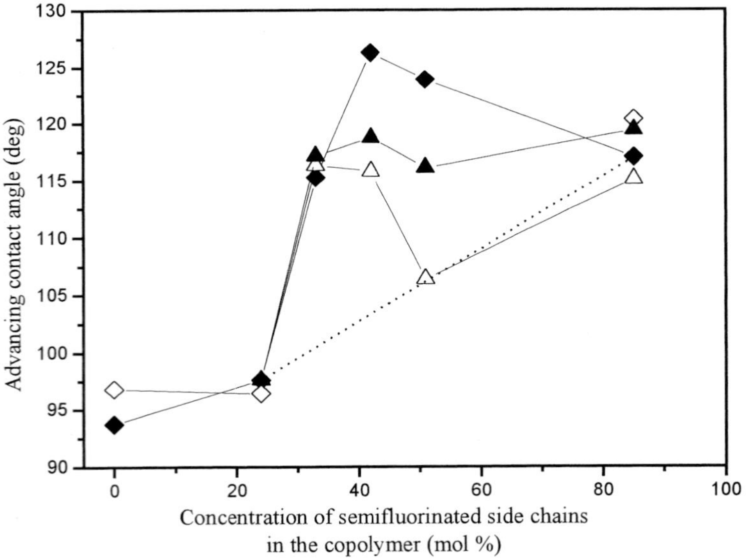

Fig. 9. Contact angles of PSU copolymers with semifluorinated side chains and their 50/50 (wt/wt) blends with PSU, dried (50 °C / 4 h / vac) and after annealing (200 °C / 4 h / vac) (♦: copolymers, dried; ◊: copolymers, annealed; ▲: blends, dried; Δ: blends, annealed).

The contact angles of 50/50 (wt/wt) blends of the copolymers with PSU are lower than those of the pure copolymers. Annealing only slightly affects the advancing contact angles of the copolymers.

Conclusion

Segmented block copolymers of polysulfone and semifluorinated polyesters with medium segment molecular weights are more effective to compatibilize blends with the PSU homopolymer than random PSU copolymers with semifluorinated substituents distributed along the polymer chain. The reason is the strong self-organization of the semifluorinated chains in the latter preventing the mixing of the phases. Partial miscibility between the segmented BCP and PSU matrix in a blend is a necessary requirement to obtain homogeneous, stable surfaces with low surface free energy because partial demixing allows surface segregation of non-self organized semifluorinated chains. Thus, control of phase separation in the bulk and at the surface is possible in such blends and can successfully be used to tailor surface properties of films.

Acknowledgements

Financial support by German Science Foundation and NATO (collaborative linkage grant) is gratefully acknowledged. The authors would like to thank Mrs. Monika Dittrich, Mrs. Kerstin Arnold and Mrs. Kathrin Pöschel for valuable technical assistance.

[1] K. Song, J. R. Twieg, J. F. Rabolt, *Macromolecules* **1990**, *23*, 3712.
[2] W. A. Zisman,. "Contact angle, Wettability and Adhesion", *Advances in Chemistry Ser.* **43**, American Chemical Society, Washington, DC 1964.
[3] E. Lindner, *Recent Dev. Biofouling Control*, **1994**, 305.
[4] J. Höpken, M. Möller, *Macromolecules,* **1992**, *25*, 1461.
[5] J. Wang, G. Mao, C. K. Ober, E. J. Kramer, *Macromolecules*, **1997**, *30*, 1906.
[6] M. Xiang, X. Li, C. K. Ober, K. Char, J. Genzer, E. Sivaniah, E. J. Kramer, D. A. Fischer, *Macromolecules*, **2000**, *33*, 6106.
[7] D. Jehnichen, D. Pospiech, A. Janke, P. Friedel, L. Häußler, A. Gottwald, S. Kummer, W. Kollig, K. Grundke, *Materials Science Forum*, **2001**, 378-381, 378.
[8] J.-M. Corpart, *Melliand Textilber.,* **1997**, *9*, 625.
[9] R. Bongiovanni, N. Pollicino, G. Gozzelino, G. Malucelli, A. Priola, B. Ameduri, *Polym. Adv. Technol.* **1996**, *7*, 403.
[10] D. Pospiech, L. Häußler, K. Eckstein, D. Voigt, D. Jehnichen, A. Gottwald, W. Kollig, A. Janke, K. Grundke, C. Werner, H.R. Kricheldorf, *Macromol. Symp.* **2001**, 163, H. Cherdron, B. Sandner, H. U. Schenk, B. Voit (Eds.), "Tailormade materials", ISSN 1022-1360, pp. 113.
[11] D. Pospiech, L. Häußler, K. Eckstein, H. Komber, D. Voigt, D. Jehnichen, P. Friedel, A. Gottwald, W. Kollig, H. R. Kricheldorf, *High Performance Polymers* **2001**, 13, 2, 275.

[12] to be publ.
[13] D. Y. Kwok, T. Gietzelt, K. Grundke, H. J. Jacobasch, A. W. Neumann, *Langmuir* **1997**, *13*, 2880.
[14] P. Friedel, D. Pospiech, D. Jehnichen, J. Bergmann, C. K. Ober, *J. Polym. Sci., Phys.* **2000**, 38, 1617.
[15] D. Pospiech, A. Gottwald, D. Jehnichen, P. Friedel, C. Harnisch, D. Voigt, G. Khimich, A. Yu. Bilibin, *Colloid & Polymer Sci.* **2002**, 280, 1027.
[16] D. W. Van Krevelen, "Properties of polymers, their correlation with chemical structure; their numerical estimation and prediction from additive group contributions", Elsevier, Amsterdam Oxford, New York, Tokyo 1990.
[17] D. Pospiech, L. Häußler, K. Eckstein, D. Jehnichen, A. Gottwald, P. Pötschke, *Macromol. Symp.* **2000,** 149, "Polymer Blends", M. Stamm, Ed., Wiley-VCH, Weinheim, ISBN 3-527-30 128-3, pp. 219.
[18] A. Gottwald, D. Pospiech, D. Jehnichen, L. Häußler, P. Friedel, J. Pionteck, M. Stamm, G. Floudas, *Macromol. Chem. Phys.,* **2002**, 203, 5-6, 854.
[19] D. Pospiech, D. Jehnichen, A. Gottwald, L. Häussler, U. Scheler, P. Friedel, W. Kollig, C. K. Ober, X. Li, A. Hexemer, E. J. Kramer, D. A. Fischer, *Polymeric Materials: Science & Engineering* 2001, 84, 314.
[20] D. Pospiech, H. Komber, D. Voigt, D. Jehnichen, L. Häußler, A. Gottwald, W. Kollig, K. Eckstein, A. Baier, K. Grundke, „Synthesis and characterization of semifluorinated polymers and block copolymers", *Macromol. Symp.* (submitted, November 2002).
[21] K. Grundke, M. Nitschke, S. Minko, M. Stamm, C. Froeck, F. Simon, S. Uhlmann, K. Pöschel, M. Motornov, „Merging Two Concepts: urahydrophobic Polymer Surfaces and Switchable Wettability", *J. Adhesion Sci. Technol.* (accepted, 2002).
[22] K. Grundke, D. Pospiech, W. Kollig, F. Simon, A. Janke, *Colloid & Polym. Sci.* 2001, 279, **8**, 727.
[23] D. Pospiech, D. Jehnichen, A. Gottwald, L. Häußler, W. Kollig, K. Grundke, A. Janke, S. Schmidt, C. Werner „Surface structure of fluorinated polymers and block copolymers", *Surface Coatings International Part B: Coatings Transactions* **2002**, 85, B4.
[24] P. Müller-Buschbaum, M. Stamm, *Macromolecules,* **1998**, 31, 11, 3686.
[25] G. Krausch, *Materials Science & Engineering,* **1995**, R14, 1, 1.

Macromol. Symp. **2003**, *198*, 435-447

Effect of Fillers on Fire Retardancy of Intumescent Polypropylene Blends

X. Almeras,[1] *M. Le Bras,**[1] *F. Poutch,*[2] *S. Bourbigot,*[3] *G. Marosi,*[4] *P. Anna*[4]

[1]Laboratoire PERF UPRES EA 1040, ENSC-Lille, BP 108, 59652 Villeneuve d'Ascq Cedex France
E-mail: Michel.Le-Bras@ensc-lille.fr
[2]CREPIM Bruay la Buissière, France
[3]ENSAIT, Gemtex, Roubaix, France
[4]Technical University of Budapest, Hungary

Summary: Addition of Ammonium Polyphosphate/Polyamide-6 system is known to provide flame retardancy in many polymers blends via an intumescent process. Particulate fillers (talc and calcium carbonate) are used in large quantities in PP. Combination of fillers in PP can modify the properties of the polymeric matrix.
This study investigates the effect of fillers (talc and calcium carbonate) on the fire performance of the Polypropylene/Ammonium polyphosphate/Polyamide-6 blend. It is shown that the fire performance strongly depends on the nature of the filler used. Talc increases and calcium carbonate decreases in the fire performance of the blend.

Keywords: ammonium polyphosphate, cone calorimetry, intumescence, polyamide-6, polypropylene, talc and $CaCO_3$

Introduction

Polypropylene (PP) is representing a big part of polymer worldwide consumption just after polyvinyl chloride and polyethylene. Addition of low cost fillers such as calcium carbonate or talc, was historically justified by cost reduction but more and more improvement of performance becomes the main driver. The low cost material have numerous advantages like a high modulus, excellent chemical resistance and easy process conversion. This explains why the polymer has grow significantly in automotive or appliances.[1]

Moreover, for safety reasons, applications of polyolefinic materials in common transport, materials require fire retardant properties. One way to increase the fire properties is to develop intumescent systems.

Intumescent technology[2],[3] has found a place in polymer science as a method of providing flame retardancy to polymeric materials.[4],[5] On heating, fire retardant (FR) intumescent materials form a foamed cellular charred layer on their surfaces, which protects the underlying materials from the action of heat flux and flame. The proposed mechanism is based on the

 DOI: 10.1002/masy.200350837

charred layer acting as a physical barrier, which slows down heat and mass transfer between the gas and the condensed phase.[6] Generally, intumescent formulations contain three ingredients: an acidic source, a carbonisation agent and a blowing agent.

It was shown that incorporating ammonium polyphosphate (APP) in polyamide 6 (PA-6) provides fire properties of interest.[7]-[8] However, the compatibility of the APP/PA6 blends obtained by direct mixing of APP in melted PA-6, is low : a migration of the mineral salt [9] occurs during solidification of the melt. In this case, the use of an interfacial agent is required. Ethylene-vinyl-acetate (EVA) copolymers, which are known to be very efficient interfacial agents[10]-[11] have thus been used as compatibilisers in the intumescent PP/APP/PA6 system. That has been revealed using calorimetry by oxygen consumption that fire properties of interest were obtained by using EVA_{24} as interfacial agent in the PP/APP/PA-6 intumescent system.[12] Moreover, some authors have shown that fillers, particularly glass fibres, have a great influence on the fire properties and lead to an increase in the fire performance.[13]

In this study, the effect of talc and calcium carbonate on the fire properties are investigated and compared with the PP/PA-6/APP/EVA_{24} system without fillers.

Experimental

Materials

Raw materials were polypropylene (homopolymer grade 3120 MN 1 MFI 8 (230°C, 2.16kg), as pellets supplied by Appryl), PA-6 (as pellets supplied by Nyltech), APP (AP 462: $(NH_4PO_3)_n$, $n \approx 700$, powder supplied by Clariant), EVA_{24} supplied as pellets by Elf-Atochem (EVATANE grades). Polypropylene talc (PPTALC, MFI Polyfort FPP 20 T as pellets supplied by Schulman S.A. hereafter called talcom).

Two fillers were added into the polypropylene : Talc (mean size of particles 15 µm supplied by Suommen takki) and Calcium carbonate (mean size particles 2.5 µm supplied by Omnia).

Processing and Formulations

The fillers (talc and $CaCO_3$ without sizing agent) were added to PP using a Brabender DSK 42/7 intermeshing counter-rotating twin-screw extruder at 190°C and 150 rpm to obtain a composite with mineral loading of 20 wt-%. Then the PP composites were mixed with PA-6, APP, and EVA_{24} using a Brabender Mixer measuring head (type 350/EH, roller blades) at 230°C and 50 rpm. We have checked the mixing conditions using the data processing torque rheometer system Brabender Plasticorder PL2000. Sheets (100 x 100 x 3 mm^3) were then obtained using a Darragon pressing machine at 200°C and at a pressure of 10^6 Pa. The different processed blends mixed are listed in Table 1.

Table 1. Composition of the PP based blends.

PP (wt-%)	PA-6 (wt-%)	APP (wt-%)	Talc (wt-%)	$CaCO_3$ (wt-%)	EVA_{24} (wt-%)	Samples label
60	8.75	26.25			5	Reference
48	8.75	26.25	12		5	PPtalcom
48	8.75	26.25	12		5	PPtalc
48	8.75	26.25		12	5	PPcarb

Fire Testing

Limiting Oxygen Index (LOI) was measured using a Stanton Redcroft instrument on sheets (100x10x3 mm^3) according to ASTM 2863.[14] UL-94 tests were carried out on 127x12.7x3 mm^3 sheets according to the UL-94 testing.[15]

Samples are exposed to a Stanton Redcroft Cone Calorimeter according to ASTM 1356-90 and ISO 5660 at an external heat flux of 50 kW/m^2 which corresponds to the heat evolved during a fire.[16] Conventional data (Rate Heat Release (R.H.R.), Total Heat Emitted (T.H.E.), CO/CO_2 evolutions and residual mass) were computed using a software developed in our laboratory. For experiments performed at 50 kW/m^2, RHR and THE values were reproducible to within ±10% and weight loss, CO, CO_2 were reproducible to within ±15%. The data presented in the following are averages from at least 3 experiments.

Thermal Analysis

Thermogravimetric analyses were carried out at heating rate 10°C/min in a flowing synthetic air (Air Liquid grade; flow rate = 5.10^{-7} Nm^3/s) using a SETARAM MTB 10-8 thermobalance. In each case, samples were 10 mg and were positioned in open vitrous silica pans. The precision on temperature measurements is ± 1.5°C in the range 50°C-850°C.

The curves of weight difference between the experimental and theoretical TG curves are computed as follows : $\Delta(T)$: curve of weight difference : $\Delta(T) = M_{exp}(T) - M_{th}(T)$. with:

$$M_{th}(T)=(0.60-X)M_{poly}(T)+XM_{fill}(T)+0.2625M_{APP}(T)+0.0875M_{PA\text{-}6}(T)+0.05M_{EVA}(T).$$

Where $M_{poly}(T)$, $M_{APP}(T)$, $M_{PA-6}(T)$, $M_{fill}(T)$, $M_{EVA}(T)$ and $M_{exp}(T)$ are respectively the polymer weight, APP, PA-6, the fillers and of the blends given by their TG curves.

The $\Delta(T)$ curves allow to point out an eventual increase or decrease in the thermal stability of the polymeric matrix related to the presence of one or every additives.[17]

Results

Fire Testing

LOI values and UL-94 rating of the different blends are given in the Table 2.

Table 2. LOI values and UL-94 ratings of the intumescent blends versus the fillers grade.

Blends	LOI (vol %)	UL-94 rating
$PP/PA-6/APP/EVA_{24}$	32	V0
$PPcarb/PA-6/APP/EVA_{24}$	29	V1
$PPtalc/PA-6/APP/EVA_{24}$	31	V0
$PPtalcom/PA-6/APP/EVA_{24}$	32	V0

Incorporating talc has no effect on the LOI nor on the UL-94 rating whereas with $CaCO_3$, the LOI value is lower than those obtained with talc or without fillers. Moreover, a V1 rating is obtained with $CaCO_3$ whereas a V0 rating is achieved with the other formulations. Using $CaCO_3$ leads to a decrease of the fire performance of the blend.

The R.H.R. curves of the formulations are presented on Fig. 1.

Incorporating APP with PA6[12] in PP and PP/fillers blends leads to the development of an intumescent structure in the cone calorimeter conditions. Whatever the blend, two peaks are observed separated by a plateau value of R.H.R. According to other studies,[6] the first peak may be assigned to the development of the intumescent protective structure. The formation of a stable efficient protective shield may explain the plateau value where the R.H.R is roughly constant. The second peak is assigned to the degradation of this protective layer.

Incorporating calcium carbonate in the blend increases the first RHR peak from 200 kW/m^2 without fillers to 350 kW/m^2 and decreases the time needed for the complete degradation of the polymeric material.

Adding talc decreases the first peak of RHR but no significant difference of the time to destruction and of the second RHR peak value is observed. However, it seems that the best results are obtained using the commercial PPtalc matrix. The stability of the shield is about 420 seconds whereas it is 350 seconds for the other blends.

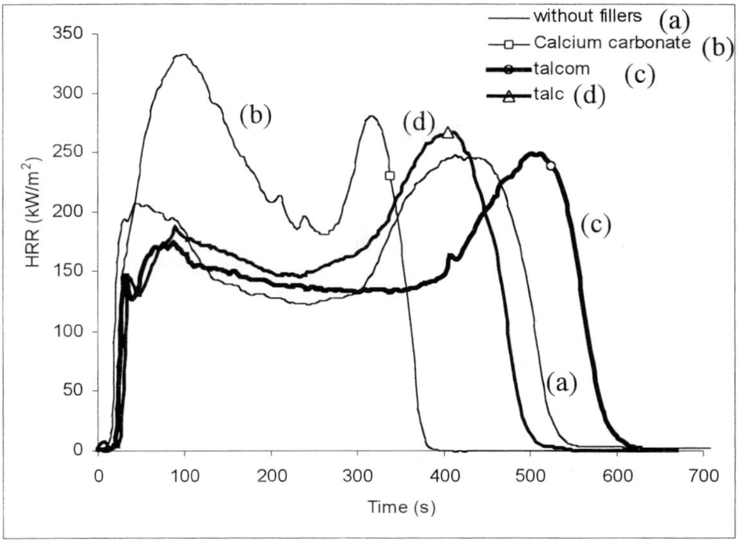

Fig. 1. R.H.R. versus time for PP/APP/PA6/EVA_{24} formulations at an heat flux of 50 kW/m^2.

The curve of the total heat evolved during the test are plotted on Fig. 2.

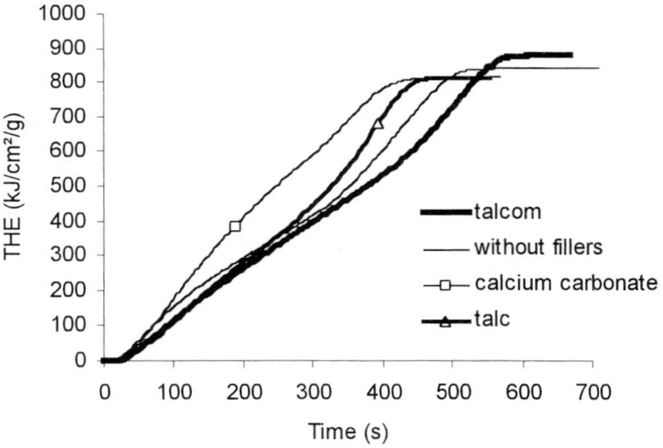

Fig. 2. THE versus time for PPfillers/APP/PA6/EVA_{24} formulations at an external heat flux of 50 kW/m^2.

Incorporating fillers does not lead to a decrease in the final THE value. However, the time to heat release depends upon the fillers and is highly decreased using $CaCO_3$. The final THE is the same whatever the blend.

CO_2 and CO emissions curves are plotted on Fig. 3 and Fig. 4.

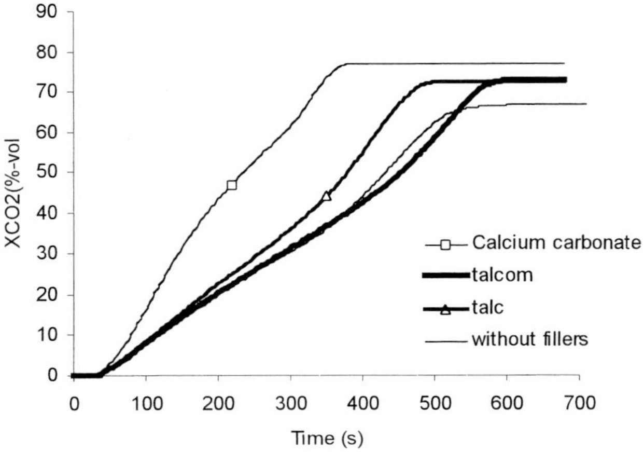

Fig. 3. CO_2 emission versus time for PP/APP/PA6/EVA_{24} formulations under an external heat flux of 50 kW/m^2.

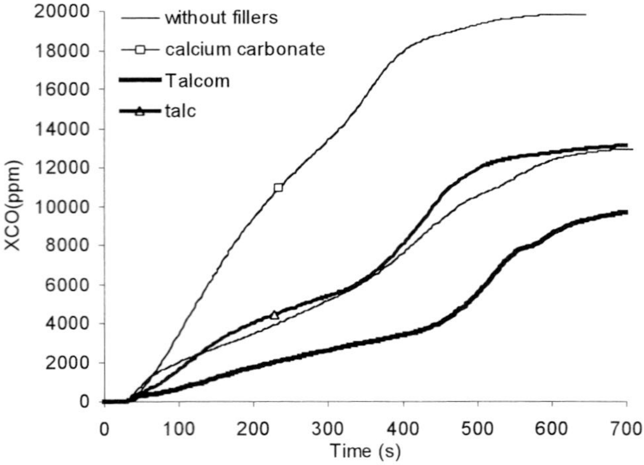

Fig. 4. CO emission versus time for PP/APP/PA6/EVA_{24} formulations under an external heat flux of 50 kW/m^2.

Incorporating fillers lead to an increase in the final among of evolved CO_2. Adding talc allows to increase the delay of the CO and CO_2 emission compared to the blend with $CaCO_3$. This delay particularly increases with the commercial product and the amount of CO evolved decreases. Moreover, the amount of CO_2 evolved does not increase. So, a part of the carbonaceous residue is preserved. Nevertheless, the efficiency of talc is reduced when the blend extruded in our lab is considered. Using $CaCO_3$ induces a decrease in the delay to obtain the maximum value. This could be interpreted as a destabilisation of the protective shield.

An increase in the amount of the evolved CO for the blend with $CaCO_3$ is noted Fig. 4, compared to the other blend. This is probably due to the degradation of $CaCO_3$. This hypothesis will be discussed in the following.

The residual weight curves for the blends versus time are plotted on Fig. 5.

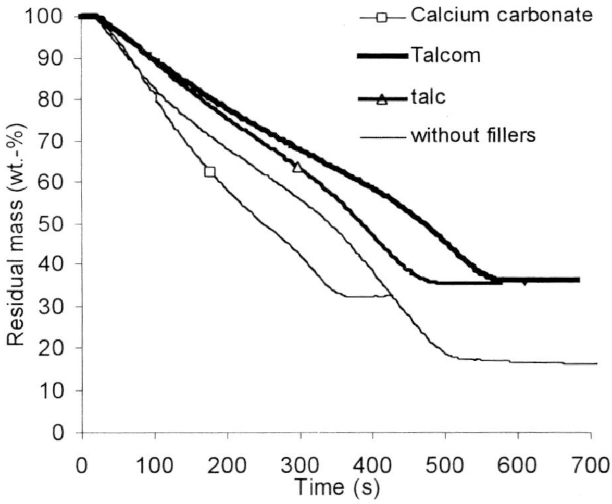

Fig. 5. Evolution of the residual mass for PP/APP/PA6/EVA$_{24}$ formulations versus time under an heat flux of 50 kW/m^2.

Incorporating talc enables to obtain a final residual weight greater than without. Moreover, as previously observed, using talc enables to get a higher stability of the material which is maximal using the commercial product.

Using $CaCO_3$ leads to a loss of the thermal/fire stability in the cone calorimeter condition. The amount of residual weight obtained is about 35 wt-% with talc and 30 wt-% with calcium carbonate whereas it is only 17 wt-% without. As there is 13 wt-% of fillers, with talc in the cone calorimeter conditions, some materials are preserved.

The curves of the thermo-oxidative degradation of the blends with their fillers and their derivatives are plotted Figure 6 and Figure 7.

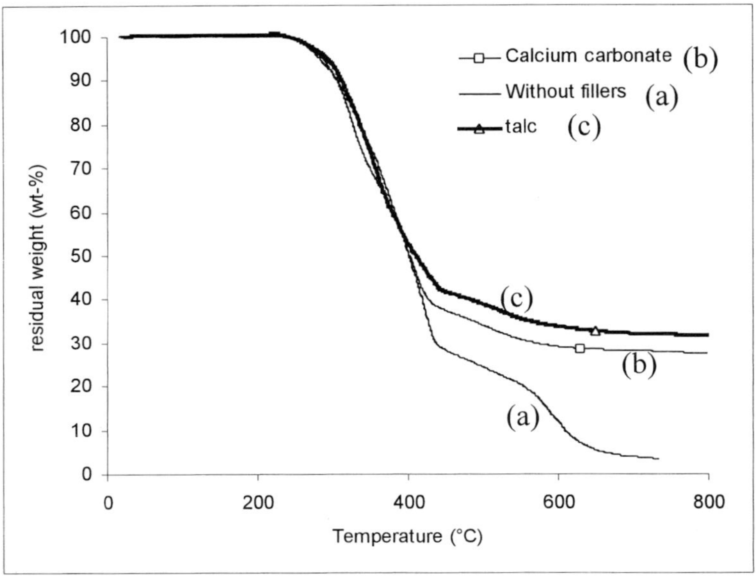

Fig. 6. TG curves under air of the PP/APP/PA-6/EVA$_{24}$ intumescent blend (heating rate: 10°C/min).

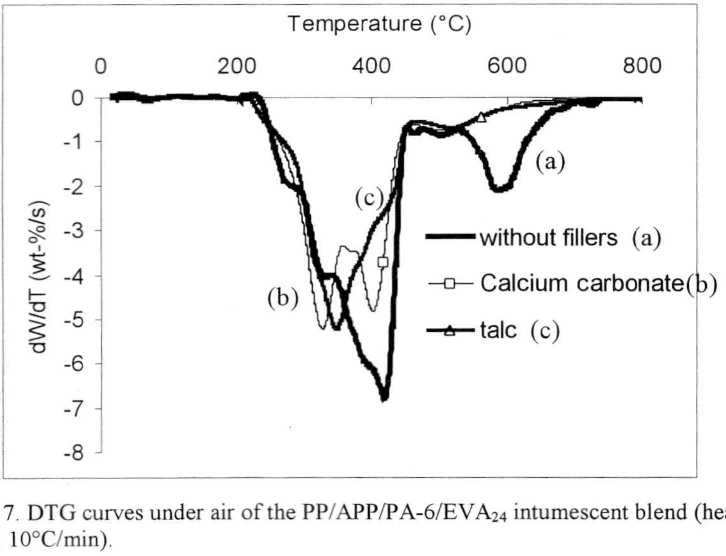

Fig. 7. DTG curves under air of the PP/APP/PA-6/EVA$_{24}$ intumescent blend (heating rate: 10°C/min).

TG and DTG curves of the PP/PA-6/APP/EVA_{24} blend (Figure 6 and Figure 7) show four significant changes in the slopes, which prove that its degradations is at least a four main steps process. Previous studies show that the carbonization process of the system occurs via 4 steps.[17] From 230°C the additives react and form phosphate esters. Between 280°C and 350°C, the development of intumescence occurs and between 350°C and 430°C, the intumescent coating degrades. At higher temperatures, there are structural changes leading to the formation of a new carbonaceous species (established in the temperature range 430°C $<$ T $<$ 560°C) up to 430°C which may be interpreted as above. Then a plateau is observed between 430-560°C. This plateau may be assigned to a particular protective "carbon". Finally, the curve decreases at higher temperature. There is modification of the carbonaceous structure which is no longer protective and the formation of a "high temperature" residue (about 4 wt.-% at 800°C).

TG and DTG curves of the PPtalc/PA-6/APP/EVA_{24} blend (Figure 6 and Figure 7) are similar to the curves of the blend without fillers until 560°C. Beyond 560 °C, a high temperature residue (about 32 wt-%) is obtained.

TG and DTG curves of the PPcarb/PA-6/APP/EVA_{24} blend (Figure 6 and Figure 7) shows three significant changes in the slopes, which prove that its degradations is at least a three steps process. The first one begins at about 250°C until 430°C. Then, the material shows a low rate degradation between 430°C and 560°C, and finally a high temperature residue (about 28 wt-%) is formed.

The ΔT curves are presented Figure 8. Under the conditions of thermo-oxidative degradation, it is observed that the protection by the intumescence phenomenon is a several steps' process.

In the case of PP/PA-6/APP/EVA_{24}, the curve of weight loss difference shows that interactions between the additive and the polymeric matrix occur. The formation of the intumescent coating occurs from 210°C which implies a comparatively more important stability of the system (experimental curve>simulated curve) from 350°C up to 500°C. In this region (210<T<500°C), a first coating is formed with a maximum of protection at 380°C. Above this temperature, the shield looses partially its efficiency. Above 420°C, a second protective process is observed which leads to the formation of a second shield. This latter acts as a relay of the first one. Finally from 560°C, we can assume degradation of the material (fast decrease of weight difference) and the formation of a residue.

In the case of PP/talc/PA-6/APP/EVA_{24}, the curve of mass difference has the same shape as PP/PA-6/APP/EVA_{24} up to 450°C which may be interpreted as above. Then a plateau is observed between 450-550°C. This plateau may be assigned to a particular protective "carbon".

Finally the curve increases until 630°C, a second protective process is observed which leads to the formation of a second shield. Then a plateau is observed at high temperature.

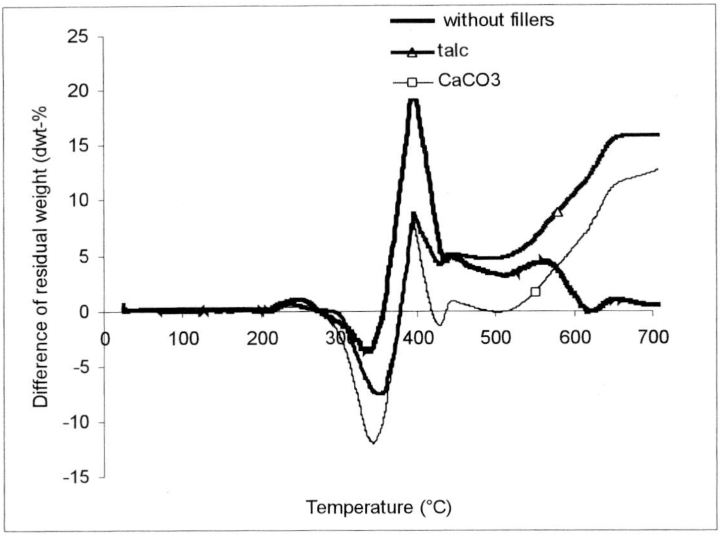

Fig. 8. Difference between experimental and theoretical TG curves of PP/PA-6/APP/EVA_{24}, PPtalc/PA-6/APP/EVA_{24} and PPcarb/PA-6/APP/EVA_{24} systems (under air flow).

In the case of PP/$CaCO_3$/PA-6/APP/EVA_{24}, the mass difference curve has the same shape as PP/PA-6/APP/EVA_{24} up to 410°C which may be interpreted as above. Then the difference between experimental residual weight and computed residual weight is null until 560°C. There is no efficient shield which is formed as in the case of other blends. Over 560°C, the curve of mass difference increases.

The comparison between the curves of mass difference shows interesting results ; particularly over 560°C. The curves of filled systems with fillers increase whereas the curve of the blend decreases. This implies that interactions and reactions between polymer, and additives are larger with fillers. So, they may allow to form a structure which may block polymer links (or its degradation products) within the coating and therefore, to decrease the degradation rate of the system. These results are to be compared with the one observed in the case of APP/PER-Zeolite systems.[19]-[20] In these systems, zeolite is a synergistic agent which strongly enhances fire proofing performances. This acts by improving the "quality" of intumescent coating and by "trapping" polymer links (via the formation of stable aluminosilicophosphate esters) into the structure which provides interesting mechanical properties to the shield. In our case, formations of calcium phosphate and magnesium phosphate are suspected.

Considering possible formation of magnesium phosphate and calcium phosphate, if only fillers and APP interact, the maximum residual weight is about 17 wt-% and 13.4 wt-% for talc and calcium carbonate. But, 32 wt-% and 28 wt-% with talc and calcium carbonate result from a thermo-oxidative degradation. This proves that a part of carbon is preserved at high temperature by the formation of a protective ceramic shield.

Another point could be pointed out. As observed by TGA, the use of $CaCO_3$ reduces the delay of stabilisation.

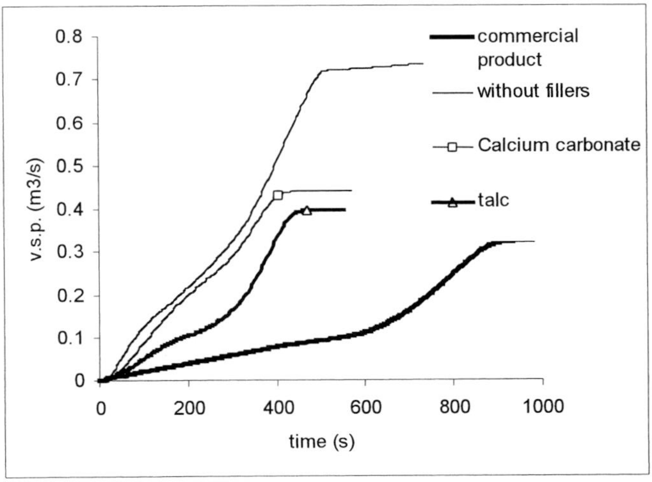

Fig. 9. Volume of Smoke Production (V.S.P.) emission versus time for PP/APP/PA6/EVA_{24} formulations under an external heat flux of 50 kW/m^2.

Incorporating the fillers enable to reduce the amount of V.S.P. (Figure 9) increasing the conversion for a high temperature residue. Moreover, using talc enables to get a higher stability of the material which is maximal using the commercial product. Considering the amount of V.S.P. evolved without fillers, this is due to the combustion of a part of polymer.

Discussion

PP/APP/PA6/EVA_{24} formulation presents enhanced fire performance in comparison with virgin PP. This improvement can be assigned to the formation of a carbonaceous shield during the combustion that slows down the degradation of formulations and that enables to form a stable residue.[12] With fillers, a similar behaviour is observed. The fire behaviour strongly depends on the nature of fillers. It was first shown by Bertelli and al[21] that fillers, in particular hydrated silica or alumina decrease fire retardancy of PP based on intumescent systems with APP. Contrary to these results, Levchik and al[22] observed an increase in the fire properties with talc and calcium carbonate in PA-6/APP system.

Incorporating talc enables to get enhanced fire properties (decrease in the first RHR peak, increase of the residual weight). On the contrary, using calcium carbonate leads to a decrease in the fire performances. Moreover, a protective material does not form in the 420-520°C range and the residual weight obtained (31 wt-%) is lower than with talc (36 wt-%) whereas fillers were added in same quantities.

First, the results obtained with calcium carbonate could be explained by its basic properties. To obtain an intumescent behaviour, three ingredients are needed [12] : a carbonific or char former, an acidic species promoter and a « spumific » or gas-generating and inflating agent. The basic calcium carbonate reacts with APP to give calcium phosphate stable until 1400°C[23] and as a consequence forbids the formation of phosphoric acid and the intumescent char could not be formed as confirmed by visual observation during the cone calorimeter experiment. With $CaCO_3$, the samples do not blow up such as with talc or without fillers. The surface of the sample has the same colour at the end but, whereas with talc, the thickness of the sample increase from 3 mm to 15 mm in the middle whereas it is 3 mm for the burning sample with $CaCO_3$.

Many differences were observed when talc was added in our lab and with a commercial product. Many reasons could explain this phenomenon : first, the structure (cristallinity, cohesion of the blend)[1] strongly depends on the morphology and on the dimensions of the talc particles. As the composition of the intumescent blends are the same, the PP and talc characteristics are suspected to play a great part in the fire properties. Absence of information about the characteristic of the commercial PPtalc do not allow any discussion about this part.

Conclusion

This work shows the importance of the nature of fillers on the fire properties of an intumescent composite materials. In fact, whereas, it decreases with calcium carbonate, the fire behaviour is preserved or increased with talc. Moreover, the residual weight increases at the end because of the development of a ceramic at the surface To follow this work, an investigation on the blend composition would be undertaken after the process and after the burning of the sample.

To conclude the PPtalc/PA-6/APP/EVA_{24} blend shows very good fire performance : LOI > 30 vol-%, $R.H.R._{max} < 250$ kW/m^2, a low level of CO, CO_2 and smoke emission which could allow to use it for some application.

Acknowledgement

This work has been financially supported by the Commission of the European Communities through contract N° G5RD-CT-1999-00120.

[1] Shields TJ, Zhang J. In: Karger-Kocsis J, Editor. Polypropylene an A-Z Reference. Dordrecht (Netherlands): Klumer Academic Pub., 1999:247.
[2] Tramm, H.L., US Patent No. 2106938, 1938
[3] Vandersall, H.L., J. Fire and Flammability, 2, 1971, 97
[4] Montaudo, G., Scamporino, E. & Vitalini, D., J. Polym. Sci., Polym. Chem., 21 (1983), 3361
[5] Camino, G., Actes du Premier Colloque Francophone sur l'Ignifugation des Polymères, ed. J. Martel, Saint Denis, France, 1985 p36
[6] Bourbigot S, Le Bras M, Delobel R. J Fire Sci 1995;13:3
[7] Bourbigot S, Le Bras M, Siat C. In: Lewin M, Editor. Recent advances in Flame Retardancy of Polymeric Materials. Norwalk (USA): BCC Pub., 1997;7:146.
[8] Bourbigot S, Le Bras M. In: Troitzsch J, Bourbigot S, Le Bras M, Editors. International Plastics Flammability Handbook. New-York : Hanser Pub., 2000, Chapter 5, under press.
[9] Le Bras M, Bourbigot S, Felix E, Pouille F, Siat C, Traisnel M. Polymer 2000;41:5283.
[10] Davis RD, Jarrett WL, Mathias LJ. Polymer 2001;42:2621.
[11] Liu NC, Baker WE. Adv Polym Tech 1992;11: 249.
[12] Almeras X, Dabrowski F, Le Bras M, Poutch F, Bourbigot S, Marosi G, Anna P, Polym Deg Stab (accepted).
[13] Dabrowski F, Le Bras M, Delobel R, Le Maguer D, Bardollet P and Aymami J, Flame retardant 2002, p 127
[14] Standard test method for measuring the minimum oxygen concentration to support candle-like combustion of plastics ASTM D2863/77 : American Society for Testings and Materials: Philadelphia, PA, 1977.
[15] Tests for flammability of plastics materials for part devices and appliances, Underwriters Laboratories, ANSI//ASTM D-635/77: Northbrook, IL, 1977.
[16] Babrauskas V. Fire and Mat 1984;8(2):81.
[17] Delobel, R., Le Bras, M., Ouassou N. and Alistiqsa, F., J. Fire Sc., 1990, 8(2):85.
[18] Bourbigot S, Le Bras M, Bréant P, Trémillon JM. Polym Deg Stab 1996;53:275.
[19] Bourbigot, S., Le Bras, M., Delobel, R., Décressain, R. and Amoureux, J.P. J. Chem. Soc, Faraday Trans, 1996 92(1):149
[20] Bourbigot, S., Le Bras, M., Delobel, R., Bréant, P. and Trémillon, J.M.. J. Chem. Soc, Faraday Trans, 1996, 92(18):3435.
[21] Bertelli G, Camino G, Marchetti E, Costa L, Casorati E and Locatelli R, Polym Deg Stab, 1989, 25, 277
[22] Levchik G.F., Levchik S.V., and Lesnikovich, Polym Deg Stab, 1996, 54, 361
[23] Van Wazer, In Interscience publishers, INC, New York, Phosphorus and its compounds, Vol I, 1958, p 528-529

Water Sorption and Electrical/Dielectric Properties of Organic-Inorganic Polymer Blends

Yevgen Mamunya,[1] Athanasios Kanapitsas,[2] Polycarpos Pissis,[2] Gisele Boiteux,[3] Eugene Lebedev[1]*

[1]Institute of Macromolecular Chemistry, National Academy of Sciences of Ukraine, 48, Kharkovskoe shausse, 02160 Kiev, Ukraine
[2]National Technical University of Athens, Department of Physics, Zografou Campus, 157 80 Athens, Greece
[3]Laboratoire des Materiaux Polymeres et des Biomateriaux, UMR-CNRS 5627, Universite Claude Bernard Lyon 1, 43 Bd du 11 Novembre 1918, 69622 Villeurbanne Cedex, France

Summary: Organic-inorganic polymer blends (OIPB) were obtained by reaction of organic and inorganic oligomers. The organic oligomer was synthesized with 2,4-toluene diisocyanate (TDI) and oligooxypropylene glycols (OPG) with various molecular weights (MW). The inorganic component was a water solution of sodium silicate. The OIPB obtained are hydrophilic and have great water sorption ability (the relative weight of sorbed water reaches 2000 %). The kinetics of water sorption and the changes of electrical conductivity during sorption were studied. Sorption ability, and mechanical, electrical and dielectric properties of OIPB depend on molecular weight of OPG: conductivity increases with increasing MW, whereas the sorption ability correlates with the mechanical properties. The influence of the inorganic phase content on the electrical and dielectric properties was studied as well.

Keywords: blends, conductivity, hydrophilic polymers, mechanical properties, water sorption

Introduction

Organic-inorganic polymer blends (OIPB) are a new class of composites which began to be studied the last ten years.[1-4] These composites are formed from mixtures of organic and inorganic components during the process of their joint polymerization. This process can be carried out both at room and higher temperatures. The structure of polymer blends is characterized by the existence of three phases, including the phase of organic polymer, the phase of inorganic polymer and the phase of interaction product between organic and inorganic components.

 DOI: 10.1002/masy.200350838

In previous research we have studied OIPB based on polyisocyanates and sodium silicates. These composites were hard materials with extremely high resistance to aggressive media (acids, alkalis) and to γ–radiation.[5] Using of an organic oligomer with free isocyanate groups instead of the low molecular polyisocyanate results in elastic properties of the synthesized polymer blends. These composites exhibit also other interesting properties, for example high water sorption ability and relatively high conductivity.

The aim of this work is the study of water sorption ability, mechanical characteristics, electrical conductivity and dielectric properties of OIPB based on the organic oligomer with various molecular weights.

Experimental

The materials were obtained on the basis of 2,4-toluene diisocyanate (TDI), oligooxypropylene glycols (OPGs) with various molecular weights (MW) and functionality equal to 2 and 3 (two or three OH-groups at the ends of a molecular chain, in the first case the chain is linear, in the second one the chain has star (branched) configuration). OPGs marked 1052, 2102, 3603 and 5003, according to MW and functionality (for example OPG-1052 means MW=1050 and functionality 2), were used. Water solution of a sodium silicate (SC) was used as inorganic component and OIPB was obtained in two stages. First, the organic oligomers were synthesized from TDI and OPGs. Depending on MW and functionality of OPG, organic oligomers were obtained with various molecular weights and branching of the molecular chains. In the second stage, OIPBs were obtained as a result of reactions between oligomers and SC and were marked OIPB-1, OIPB-2, OIPB-3, OIPB-5, following the designation of OPGs. For comparison with OIPB, samples of pure polyurethane without inorganic phase were obtained by hardening of the organic oligomers with atmosphere moisture and marked PU-1, PU-2, PU-3 and PU-5.

After mixing of the organic and inorganic components, the mixture was put in a teflon die, where samples of OIPBs with 30 mm diameter and 1 mm thickness were formed at room temperature. The samples of PUs were obtained in the same way.

The kinetics of water sorption was studied by a gravimetric method with samples, which were immersed in water. For evaluation of volume changes of OIPG during sorption, the samples of OIPB were taken out of water for short time at predetermined times and their thickness and diameter were measured. At the same time electrical dc conductivity of the samples was also measured by using the teraohmmeter E6-13 under constant voltage 10 V and flat brass electrodes.

For the investigation of the dielectric properties dielectric relaxation spectrometry (DRS) in wide ranges of frequency and temperature and thermally stimulated depolarization current (TSDC) techniques were used. For DRS measurements the complex dielectric permittivity, $\varepsilon^*=\varepsilon'-i\varepsilon''$, was determined as a function of frequency (10^{-2}-10^{6} Hz) in the temperature range –60 to 80 °C. The sample was sandwiched between gold-coated brass electrodes. A Schlumberger Frequency Response Analyzer (FRA SI 1260) supplemented with a buffer amplifier of variable gain (Chelsea Dielectric Interface) in combination with the Novocontrol Quatro Cryosystem was used. TSDC corresponds to measuring dielectric losses as a function of temperature at a fixed frequency in the range 10^{-2}-10^{-4} Hz. A home-made TSDC equipment was used for measurements in the range –185 to 30 °C.

Mechanical characteristics of OIPB were obtained with standard stress-strain measurements.

Results and Discussion

The kinetics of water sorption was studied for OIPBs based on various OPGs. Increase of OPG molecular weight (1050 to 2100) at the same functionality 2 results in the increase of the amount of water sorbed and of the rate of sorption (Fig. 1), whereas the kinetics is strongly different for OIPB based on OPG 3603 with functionality 3. The kinetics curve of OIPB-3, with the presence of a maximum which corresponds to equilibrium sorption, allows to calculate various sorption parameters, such as diffusion D, sorption S, permeability N coefficients and relative weight of sorbed water ΔP_{max}. They were calculated as:[6]

$$D = 0{,}0494\ (d^2/\tau_{1/2}); \qquad S = P_w / V_{max}; \qquad N = D\cdot S; \qquad \Delta P_{max} = 100\cdot P_w / P_0;$$

where d is the sample thickness, $\tau_{1/2}$ is the time at which $1/2(\Delta P_{max})$ is achieved, P_w is the maximal weight of sorbed water, P_0 is the weight of the dry sample, V_{max} is the sample volume when ΔP_{max} is achieved. The values of the sorption parameters are given in a Table 1.

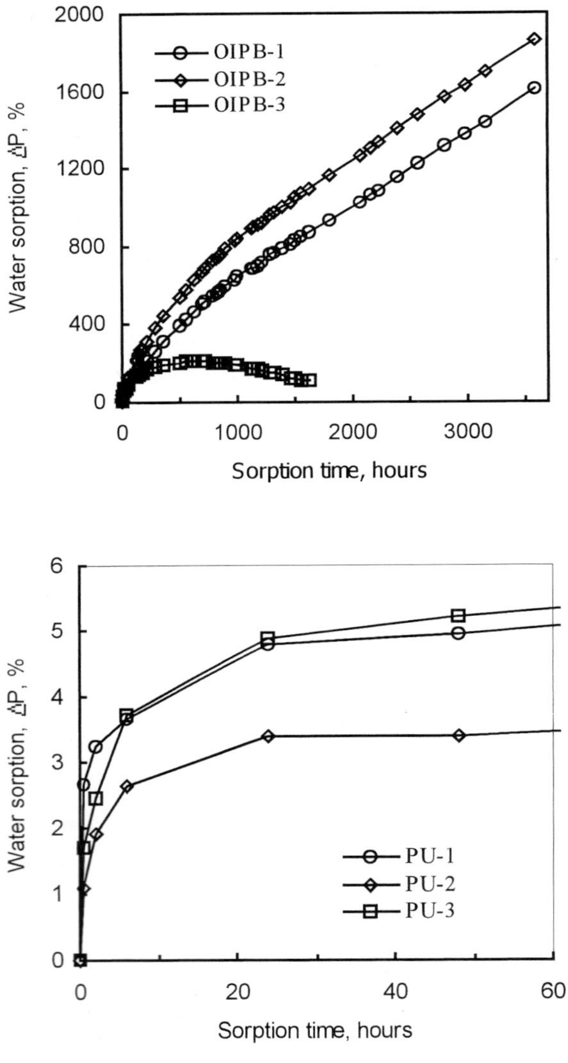

Fig. 1. Kinetics of water sorption of OIPBs and PUs based on OPG with warious molecular weights and functionalities. The weight ratio of organic/inorganic phases is 80/20 for all composites. ΔP was calculated as $100 \cdot (P-P_0)/P_0$, where P is the weight of the sample after exposition to water.

Table 1. Sorption parameters of OIPB-3 and PU-3.

Composite	Diffusion coefficient, D	Sorption coefficient, S	Permeability coefficient, N	Relative weight of sorbed water, ΔP_{max}
	cm^2/s	g/cm^3	$g/(cm \cdot s)$	%
OIPB-3 with 20 wt. % of the inorganic phase	$3.06 \cdot 10^{-9}$	0.609	$1.86 \cdot 10^{-9}$	210
PU-3	$1.80 \cdot 10^{-7}$	0.041	$7.38 \cdot 10^{-9}$	4.6

It is seen from Table 1 that the diffusion coefficient D of PU-3 is two orders of magnitude higher than D of OIPB-3 because of the large differences in time $\tau_{1/2}$, the value of equilibrium sorption is achieved much faster in PU-3 than in OIPB-3 (see Fig. 1). Sorption coefficient S and weight of sorbed water ΔP_{max} are much higher for OIPB-3 than for PU-3, indicating great sorption capacity of OIPB-3 due to the hydrophilic properties of the inorganic phase. Fig. 1 shows that the sorption capacity of OIPB-1 and OIPB-2 (based on OPG with functionality 2) is much higher than for OIPB-3 and reaches nearly 2000 % for the exposition time used. Consequently, the volume of water sorbed by the inorganic phase is controlled by the mechanical characteristics of the organic matrix.

These data are in agreement with the results obtained for the mechanical properties of OIPB. Fig. 2 shows that OIPB-2 has the highest elongation, whereas OIPB-3 has the lowest elastic characteristics.

These results show that the inorganic component of the blend is responsible for water sorption. Swelling of the particles of the hydrophilic inorganic phase during water sorption is controlled by the expansion of the organic matrix, in which the microparticles of mineral component are distributed. The most rigid organic matrix of OIPB-3 allows to swell the inorganic particles to a definite limit during the sorption process, then the mineral gel structure is destroyed under the compressing stress of the matrix and decreasing of the sorption is observed. As a result, IOPB-3 exhibits a sorption curve with maximum (see Fig. 1). In the elastic organic matrix of OIPB-1 and OIPB-2 such an effect is not observed and the increase of composite weight (and of the volume, consequently) reaches nearly 2000 % for the sorption time used.

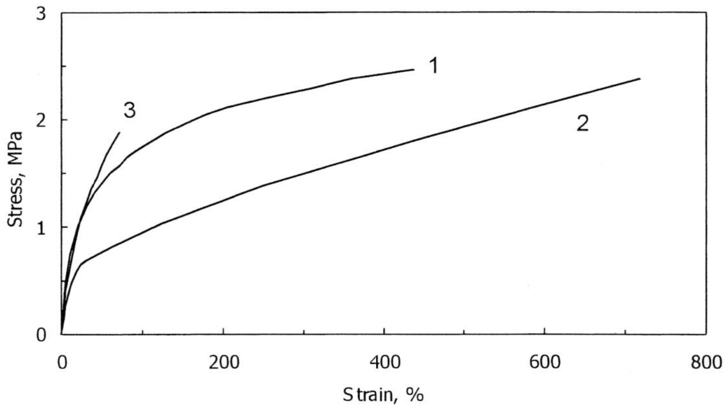

Fig. 2. Mechanical properties of OIPB-1 (1), OIPB-2 (2) and OIPB-3 (3) with inorganic phase content 20 wt.%.

Investigation of the cycling swelling of OIPB in water showed that the process of water sorption is reversible and the elastomer maintains its sorption and mechanical properties after water desorption (Fig. 3, cycles 1). A substitution of water by the electrolyte (aqueous solution of $CuCl_2$) greatly reduces the value of sorption (cycles 2). Probably this effect is a result of ionic reactions in the inorganic phase and of substitution of Na^+ by Cu^{++} on the particles surface of inorganic phase. The next cycle of water sorption (cycle 3) recovers the hydrophilic properties of OIPB.

Electrical dc conductivity of OIPB-3 during the process of cycling sorption is shown in Fig. 4. A sharp increase of conductivity can be seen (of two orders of magnitude) in the first moments of sorption, when the level of sorption is still less than 10 %, whereas afterwards the increase of conductivity is very slow (Fig. 4, curves 1). The conductivity of the composites during sorption of the electrolyte is enhanced, as compared with the cycles of water sorption (curves 2). The dependence of conductivity during the third cycle of water sorption is similar to curves 1, but the absolute values higher (curves 3). Consequently, immersion of the composite in the electrolyte makes the hydrophilic structure more sensitive to water sorption and increases the ionic conductivity.

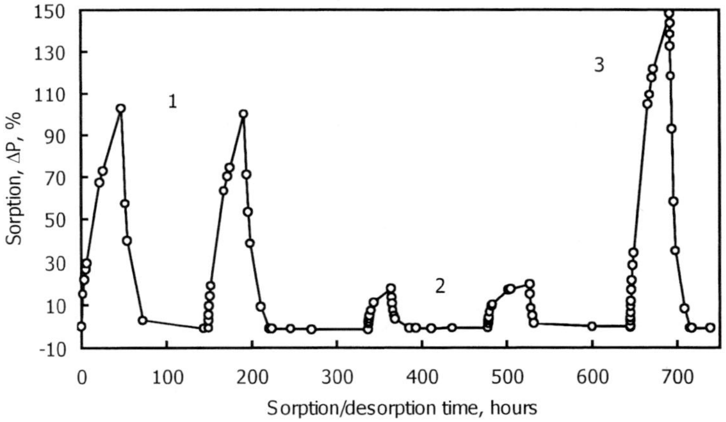

Fig. 3. Cycling regime of the sorption of OIPB-3 (with inorganic phase content 20 wt.%) in water (1, 3) and in aqueous solution of $CuCl_2$ (2).

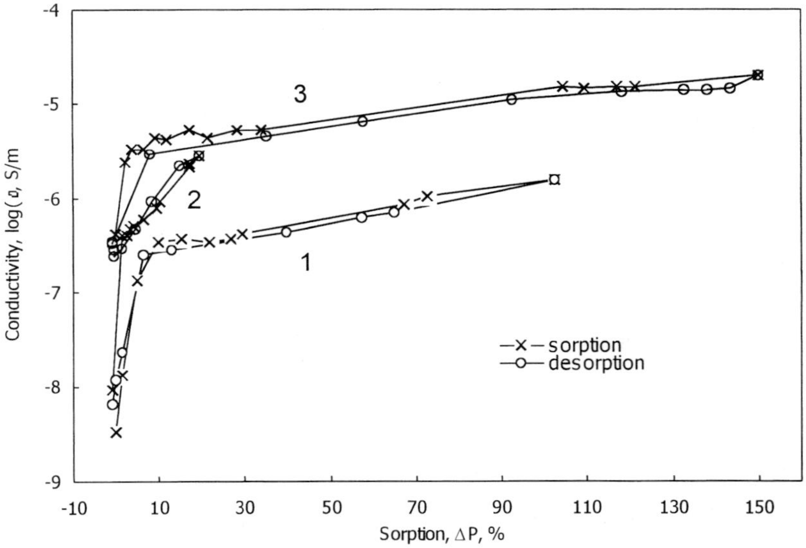

Fig. 4. Dependence of dc conductivity on water sorption during the process of cycling sorption-desorption. The numbers near the curves correspond to the numbers of the sorption/desorption cycles in Fig. 3.

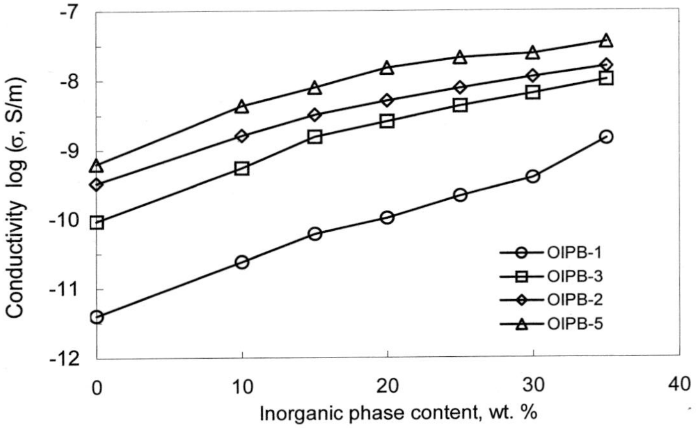

Fig. 5. Influence of the inorganic phase content on conductivity of OIPBs.

Fig. 5 demonstrates the influence of inorganic phase content and kind of OIPB on dc conductivity of these systems. Conductivity increases with the molecular weight of OIPBs, by two orders of magnitude with increase of the molecular weight of OPG from 1050 to 5000. Probably increase of the molecular chain length facilitates the transport of a charge along a chain. Functionality of OIPBs and mechanical characteristics of the polymer matrix do not affect conductivity. The dependence of conductivity on the inorganic phase content is similar for the OIPBs based on various OPGs. Increase of the inorganic phase content to 30 % leads to the rise of conductivity by almost two orders of magnitude.

DRS results show that the relaxation spectra of OIPBs are changed under variation of composition and temperature. Fig. 6 shows comparative plots of dielectric permittivity ε′ and dielectric losses ε′′ as a function of frequency, at room temperature, for OIPB-1. Similar results were obtained also with the other OIPBs. The most striking result in Fig. 6 is the overall increase of mobility in the composites as compared to the pure matrix. The high values of ε′ at low frequencies in the composites are due to conductivity effects, as indicated also by the increase of ε′′ in the same frequency range. In ε′′(f) a loss peak is observed at about 1 kHz in the pure PU-1 matrix. Measurements at several temperatures show, in agreement with TSDC

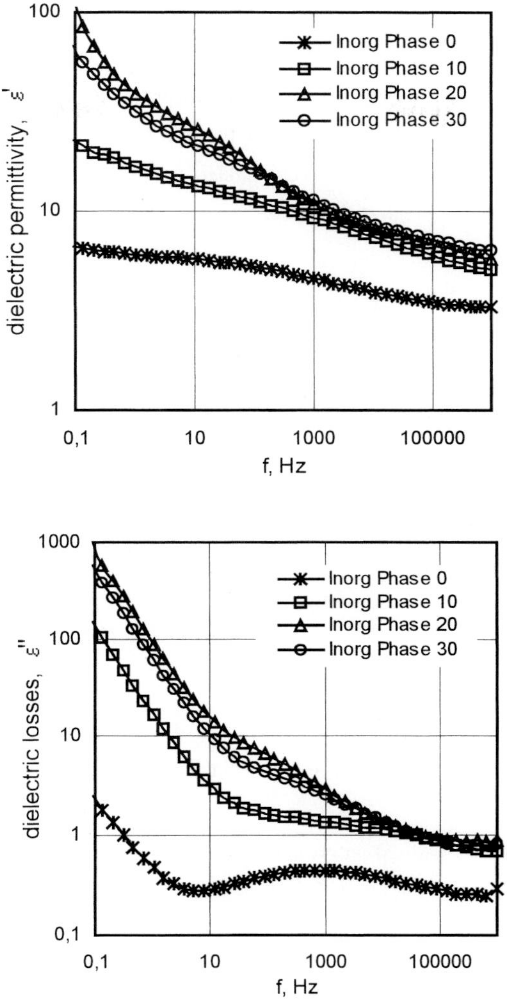

Fig. 6. Dependence of dielectric permittivity and dielectric losses on frequency at room temperature for PU-1 and OIPB-1 with various content (in wt.%) of the inorganic phase.

measurements and preliminary DCS measurements, that this peak is due to the α relaxation associated to the glass transition of the matrix. The peak is present also in the OIPBs, but it is masked by a larger peak at lower frequencies, tentatively attributed to interfacial Maxwell-Wagner-Sillars polarization. It is interesting to note that the overall dielectric response of the OIPBs with 20 and 30 wt. % inorganic phase is very similar, suggesting a kind of saturation. This is reflected also in the TSDC plots not shown here.

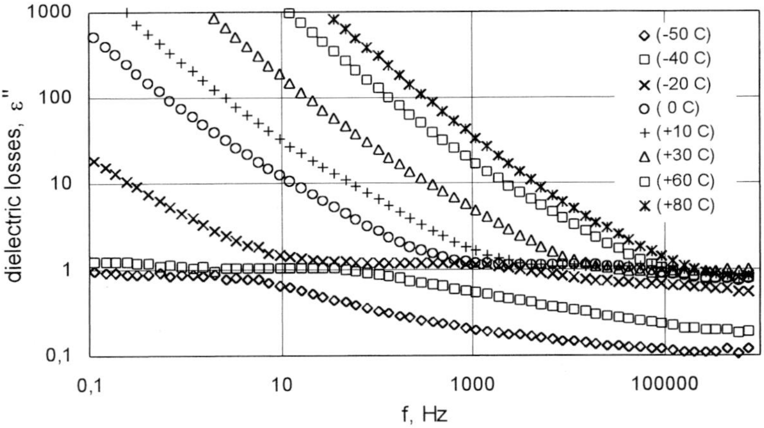

Fig. 7. Dependence of dielectric losses on frequency at various temperatures for OIPB-3 with inorganic phase content 20 wt.%.

Fig. 7 shows ε´´(f) for the OIPB-3 with 20 wt. % inorganic phase. A change of the response is observed at temperatures higher than about –40 °C, associated to the glass transition. The α loss peak shifts to higher frequencies with increasing temperature. At high temperatures/low frequencies high values of ε´´ are measured related to conductivity effects. Analysis of these and similar dielectric data is now in progress to extract quantitative information on the various mechanisms of molecular mobility.

Conclusion

Organic-inorganic polymer blends (OIPBs), obtained by polymerization of urethane oligomers with various molecular weights and aqueous solution of sodium silicate, demonstrate elastic mechanical properties which are typical for elastomers. Due to the presence of the inorganic mineral phase in the organic matrix, OIPBs possess hydrophilic properties with very high sorption capacity (about two 2000%). Cycling process sorption/desorption is reversible and shows that the electrical conductivity of OIPBs is changed during the sorption/desorption process by several orders of magnitude due to ionic conductivity in the inorganic phase in the presence of water. The electrical conductivity of OIPBs depends on the molecular weight of oligomers and content of the inorganic phase and is determined by formation of the conductive inorganic structure in the polymer matrix. Dielectric characteristics depend on the inorganic phase content and temperature and show increase of the polymer chains mobility in OIPB, as compared with pure PU.

Acknowledgement

This work was supported by a grant of the Greek Ministry on National Economy for Y.P.M and by the Exchange Programme (Project 12710) of the CNRS of France and of the National Academy of Sciences of Ukraine.

[1] S.S.Ischenko, A.B.Pridatko, *Compos. Polym. Mater.*, **1993**, N55, 53.
[2] E.V.Lebedev, S.S.Ischenko, A.B.Pridatko, *Compos. Polym. Mater.*, **1999**, 21, N1, 3.
[3] E.C.O. Lima, F.Galembeck, *J. Colloid Interf. Sci.*, **1994**, 166, 309.
[4] E.F. de Souza, C.C.Bezerra, F.Galembeck, *Polymer*, **1997**, 38, 6285.
[5] S.S.Ischenko, E.V.Lebedev, *Ukr. Chem. J.*, **2001**, 67, N8, 116.
[6] A.J.Malkin, A.E.Tchalykh, "*Diffusion and viscosity of polymers. Methods of meausurements*", Khimia, Moscow 1979.

Structure and Dielectric Properties of a Thermoplastic Blend Containing Dispersed Metal

Haralampos Zois,[*1] Yevgen P. Mamunya,[2] Lazaros Apekis[1]*

[1]National Technical University of Athens, Department of Physics, Zografou Campus, 157 80 Athens, Greece
[2]Institute of Macromolecular Chemistry, National Academy of Sciences of Ukraine, Kharkovskoe shausse 48, Kiev 02160, Ukraine

Summary: In the present work broadband dielectric relaxation spectroscopy measurements were employed to investigate the dielectric properties of polymer composites. A polyethylene/polyoxymethylene (PE/POM) thermoplastic blend was used as a matrix, while the inclusions were iron (Fe) particles. For comparison, the two pure polymers- PE and POM- were used as a matrix, too. In the PE/POM-Fe composites, the polymer matrix is two-phase and the filler particles are localized only in the POM phase, resulting in an ordered distribution of the dispersed filler particles within the blend. In PE-Fe and POM-Fe composites, the filler spatial distribution is random. The behaviour of all the composites studied is described in terms of the percolation theory. The PE/POM-Fe composites, based on the PE/POM blend, demonstrate different electrical behaviour compared to that of POM-Fe and PE-Fe systems. The percolation threshold value of the PE/POM-Fe composites was found much lower than that of the other two systems. The results were related to the microstructure of the composites. A schematic model for the morphology of the composites studied has been proposed. This model explains the peculiar behaviour of the PE/POM-Fe composites by taking into account the ordered distribution of the filler particles in a binary polymer matrix. Optical microscopy photographs confirm this model.

Keywords: composite materials, dielectric properties, morphology, percolation

Introduction

Composite materials, consisting of a polymer matrix and dispersed conductive inclusions, is a relatively new, interesting class of materials. The electrical characteristics of such composites are close to the properties of the filler, whereas their mechanical properties and processing methods are typical for plastics.[1] These composites have several advantages over the conventional conductive materials.[2] The electrical and dielectric properties of polymer composites have been the subject of both theoretical and experimental studies over the last decades, due to their wide range of industrial applications.[1-3] At the same time, these materials hold interest from fundamental point of view. They can be considered as typical disordered systems, consisting of randomly distributed inclusions within a polymer matrix.

 DOI: 10.1002/masy.200350839

Metal-filled polymer composites have been the subject of extensive research.[1-3] Considerable effort has been expended in the investigation of the structure-properties relationship,[4-7] at molecular and morphological level, which is necessary for designing new materials with desirable properties for specific technological applications.[8-11]

In this work our experimental results on the dependence of the dielectric characteristics on the filler volume fraction and the frequency are reported. A thermoplastic polymer blend (PE/POM) was used as a matrix, while the dispersed filler was iron (Fe) particles. For comparison, the two pure polymers- PE and POM- were used as a matrix, too. Broadband dielectric relaxation spectroscopy measurements were carried out. The behaviour of such systems is described on the basis of the percolation theory.[12-14] The main objective of this work is to study the effect of the polymer matrix on the dielectric properties of the composites. The results were related to the microstructure of the systems and a schematic model was proposed. This model is in good agreement with results of optical microscopy investigations.

Fundamental Aspects

Polymers filled with metal additives have very interesting properties and behaviour. When the filler volume fraction is low, the samples behave as insulators or semiconductors, while for higher metal concentration these materials exhibit conducting behaviour. In order to achieve conduction in such filled polymer systems, conductive pathways of metal particles throughout the polymer matrix are required. This situation is very well known as percolation.[12-14] The electrical percolation phenomenon has been thoroughly studied during the last years. Although many experimental works study this phenomenon, no general theory exists, that can satisfactorilly describe all experimental results. Several theoretical models and mixing rules have been proposed to describe the percolation phenomenon in binary mixtures. They include statistical percolation, thermodynamic, geometrical and structure-oriented percolation models. A detailed discussion on these models can be found in the review paper published by Lux.[15]

By gradually increasing the filler content in a metal-polymer composite, the most significant changes on the electrical and dielectric properties occur in a certain, relative narrow critical region of the filler volume fraction; the so-called percolation threshold, P_C. Percolation theory[12-14] predicts that the dielectric characteristics fulfill, in the critical region near the threshold, the following universal power-laws:

$$\varepsilon' \propto (P_C - P)^{-s} \qquad \text{for } P<P_C \qquad (1)$$

$$\text{and} \qquad tan\,\delta \propto (P_C - P)^{-r} \qquad \text{for } P<P_C \qquad (2)$$

where s, r are the dielectric permittivity and dielectric loss tangent critical exponents, P is the filler volume concentration and P_C is the percolation threshold.

The values of the critical exponents, s and r, depend only on the lattice dimensionality and not on its geometrical characteristics.[14] In particular, for three-dimensional structures, values of s=0.7-0.9 have been reported.[12, 13] Divergence of the critical exponents values from the theoretical ones can be attributed to the non-random (ordered) distribution of the filler particles within the host matrix.[16] On the other hand, percolation threshold value depends on many factors, such as the size, the shape and the spatial distribution (topology) of the filler particles within the polymer matrix, the adhesion and the interactions between the two phases and the processing method.[5, 17-20] The influence of these factors on the percolation threshold values has been discussed in details elsewhere.[1, 5, 11, 17-20]

Experimental

Composite materials, consisting of polyethylene/polyoxymethylene (PE/POM) polymer blend as a matrix and dispersed iron (Fe) particles as inclusions, were studied. For comparison, the two pure polymers- PE and POM- were used as a matrix, too. The iron particles were of R-10 type with an average size of 3.5 μm and shape close to spherical. The filler volume fraction varying in a wide range (0-40 % vol. Fe). PE/POM-Fe composites were prepared in a two-step procedure. First, a master batch of the less viscous polymer (POM) with iron particles was prepared. The filler content in the master batch was 32 % vol. Fe. Then the grinded master of POM-Fe was mixed with the other polymer (PE) at the necessary ratio and extruded. The samples were obtained by hot pressing of the grinded extrudates. PE-Fe and POM-Fe composites were prepared by extrusion of the mechanical mixture of the thermoplastic polymer and the iron powder. The samples were disc-shaped with diameter of 30 mm and thickness of about 1.5 mm.

Broadband dielectric relaxation spectroscopy measurements[21] of the complex dielectric permittivity: $\varepsilon^*=\varepsilon'-i\cdot\varepsilon''$ in the frequency range 10^{-2}-10^{6} Hz were carried out at room temperature using a Frequency Response Analyzer (Schlumberger, FRA 1260) supplemented by a buffer amplifier of variable gain (Chelsea Dielectric Interface). A two-terminal parallel-plate capacitor dielectric cell with gold-coated metal electrodes (Novocontrol) was used.[22]

Optical microscopy investigations of the composites were carried out by means of a BIOLAR microscope via light transmission through thin sections of the samples, which were cut with thickness above 10-20 μm.

Results and Discussion

Figure 1 presents the dependence of the dielectric permittivity, ε′, at the constant frequency of 1 kHz on the filler concentration, P, measured at room temperature for all the composites studied. The sharp increase of the ε′ values when the filler content reaches the critical P_C value is clearly observed. This behaviour is similar for the concentration dependence of the dielectric loss tangent, too (not presented here). This increase is characteristic for the existence of percolation phenomenon in composite materials.

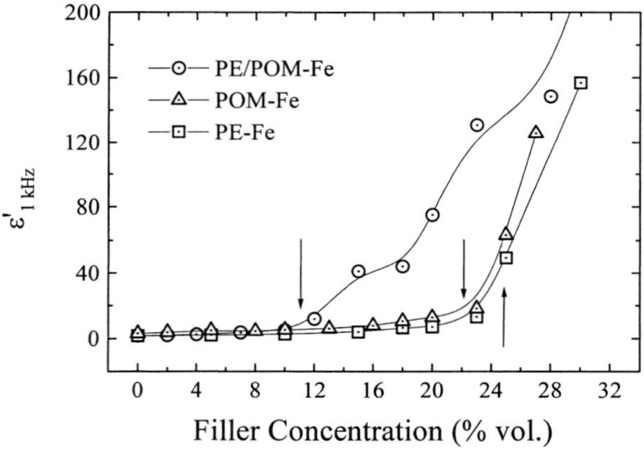

Fig. 1. Dielectric permittivity, ε′, obtained at a constant frequency equal to 1 kHz and at room temperature as a function of the filler concentration, P, for all the composites studied. The arrows indicate the percolation threshold values. The lines are to guide the eye.

Another interesting point, which is worth mentioning, is that, for filler concentration higher than 10 % vol., the PE/POM-Fe composites have higher values of the dielectric permittivity, compared to that of PE-Fe and POM-Fe composites. This means that, in the PE/POM-Fe composites, direct contacts between adjacent metal particles exist at lower filler concentration and at lower conductivity values, than in the other two composites.

Eqs. (1) and (2) suggest that plots of logε′ (or logtanδ) as a function of the logarithm of the reduced filler concentration, $\log[(P_C-P)/P]$, should be straight lines with slopes equal to the critical exponents s and r, respectively. These linear relations are shown in Figure 2. For the PE/POM-Fe composites, least-square best fitting lines give: P_C=11 % vol. Fe, s=0.75 and r=5.1. The results for the other two systems (PE-Fe and POM-Fe) are listed in Table 1 below. The most striking result is that, the percolation threshold values, for PE-Fe and POM-Fe composites, were calculated equal to 22 and 25 % vol. Fe, respectively, which are much higher compared to that of the PE/POM-Fe system. This will be discussed in details later.

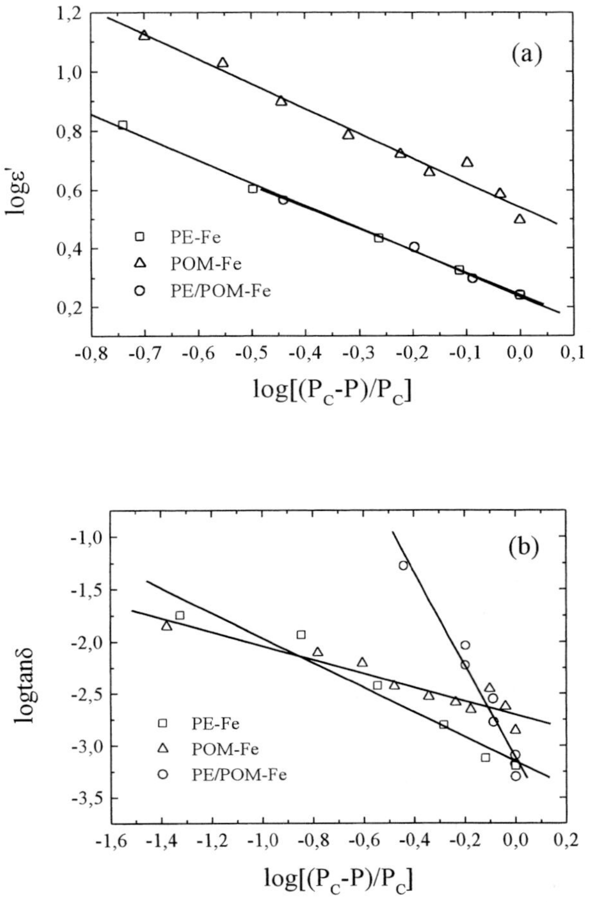

Fig. 2. Linear plots of logε´ (a) and logtanδ (b) as a function of $\log[(P_C-P)/P_C]$ for all the composites studied. The slopes of these lines give the values of the critical exponents s and r, respectively.

For all the composites studied, the values of the critical exponent s (0.75-0.84) were estimated in good agreement with those predicted by the theory ($s_{theor}\cong0.8$)[12, 13] for systems having random filler distribution, as well as with other experimental values reported for metal-filled polymer systems.[23, 24]

The values of the critical exponent r, for the composites studied, differ each other, but they are in good agreement with the values obtained by dc conductivity measurements.[25] The peculiar micromorphology and the complex structure of the created infinite conductive cluster of the composites based on the PE/POM blend may be responsible for this effect, to be discussed later.

Table 1. Dielectric parameters for all the composites studied.

Composite System	P_C (% vol. Fe)	s	r
PE-Fe	22	0.77	1.6
POM-Fe	25	0.84	0.8
PE/POM-Fe	11	0.75	5.1

The filler volume concentration dependence of the dielectric permittivity, ε′, for PE-Fe and PE/POM-Fe composites at several frequencies is presented in Figure 3. At a given frequency and for filler contents below the percolation threshold, only moderate changes to the dielectric permittivity values occur. A sudden increase to the dielectric permittivity values over 2-3 orders of magnitude is clearly observed, when the iron concentration aproaches the percolation threshold, P_C. This increase is more intense for the lower frequencies, indicating the strong contribution of the metal particles (and the created conductive networks) to the dielectric permittivity of the composites.[24]

At a given filler content below the percolation threshold ($P<P_C$), dielectric permittivity practically does not depend on frequency (ε′ is almost constant for all the frequencies shown). Near the percolation threshold, only negligible changes to the dielectric permittivity values, at the various frequencies studied, are observed. Finally, above the percolation threshold ($P>P_C$), great differences to the dielectric permittivity values, at the various frequencies, are observed. The lower the frequency is, the higher are the dielectric permittivity values. This behaviour is also similar for the POM-Fe composites and the same holds for the dependence of the dielectric loss tangent on the filler volume fraction (not presented here).

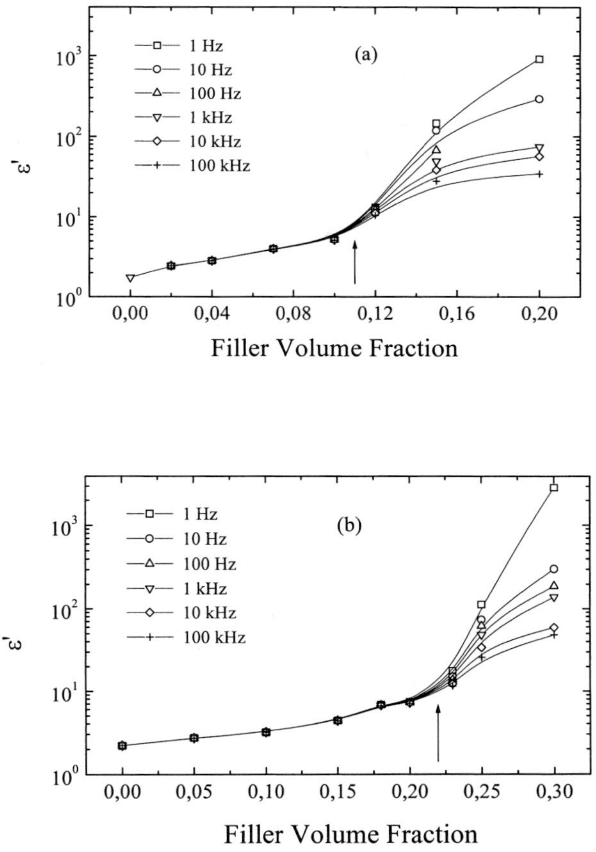

Fig. 3. Dielectric permittivity, ε′, as a function of the filler volume fraction for PE/POM-Fe (a) and PE-Fe (b) composites at several frequencies indicated on the plots. The arrows show the percolation threshold values. The lines are guides to the eye.

We also observe that, for the PE/POM-Fe composites, a specific value of the dielectric permittivity is achieved at lower filler volume fraction, compared to that of PE-Fe and POM-Fe systems. Moreover, thermal conductivy studies in these composites showed that the PE/POM-Fe composites possess better conditions for heat transport through its volume, as compared to the other two composites of the same filler.[25] This result is interesting from the point of view of applications of these materials, since PE/POM-Fe composites can be used for absorption of electromagnetic radiation.

It is well known that, the appearance of conductivity in metal-filled polymer systems is determined by the structural features of the composites. One of the most important factors affecting the percolation threshold value is the interactions between the polymer and the metal inclusions.[17] In general, the stronger the interaction is, the slower becomes the creation of the conductive network and, consequently, the higher is the percolation threshold value. This is a possible reason for the different percolation threshold value of PE-Fe and POM-Fe composites. The percolation threshold value is also affected by the spatial distribution (topology) of the metal particles within the host polymer matrix. The percolation threshold value for composites having ordered filler distribution is lower compared to that of systems having random filler distribution.

It follows, from the results listed in the Table, that the percolation threshold value for the PE/POM-Fe composites, which have the PE/POM blend as a matrix, is much lower (P_C=11 % vol. Fe) compared to that of the PE-Fe and POM-Fe systems. The differences observed can be attributed to the peculiar morphology of the PE/POM-Fe composites.

As mentioned above, PE/POM-Fe composites have been prepared in a two-step procedure. First, the metal particles were introduced into one polymer (POM) and then this mixture was diluted by pure PE. Since POM is less viscous than PE, iron particles are localized only in POM and, consequently, they are not randomly distributed within the whole volume of the blend matrix. In that case, PE forms a separate, non-conductive phase, which does not contain iron particles and the filler distribution is now ordered. As a result, the two distinct phases (POM-Fe and PE) create an interpenetrating structure, at comparable concentrations, and the filler distribution is now ordered. Similar morphology with non-random distribution of the filler particles (carbon black) and unusual electrical characteristics have also been observed in composites based on various polymer blends.[20, 26-29]

During the preparation of PE/POM-Fe composites, the iron concentration within the POM phase was always kept constant (32 % vol. Fe), which is higher than the percolation threshold of POM-Fe system (25 % vol. Fe). So, for low filler concentration within the PE/POM blend ($P<P_C$), the dispersed phase is the POM-Fe, creating isolated conductive aggregates within the non-conductive PE phase. For high filler content ($P>P_C$), the conductive POM-Fe clusters come in physical contact each other, creating a continuous conductive network.

In the case of composites having ordered filler distribution, two different filler concentrations can be specified; a local and an average one. In our case, the local filler concentration in the

POM phase is high, so that POM-Fe phase becomes conductive, even if the average filler content within the PE/POM blend matrix may be low.

In order for conductivity to exist in the PE/POM-Fe composites, it is necessary that two conditions are fulfilled: existence of a conductive network of dispersed iron particles into the POM phase and second, continuity of the conductive POM-Fe network within the PE/POM matrix. Sumita and coworkers[28] first proposed the term ''double percolation'' for such an effect in composites with a two-phase polymer matrix. Since, in our case, the POM-Fe phase is conductive, the percolation threshold value of the PE/POM-Fe composites will be defined only by the continuity of the conductive POM-Fe phase.

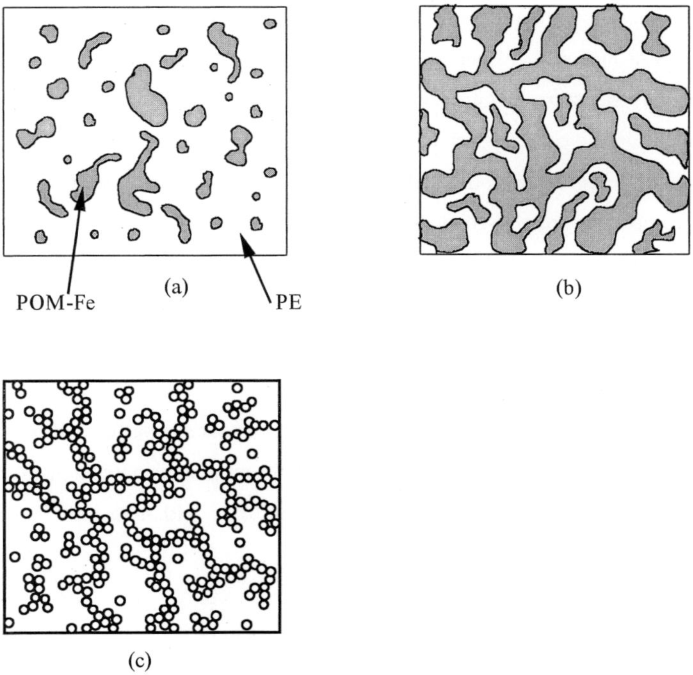

Fig. 4. The proposed model for the morphology of the composites having ordered (PE/POM-Fe) (a, b) and random filler distribution (PE-Fe or POM-Fe) (c). The filler volume fraction is: $P<P_C$ (a) and $P>P_C$ (b, c).

A schematic model, based on the above-mentioned considerations, has been proposed[25] for the microstructure of the composites studied. The proposed structural model compares the morphologies of the composites having ordered (PE/POM-Fe) and random filler distribution (all the other systems) and is schematically shown in Figure 4. This model explains the much lower percolation threshold value of the PE/POM-Fe composites. As it is seen, in the range $P<P_C$, the filler particles are incorporated in small aggregates located in the volume of one polymer (POM) and these capsules of iron-filled POM are distributed randomly in the PE phase (Figure 4a). When the filler content takes values higher than the percolation threshold ($P>P_C$), the conductive clusters of POM-Fe merge together and a branched structure is created (Figure 4b). On the other hand, PE-Fe and POM-Fe composites have distinctly different structures. For these systems, the filler particles are randomly distributed within the volume of the polymer matrix, both below and above the percolation threshold (Figure 4c).

This structural model is in good agreement with results of optical microscopy investigations, shown in Figure 5. For the PE/POM-Fe composites and at low filler content ($P<P_C$), these composites have ''island'' structure, since the iron particles are localized only in the POM phase (Figure 5a). At higher filler concentration, the conductive POM-Fe clusters merge in a branched continuous structure (Figure 5b) and the PE/POM-Fe composites become now conductive ($P>P_C$). On the contrary, for the PE-Fe composites, the iron particles are situated separately, or in small aggregates, both at low and high filler concentrations (Figure 5c). In both cases, the composites are non-conductive, since these filler contents are below the value of the percolation threshold. This behaviour is similar for the POM-Fe composites having random filler distribution, too. These pictures justify the microstructures proposed by the model.

Percolation theory predicts the values of the percolation threshold, P_C, and the critical exponents for composites with statistical distribution of the conductive particles. The proposed model explains the low values of the percolation threshold by assuming ordered distribution and high local concentration of the conductive filler particles within one of the polymer components.

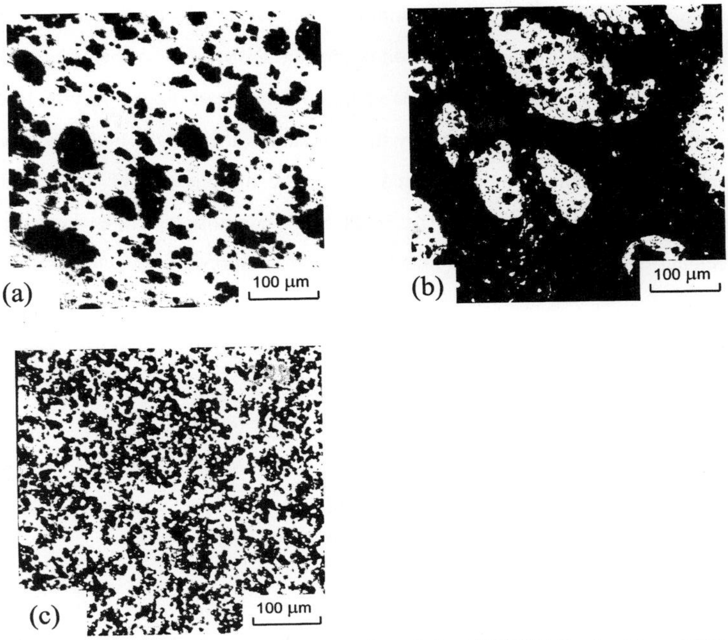

Fig. 5. Optical microscopy photographs of the PE/POM-Fe (a, b) and PE-Fe (c) composites. The filler volume concentration is below (a) and above (b, c) the percolation threshold. These photographs are in agreement with the schematic model presented in Figure 4.

Conclusions

The dielectric properties and the microstructure of various composite materials based on a thermoplastic polymer blend (PE/POM) as a matrix, and iron (Fe) particles as inclusions were investigated in this work. For comparison, the two pure polymers- PE and POM- were used as a matrix, too. In PE-Fe and POM-Fe composites the filler particles are randomly distributed within the polymer matrix, while in the PE/POM-Fe system the filler spatial distribution is ordered. For all the systems studied, the values of the dielectric permittivity critical exponent, s, are in good agreement with the values predicted by the general power-laws of the percolation theory. On the other hand, the percolation threshold value depends on the type of the polymer matrix (PE, POM or PE/POM). This fact is a result of the polymer-filler interaction. For the composites based on the PE/POM polymer blend, the percolation threshold value is much lower, as a result of the peculiar structure of these composites. For such a system, the polymer matrix is two-phase and the dispersed filler particles are localized within one (POM) of the two polymer components. The two distinct phases (the conductive POM-Fe and the nonconductive pure PE) create an interpenetrating network and double percolation phenomenon exists. The local iron concentration within the POM phase is high, whereas the average metal content, on account of the whole volume of the blend matrix, may be low. So, the percolation threshold value is determined by the continuity of the POM-Fe network within the PE/POM blend. A schematic model, proposed for the morphology of the composites studied, explains the different behaviour of the PE/POM-Fe composites. The proposed model is in good agreement with results of optical microscopy investigations.

[1] S.K. Bhattacharya (Ed.), *"Metal-Filled Polymers, Properties and Applications"*, Marcel Dekker, New York, 1986.
[2] R.H. Norman, *"Conductive rubbers and plastics"*, Elsevier, New York, 1970.
[3] J. Delmonte, *"Metal/Polymer Composites"*, Van Nostrant Reinhold, New York, 1990.
[4] R. Pelster, U. Simon, *Colloid Polym. Sci.* **1999**, *277*, 2.
[5] E.P. Mamunya, V.V. Davidenko, E.V. Lebedev, *Polym. Compos.* **1995**, *16*, 319.
[6] H.S. Gokturk, T.J. Fishke, D.M. Kalyon, *IEEE Trans. Magn.* **1993**, *29*, 4170.
[7] J. Yacubowicz, M. Narkis, S. Kenig, *Polym. Eng. Sci.* **1990**, *30*, 469.
[8] G. Boiteux, J. Fournier, D. Issotier, G. Seytre, G. Marichy, *Synth. Met.* **1999**, *102*, 1234.
[9] R. Strumpler, *J. Appl. Phys.* **1996**, *80*, 6091.
[10] R.A. Crossman, *Polym. Eng. Sci.* **1985**, *25*, 507.
[11] D.M. Bigg, D.E. Stutz, *Polym. Compos.* **1983**, *4*, 40.
[12] S. Kirkpatrick, *Rev. Mod. Phys.* **1973**, *45*, 574.
[13] D. Stauffer, *"Introduction to Percolation Theory"*, Taylor and Francis, London 1985.
[14] R. Zallen, *"The Physics of Amorphous Solids"*, Wiley, New York, 1973.
[15] F. Lux, *J. Mater. Sci.* **1995**, *28*, 285.
[16] Ye.P. Mamunya, V.V. Davidenko, H. Zois, L. Apekis, A.A. Snarskii, K.V. Slipchenko, *Polym. and Polym. Compos.* **2002**, *10*, 219.
[17] E.P. Mamunya, V.V. Davidenko, E.V. Lebedev, *Compos. Interf.* **1997**, *4*, 169.
[18] Ye.P. Mamunya, *Macromol. Symp.* **2001**, *170*, 257.
[19] Ye.P. Mamunya, Yu.V. Muzychenko, P. Pissis, E.V. Lebedev, M.I. Shut, *J. Macrom. Sci.-Phys.* **2001**, *B40*, 591.
[20] Ye.P. Mamunya, *J. Macrom. Sci.-Phys.* **1999**, *B38*, 615.
[21] J.P. Runt, J.J. Fitzgerald (Eds.), *"Dielectric Spectroscopy of Polymeric Materials, Fundamentals and Applications"*, American Chemical Society: Washington DC, 1997.
[22] H. Zois, L. Apekis, M. Omastová, *Macromol. Symp.* **2001**, *170*, 249.
[23] In-Gann Chen, W.B. Johnson, *J. Mater. Sci.* **1991**, *26*, 1565.
[24] Y. Baziard, S. Breton, S. Toutain, A. Gourdenne, *Eur. Polym. J.* **1988**, *24*, 633.
[25] Ye.P. Mamunya, Yu.V. Muzychenko, P. Pissis, E.V. Lebedev, M.I. Shut, *Polym. Eng. Sci.* **2002**, *42*, 90.
[26] R. Tchoudakov, O. Breuer, M. Narkis, A. Siegmann, *Polym. Eng. Sci.* **1996**, *36*, 1336.
[27] O. Breuer, R. Tchoudakov, M. Narkis, A. Siegmann, *J. Appl. Polym. Sci.* **1999**, *73*, 1655.
[28] M. Sumita, K. Sakata, Y. Hayakawa, S. Asai, K. Miyasaka, M. Tanemura, *Colloid Polym. Sci.* **1992**, *270*, 134.
[29] F. Gubbels, S. Blacher, E. Vanlathem, R. Jerome, R. Deltour, F. Brouers, Ph. Tessyie, *Macromolecules* **1995**, *28*, 1559.

Characterization of Extruded Ethylene-Vinyl Alcohol Copolymer Based Barrier Blends with Interest in Food Packaging Applications

J. M. Lagarón,[*1] *E. Giménez,*[2] *B. Altava,*[3] *V. Del-Valle,*[1] *R. Gavara*[1]

[1]Institute of Agrochemistry and Food Technology, CSIC, Apdo. Correos 73, 46100 Burjassot, Spain
E-mail: lagaron@iata.csic.es
[2]Department of Technology, University Jaume I, Campus Riu Sec, Castellón 12071, Spain
[3]Department of Organic and Inorganic Chemistry, University Jaume I, Campus Riu Sec, Castellón 12071, Spain

Summary: Based on previous work a number of optimum extruded blends with high contents of a high barrier ethylene-vinyl alcohol copolymer were selected and characterized in terms of phase morphology, water sorption and barrier properties. Blend components were an ethylene vinyl-alcohol copolymer (EVOH with 32 mol% ethylene), an amorphous polyamide (aPA) and a nylon-containing ionomer. A fine two phase structure was found for these blends in all cases. However, Raman spectroscopy results indicated a poor interface interaction between the blend components in the case of the EVOH/aPA blends. Higher interface interaction had been previously found in the dry EVOH/ionomer blends. Equilibrium moisture solubility and diffusion were found to be higher than expected from simple additivity. However, the oxygen transmission rate was found to be clearly lower than expected from the rule of mixtures, particularly under dry conditions, fitting closely a simple Maxwell model.

Keywords: blends, ethylene-vinyl alcohol copolymer, high barrier packaging

Introduction

Ethylene-vinyl alcohol copolymers are a family of random semicrystalline materials with excellent barrier properties to gases and hydrocarbons, and with outstanding chemical resistance.[1] These copolymers are therefore being increasingly used in the food packaging industry as barrier layers to protect foods from the ingress of oxygen and losses of flavours and

 DOI: 10.1002/masy.200350840

consequently to increase package shelf-life.[2] In spite of that, EVOH copolymers are also largely affected by water and possess a number of processing associated problems, like deficient thermoformability, due to high rigidity and crystallization kinetics.[3] In addition to high water sensitivity, EVOH copolymers do not have a good compatibility (adhesion and miscibility) with other polymers whether of polar or non-polar nature.[4] The lack of good compatibility with other polymers is thought to be a consequence of the fact that EVOH copolymers are strongly self-associated, while the inter-association of the hydroxyl groups of EVOH with, for instance, the carbonyl groups of complementary polymers is comparatively weak. Even miscible systems have low LCST (Lower Critical Solution Temperatures) and phase separation is anticipated at processing temperatures. In order to avoid the deterioration of the physical and barrier properties by the uptake of water, these materials are usually sandwiched by coextrusion in multilayer structures where and inner and outer layers of hydrophobic materials, e.g. polyethylene, wrap up the barrier layer made up of EVOH. Blends of EVOH with other materials have been studied over the last years with the aim of boosting the barrier properties of low barrier performance materials, improving mechanical properties of EVOH and/or adhesion, etc.[5,6,7,8,9]

It is well known that the EVOH solid phase formability window, usually between 90ºC and 110ºC, does not overlap with that of many polymers used in coextrusion to form multilayer structures in the packaging industry. Blends of EVOH with low contents of amorphous PA have been claimed by manufacturers to broaden the forming capacity of EVOH with respect to temperature and draw ratios without sacrifice in gas barrier properties[3]. This is because the amorphous PA exhibits a much better forming capacity and its gas barrier performance[10] improves with increasing relative humidity as opposed to EVOH, which barrier properties deteriorate with increasing relative humidity.[11,12] Nylon containing, partially neutralized, ethylene-methacrylic acid copolymers (ionomer) are a family of materials highly transparent and with low hygroscopicity. These ionomers are thermoplastic materials ionically crosslinked and their properties vary with molecular weight and the amount of crosslinking. These materials are designed to be tumble-blended with EVOH in conventional equipment and are claimed[13] to provided a soft, clear, flexible film or sheet with improved flex resistance, excellent thermoformability, and gas barrier at high humidity, compared to films made of EVOH alone.

Previous characterization studies carried out in melt-mixed compresion moulded specimens indicated a two phase structure with a fine distribution of the disperse phase within the EVOH matrix.[14,15] Better compatibility was observed in EVOH/ionomer blends, particularly for blends with higher contents of ionomer. Moreover, a thermofomability study[16] of the blends indicated that EVOH/aPA extruded blends did not improve the inherently poor formability of EVOH alone. In contrast, significant improvements in thermoformability were achieved by blending EVOH with a compatibilized-ionomer in the range 100-150°C. Optimum forming capacity was achieved in a ternary blend by addition of a compatibilized-ionomer to EVOH/aPA blends in the range of 140°C-150°C.

In this study, some additional results based on SEM, vibrational spectroscopy, moisture sorption and oxygen permeability are presented on extruded films of the blends, which permit us to characterize the properties of these barrier blends in the usual form in which they will be processed into food containers.

Experimental

Polymer sheets were obtained using conventional film co-extrusion equipment by extruding pelletised pre-extruded blends between polyethylene layers without adhesive. This allowed for an easy peeling off of the barrier blend film. The blends comprised EVOH-32 as main component and an amorphous PA (Selar PA UX-2034) and/or a Nylon-containing ionomer (Surlyn AM-7938) as minor components. The EVOH copolymer used was a commercial product (Soarnol DC3203) of Nippon-Gohsei (Japan) with 32 mol% of ethylene. The binary and ternary blends prepared were: EVOH/aPA (80/20), EVOH/ionomer (80/20) and EVOH/aPA/ionomer (80/10/10).

FT-Raman spectra were obtained using a FT-IR/NIR-FT-Raman Perkin-Elmer Spectrum 2000 instrument equipped with a diode pumped Nd:YAG laser PSU. Curve-fitting of the experiments was carried out with the Grams Research 2000 software package (Galactic

Industries). Voigt line shapes (convolution of Lorentzian and Gaussian functions), no constraints, and linear baselines were used.

Scanning electron microscopy experiments were carried out in a Jeol JMS-6300. Samples were cryofractured and gold coated prior to SEM observations.

The kinetics of water sorption at 100% RH were obtained for the different samples by immersing dry specimens in distilled water and follow water uptake gravimetrically until saturation (equilibrium sorption).

Oxygen transmission rate was measured in an Oxtran MS (Modern Control Inc., Minneapolis, US) equipment with humidity control.

Results and Discussion

Previous characterization work in these blends by DSC, WAXS, DMA, SEM, microhardness and tensile testing indicated that the miscibility of this high barrier EVOH grade (with T_g at 59°C and T_m at 183°C) with the amorphous PA is very poor, and clear phase segregation throughout composition was shown by DSC, DMA and SEM.[14] Factors like geometric hindrance and chain stiffness of the amorphous PA (with a T_g around 127°C) could be responsible for this behaviour. A lack of good interaction between EVOH/aPA blend components was further supported by the negative deviation from the simple additive rule seen in the mechanical properties of these blends. A two phase structure was also observed in the EVOH/ionomer blends, but from the results a better phase compatibility was inferred. This compatibility increased in the ionomer rich blends and was thought to be enhanced by the presence of crystalline Nylon in the formulation of the ionomer.

Figure 1 shows SEM results carried out on extruded films perpendicular to the flow direction. From this figure, a fine particle dispersion with some debonding at the particles interface is observed. Similar observations were already reported in melt-mixed compression molded specimens.[14] However, while in the EVOH/aPA blends the dispersed phase is rather spherical in shape, this is clearly more elongated in the case of the EVOH/ionomer blend. The latter observation caused by the shear stress imposed during the extrusion procedure is probably a consequence of the lower viscosity and maybe of the phase interaction exhibited by the ionomer in this particular blend.

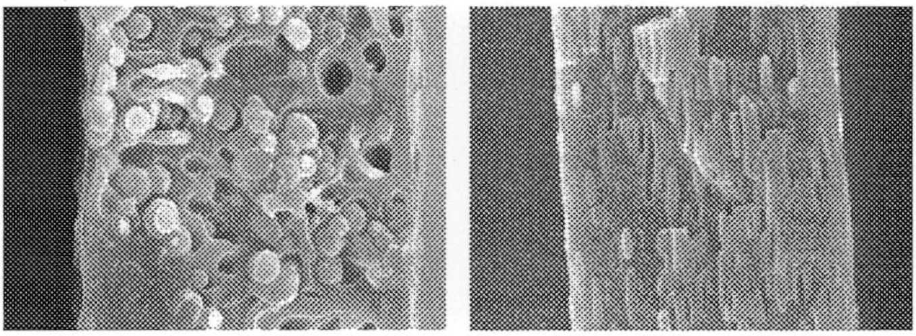

Fig. 1. SEM micrographs of films of EVOH/aPA (left) and EVOH/ionomer (right).

Figure 2 shows the Raman region centered at 1600 cm^{-1} containing some ring vibrations and the amide I mode at 1638 cm^{-1} for the aPA, water equilibrated aPA and EVOH/aPA 80/20 blend. This amide I mode is mainly due to CO stretching vibrations and contains two contributions one small due to free CO groups at 1671 cm^{-1} and another one at 1638 cm^{-1} originating from hydrogen bonded CO groups. Depending on the severity of the drying process the relative amount of free CO groups has been found to vary between ca. 10% and 16%.[17] In the presence of moisture however, the free CO band vanishes due to association of these free carbonyl groups with water molecules (see Figure 2, center). Therefore, the presence or absence of these free carbonyl groups, as well as the position of both of these contributions, can tell about potential interactions at a molecular scale. As EVOH is also capable of forming hydrogen bonding interactions via the hydroxyl groups present in the polymer chain, the disappearance of this free CO groups could then indicate phase interaction between EVOH and the aPA. Figure 2 also shows the amide I range in the EVOH/aPA 80/20 blend. From this Figure, it can be seen that free CO groups are still present, and therefore, there is no clear evidence of interaction between EVOH groups and the amide groups of the aPA. However, the position of the free CO contribution appears shifted towards lower wavenumber in the blend compared to the neat component. This could be attributed to a potential influence of the EVOH matrix over the free amide moieties. Consequently, Raman confirms the lack of a strong interaction between blend components in agreement with results gathered by other techniques for this particular blend. The spectroscopic analysis of the EVOH/ionomer blend is more complicated due to the compositional complexity of the ionomer, i.e. already a blend of an ethylene-methacrylic acid copolymers and a crystalline Nylon.

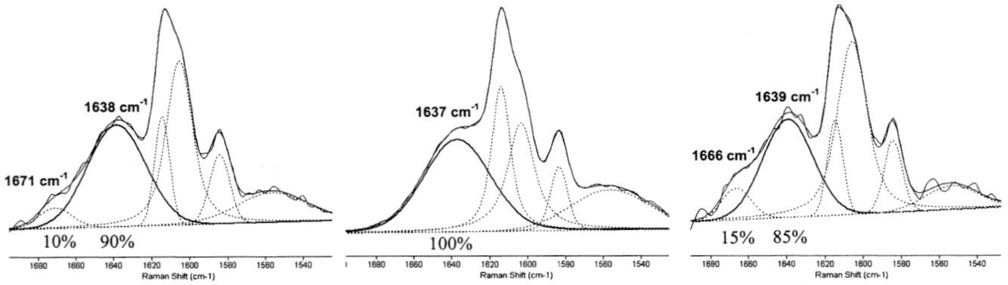

Fig. 2. Raman spectra in the range of the amide I mode for aPA (left), water equilibrated aPA (center) and the EVOH/aPA 80/20 blend (right). Percentages of the different contributions to the amide I envelop and positions are given.

The effect of moisture uptake was previously investigated in a number of melt-mixed compression molded blends by TGA, DMA, microhardness and tensile testing.[15] The study of the influence of water sorption on these blends is relevant to understand the gas barrier behavior of these materials under high relative humidity conditions; of concern, for instance, during typical industrial processes such as retorting of packaged food. From the results, water sorption was found to be in the order EVOH>aPA>>ionomer. The moisture was found to bind more strongly (higher temperature desorption by TGA) with EVOH than with aPA or ionomer. The T_g of water equilibrated aPA was still above room temperature, whereas it was well below room temperature in water equilibrated EVOH. Accordingly, EVOH showed increased plasticity and toughness at high relative humidity conditions. An intrinsic high stiffness and brittleness was measured for the aPA irrespective of the relative humidity. The ionomer showed low moisture dependence in the mechanical properties and much higher flexibility than aPA. In EVOH/aPA and EVOH/ionomer blends, the EVOH fraction was found to be fully plasticized. The phase compatibility suggested from earlier work for some EVOH/ionomer blends appeared to be eliminated by the effect of sorbed moisture on the potential interfacial adhesion via hydrogen bonding. In the mechanical properties of the blends exposed to moisture, EVOH/aPA blends showed an increase in stiffness, and in turn, of the fragility compared to EVOH alone, whereas high flexibility and toughness was still observed in EVOH/ionomer blends at high relative humidity conditions.

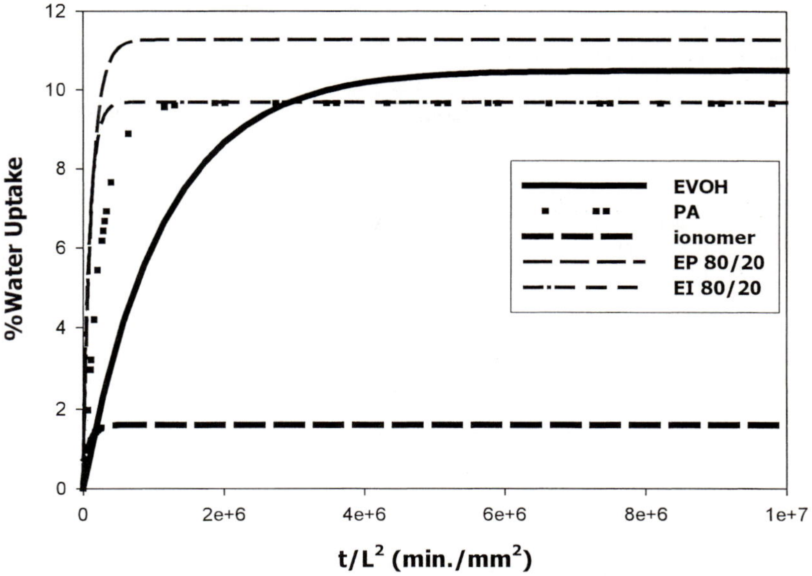

Fig. 3. Kinetics of water sorption at 100%RH for neat components and blends.

Figure 3 shows water sorption curves as a function of time at room temperature and at 100%RH for extruded specimens of the neat components and blends. From this figure, it can be seen that equilibrium sorption for the neat components is, as previously reported, in the order EVOH>aPA>>ionomer in the extruded specimens. However, from the isotherms it is also possible to observe that the diffusion of water is faster in the ionomer and slower for EVOH. In the blends however, it can be seen that solubility is higher than expected from simple additivity of those of the neat components; and so is the diffusion rate. A contribution to this observation could come from a slight decrease in crystallinity observed for the EVOH fraction in the blends during WAXS experiments; however this was not seen by DSC. Nevertheless, the cited effect must be largely attributable to lack of a strong interaction between components in the blends, which makes then the plasticized interface a preferential pathway for diffusion of moisture. In fact, if a more strong interaction was going to occur at the particles interface, that would have probably been disrupted by a more favorable interaction with moisture molecules, as was previously reported in EVOH/ionomer blends. Consequently, transport of moisture is observed to be higher in these blends than expected from simple additivity of the neat components.

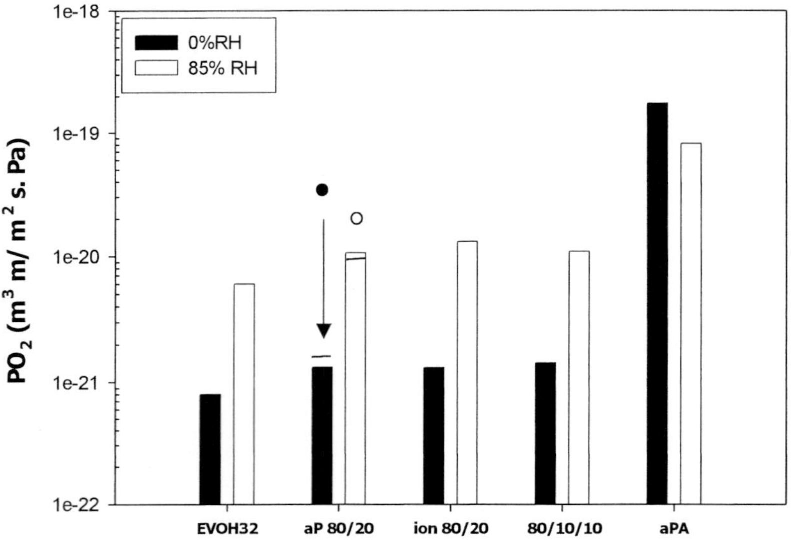

Fig. 4. Oxygen transmission rate at 0%RH and at 85%RH. Circles in the EVOH/aPA 80/20 blend correspond to parallel alignment of the two phases to permeant flow (simple additivity) and dashes to the Maxwell equation.[18]

Figure 4 shows the oxygen transmission rate results at 0%RH (dry) and at 85%RH in binary and ternary blends. A full map of the oxygen barrier properties across relative humidity is being currently measured and will be reported elsewhere. From the results, it can be seen that the oxygen permeability at 0%RH is lower for EVOH but is also very low for the blends. At high relative humidity (85% RH) the barrier properties of the materials are much higher in all cases, except for the amorphous PA. The latter material is known to have better barrier properties at high relative humidity due to its particular interaction with moisture. It has been recently found that moisture sorption does not disrupt the originally existing hydrogen bonding structure, but rather links to free amide moieties being the majority of the sorbed water under the form of clusters.[17] Clustered water is thought to fill the available polymer free volume, hence leading to a decrease in oxygen permeability on the bases of a competing mechanism. The barrier properties of the neat ionomer could not be measured because it could not be peeled off due to strong adhesion to the structural polyolefinic layers of the co-extruded sheets supplied. The deterioration in barrier properties of the blends at high humidity is consequence of the plastification of the EVOH continuous phase.

The barrier properties of the blends seem to follow a relationship (see equation 1) close to that proposed by Maxwell and extended by Roberson (see equation 1) for spheres of a low barrier phase dispersed in a high barrier continuous matrix (see dashes in Figure 4).[18] This simple model would appear to closely reflect, albeit with a slight positive deviation, the case of the dispersed morphology found for the EVOH/PA blend in Figure 1. The EVOH/ionomer blend could even present considerable better barrier than predicted from equation 1 due to the fact that particles are elongated (higher aspect ratio) in a direction normal to the permeation direction.

$$P_{EVOH/aPA} = P_{EVOH}\left[\frac{P_{aPA} + 2P_{EVOH} - 2V_{aPA}\left(P_{EVOH} - P_{aPA}\right)}{P_{aPA} + 2P_{EVOH} + V_{aPA}\left(P_{EVOH} - P_{aPA}\right)}\right] \quad (1)$$

The permeability of the blends consider here would then approach the permeability of a coextruded multilayer (series laminate normal to permeant flow, see equation 2) system comprising two layers, one made of a lower barrier disperse phase and the other of high barrier matrix; being therefore, the overall permeability close to the permeability of the neat high barrier matrix for a sufficiently high volume fraction of the matrix (V_{EVOH}). Equation 2 is a very favorable situation in terms of permeability for a non miscible blend.

$$P_{EVOH/aPA} = \frac{P_{EVOH}P_{aPA}}{V_{aPA}P_{EVOH} + V_{EVOH}P_{aPA}} \quad (2)$$

The circles on the graph in Figure 4 represent the values of permeability obtained by application of a simple additive rule (layers parallel to permeant flow, see equation 3). This case would clearly be a very disfavorable situation in terms of permeability for these blends.

$$P_{EVOH/aPA} = P_{EVOH}V_{EVOH} + P_{aPA}V_{aPA} \quad (3)$$

At 85% RH the barrier properties of the blends are higher and approach closer the value for the simple additivity (equation 3) in the case of the EVOH/aPA blend. Nevertheless, they still show a reasonable close fit to equation 1, albeit with negative deviation. This observation may appeal for the application of not so simplified Maxwell models. This particular behavior is thought to be affected by the higher solution and diffusion of moisture at high humidity levels in these blends. Results for a ternary blend are also included. This blend has been found to provide optimum thermoforming performance.[16]

Acknowledgments

The authors would like to acknowledge CICYT (program IFD97-0813-C03) for financial support and wish to thank the assistance of AIMPLAS (Institute Technologic of Plastic) for supplying co-extruded sheets. The authors would also like to thank the program "Ramon y Cajal" of the Spanish Ministry of Science and Technology for financial support.

[1] Lagaron JM, Powell AK, Bonner JG., Polymer Testing 2001; 20/5:569.
[2] Foster RH. Polymer News 1986; 11:264.
[3] Chou R, Lee IH. J of Plastic Film & Sheeting 1997; 13:74.
[4] Coleman MM, Yang X, Zhang H, Painter PC. J Macromolecul Sci-Phys 1993; B32(3):295.
[5] Samios CK, Kalfoglou NK, Polymer 1998; 39:3863.
[6] Petris S, Laurienzo P, Malinconico M, Pracella M, Zendron M, J Appl Polym Sci 1998; 68:637.
[7] Lohfink GW, Kamal MR. Polym Ing Sci 1993; 33:1404.
[8] Lee SY, Kim SC. J Appl Polym Sci 1998; 67:2001.
[9] Uno M, Norton LJ, Kramer EJ, Oda H. J Mater Sci 1998; 33:853.
[10] Hernandez RJ, Giacin JR, Grulke EA. J Membrane Sci 1992;65:187.
[11] Aucejo S, Marco C, Gavara R. J Appl Polym Sci 1999;74:1201.
[12] Zhang Z, Britt IJ, Tung MA. J Polym Sci: Part B: Polym Phys 1999 ; 37:691.
[13] Fetell AI. 1997 TAPPI Polymers, Laminations and Coatings Conference, August 24-28 , Toronto, Canada 1997.
[14] J.M. Lagaron, E. Gimenez, J.J. Saura, R. Gavara, Polymer, 42, 7381, 2001.
[15] J.M. Lagaron, E. Gimenez, R. Gavara, J.J. Saura, Polymer, 42, 9531, 2001.
[16] Gimenez E., Lagaron J.M., Cabedo L., Gavara R., Saura J. in press
[17] Lagaron J.M., Gimenez E., Gavara R., Catala R., Macrom. Chem. Phys., accepted for publication.
[18] Hopfenberg H.B., Paul D.R.. In Polymer Blends; Paul D.R., Newman S.; Academic Press: New York, 1978.